Progress in Botany 68

68 PROGRESS IN BOTANY

Review
Genetics
Physiology
Ecology

Edited by

K. Esser, Bochum
U. Lüttge, Darmstadt
W. Beyschlag, Bielefeld
J. Murata, Tokyo

Springer

With 48 Figures and 14 Tables

ISSN 0340-4773
ISBN 978-3-642-07193-5 e-ISBN 978-3-540-36832-8

Springer is a part of Springer Sciences+Business Media
springer.com

Softcover reprint of the hardcover 1st edition 2007

Cover design: Design & Production, Heidelberg

31/3150/11541097/SPi - 5 4 3 2 1 0 - Printed on acid-free paper

Editorial

FORTSCHRITTE DER BOTANIK was founded in 1932 by Fritz von Wettstein. The aim of this series, which today is published annually, is clearly stated in the preface to the first volume: "The large number of journals, books and monographs currently being published, prohibits individuals from maintaining an overview of the progress in all the specialised areas of botany, let alone from keeping abreast of the results from related areas. The necessary interconnections between different specialised areas can only thrive, or be revealed, if one is able to maintain an overview of the whole field of botany."

Since its first appearance more than 70 years ago, FORTSCHRITTE DER BOTANIK or PROGRESS IN BOTANY, as it was later called, has made an effort to meet the high standards it set for itself. As a result, the series stands out from other Annual Reviews in the English-speaking world, in which articles pertaining to a particular theme just tend to be listed and briefly summarised.

Following the untimely death of von Wettstein in 1945, the series was brought back to life by Erwin Bünning (Tübingen) and Ernst Gäumann (Zürich) in 1948. After Ernst Gäumann's death in 1963, Heinz Ellenberg took his place in 1964 (volume 26). As the field of botany was being transformed more and more into an experimental science, it seemed appropriate for this to be reflected in the series. It was with this in mind that Erwin Bünning, and the publishers Dr. Heinz Götze and Dr. Konrad Springer invited me to take charge of the area of Genetics as an editor. In order to be able to follow the whole spectrum of botanical research, Dr. Heinz Götze and Dr. Konrad Springer followed my suggestion of expanding the editorial board to include 5 members. Thus, from volume 27 onwards, Peter Sitte (Freiburg) was responsible for Anatomy and Morphology, Erwin Bünning (Tübingen) for Physiology, Karl Esser (Bochum) for Genetics, Hermann Merxmüller (Munich) for Systematics, and Heinz Ellenberg (Göttingen) for Ecology.

The idea of structuring the series into 5 sections not only revitalised the series but also proved successful over the following decades. There was of course turnover of the members of the editorial board. It was not until 1998 (volume 60) that the editorial board was reduced to 4 members. Over the last few years, we have broadened the scope of PROGRESS IN BOTANY by giving retired colleagues the opportunity to publish overview articles about their life's work.

In order to communicate the botanical nature of the series to the public, the chlorophyll molecule was used as a logo on the cover from volume 28 onwards. To keep pace with the increasing use of English in the field of natural sciences, from 1974 (volume 36) more and more articles were published in English. This trend was further reflected by adopting the English title "Progress in Botany". From volume 47 (1985), all articles were published in English. As a consequence, the German subtitle "Fortschritte der Botanik" was completely removed. From volume 59 (1997), the cover illustration was "modernised".

After 42 years of service to Fortschritte der Botanik/Progress in Botany and having reached almost 83 years of age, I feel the time has come for me to retire and to make way for a younger editor.

I would like to take this opportunity to thank the countless authors who supported me over the last decades with their contributions to the series. Moreover, I would like to thank my many colleagues who served with me on the editorial board for their cooperation. I would also like to acknowledge Dr. Czeschlik and Mrs. Gramm who, over the last few years, were involved with the series on behalf of the publisher.

Bochum, autumn 2006 Karl Esser

Contents

List of Editors

Professor Dr.Dr.h.c. mult. K. Esser
Lehrstuhl für Allgemeine Botanik, Ruhr Universität
Postfach 10 21 48
44780 Bochum, Germany

Phone: +49-234-32-22211; Fax: +49-234-32-14211
e-mail: karl.esser@rub.de

Professor Dr. U. Lüttge
TU Darmstadt, Institut für Botanik, FB Biologie (10)
Schnittspahnstraße 3-5
64287 Darmstadt, Germany

Phone: +49-6151-163200; Fax: +49-6151-164630
e-mail: luettge@bio.tu-darmstadt.de

Professor Dr. W. Beyschlag
Fakultät für Biologie
Lehrstuhl für Experimentelle Ökologie
und Ökosystembiologie
Universität Bielefeld, Universitätsstraße 25
33615 Bielefeld, Germany

Phone: +49-521-106-5573; Fax: +49-521-106-6038
e-mail: w.beyschlag@uni-bielefeld.de

Professor Dr. Jin Murata
Botanical Gardens
Graduate School of Science
University of Tokyo
3-7-1 Hakusan
Bunkyo-ku, Tokyo 112-0001
Japan
Phone: +81-3-3814-2625; Fax: +81-3-3814-0139
e-mail: murata@ns.bg.s.u-tokyo.ac.jp

Curriculum Vitae

Charles Barry Osmond was born 20 September 1939 in Cooranbong, a small town in the foothills of the central coastal range New South Wales (NSW), Australia, to Joyce Daphne (nee Krauss) and Edward (Edmund) Charles Osmond, carpenter.

1946–55	Schooling at Morisset Central and Wyong High Schools, NSW
1956–62	BSc (Hons) and MSc, Department of Botany, University of New England, Armidale NSW, with plant ecophysiologist, Noel C.W. Beadle.
1963–65	PhD, Department of Botany, University of Adelaide, South Australia, with plant physiologist-biochemist Sir Rutherford Ness (Bob) Robertson.
1965–66	Postdoctoral research associate with plant physiologist George G Laties, UC Los Angeles USA.
1966–67	CSIRO overseas postdoctoral scholarship with plant biochemist Tom ap Rees, Botany School, University of Cambridge, UK.
1967–87	Successively, non-tenured Research Fellow, tenured Fellow, Senior Fellow and Professor (1978) Department of Environmental Biology, Research School of Biological Sciences (RSBS), Australian National University (ANU), Canberra Australia.
1980–86	Director, Biosciences Center, Desert Research Institute, University of Nevada-Reno NV USA (joint appointment with ANU)
1987–91	Arts and Sciences Distinguished Professor, Botany Department, Duke University, Durham NC, USA.

1991–2001	Director, RSBS (until 1998) then Professor, Photobioenergetics Group, RSBS ANU.
2001–03	President and Executive Director, Biosphere 2 Center, Columbia University, Oracle AZ, USA.
2003–	Visiting Fellow, School of Biochemistry and Molecular Biology, ANU, and Research Professor (adjunct) Nicholas School of the Environment, Duke University
2005	Visiting scientist, Institut Phytosphäre (ICG III) Forschungszentrum Jülich, Germany.

Elected Fellow, Australian Academy of Science (1978), Royal Society of London (1984) and Mitglied, Deutsche Akademie der Naturforscher, Leopoldina (2001). Edgeworth David (1972) and Clarke (1997) Medals, Royal Society of NSW. Fulbright Senior Scholar UC Santa Cruz and Carnegie Institution Fellow, Department of Plant Biology Stanford; Richard Merton Gastprofessor, Institut für Botanik, Technische Universität München (1974). Overseas Fellow, Churchill College, Cambridge (1980); Inaugural Distinguished Alumnus, University of New England (1994); Forschungspreis, Alexander von Humboldt Stiftung (1996).

Crassulacean Acid Metabolism: Now and Then

Charles Barry Osmond

Previous title chapters in Progress in Botany, from giants of European botanical research in the latter half of the twentieth century, have explored significant areas of plant science. I am honoured, and more than a little over-awed, by the Editor's invitation to contribute in this context (and alarmed to discover that I am only, but precisely, a decade younger than the previous contributor!!). Although this chapter may not present the long view of the discipline offered by others, any perspective on crassulacean acid metabolism (CAM), a pathway of photosynthetic carbon metabolism that occurs in about 5% of vascular plants (Winter and Smith 1996; Lüttge 2004), reveals much of wider significance in plant physiology and biochemistry. When Clanton Black and I prepared a brief historical overview of CAM (Black and Osmond 2003), we emphasized the close relationship between these succulent plants and humans through the romantic paintings of Carl Spitzweg. We noted that the taste–test diagnostic of CAM may have been known to the Romans, and that accounts of morning acidity in leaves of succulents that disappeared by evening were published by Grew in the seventeenth century and Heyne in the early nineteenth century. The early literature on CAM into the 1960s was highlighted, but space constraints relegated most of the influential studies of the last 40 years to little more than a few citations from more than 40 mentors, former students and colleagues.

This chapter provides an opportunity to make amends by recording my indebtedness to very many companions in CAM research. It is a personal view of an active and exciting area of plant biology since about 1970. Indeed, Lüttge (2004) cited a selection of more than 20 reviews, edited volumes and books on CAM (notably Kluge and Ting 1978; Winter and Smith 1996) over the last 2–3 decades. My reminiscences will be largely confined to areas in which my companions and I have published, but the temptation to range more broadly sometimes will be difficult to resist. Throughout, I will link to important current developments, and emphasize some broader implications that have emerged. As will become evident, my peripatetic research on photosynthesis in succulent plants with CAM continues to depend on stimulus from many colleagues in plant science, particularly those in Germany.

Progress in Botany, Vol. 68

1.1 A pathway to CAM via oxalate and malate in *Atriplex*

There was little in my family or educational background to suggest any particular scholastic ability or affinity with plant biology. Alfred Kurtz, a distant relative of my mother, and a well known second generation viticulturist in the Mudgee region of central western NSW, was the only strong family connection to things botanical. His vineyard is generally credited as the source (in the 1950s) of robust Chardonnay root stocks that supported the worldwide expansion of this variety in the last half of the twentieth century (Halliday 1985). However, since both sides of my family were teetotal for two generations, I did not think to explore viticulture or enology. Rather, my botanical career emerged accidentally. It was made more likely by a spectacular collapse in mathematical ability between high school and university that terminated my aspiration to qualify as a teacher of math and science. After 3 undistinguished years, I began again, and in 1960 did well enough in botany and in natural products organic chemistry to commence graduate research in botany at the University of New England, Armidale NSW, then the only university in Australia outside a State capital city.

I was influenced by the ecological focus of the Armidale Botany Department especially its interest in halophytes of the genus *Atriplex*, the "saltbushes" from semi-arid ecosystems in southern Australia. Physiological plant anatomy featured in the curriculum, and *Atriplex* leaves were fascinating for their large crystals of calcium oxalate, their huge epidermal bladders that proved to be salt secreting systems, and their "Kranz" arrangement of mesophyll and bundle sheath tissues that proved to be the foundation of C_4 pathway of photosynthetic metabolism. One could not have been presented with a more fascinating complex of leaf physiological anatomy, and all three features were to provide this starting graduate student with significant opportunities for original research. My first encounter with a CAM plant in the wild (a huge specimen of "tree pear" introduced *Opuntia tomentosa* Salm-Dyck) took place about 1960 during an excursion to the arid shrublands of south west Queensland as a field assistant to Professor Noel Beadle, a pioneering Australian plant ecophysiologist. Sadly, it was a decade or more before I rediscovered the impact of CAM on the Australian landscape.

Presented with equipment for ether extraction of organic acids, I found oxalate to be the balancing anion for the inorganic cation excess in *Atriplex* leaves. Subsequently, as a PhD student in the laboratory of Professor "Bob" Robertson in the University of Adelaide, it was possible to explore the synthesis of oxalate following $^{14}CO_2$ fixation in the light and dark, using ion-exchange and paper chromatography. Malic acid, exclusively labelled in the 4-C carboxyl was the most abundant early labelled product in the dark (the terms malic acid and malate will be used interchangeably throughout this chapter). Unexpectedly, 4-C labelled malic and aspartic acids, were also the most abundant initial products of $^{14}CO_2$ fixation in the light in *Atriplex* leaves. As related elsewhere (Osmond 1997), I had stumbled across the "β-carboxylation" pathway of primary CO_2 fixation, subsequently associated with "Kranz" anatomy, in this large genus of C_4 plants.

These early adventures in intermediary metabolism were stimulated by P.N. ("Danny") Avadhani, who was visiting Adelaide from the University of Singapore. Danny considered himself an "ideas man" and he occupied the chalkboard in the Departmental tearoom for days on end with a frequently amended forerunner of the metabolic wall charts that Boehringer-Mannheim later supplied to decorate laboratories throughout the world. Our interpretation then of the pathway of oxalate synthesis in *Atriplex* leaves (Osmond and Avadhani 1968) was based on analogies with the isocitrate cycle and was probably incorrect. Danny had taken his PhD in the University of Newcastle upon Tyne, where Thomas and Beevers (1949) had introduced the term Crassulacean acid metabolism. In the same laboratory, David Walker (1956) had demonstrated that phosphoenolpyruvate carboxylase (PEPCase), the legendary "wouldn't work man!" reaction (Wood and Werkmann 1938), was involved in the pathway to malic acid in CAM. It is a particular pleasure now to observe that another generation of researchers has been "taking coal to Newcastle" in the form of highly original physiological biochemical and unequivocal molecular evaluations of CAM (Griffiths et al. 1989; Borland et al. 1999; Borland and Dodd 2002).

I had been introduced to plant physiology through the 1956 edition of Thomas' textbook but its description of CAM in terms of respiratory quotients probably explains why I did not readily connect to the photosynthetic implications of this pathway. Even now, with O_2- and CO_2-specific electrodes (Osmond et al. 1996) and mass spectrometers (Maxwell et al. 1998), the stoichiometries of net CO_2 and O_2 exchanges in CAM are difficult to interpret. Ranson and Thomas (1960) provided the authoritative source on CAM in English, but it had been reported that malic acid accumulating in CAM in the dark was labelled in both 1-C and 4-C carboxyl positions, in the ratio 1:2 (Bradbeer et al. 1958). At the time, the "Newcastle overall hypothesis" seemed rather perplexing and only remotely connected to my observations in *Atriplex*. Nevertheless, Danny led me through the simple diel routine of acid extraction in boiling water and titration to phenolphthalein end-points that even now draws me through nights of interrupted sleep. There is no escape from this fundamental reference for the temporal expression of CAM in different conditions, a simple reference that could bring greater rigor to contemporary studies of regulatory cascades of gene expression.

Evidently a slow learner, I should have recognized the research potential of CAM during a postdoctoral year in George Laties' laboratory at UC Los Angeles exploring ion transport and malate compartmentation in beet discs (Osmond and Laties 1967). Ulrich Lüttge occupied the other side of the lab bench, and although I doubt we spoke of it then, our lifelong friendship later came to be entwined with CAM. A second post-doc year with Tom ap Rees in Cambridge introduced me to enzymology and the use of specifically labelled substrates for evaluation of metabolic pathways. I then had the good fortune to join Ralph Slatyer's Department of Environmental Biology in the new Research School of Biological Sciences (RSBS) at the Australian National University (ANU) in Canberra, and returned to Australia in 1967. The "research only" appointments in the Max Planck-like Research Schools embedded in a university environment, with limited opportunities for tenure but access to front-line equipment of the day, provided privileged starts for many research careers in Australia at the time. I had been hired to work on starch to malate metabolism in stomatal guard cells, but any links to CAM research that may have occurred to me at the time were soon put to one side by the wave of interest in C_4 photosynthesis. The Research School was just across the street from CSIRO Plant Industry, the nation's strongest concentration of plant physiologists and biochemists, notably Jan Anderson, Keith Boardman and Hal Hatch.

Ralph's prestige, and the popularity of environmental science at the time, may have conspired to grant us the opportunity to organize the first workshop sponsored by the US–Australia bilateral programme in science and technology. With Ralph's deft handling, the programme was expanded to include some leading scientists from the UK, Germany and Japan. The ensuing workshop on photosynthesis and photorespiration was most timely and evidently of lasting impact (Sage and Monson 1999). It was a very exciting time in photosynthetic metabolism, and the meeting afforded excellent early career opportunities to build enduring networks. My latent interest in CAM was stimulated by this meeting when a plant anatomist (Laetsch 1970) provocatively declared the C_4 pathway to be "CAM mit Krantz". A better understanding of CAM in relation to C_3 and C_4 pathways of metabolism was obviously needed, especially after the surprising observation of Klaus Winter that the ice-plant *Mesembryanthemum crystallinum* L. could be converted from C_3 to CAM patterns of CO_2 fixation by salt stress (Winter and von Willert 1972). Hal Hatch was providing inspirational research leadership in C_4 metabolism in CSIRO, and the time seemed ripe to explore CAM as a photosynthetic process.

1.2 Sorting the phases of CAM

We now know that C_4 and CAM pathways of photosynthetic carbon metabolism are both based on largely analogous preliminary CO_2 concentrating mechanisms (CCMs) in which primary carboxylation leads to 4C acids (and amino acids) that serve as intermediate, internal stores of carbon. These substrates are subsequently decarboxylated to generate internal CO_2 concentrations of 1000–25,000 ppm (Cockburn et al. 1979) that largely mitigate the oxygenase activity of Rubisco (Leegood et al. 1997). From an evolutionary perspective, these CCMs recreate the atmospheric CO_2 concentrations of the Cretaceous, a time of grand expansion of terrestrial plants under conditions of CO_2 saturation in which O_2 fixation by Rubisco oxygenase and subsequent C recycling in photorespiration would not have carried the same penalty, in energetic terms, as it does for C_3 photosynthesis today. Simply put, CCMs of C_4 plants are based on small (about 1–10 mM), spatially separated cytoplasmic pools of 4C acids that turn over rapidly ($t_{1/2}$ about 1–10 s). These CCMs can be distinguished from those of CAM plants which are based on larger (100–500 mM) pools of 4C (and 6C) acids in the vacuoles, that turn over much more slowly ($t_{1/2}$ about 5000–50,000 s) with complex, temporally separated, patterns of acid synthesis and degradation.

With a lot of help from colleagues, I set out to impose some order on the carbon metabolism of these temporally separated processes (so-called phases I–IV) and to place the curiosity of CAM into the context of other pathways of photosynthetic metabolism. The 1970 workshop stimulated two CAM enthusiasts, Manfred Kluge and Irwin Ting to spent sabbatical periods in RSBS where they successfully demonstrated the presence of pyruvate Pi dikinase and the distinctive kinetic properties of PEPCase in extracts of these plants (see below). Bruce Sutton, my first PhD student, undertook a reassessment of the labelling patterns of malic acid in CAM plants exposed to $^{14}CO_2$ in the light by comparing the previously employed *Lactobacillus* culture degradation method and degradation with purified malic enzyme to remove the 4-C carboxyl of specifically labelled malic acid preparations (Sutton and Osmond 1972). These experiments strongly suggested that fumarase activity in *Lactobacillus arabinosus* (synonym for *L. plantarum* WCFS1) led to randomization of label from 4-C to 1-C in ^{14}C-malic acid prior to or during decarboxylation, especially when old cultures were used to degrade large amounts of malic acid. Indeed, using the purified enzyme, we found that malic acid from dark $^{14}CO_2$ fixation in CAM plants was initially and predominantly 4-C labelled, consistent with primary CO_2 fixation of unlabelled PEP by PEPCase. On the other hand, malate labelling in the light was closer to the 1-C to 4-C ratio of 1:2 observed by Bradbeer et al. (1958), consistent with PEP formation from two molecules of PGA, one of which had been previously labelled as a result of prior $^{14}CO_2$ fixation by Rubisco (Osmond and Allaway 1974).

Generous sabbatical provisions in ANU (1 year in 4 for tenured staff; a legacy of the postwar sense of isolation down-under) enabled me to work in UC Santa Cruz and the Technische Universität, München in 1973–1974. Harry Beevers evidently had a soft spot for CAM from his days in Newcastle upon Tyne and was a most generous host in Santa Cruz. Although germinating castor beans were an ideal system for investigation of Rubisco in proplastids, like most others in the lab, I welcomed opportunities to escape the nauseous extraction process. It proved possible to commute over the coast range for nocturnal gas exchange experiments with CAM plants in Olle Björkman's lab at Carnegie Plant Biology, Palo Alto. By the time I joined Professor Hubert Ziegler in München, it was clear that CO_2 fixation in the dark in CAM plants was insensitive to O_2, whereas CO_2 fixation in the light was inhibited by O_2 (Björkman and Osmond 1974), further confirming that C_4- and C_3-like carboxylation systems were functioning in a temporally separated fashion.

I went to München because Professor Ziegler had excellent access to natural abundance ratio mass spectrometers. We and others had earlier speculated that the variable natural abundance $\delta^{13}C$ values of CAM plants might reflect the variable contributions of C_4- and C_3-like carboxylations in the dark and light (Bender et al. 1973; Osmond et al. 1973). With time to think and colleagues to challenge, it now seems natural that notions of the "phases of CAM" should have matured in München, to emerge then in Naturwissenschftliche Rundschau (Osmond and Ziegler 1975; Fig. 1), some time before their most commonly cited source (Osmond 1978). Much more comprehensive studies have subsequently refined the above simple interpretation of $\delta^{13}C$ values in different taxa in different environments (Winter and Holtum 2002; Holtum et al. 2005a).

Unequivocal and independent confirmation of the labeling patterns in Fig. 1 followed later from gas chromatograph-mass spectrometry (GCMS) analysis of ^{13}C-malate extracted from CAM plants after exposure to $^{13}CO_2$. Only singly labelled malic acid molecules were detected in the dark (Cockburn and MacAuley 1975), with doubly and multiply labelled molecules appearing during $^{13}CO_2$ fixation in the light (Ritz et al. 1986; Osmond et al. 1988). Griffiths et al. (1990) provided the ultimate proof of the shifting carboxylation activities in the phases of CAM in-vivo with elegant on-line natural abundance isotope discrimination studies, and these also sealed the interpretations of shifting $\delta^{13}C$ values discussed below. Subsequent GCMS and nuclear magnetic resonance (NMR) studies facilitated assessment of the extent of fumarase randomization in CAM itself, and suggested that the

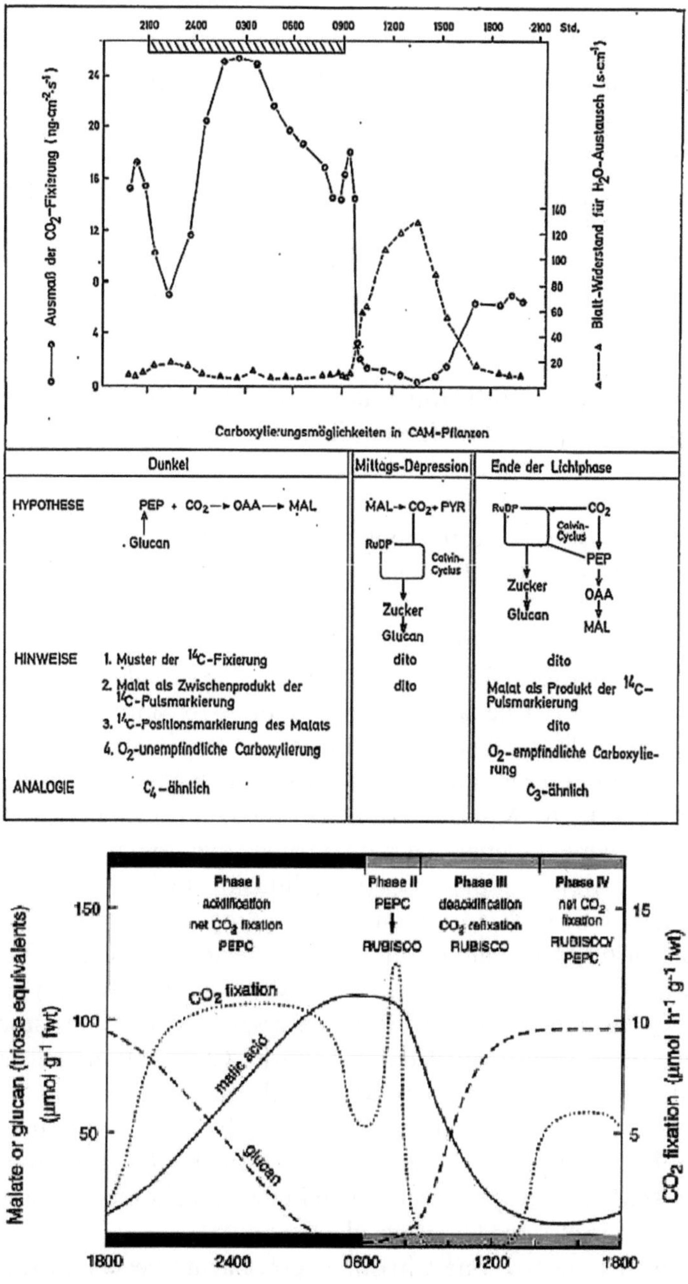

Fig. 1. Origins of the phases of CAM concept. An early summary of evidence (above) for C_4-like and C_3-like carboxylation events separated by deacidification of malic acid (Osmond and Ziegler 1975) corresponding to phases I, IV and III respectively, and (below) from a textbook chapter (Leegood et al. 1997). Phase II, the transition in carboxylation events early in the day, emerged from studies of plants in growth chambers exposed to sudden transitions in light. Perhaps one should also observe phase V, another period of transition in carboxylation events at the end of the day, especially in CAM plants exposed to slowly declining light under natural conditions. Diagrams reproduced by permission of the publishers; Wissenschaftliche Verlagsgesellschaft mbH and Pearson Education Ltd, respectively

equilibration of vacuolar, cytoplasmic and mitochondrial pools of malate changed during long-term exposures to $^{13}CO_2$ in the dark (Osmond et al. 1988). Much still remains to be done to convincingly evaluate these relationships. Although specification of the phases of CAM provided a helpful framework for a better understanding of CAM, it is important to recognize now that CAM can be much more plastic, indeed more fantastic than one then could have imagined (Dodd et al. 2003; Lüttge 2004).

1.3 Biochemistry and diffusion as determinants of the $\delta^{13}C$ value in CAM plants; improved understanding of water use efficiency in C_3 plants

Perhaps the most significant application arising from these studies of labelling patterns in CAM emerged from Marion O'Leary's interest in PEPCase in vivo. It is fair to say that interpretations of $\delta^{13}C$ values in C_3 and C_4 plants prior to 1980 were empirical, and lacked a rigorous mechanistic insight. When we were able to move the $\delta^{13}C$ value of *Kalanchoë daigremontiana* Hamet et Perrier de la Bâthie from about −16‰ to −29‰ by simply changing day-night temperature regimes and water stress exposures of plants in the same cross-gradient growth room of the Madison Biotron (Osmond et al. 1976), Marion became curious. An expert in heavy-isotope effects on enzyme kinetics, he immediately saw the merit of detailed evaluation of component processes (CO_2 diffusion, hydration to HCO_3^- and enzyme catalysis by PEPCase and malate dehydrogenase) contributing to the $\delta^{13}C$ value of CAM malate. From Marion's perspective of enzyme kinetic analyses, the carboxylation process in CAM in the dark was reporting in-vivo, as close as it gets a coupled PEPCase assay in-vitro. Many studies had shown that little else was labelled during dark $^{14}CO_2$ fixation, that there was little further metabolism of the product in the dark, and the isotopic composition of all C atoms in the product could be examined. Furthermore, because most CAM plants show substantial stomatal limitation to CO_2 diffusion, even when stomata are wide open in the dark, isotopic signatures due to diffusion were also readily detected.

Arriving in Canberra one Christmas eve, having been rained-out of a camping and walking tour of New Zealand, Marion went to work on the natural abundance ^{13}C of carbons in malate accumulated in *K. daigremontiana* and *B. tubiflorum* Harvey. His analysis of dark CO_2 fixation in CAM showed us how to sum the biophysical and biochemical components of in-vivo isotope fractionation (O'Leary and Osmond 1980). These insights soon led to a new understanding of carbon isotope fractionation in C_3 and C_4 plants with much more important consequences. It was recognized that integrated average stomatal conductance could be inferred from changes in $\delta^{13}C$ values in C_3 plants and correlated with water use efficiency (Farquhar et al. 1982). As a result, $\delta^{13}C$ values have been used in breeding programmes to select more water use efficient cultivars of wheat (Condon et al. 1990) and other crops, adding much to the value of marginal agriculture in Australia and elsewhere. At the time we were also engaged in ecophysiological studies on "prickly pear" [*Opuntia stricta* (Haw.) Haw.] that, by the 1930s, had denied access to otherwise productive land over large areas central-eastern Australia (see below). In retrospect, it is a delightful irony that, half a century later, insights from CAM should help advance cereal agriculture in the very same regions.

These insights into the importance of diffusion later led us to predict that the $\delta^{13}C$ value of malate should be somewhat less negative towards the centre of thick CAM tissues (Robinson et al. 1993). Indeed, slow diffusion of CO_2 in thick leaves of CAM plants, with low stomatal frequencies and intercellular airspaces often below 5% (Smith and Heur 1981), is manifest in photosynthetic metabolism in other interesting ways. Maxwell et al. (1997) estimated that intercellular CO_2 concentrations at the sites of Rubisco carboxylation were only 108 μbar in *Kalanchoë daigremontiana* with open stomata in air (380 μbar CO_2) during C_3 carbon assimilation in the afternoon. It should be no surprise then, that we had earlier noted clear labelling of intermediates of photorespiration during $^{14}CO_2$ feedings in phase IV (Osmond and Allaway 1974).

We now have other evidence that the high internal resistance to CO_2 diffusion also seems to manifest itself during decarboxylation in phase III. Chlorophyll fluorescence images of the efficiency of PSII are extremely heterogeneous in phase III (Rascher et al. 2001; Siebke and Osmond, unpublished), and are characterized by randomly arising and fading patches (or fronts) of higher efficiency. The heterogeneity persists during endogenous rhythms in continuous light. These suggest spatial and temporal differences in CO_2 concentration arise behind closed stomata as deacidification in some areas proceeds faster than in others. Remembering that CO_2 diffusion in wet cell walls is likely to be 3–5 orders of magnitude slower than in intercellular air spaces, the interpretation seems reasonable. Although Duarte et al. (2005) have recently demonstrated that lateral diffusion of CO_2 occurs in leaves of *K. daigremontiana* over periods of hours, the patterns observed during deacidification change with time constants of minutes. It seems likely that the smooth curves of deacidification are a product of local heterogeneity in the implementation of the regulatory cascade that control PEPCase sensitivity to malate and other elements of this part of the CAM cycle.

1.4 Regulation of CAM PEPCase in the dark and light; its role in the diurnal rhythms of CAM and in C_4 plants

The desensitization of CAM PEPCase to end product inhibition by malate in the dark and its sensitization to the same process in the light is one of the most elegant, thoroughly and creatively documented, reversible regulatory cascades of a core physiological function in plant metabolism (Nimmo 2000). Early studies by Manfred Kluge (Kluge and Osmond 1971a,b) and Irwin Ting established the distinctive K_m, high V_{max} form of PEPCase in CAM (Ting and Osmond, 1973a) in which G-6-P desensitized the enzyme to the inhibitor malic acid (Ting and Osmond, 1973b). Thanks to Manfred and others, feedback inhibition of PEPCase by malic acid was soon implicated in the regulation of dark CO_2 fixation which tended to decline as malic acid accumulated in the vacuole towards the end of the dark period (Kluge et al. 1980). However, it was the experiments of Klaus Winter during his postdoc in Canberra (Winter 1981, 1982), and those of Jones et al. (1981) in Wilkins' laboratory in Glasgow, that stimulated the search for the PEPCase regulatory cascade. We now know that in CAM high affinity CO_2 fixation in the dark continues in the face of high malic acid contents because PEPCase is phosphorylated and desensitized to inhibition by the accumulating malic acid. In the light, PEPCase is de-phosphorylated, becomes more malic acid sensitive, and is largely prevented from competing with Rubisco in a futile carboxylation cycle during deacidification (Nimmo et al. 1986, 1987).

Damped diurnal rhythms of CO_2 evolution in CO_2-free air in continuous dark, of CO_2 exchange in air in continuous light, and their temperature responses, have been distinctive and enduring features of CAM research (Wilkins 1959; Nuernbergk 1961). The early acceptance of an overriding controller was best summed up by Queiroz (1974). He noted then that *"all the available data on CAM rhythms suggest that even if malate feedback inhibition operates under certain conditions in vivo, this effect should be superimposed on a basic oscillator (of unknown nature) which underlies the coherent operation of several enzymes of the pathway"* and that "control by feedback could be more efficient if applied to an already oscillating system". As is evident above, these thoughtful assertions provoked a determined and remarkably successful assault on PEPCase regulation from the likes of Klaus Winter and the biochemists in Glasgow. However, before these studies changed the way we think about the "CAM clock", a moment of levity intruded into this otherwise serious discussion of matters circadian. The first transmission EM pictures of the photosynthetic organelles in CAM plants were published from Eldon Newcomb's laboratory in 1975. These outstanding pictures also seemed to show a "CAM clock" in elegant physical reality. The stained, ultra-thin sections revealed an "anomalous microcylinder" that presented a perfectly circular multi-point array in transverse section; quite clearly the 24-point cog of the "CAM clock" in *K. daigremontiana* (Kapil et al. 1975; see insert in their Fig. 13). In fact, the plant material had been maintained by vegetative propagation in the Madison greenhouses for decades, and presumably accumulated a virus (possibly a potyvirus; R. Milne personal communication) the structural proteins of which might have produced the "rifled" cylindrical structure responsible for the multi-pointed (18–24) cogwheel in transverse section.

In reality, explanations of the endogenous rhythms of CAM require nothing less than the careful interpolation and interpretation of the whole of CAM physiology and biochemistry. In those CAM plants that display damped diurnal rhythms of CO_2 exchange (only a handful are known in detail), we must embrace not only the coherent regulation of CO_2 uptake by PEPCase, deacidification and CO_2 release by malic enzyme, but also the refixation of CO_2 by PEPCase and Rubisco (Griffiths et al. 2002; Wyka and Lüttge 2003), as well as compartmentation dominated by malic acid fluxes into and out of the vacuole (Hafke et al. 2003), its relationships to metabolite fluxes among smaller organelles (Kore-eda et al. 2005) and of course, the even more complex coherent regulation of carbohydrate metabolism.

I confess to having long favoured the system view; that if left alone, the intricate network of physiology and biochemistry of CAM will oscillate of its own accord in continuous light and dark, so long as C-resources allow. The rather rapid dampening of the rhythm in CO_2-free air in the dark is almost certainly limited by carbohydrate reserves and respiration, and the numerous oscillations in continuous light in a variety of CAM plants obviously reflect the interactions of 2 carboxylases and differing decarboxylation options. The system view has taken strength from elegant temperature shift analyses augmented by on-line stable isotope discrimination and modelling (Grams et al. 1997), from biochemical and molecular evidence that malate overrides the circadian regulation of the PEPCase kinase (Carter et al. 1991; Borland et al. 1999), and from images of the entrainment of areas of low and high PSII efficiency during oscillations in continuous light (Rascher et al. 2001). Indeed, interpretation of these images in terms of the "*the biological clock as an assembly of coupled individual oscillators*" simply refers to independent nodes of the CAM system isolated by slow diffusion of internally generated CO_2 in a tissue with little intercellular air space connectivity.

The remarkable long-term commitment of Bohnert and Cushman to the molecular genetics of CAM in *M. crystallinum* (Cushman and Bohnert 1999) is now facilitating great progress on the circadian regulation of gene expression behind key components of the regulatory cascade (Hartwell et al. 2002;

Boxall et al. 2005; Hartwell 2005). To paraphrase Orlando Queiroz then, perhaps we really are closing on the nature of the basic oscillator that directs the coherent operation of the regulatory cascades of PEPCase and other enzymes of the pathway. To quote Boxall et al. (2005) now *"these experiments will allow us to finally resolve the nature of the circadian oscillator that controls CAM"*. But it brings little comfort to a committed CAM physiologist to learn that analogues of circadian clock genes from a widely studied, but physiologically undistinguished weed (*Arabidopsis*) can now be recognized in *M. crystallinum* and *K. daigremontiana*. Whatever, it remains a matter of some pride that the PEPCase regulatory cascade discovered in CAM plants subsequently has been shown to apply also in C_4 plants (Jiao and Chollet 1991). It works in reverse in C_4, with de-phosphorylated PEPCase less active and more malate sensitive in the dark, and the primary carboxylase more active in the light following the attention of a light activated protein kinase (McNaughton et al. 1991). Some of the debt for insights into CAM inherited from research into the C_4 pathway in the 1960–1970s, has been repaid.

1.5 Malic acid compartmentation in CAM

Since the study of Kluge (1968), it has been widely accepted that accumulation of malic acid into the vacuole of CAM plants, and its concentration in the cytosol, play central roles in regulation of CAM photosynthesis. I remember wondering why Kenyon et al. (1978) went to the trouble of isolating malate laden vacuoles from *Sedum*; where else could one put 0.5 M free organic acid? Although compartmental analysis using specific activity arguments and isotope exchange methods indicated separate cytoplasmic and vacuolar pools of malate in storage tissue discs (MacLennan et al. 1963; Osmond and Laties 1968; Cram and Laties 1974), direct attack on the compartmentation and tonoplast transport systems for malic acid in CAM was surprisingly slow to emerge (Nishida and Tominaga 1987). Not personally engaged in pursuit of these objectives, my background in organic acids and ion transport made me a keen spectator.

Early physiological experiments (Lüttge et al. 1975) explored turgor and tonoplast fluxes of malic acid as key components of the diel CAM system. At a meeting in Toronto, Canada, Ulrich and I sketched some of the implications of an ATP dependent proton pump for the energy metabolism of CAM on the back of a napkin over a meal in a Chinese restaurant. Surprised by the bioenergetic demands of malic acid compartmentation in CAM in the dark, Ulrich engaged expert opinion and published a speculative paper (Lüttge et al. 1981) that set him on a path to deep engagement with these processes that continues today. A decade later, and in the days before e-mail, Ulrich and I used a facsimile machine to hastily assemble a progress report on tonoplast fluxes for Harry Beever's retirement symposium. We made light of the notion that Harry may have lost his way in choosing not to pursue his early interest in CAM (Thomas and Beevers 1949), and had in fact overlooked the remarkable transporter properties of the largest organelle of all. Being realists, Harry's friends presented him with an original print from Beseler's *Hortus Eichstettensis*, a print of *Ricinus communis* L.

Holtum et al. (2005b) have provided an excellent summary of the present understanding of proton and malic acid fluxes into and out of the vacuole of CAM plants. Quantitative freeze-fracture analyses of membrane particles and of protein subunit patterns produced convincing evidence for the induction of vacuolar H^+-ATPase and H^+-PP_iase correlated with the induction of CAM (Lüttge and Ratajczak 1997). The picture emerges of a CAM tonoplast with a high density of H^+-ATPases (plus H^+-PP_iase) and low density of anion-selective ion channels (Hafke et al. 2003) with an apparent $K_{1/2}$ of 2.5 mM for malate^{2-} that facilitates passive movement from the cytoplasm down the electrochemical potential gradient at rates adequate to account for the high influx of malic acid into the vacuole. Efflux during phase III is less well understood, but Holtum et al. (2005) cautiously accept that the vacuolar malate carrier in *Arabidopsis*, a homolog of the human Na/dicarboxylate cotransporter (Emmerlich et al. 2003), might account for H^+ and malate efflux from the vacuole.

1.6 Light use efficiency and photoinhibition in CAM plants; the role of CO_2 supply in the avoidance of photoinactivation

If a preoccupation with CAM came to define my interest in the dark reactions of photosynthesis, early collaborations with Olle Björkman, and in the Anderson-Boardman lab in CSIRO, led me to seek links with the light reactions. As described elsewhere (Osmond and Förster 2006), Olle and I had postulated that when stomata close in response to water stress, photorespiratory CO_2 cycling in C_3 plants might continue to generate a sink for electron transport, and thus mitigate photoinhibition. It was a short step to ask whether the high internal CO_2 concentration during deacidification in phase III of CAM contributed to mitigation of photoinhibition, at least during the first part of the day. The application to CAM may have been consolidated during the long night experiments on gas exchange at Carnegie in 1974. These two streams of research came together a decade later when William Adams III became a graduate student at ANU with the opportunity to undertake a good part of his research in the Desert Research Institute, Reno, Nevada.

The best criterion then for photoinhibition in vivo was the initial slope of the light response curve of photosynthesis at CO_2 saturation. Quantum yield measurements at CO_2 saturation during phase III of CAM proved to be a straightforward matter using David Walker's O_2 electrode system. It was our good fortune that Professor K. Nishida spent some months in RSBS. A Japanese plant physiologist highly respected for his demonstration of nocturnal opening and diurnal closure of stomata in CAM using a viscous flow porometer (Nishida 1963), three generations of researchers collaborated to estimate quantum yields of O_2 evolution in phase III in CAM. By chance, gardeners had planted the highly reflective CAM plant *Cotyledon orbiculata* L. adjacent to the entrance of RSBS in Canberra. With this plant we not only had an opportunity to evaluate the role of internal CO_2 supply, but the ability to decrease reflectance (and hence increase absorption) by simply brushing the wax surface. William's first paper on quantum yield in CAM (Adams et al. 1986) appeared practically simultaneously with the wider survey of O_2 evolution in C_3 plants by Demmig and Björkman (1986). What happened next is now history as the Adams-Demmig-Adams partnership emerged as one of the most impressive teams in modern plant ecophysiology.

William also used *K. daigremontiana* to clearly show that if malic acid pools in CAM were reduced by exposure to N_2 atmospheres in the dark, the extent of photoinhibition was exaggerated (Adams and Osmond 1988). The role of CAM in conserving respiratory CO_2 and mitigating photoinhibition in tropical ferns was soon confirmed by Griffiths et al. (1989). Meanwhile, the extreme light environment of the Mohave Desert and the extensive work of Irwin Ting on the beautiful, but optically opaque beaver-tail cactus *Opuntia basilaris* Engelm, and Bigelow made this plant a natural choice for investigation of photoinhibition on sun and shade exposed sides of the cladodes. I was not much help to the project, once requiring repatriation from the field after falling victim to severe sunstroke while working in Grapevine Canyon, Death Valley. William's 77K chlorophyll fluorescence and quantum yield data (Adams et al. 1987) were interpreted in terms of our then understanding of photoinhibition. Within 2–3 years our understanding of photoinhibition changed rapidly as the importance of photoprotection associated with the xanthophyll cycle became evident. This is perhaps most clearly shown in the later study (Adams et al. 1989) of *Nopalea cochinillifera* (L.) Salm-Dyck (synonym for *Opuntia cochenillifera* (L.) P. Mill.) in Venezuela.

We came full circle in this aspect of CAM when Sharon Robinson showed that removal of the reflective wax from *C. orbiculata* increased the depth within the tissue to which conversion of violaxanthin to zeaxanthin could be detected (Robinson and Osmond 1994). Her study neatly integrated the concepts of external photoprotection by reflectance, and internal photoprotection by the xanthophyll cycle. Nevertheless, there is no doubt that the CO_2 concentrating mechanism of CAM provides an excellent natural model for the importance of access to CO_2 in sustaining high photosynthetic efficiency in strong light. It is clearly a less ambiguous general model for the role of internal CO_2 generation in mitigation of photoinactivation than our earlier hypothesis for the role of photorespiratory carbon cycling in C_3 plants when stomata close in response to water stress.

1.7 Recycling of respiratory CO_2 in CAM and diel variation in the engagement of cyanide insensitive respiration

Recycling of respiratory CO_2 in phase I is a feature of CAM cycling and CAM idling (Kluge and Ting 1978), but engagement of the CAM pool of malate in tricarboxylic acid (TCA) cycle metabolism and mitochondrial electron transport remains to be assessed in detail. Mitochondria isolated from CAM plants have a high capacity of the alternative oxidase in vitro (Rustin and Quiroz-Claret 1985), but as with all such studies, evaluation of activity and function in vivo remained elusive. The creative insights of Joe Berry, Jim Siedow and Dan Yakir, the technical excellence of Larry Giles and the post-doctoral confidence of Sharon Robinson and Miguel Ribas-Carbo, were directed to meddling with an O_2 electrode attached to a mass-spectrometer made available by a grant from the broadminded North Carolina Biotechnology Center. The team discovered that alternative oxidase activity peaked in phase III of CAM, just as predicted to assure the least adenylate control and maximum TCA cycle flexibility during deacidification (Robinson et al. 1992). Their method proved generally applicable, and has now given better insights into the functional significance of cyanide-insensitive respiration in other plants (Robinson et al. 1995).

When fitted with a gas phase CO_2 electrode, David Walker's O_2 electrode system revealed that stoichiometric nightmares of net O_2 and CO_2 exchanges prevail in all phases of CAM (Osmond et al. 1996). Again, Maxwell et al. (1998) showed that these can only begin to be resolved when the CAM tissues are attached to a mass spectrometer tuned for tracer experiments with $^{13}CO_2$ and $^{18}O_2$. In the course of these experiments, we were confronted with a simple but poorly understood feature of CAM; the extraordinarily high concentrations of O_2 that accumulate in the closed system during phase III conversion of a soluble CO_2 source (malic acid) to soluble and insoluble products (sugars and starch) accompanied by photosynthetic O_2 evolution. The process could be observed by allowing the CO_2 concentration to oscillate over a narrow range in short light-dark cycles while O_2 concentration increased in each light period (Osmond et al. 1996, 1999). The over-pressure of O_2 was clearly audible when the chamber was opened at the conclusion of each experiment.

We estimated that deacidification of 100 μmol malic acid g^{-1} fwt could generate up to 9 ml O_2 in a tissue with a gas space of about 50 μl ml^{-1}. The high internal CO_2 concentrations generated during deacidification seem adequate to mitigate the oxygenase activity of Rubisco and minimize photorespiration during CO_2 fixation (Osmond et al. 1999; Lüttge 2002), but the O_2 partial pressures that build up behind closed stomata in phase III of CAM are probably higher than in any other living system. One had encountered anecdotes about this phenomenon from time to time, but now we can be grateful to Krätz (2001) Ulrich Lüttge and Otto Lange for pointing us to the discovery of O_2 over-pressure in *Clusia rosea* Jacq. by Alexander von Humboldt, in the field, in 1800! When a cut leaf was illuminated with its petiole in water, the gas bubbles released from the petiole were found to contain 30–35% O_2. The implications for oxidative stress are obvious, but still need to be probed in depth (Broetto et al. 2002). Many investigations of the potential for photo-oxidative stress associated with photosynthetic O_2 evolution in the chloroplast have been frustrated by problems of access, detection of reactive oxygen species, and the remarkable multiplicity of metabolic pathways. Frankly, why would one look beyond the CAM system to discover what really matters in oxidative stress?

1.8 Field CAMpaigns

From the above, it is clear that my CAM research has been largely lab-bound. I was much aware of the advances arising from Irwin Ting's fieldwork, and while in München, made brief excursions to dry sites in the Alps with Professor Ziegler. We were also impressed by the comprehensive ecophysiological and biophysical evaluation of massive CAM succulents being undertaken by Park Nobel. His book (Nobel 1988), complete with cover illustrations provided by the US Postal Service, made it clear that field CAMpaigns should be directed elsewhere. However, there was a time when visitors to the RSBS laboratory were pressed into brief field trips over long distances to study "prickly pear" (Osmond et al. 1979a). Unfortunately, our primitive hands-on equipment made it difficult to avoid the annoying, tiny barbed glaucids of this otherwise benign succulent; one hopes now all is forgiven!

Introduced as a hedge for vineyards in 1846, today's highly dispersed remnant populations of *Opuntia stricta* give little clue to the impact of this invasive CAM plant in central-eastern Australia. By 1930 this succulent had become a noxious weed of immense proportions, having expanded to deny agricultural and pastoral access to an area as large as that of the British Isles, often attaining 500 or more tonnes fresh weight per hectare (Osmond and Monro 1981). As early as 1915 investigators knew they were dealing with something different. A chemical control agent (arsenic pentoxide in 15% sulphuric acid) proved more effective when sprayed at night (presumably because stomates were more open at night), and thousands of farmers knew that cladodes of prickly pear seemed to survive "on air", without roots. However, nocturnal malic acid synthesis and the water use efficiency of CAM in *Opuntia* did not attract much interest among Australian plant physiologists of my generation, possibly because the "prickly pear" problem had been resolved by one of the most spectacular biological control regimes ever established.

Research by Gert Stange, a visual neurophysiologist now working on bio-inspired autopilots for flying robots, revealed that the continued highly effective biological control of this noxious invasive CAM plant depends to a significant extent on exquisitely sensitive CO_2 detectors in the mouth parts of female moths *Cactoblastis cactorum* Berg (Stange et al. 1995; Stange 1997). Ever since their introduction in the late 1920s, female *C. cactorum* have been making boundary layer CO_2 profiles 1–5 mm above any surface upon which they alight during their early evening perambulations. Possibly the most persistent and numerous CAM researchers on the planet, after detecting a surface with net CO_2 influx in the dark, the female *C. cactorum* proceed to deposit eggs on the glaucids in the aureoles of *Opuntia* cladodes, the most uncomfortable but most secure spot on the cladode surface.

There is probably a lot more involved in host plant detection (Pophof et al. 2005) but measurement of nocturnal CO_2 influx evidently remains a good targeting system throughout the range of *O. stricta* in south-eastern Australia (there is only one native CAM plant in the region; *Sarcostemma australe* R.Br.). Larvae of *C. cactorum* simply burrow into the cladodes and totally devour them from within. Millions of tonnes of *O. stricta* were consumed by trillions of larvae and after three damped cycles of devastation and recovery over a decade, *O. stricta* ceased to be a problem (Osmond and Monro 1981). Half a century later, when ecophysiologists began to study the dispersed remnant populations of the CAM invader, we quickly gained an appreciation of the tight population dynamics underlying this unobtrusive, highly effective biological control. Unless clumps of *O. stricta* were isolated by insect netting, *C. cactorum* usually found and devoured its host before we could return to complete seasonal measurements of CAM (Osmond et al. 1979a,b).

I may have been cured of CAM fieldwork by the *Stylites* expedition of 1982. In an unguarded moment during a discourse on the stomate free cuticles of fossil plants at the Sydney Botanical Congress in 1981, John Raven leaned over and muttered "they must have been CAM then". He was overheard by Jon Keeley, who retorted "I know just where to find such CAM plants". The cover article in *Nature* that followed should have been headlined with the best insight of the trip ("How does Stylites CAMpeat?"; courtesy of Sterling Keeley). However, the remarkable capacity of the sporophylls of this taxon to recycle CO_2 from its peaty root zone via CAM in chloroplast containing cells surrounding the air canals connected to gas columns in living roots was a good story in itself. The fieldwork cure took the form of excruciating headaches (from inattention to altitude in the Peruvian Andes), and a really debilitating stomach problem (from inattention to ethnic culinary style in restaurants) that almost led to a JAL cabin crew being quarantined at Tokyo airport over Christmas.

As it was, Klaus Winter organized my first and only field encounter with CAM in the tropics (Winter et al. 1986), and I regret having been distracted by other things during the "grand era" of tropical CAM ecophysiology. One cannot but admire the results and impact of expeditions to tropical CAMscapes; to the Bromeliads in Trinidad (Griffiths and Smith 1983), to the Clusias in the Caribbean and tropical Americas (Ting et al. 1985) and to the diverse stem succulents of Madagascar (Kluge et al. 1991). These expeditions have stimulated our wider appreciation of CAM plants well beyond the Fensterpflanzen of Carl Spitzweg. The recent history of CAM ecophysiology is remarkable for the ways fieldwork has led to creative laboratory investigations, ranging from use of isolated vacuoles from orchids for transport studies (White and Smith 1992) to the establishment of records such as 1.4 M titratable protons in vacuoles of *Clusia* (Borland et al. 1992) and to records in xanthophyll photoprotection set by *Guzmania* (Maxwell et al. 1994).

As one may sense from Black and Osmond (2003), I thought that my dreams for field work with CAM plants might be realized after accepting the challenge to lead the Biosphere 2 Laboratory to fulfil its potential for experimental ecosystem and climate change research. This was by no means the highest priority for research in the beautifully situated, extraordinary facility in the higher Sonoran Desert, but we set out to examine some system level impacts of CAM under controlled conditions. An enclosed environmentally controlled facility, retrofitted and named in honour of Manfred Kluge and Ulrich Lüttge, Nobel and Bobich (2003) explored the carbon sources for new root growth of *Opuntia* spp. following summer precipitation, and Rascher et al. (2006) demonstrated that the imprints of the phases of CAM can be discerned in net ecosystem CO_2 exchanges.

The premature termination by Columbia University of its 10-year commitment to the transformation of this unique apparatus, just as its potential for global change research was becoming so clearly evident, will be remembered by many as one great opportunity lost. In a lighter vein, I regret that we did not get to evaluate the havoc that must be wrought on CAM in *Carnegia gigantea* (Engelm.) Britt. & Rose draped each night in suburban gardens of Arizona with garlands of red, white or blue LEDs for weeks at a time during the festive season. Although I pointed out the misinformation on a panel at the Arizona Sonora Desert Museum (that asserts CAM plants have the remarkable ability to split H_2O to O_2 in the dark!) nothing had changed on my last visit.

1.9 Origins of CAM and its future prospects in a high CO_2 world

The evolution of CAM is a question that is best left to the experts. Aside from noticing that the CCM upon which CAM depends simply involves internalizing the CO_2 atmospheres of the Cretaceous, I have not paid much attention to the problem. In an unguarded moment at a photosynthesis congress in Brussels, shortly after the encounter with *Stylites*, and no doubt while still overly impressed with contemporary progress in CAM, I was drawn to speculate that it might represent an ancestral, terrestrial photosynthetic metabolism for all seasons. Now in one's dotage even a firm believer in the principle that one should either put up (experimental evidence) or shut up, may be allowed a little fanciful speculation. If one contemplates the primal cellular requirements of cation–anion balance, pH stasis and turgor in a vacuolated autotroph during metabolism of nitrate (a shadow of *Atriplex* here!), then the anaplerotic, cytoplasmic synthesis and vacuolar accumulation of malate in CAM seems an admirably comprehensive compromise. If recycling of respiratory CO_2 and conservation of water are of further selective advantage to the C-balance of a multi-cellular terrestrial system, we seem presented with the key design elements for CAM. Lüttge et al. (2000) took a step towards evaluation of these possibilities in their demonstration of linkages between the housekeeping function of vacuolar malate transport for cation balance in tobacco under nitrate (but not ammonium) nutrition, and the special role of malic acid fluxes in CAM.

Some more thoughtful commentators now seem not to dismiss the above notions out of hand in their discussions of the physiological and biochemical realities of C-acquisition in aquatic, and subsequently in water limited terrestrial habitats. Keeley and Rundel (2003) entertained the possibility that the temporally regulated survival attributes of CAM in aquatic and terrestrial habitats may well have predated the more productive, spatially regulated attributes of C_4 metabolism. Sage (2002) made an outstanding case that the developmental, physiological and biochemical analogies between the CCMs in CAM and C_4 metabolism are probably mutually exclusive in functional terms, except perhaps if spatially and temporally separated in *Portulacca* (Guralnick et al. 2002). Unlike C_4 photosynthesis where the CCM is associated with the vasculature and defined by suberin, the CCMs of CAM plants cannot be expected to leave much of an anatomical signature during fossilization. On the other hand, as Uwe Rascher has pointed out, woody CAM plants that rarely if ever engage in CO_2 assimilation in phase IV, might provide long lasting less negative $\delta^{13}C$ signatures that are relatively unresponsive to environment variables.

Whatever, the safe position in these matters is to retreat to the notion that CAM is a remarkably flexible, niche-filling form of photosynthetic metabolism originating in many families that is found now in some 16–20,000 ancient and modern aquatic and terrestrial species. Crayne et al. (2004) make the point that terrestrial and epiphytic niche differentiation seems to have been important during evolution in different genera in the Bromeliaceae. In this "Arabicentric era" of plant biology, Hartwell (2005) speaks for all in this field when he asserted that "*we overlook such valuable adaptations (as CAM) at our own peril in the face of current predictions of global warming*". In spite of the efforts of the dominant mammal (humankind) to restore Cretaceous-like atmospheres on Earth by profligate combustion of fossil photosynthates, the CCM of CAM may confer advantages in many hot tropical and seasonally arid climates for some time to come. Perhaps one should be concerned for future biological control of "prickly pear" because, as atmospheric [CO_2] continues to increase, the ability of *Cactoblastis* to detect inwardly directed CO_2 gradients over *Opuntia* cladodes at night is compromised (Stange 1997). Perhaps we should be more concerned with the possibility that promotion of profligate use of fossil fuels may be seen by some administrations as a convenient way of burning much of the evidence for biological evolution itself!

1.10 A view from over the hill

I concluded an earlier review of CAM with a quote from Nehemiah Grew that promised research would continue to be an uphill quest (Osmond 1978). Fortunately perhaps, I seem to find myself wandering the foothills still, confronted with so many questions in CAM demanding to be addressed that one scarcely knows where to turn. A new set of generic questions seems to turn up with each forward step, and we remain somewhat blind to several major issues. For example, we have been preoccupied with elucidation of the signature metabolism of CAM, with malic acid synthesis and degradation, and have paid less attention to what may be even more sophisticated regulation of carbohydrate metabolism and its relation to growth. The fundamental question of how CAM plants preserve carbohydrate reserves in the light for acid synthesis in the dark, while at the same time providing carbon for growth remains enigmatic. In particular, the natural abundance stable isotope evidence for discreet pools of carbohydrates engaged in dark CO_2 fixation and growth (Deleens et al. 1979) challenges all present models of metabolic compartmentation. It is astonishing, but the first detailed diel carbon allocation budgets of CAM plants seem to have been published by Borland (1996) and Borland and Dodd (2002).

Two of my earliest PhD students made important contributions to carbohydrate stoichiometry in CAM (Sutton 1975a,b) and to gluconeogenesis (Holtum and Osmond 1981); studies that have continued with subsequent students (Christopher and Holtum 1996, 1998). Carbohydrate metabolism in CAM became a cornerstone in Clanton Black's lab, with discovery of the novel pyrophosphate dependent 6-phosphofructokinase (Carnal and Black 1979) and recognition that sugars in the vacuole, rather than starch in the chloroplast, are the source of substrates for acidification in pineapple and other species (Black et al. 1996). Clearly, major issues of carbohydrate compartmentation and transport in CAM, such as vacuolar sucrose fluxes (McRae et al. 2002), glucose and glucose-6-phosphate fluxes into and out of chloroplasts, and the unusual regulatory relationships with fructose 2,6 bisphosphate (Fahrendorf et al. 1997), need close attention. Given the revolution in our understanding of starch metabolism in leaves (Smith et al. 2005), the unusual demands of CAM may soon be placed in context (Holtum et al. 2005).

When do CAM plants actually grow? Years ago, Bill Allaway weighed *K. daigremontiana* grown in controlled environments in the Canberra Phytotron that were designed to vary the amount of CO_2 fixed in phase IV, and as expected these plants grew faster than those confined to CO_2 uptake in phase I. Ignoring for the moment that we do not know much about the diel growth properties of leaves from more than a handful of tame C_3 plants that mostly grow at night, we have been surprised to find that CAM plants grow in the day during phase III (Gouws et al. 2005). A few moments of reflection were enough to convince us that this made a lot of sense. During phase III carbon skeletons need to be conserved in starch for the next night of malic acid synthesis, but in the absence of phase IV net CO_2 fixation, 25% of malate carbon could become available for growth and maintenance. Moreover, because turgor peaks in phase III, and cytoplasmic pH declines, this phase of CAM seems the most propitious for growth of leaves and cladodes.

The ecophysiological literature suggested a further test, using the conveniently planar leaves of *Clusia minor* L. that switch from CAM to C_3 and back again in response temperature and vapour pressure manipulations (Schmitt et al. 1988). It was at this point we were reminded that ecophysiologists in general are prone to study fully expanded (non-growing) mature leaves, that not much is known of CO_2 fixation in young expanding leaves, and that we now need to revisit the whole mysterious citric acid metabolism of *Clusia*. A recent comprehensive analysis of nocturnal acid synthesis and carbohydrate consumption in strongly CAM *Clusia hilariana* Schlecht demonstrated a remarkably tight sugar-malate dominated stoichiometry (Berg et al. 2004). However, the sole pulse-chase study of these processes (Olivares et al. 1993) leaves open many questions of citrate metabolism that still need to be explored C atom by C atom using $^{13}CO_2$ and GCMS (Osmond et al. 1996). We also now need to study source-sink relationships in growing photosynthetic tissues in CAM plants, taking cues from *Opuntia ficus-indica* (L.) P. Mill. (Wang et al. 1988) and the carbon budget experiments of Borland and Dodd (2002).

Not much of the above research has appeared in generalist journals now held to be of high impact by the "accountants" who now define the parameters for academic promotions and award of competitive grants. Rather, stalwart plant biology journals, served by broadminded editors and reviewers, have facilitated the reporting of CAM research. Consequently, one is no longer surprised to find the feats of CAM physiology and biochemistry cited in headlines in higher impact journals justifying attention to particular genes in *Arabidopsis* (Emmerlich et al. 2003). Preparation of this chapter has reminded me again of just how much there is in CAM that is new, sometimes entirely unexpected and highly specific (Epimashko et al. 2004), but also much that is relevant and indispensable to mainstream plant biology (Hafke et al. 2003; Boxall et al. 2005). This functionally distinctive but economically undistinguished sector of plant biodiversity has added significantly to our genetic, biochemical, biophysical, physiological and ecological understanding of plant biology in general.

Some 40 years of peripatetic engagement in CAM research, from privileged research positions in RSBS at ANU, in Reno, at Duke and briefly in Columbia, have left me with several enduring perspectives:

- First, in all the glorious functional biodiversity of plant systems, one can rarely predict where the next significant insight may arise. The huge impact of ecophysiological studies of tropical epiphytes and stranglers on

widening our understanding of CAM beyond the Crassulaceae is a case in point. Although it was clear that research in CAM would contribute much to the understanding of PEPCase regulation, tonoplast malate translocators and circadian rhythms, it was less obvious that one could have anticipated its contributions to breeding water-use efficient crops, assessment of alternative oxidase in-situ, or the still to be exploited potential of CAM in studies of oxidative stress.
- Second, the collegial networks so critical to progress in research know no boundaries. Even at times when it seemed that different views of the same phenomenon had become polarized, common ground ultimately emerged. Witness for example, the insights that continue to emerge from application of diverse biophysical and molecular genetic approaches to circadian rhythms in CAM (Borland et al. 1999; Lüttge 2000; Nimmo 2000; Hartwell 2005).
- Third, it becomes clear with the passage of time that one's efforts must be directed to sustaining individual creativity, collaborative activity and achievement among younger colleagues.

Is there a bottom line? Several of my mentors in research offered outstanding proof, in their time, of Medawar's axiom; "*If politics is the art of the possible, research is surely the art of the soluble. Both are immensely practical-minded affairs*". They bridged the apparent divide between research and politics with spectacular solutions that enhanced the possibilities for science. Clearly, research politics have not been my forte. Although I have striven, I have not prevailed, in spite of outstanding opportunities to change the ways we think about things, and go about them. In the end, one remains simply grateful to many companions in CAM, and in other pathways, whose integrity and creativity have made his career in botanical research so much more fun, so much more fascinating, and possibly somewhat more enduring, than all the other efforts.

Acknowledgements

In addition to the Editors, and to the authors and colleagues cited above, I would like to express particular thanks to Drs. Britta Förster, Howard Griffiths, Joe Holtum and Uwe Rascher for reading drafts of the manuscript. The chapter evolved in the summer of 2005 thanks to the hospitality, support and stimulation in research from Dr. Uli Schurr and colleagues in the Institut Phytosphäre (ICG III), Forschungszentrum Jülich.

References

Adams WW III, Osmond CB (1988) Internal CO_2 supply during photosynthesis of sun and shade grown CAM plants in relation to photoinhibition. Plant Physiol 86:117–123

Adams WW III, Nishida K, Osmond CB (1986) Quantum yields of CAM plants measured by photosynthetic O_2 evolution. Plant Physiol 81:297–300

Adams WW III, Smith SD, Osmond CB (1987) Photoinhibition of the CAM succulent *Opuntia basilaris* growing in Death Valley; evidence from 77K fluorescence and quantum yield. Oecologia 71:221–228

Adams WW III, Diaz M, Winter K (1989) Diurnal changes in photochemical efficiency, the reduction state of Q, radiationless energy dissipation and non-photochemical fluorescence quenching in cacti exposed to natural sunlight in the field in northern Venezuela. Oecologia 80:553–561

Bender MM, Rouhani I, Vines HM, Black CC (1973) $^{13}C/^{12}C$ ratio changes in Crassulacean acid metabolism plants. Plant Physiol 52:427–430

Berg A, Orthen B, Arcoverde de Mattos E, Duarte HM, Lüttge U (2004) Expression of crassulacean acid metabolism in *Clusia hilarana* Schlechtendal in different stages of development in the field. Trees 18:553–558

Björkman O, Osmond CB (1974) Effect of oxygen on carbon dioxide exchange in *Kalanchoë daigremontiana*. Carnegie Institution of Washington Yearbook 73:852–859

Black CC, Chen J–Q, Doong RL, Angelov MN, Sung SJS (1996) Alternative carbohydrate reserves used in the daily cycle of crassulacean acid metabolism. In: Winter K, Smith JAC (eds) Crassulacean acid metabolism. Ecological studies vol 114. Springer Verlag, Berlin, pp 31–45

Black CC, Osmond CB (2003) Crassulacean acid metabolism photosynthesis; "working the night shift". Photosynthesis Res 76:329–341

Borland AM (1996) A model for the partitioning of photosynthetically fixed carbon during the c3-CAM transition in Sedum telephium. New Phytol 134:433–444

Borland AM, Dodd AN (2002) Carbohydrate partitioning in crassulacean acid metabolism plants: reconciling potential conflicts of interest. Funct Plant Biol 29:707–716

Borland AM, Griffiths H, Maxwell K, Broadmeadow MGJ, Griffiths NM, Barnes JD (1992) On the ecophysiology of Clusiaceae in Trinidad: expression of CAM in *Clusia minor* L. during the transition from wet to dry season and characterization of the endemic species. New Phytol 122:349–357

Borland AM, Hartwell J, Jenkins GI, Wilkins MB, Nimmo HG (1999) Metabolite control overrides circadian regulation of phosphoenolpyruvate carboxylase kinase and CO_2 fixation in crassulacean acid metabolism. Plant Physiol 121:889–896

Boxall SF, Foster JM, Bohnert HJ, Cushman JC, Nimmo HG, Hartwell J (2005) Conservation and divergence of circadian clock operation in a stress-inducible crassulacean acid metabolism species reveals clock compensation against stress. Plant Physiol 137:969–982

Bradbeer JW, Ranson SL, Stiller M (1958) Malate synthesis in Crassulacean leaves. I. The distribution of ^{14}C in malate of leaves exposed in $^{14}CO_2$ in the dark. Plant Physiol 33:66–70

Broetto F, Lüttge U, Ratajczak R (2002) Influence of light intensity and salt treatment on mode of photosynthesis and enzymes of the antioxidative response system of *Mesembryanthemum crystallinum* L. Funct Plant Biol 29:13–23

Carnal NW, Black CC (1979) Pyrophosphate-dependent phosphofructokinase, a new glycolytic enzyme in pineapple leaves. Biochem Biophys Res Commun 86:20–26

Carter PJ, Nimmo HG, Fewson CA, Wilkins MB (1991) Circadian rhythms in the activity of a plant protein kinase. EMBO J 10:2063–2068

Christopher JT, Holtum JAM (1996) Patterns of carbohydrate partitioning in leaves of Crassulacean acid metabolism species during deacidification. Plant Physiol 112:393–399

Christopher JT, Holtum JAM (1998) Carbohydrate partitioning in the leaves of Bromeliaceae performing C_3 photosynthesis or Crassulacean acid metabolism. Aust J Plant Physiol 25:371–376

Cockburn W, McAuley A (1975) Pathway of dark CO_2 fixation in CAM plants. Plant Physiol 55:87–89
Cockburn W, Ting IP, Sternberg LO (1979) Relationships between stomatal behaviour and internal carbon dioxide concentration in crassulacean acid metabolism plants. Plant Physiol 63:1029–1032
Condon AG, Farquhar GD, Richards RA (1990) Genotypic variation in carbon isotope discrimination and transpiration efficiency in wheat. Leaf gas exchange and whole plant studies. Aust J Plant Physiol 17:539–552
Cram WJ, Laties GG (1974) Kinetics of bicarbonate and malate exchange in carrot and barley root cells. J Exp Bot 25:11–27
Crayne DM, Winter K, Smith JAC (2004) Multiple origins of crassulacean acid metabolism and the epiphytic habit in the Neotropical family Bromeliaceae. Proc Natl Acad Sci USA 101:3703–3708
Cushman JC, Bohnert HJ (1999) Crassulacean acid metabolism: molecular genetics. Annu Rev Plant Physiol Plant Mol Biol 50:305–332
Deléens E, Garnier-Dardart J, Querioz O (1979) Carbon-isotope composition of intermediates of the starch-malate sequence and the level of crassulacean acid metabolism in leaves of *Kalanchoë blossfeldiana* Tom Thumb. Planta 146, 441–449
Dodd AN, Borland AM, Haslam RP, Griffiths H, Maxwell K (2002) Crassulacean acid metabolism; plastic, fantastic. J Exp Bot 53:569–580
Duarte HM, Jakovljevic I, Friedemann Kaiser F, Lüttge U (2005) Lateral diffusion of CO_2 in leaves of the crassulacean acid metabolism plant *Kalanchoë daigremontiana* Hamet et Perrier. Planta 220:809–816
Emmerlich V, Linka N, Reinhold T, Hurth, MA Traub, M Enrico Martinoia, M, Neuhaus HE (2003) The plant homolog to the human sodium/dicarboxylic cotransporter is the vacuolar malate carrier. Proc Natl Acad Sci USA 100:11122–11126
Epimashko S, Meckel T, Fischer–Schliebs E, Lüttge U, Thiel G (2004) Two functionally different vacuoles for static and dynamic purposes in one plant mesophyll leaf cell. Plant J 37:294–300
Fahrendorf T, Holtum JAM, U. Mukherjee U, Erwin Latzko E (1997) Fructose 2,6-bisphosphate, carbohydrate partitioning, and crassulacean acid metabolism. Plant Physiol 84:182–187
Farquhar GD, O'Leary MH, Berry JA (1982) On the relationship between carbon isotope discrimination and the intercellular carbon dioxide concentration in leaves. Aust J Plant Physiol 9:121–137
Gouws LM, Osmond CB, Schurr U, Walter A (2005) Distinctive diel growth cycles in leaves and cladodes of CAM plants: complex interactions with substrate availability, turgor and cytoplasmic pH. Funct Plant Biol 32:421–428
Grams TEE, Borland AM, Roberts A, Griffiths H, Beck F, Lüttge U (1997) on the mechanism of the re-initiation of endogenous crassulacean acid metabolism rhythm by temperature changes. Plant Physiol 113:1309–1317
Griffiths H, Smith JAC (1983) Photosynthetic pathways in the Bromeliaceae of Trinidad: relations between life-forms, habitat preference and the occurrence of CAM. Oecologia 60:176–184
Griffiths H, Ong BL, Avadhani PN, Goh CJ (1989) Recycling of respiratory CO_2 during crassulacean acid metabolis: alleviation of photoinhibition in Pyrossia piloselloides. Planta 179:115–122
Griffiths H, Broadmeadow MSJ, Borland AM, Hetherington CS (1990) Short-term changes in carbon-isotope discrimination identify transitions between C_3 and C_4 carboxylation during crassulacean acid metabolism. Planta 186:604–610
Griffiths H, Maxwell K, Helliker B, Roberts A, Haslam RP, Girnus J, Robe WE, Borland AM (2002) Regulation of Rubisco activity in crassulacean acid metabolism plants: better late than never. Funct Plant Biol 29:689–696

Guralnick LJ, Edwards G, Ku MSL, Hockema B, Franceschi VR (2002) Photosynthetic and anatomical characteristics in the C_4-crassulacean acid metabolism plant *Portulaca grandiflora*. Funct Plant Biol 29:726–773

Hafke JB, Hafke Y, Smith JAC, Lüttge U, Thiele G (2003) Vacuolar malate uptake is mediated by an anion-selective inward rectifier. Plant J 35:116–128

Halliday J (1985) The Australian wine compendium. Angus and Robertson, North Ryde, p 53

Hartwell J (2005) The circadian clock in CAM plants. In: Hall A, McWatters H (eds) Endogenous plant rhythms. Blackwell, Oxford, pp 211–236

Hartwell J, Nimmo GA, Wilkins MB, Jenkins GI, Nimmo HG (2002) Probing the circadian control of phosphoenolpyruvate carboxylase kinase expression in *Kalanchoë fedtschenkoi*. Funct Plant Biol 29:663–666

Holtum JAM, Osmond CB (1981) The gluconeogenic metabolism of pyruvate during deacidification in plants with crassulacean acid metabolism. Aust J Plant Physiol 8:31–44

Holtum JAM, Aranda J, Virgo A, Gehrig HH, Winter K (2005a) $\delta^{13}C$ values and crassulacean acid metabolism in *Clusia* species from Panama. Trees Struct Funct 18:658–668

Holtum JAM, Smith JAC, Neuhaus HE (2005b) Intercellular transport and pathways of carbon flow in plants with crassulacean acid metabolism. Funct Plant Biol 32:429–449

Jiao JA, Chollet R (1991) Post translational regulation of phosphoenolpyruvate carboxylase in C_4 and CAM plants. Plant Physiol 95:981–985

Jones R, Buchanan IC, Wilkins MB et al. (1981) Phosphoenolpyruvate carboxylase from the crassulacean plant Bryophyllum-fedtschenkoi Hamet et Perrier—activity changes and kinetic–behavior in crude extracts. J Exp Bot 32:427–441

Kapil RN, Pugh TD, Newcomb EH (1975) Microbodies and an anomalous "microcylinder" in the ultrastructure of plants with Crassulacean acid metabolism. Planta 124:231–244

Keeley JE, Rundel PW (2003) Evolution of CAM and C_4 carbon-concentrating mechanisms. Int J Plant Sci 164, S55–S77

Kenyon WH, Kringstad R, Black CC (1978) Diurnal changes in the malic acid content of vacuoles isolated from the Crassulacean acid metabolism plant, Sedum telephium. FEBS Lett 94:281–283

Kluge M (1968) Untersuchungen über den Gaswechsel von Bryophyllum während der Lichtperiode II. Beziehungen zwischen dem Malatgehalt des Blattgewebes und der CO_2–Aufname. Planta 80:358–377

Kluge M, Osmond CB (1971a) Pyruvate, Pi dikinase in Crassulacean acid metabolism. Naturwissenschaften 58:514–515

Kluge M, Osmond CB (1971b) Studies on phosphoenolpyruvate carboxylase and other enzymes of Crassulacean acid metabolism of Bryophyllum tubiflorum and Sedum praealtum. Zeitschr Pflanzenphysiol 66:97–105

Kluge M, Ting IP (1978) Crassulacean acid metabolism. analysis of an ecological adaptation. Springer, Berlin, Heidelberg, New York

Kluge M, Böcher M, Jungnickel G (1980) Metabolic control of crassulacean acid metabolism; evidence for diurnally changing sensitivity against inhibition by malate of PEP carboxylase in Kalanchoë tubiflora. Zeitschr Pflanzenphysiol 97:197–204

Kore-eda S, Naoke C, Ohishi M, Ohnishi J-I, Cushman JC (2005) Transcriptional profiles of organellar metabolite transporters during the induction of crassulacean acid metabolism in *Mesembyranthemum crystallinum*. Funct Plant Biol 32:451–446

Krätz O (2001) Alexander von Humboldt (1769–1859) auf Pflanzenjagd. Palmengarten 53:33–46

Laetsch WM (1970) Chloroplast structural relationships in leaves of C_4 plants. In: Hatch MD, Osmond CB, Slatyer RO (eds) Photosynthesis and photorespiration. Wiley–Interscience, NY, pp 323–349

Leegood RC, von Caemmerer S, Osmond CB (1997) Metabolite transport and photosynthetic regulation in C4 and CAM plants. In: Dennis DT, Turpin DH, Leferbvre DD, Layzell DB (eds) Plant metabolism. Addison Wesley Longman, Harlow, pp 341–369

Lüttge U (2000) The tonoplast functioning as the master switch for circadian regulation of crassulacean acid metabolism. Planta 211:761–769

Lüttge U (2002) CO_2-concentrating: consequences in crassulacean acid metabolism. J Exp Bot 53:2131–2142

Lüttge U (2004) Ecophysiology of Crassulacean acid metabolism (CAM). Ann Bot 93:629–652

Lüttge U, Ratajczak R (1997) The physiology, biochemistry, and molecular biology of the plant vacuolar ATPase. Adv Bota Res 25:253–296

Lüttge U, Kluge M, Ball E (1975) Effects of osmotic gradients on vacuolar malic acid storage. A basic principle in oscillatory behaviour of of crassulacean acid metabolism. Plant Physiol 56:613–616

Lüttge U, Smith JAC, Margio G, Osmond CB (1981) Energetics of malate accumulation in the vacuoles of CAM cells. FEBS Lett 126:81–84

Lüttge U, Pfeifer T, Fischer–Schliebs E, Ratajczak R (2000) The role of vacuolar malate–transport capacity in Crassulacean acid metabolism and nitrate nutrition. Higher malate–transport capacity in ice-plant after crassulacean acid metabolism-induction and in tobacco under nitrate nutrition. Plant Physiol 124:1335–1347

MacLennan DH, Beevers H, Harley JH (1963) "Compartmentation" of acids in plant tissues. Biochem J 89:316–327

McNaughton GAL, MacIntosh C, Fewson CA, Wilkins MB, Nimmo HG (1991) Illumination increases the phosphorylation state of maize phosphoenolpyuvate carboxylase by causing an increase in the activity of a protein kinase. Biochim Biophys Acta 1093:189–195

McRae SR, Christopher JT, Smith JAC, Holtum JAM (2002) Sucrose transport across the vacuolar membrane of *Ananas comosus.* Funct Plant Biol 29:717–724

Maxwell K, Griffiths H, Young AJ (1994) Photosynthetic acclimation to light regime and water stress by the C_3-CAM epiphyte *Guzmania monostachyia*; gas-exchange characteristics, photochemical efficiency and the xanthophyll cycle. Funct Ecol 8:746–754

Maxwell K, von Caemmerer S, Evans JR (1997) Is low internal conductance to CO_2 a consequence of succulence in plants with crassulacean acid metabolism? Aust J Plant Physiol 25:777–786

Maxwell K, Badger MR, Osmond CB (1998) A comparison of CO_2 and O_2 exchange patterns and the relationship with chlorophyll fluorescence during photosynthesis in C_3 and CAM plants. Aust J Plant Physiol 25:45–52

Maxwell K, Borland AM, Haslam RP, Helliker B, Roberts A, Griffiths H (1999) Modulation of Rubisco activity during the diurnal phases of the crassulacean acid metabolism plant *Kalanchoë daigremontiana.* Plant Physiol 121:849–856

Nimmo HG (2000) The regulation of phosphoenolpyuvate carboxylase in CAM plants. Trends Plant Sci 5:75–80

Nimmo GA, Nimmo HG, Hamilton ID, Fewson CA Wilkins MB (1986) Purification of the phosphorylated night form and dephosphorylated day form of phosphoenolpyuvate carboxylase from Bryophyllum fedtschenkoi. Biochem J 239:213–220

Nimmo GA, Wilkins MB, Fewson CA, Nimmo HG (1987) Persistent circadian rhythms in the phosphorylation state of phosphoenolpyruvate carboxylase from *Bryophyllum fedtschenkoi* leaves and in its sensitivity to inhibition by malate. Planta 170:408–415

Nishida K (1963) Studies on the reassimilation of respiratory CO_2 in illuminated leaves. Plant Cell Physiol 3:111–124

Nishida K, Tominaga O (1987) Energy-dependent uptake of malate into vacuoles isolated from CAM-plant *Kalanchoë daigremontiana.* J Plant Physiol 127:385–393

Nobel PS (1988) Environmental biology of agaves and cacti. Cambridge University Press, Cambridge

Nobel PS, Bobich EG (2002) Initial net CO_2 uptake responses and root growth for a CAM community placed in a closed environment. Ann Bot 90:593–598

Nuernbergk EL (1961) Endogener Rhythmus und CO_2-Stoffwechsel bei Pflanzen mit diurnalem Säurerhythmus. Planta 56:28–70

O'Leary MH, Osmond CB (1980) Diffusional contribution to carbon isotope fractionation during dark CO_2 fixation in CAM plants. Plant Physiol 66:931–934

Olivares E, Faist K, Kluge M, Lüttge U (1993) $^{14}CO_2$ pulse-chase labeling in Clusia minor L. J Exp Bot 44:497–501

Osmond CB (1978) Crassulacean acid metabolism: a curiosity in context. Annu Rev Plant Physiol 29:379–414

Osmond CB (1997) C_4 photosynthesis: thirty (or forty?) years on. Aust J Plant Physioly 24:409–412

Osmond CB, Avadhani PN (1968) Acid metabolism in Atriplex. II. Oxalate synthesis during acid metabolism in the dark. Aust J Biol Sci 21:917–927

Osmond CB Allaway WG (1974) Pathways of CO_2 fixation in the CAM plant *Kalanchoë daigremontiana* I. Patterns of $^{14}CO_2$ fixation in the light. Aust J Plant Physiol 1:503–511

Osmond CB, Förster B (2006) Photoinhibition: then and now. In: Demmig-Adams B, Adams W W III, Mattoo A (eds) Photoprotection, photoinhibition, gene regulation, and environment. Springer, Berlin (in press)

Osmond B, Laties GG (1969) Compartmentation of malate in relation to ion absorption in beet. Plant Physiol 44:7–14

Osmond CB, Monro J (1981) Prickly pear. In: Carr DJ, Carr SJ (eds) Plants and man in Australia. Academic Press, Sydney, pp 194–222

Osmond CB, Ziegler H (1975) Schwere Pflanzen und leichte Pflanzen: Stabile Isotope im Photosynthesestoffwechsel und in der Biochemischen Ökologie. Naturwissenschaft Rundschau 28:323–328

Osmond CB, Allaway WG, Sutton BG, Troughton JH., Queiroz O, Lüttge U, Winter K (1973) Carbon isotope discrimination in photosynthesis in CAM plants. Nature 246:41–42

Osmond CB, Bender MM, Burris RH (1976) Pathways of CO_2 fixation in the CAM plant Kalanchoë daigremontiana III. Correlation with $\delta^{13}C$ value during growth and water stress. Aust J Plant Physiol 3:787–799

Osmond CB, Ludlow MM, Davis RL, Cowan IR, Powles SB, Winter K (1979a) Stomatal responses to humidity in *Opuntia inermis* in relation to control of CO_2 and H_2O exchange patterns. Oecologia 41:65–76

Osmond CB, Nott DL, Firth PM (1979b) Carbon assimilation patterns and growth of the introduced CAM plant *Opuntia inermis* in Eastern Australia. Oecologia 40:331–350

Osmond CB, Holtum JAM, O'Leary MH, Roeske C, Wong OC, Summons RE, Avadhani PN (1988) Regulation of malic acid metabolism in CAM plants in the dark and light: in–vivo evidence from ^{13}C-labeling patterns after $^{13}CO_2$ fixation. Planta 175:184–195

Osmond CB, Popp M, Robinson SA (1996) Stoichiometric nightmares: studies of O_2 and CO_2 exchanges in CAM plants. In: Winter K, Smith JAC (eds) Crassulacean acid metabolism. Ecological studies vol 114. Springer Verlag, Berlin, pp 19–30

Osmond CB, Maxwell K, Popp M, Robinson S (1999) On being thick: fathoming apparently futile pathways of photosynthesis and carbohydrate metabolism in succulent CAM plants. In: Burrell M, Bryant J, Kruger N (eds) Carbohydrate metabolism in plants. Bios Science Publications, Oxford, pp 183–200

Pophof B, Stange G, Abrell L (2005) Volatile organic compounds as signals in a plant–herbivore system: electrophysiological responses in olfactory sensilla of the moth Cactoblastis cactorum. J Chem Ecol 30:51–68

Quieroz O (1974) Circadian rhythms and metabolic pathways. Annu Rev Plant Physiol 25:115–134

Ranson SL, Thomas M (1960) Crassulacean acid metabolism. Annu Rev Plant Physiol 11:81–110

Rascher U, Hütt M–T, Siebke K, Osmond CB, Beck F, Lüttge U (2001) Spatio-temporal variation of metabolism in a plant circadian rhythm: the biological clock as an assembly of coupled individual oscillators. Proc Natl Acad Sci USA 98:11801–11805

Rascher U, Bobich EG, Osmond CB (2006) The "Kluge–Lüttge Kammer": preliminary evaluation of an enclosed Crassulacean acid metabolism (CAM) mesocosm that allows separation of synchronized and desynchronized contributions of plants to whole system gas exchange. Plant Biol 8:167–174

Ritz D, Kluge M, Veith HJ (1986) Mass-spectrometric evidence for the double carboxylation pathway of CO_2 fixation in crassulacean acid metabolism plants in the light. Planta 167:284–291.

Rustin P, Queiroz–Claret C (1985) Changes in the oxidative properties of *Kalanchoë blossfeldiana* leaf mitochondria during the development of crassulacean acid metabolism. Planta 164:415–422

Robinson SA, Osmond CB (1994) Internal gradients of chlorophyll and carotenoid pigments in relation to photoprotection in thick leaves of plants with Crassulacean acid metabolism. Aust J Plant Physiol 21:497–506

Robinson SA, Osmond CB, Giles L (1993) Interpretations of gradients in $\delta^{13}C$ value in thick photosynthetic tissues of plants with Crassulacean acid metabolism. Planta 190:271–276

Robinson SA, Ribas–Carbo M, Yakir D, Giles L, Reuveni Y, Berry JA (1995) Beyond sham and cyanide: opportunities for studying the alternative oxidase in plant respiration using oxygen isotope discrimination. Aust J Plant Physiol 23:487–496

Robinson SA, Yakir D, Ribas-Carbo M, Giles L, Osmond CB, Siedow JN, Berry JA (1992) Measurements of the engagement of cyanide-resistant respiration in the crassulacean acid metabolism plant *Kalanchoë diagremontiana* with the use of online oxygen isotope discrimination. Plant Physiol 100:1087–1091

Sage RF (2002) Are crassulacean acid metabolism and C_4 photosynthesis incompatible? Funct Plant Biol 29:775–785

Sage RF, Monson RK (eds) (1999) C_4 Plant biology. Academic Press, San Diego

Schmitt AK, Lee HSJ, Lüttge U (1988) The response of the C_3-CAM tree Clusia rosea to light and water stress. J Exp Bot 39:1581–1590

Smith JAC, Heuer S. (1981) Determination of the volume of intercellular spaces in leaves and some values for CAM plants. Annals of Botany 48:915–917

Smith AM, Zeeman SC, Smith SM (2005) Starch degradation. Annu Rev Plant Biol 56:73–98

Stange G (1997) Effects of changes in atmospheric carbon dioxide concentration on the location of hosts by the moth, *Cactoblastis cactorum*. Oecologia 110:539–545

Stange G, Munro J, Stowe S, Osmond CB (1995) The CO_2 sense of the moth *Cactoblastis cactorum* and its probable role in the biological control of the CAM plant *Opuntia stricta*. Oecologia 102:341–352

Sutton BG (1975a) The path of carbon in CAM plants at night. Aust J Plant Physiol 2:377–387

Sutton BG (1975b) Glycolysis in CAM plants. Aust J Plant Physiol 2:389–402

Sutton BG, Osmond CB (1972) Dark fixation of CO_2 by Crassulacean plants; evidence for a single carboxylation step. Plant Physiol 50:360–365

Thomas M, Beevers H (1949) Physiological studies on acid metabolism in green plants. II. Evidence of CO_2 fixation in Bryophyllum and the study of diurnal variation of acidity in this genus. New Phytol 48:421–447.

Ting IP, Osmond CB (1973a) Multiple forms of plant P-enolpyruvate carboxylase associated with different metabolic pathways. Plant Physiol 51:448–453

Ting, IP Osmond CB (1993b) Activation of plant P-enol-pyruvate carboxylase by glucose–6–phosphate. A particular role in Crassulacean acid metabolism. Plant Sci Lett 1:123–128

Ting IP, Sternberg LSL, De Niro MJ (1985) Crassulacean acid metabolism in the strangler, Clusia rosea Jacq. Science 229:969–971

Walker DA (1956) Malate synthesis in a cell free extract of a Crassulacean plant. Nature 178:593–594

Wang N, Zhang H, Nobel PS (1998) Carbon flow and carbohydrate metabolism during sink–to–source transition for developing cladodes of *Opuntia ficus*-indica. J Exp Bot 49:1835–1843

White PJ, Smith JAC (1992) Malate-dependent proton transport in tonoplast vesicles isolated from orchid leaves correlates with the expression of crassulacean acid metabolism. J Plant Physiol 139:533–538

Wilkins MB (1959) An endogenous rhythm in the rate of dark fixation of carbon dioxide in leaves of Bryophyllum. II. The effects of light and darkness on the phase and period of the rhythm. J Exp Bot 10:377–390

Winter K (1981) Changes in the properties of phosphoenolpyruvate carboxylase from the Crassulacean acid metabolism plant *Mesembryanthemum crystallinum* after isolation. Aust J Plant Physiol 8:115–119

Winter K (1982) Properties of phosphoenolpyruvate carboxylase in rapidly prepared, desalted leaf extracts of the Crassulacean acid metabolism plant *Mesembryanthemum crystallinum*. Planta 154:298–308

Winter K, Holtum JAM (2002) How closely do the $\delta^{13}C$ values of Crassulacean acid metabolism plants reflect the proportion of CO_2 fixed during day and night? Plant Physiol 129:1843–1851

Winter K, Smith JAC (1996) Crassulacean acid metabolism; biochemistry, ecophysiology and evolution. Ecological studies vol 114. Springer, Berlin, Heidelberg, New York

Winter K, Osmond CB, Hubick KT (1986) Crassulacean acid metabolism in the shade. Studies on an epiphytic fern, *Pyrossia longifolia* and other rainforest species from Australia. Oecologia 68:224–230

Winter K, von Willert DJ (1972) NaCl-induzierter Crassulaceen-Saurestoffwechsel bei *Mesembryanthemum crystallinum*. Zeitschr Pflanzenphysiol 67:166–170

Wood HG, Werkmann CH (1938) The utilization of carbon dioxide by propionic acid bacteria. Biochem J 32:1262–1271

Wyka TP, Lüttge U (2003) Contribution of C_3 carboxylation to the circadian rhythm of carbon dioxide uptake in a Crassulacean acid metabolism plant *Kalanchoë daigremontiana*. J Exp Bot 54:1471–1479

Charles Barry Osmond
Visiting Fellow
School of Biochemistry and Molecular Biology
Australian National University
Canberra ACT 0200 Australia
Fax: +61 2 6125 0313
Barry.Osmond@anu.edu.au

Address for correspondence:
PO Box 3252, Weston Creek
ACT 2611, Australia
Ph: +61 2 6287 1487

Genetics

Function of Genetic Material: Progressive Insight into Antimicrobial Peptides and their Transcriptional Regulation

Silke Hagen and Ulf Stahl

1 Introduction

Antimicrobial peptides (AMPs) are small proteins that display growth inhibitory effects on a multitude of organisms. They are constituents of the innate defence mechanism of multicellular organisms (Zasloff 2002), in which they are essential factors in repelling pathogen attack. Interestingly, the host range of AMPs is largely restricted to a specific group of pathogens. Furthermore, they often possess high antimicrobial potential. Due to these two aspects, AMPs are considered to be "nature's antibiotics" (Wang and Wang 2004), representing attractive alternatives to chemical antimicrobial agents presently in use (Reddy et al. 2004a). However, before research can take full advantage of AMP characteristics, more detailed information has to be accumulated concerning their expression and regulation, respectively.

During the last few years, dramatic progress has been made in understanding how expression of AMP-encoding genes is induced and regulated. It appears that many organisms, such as *Arabidopsis thaliana*, possess highly sophisticated defence-related signalling pathways (Nimchuk et al. 2003). These enable the host to discriminate between different qualities of pathogen infection and other stress-evoking factors, and to react accordingly.

In this review, the interested reader is given an overview on the recent progress achieved in understanding AMP expression and regulation of AMP-encoding genes. For further leading in-depth information, the reader is referred to a selection of comprehensive review articles from Lemaitre (2004), Zhao (2005) and Zipfel and Felix (2005).

2 What are antimicrobial peptides?

Antimicrobial peptides, comprising of antibacterial or antifungal activity, have been identified in many species, ranging from bacteria and fungi to insects, mammals and plants. In evolutionary terms, they are thought to be ancient

Progress in Botany, Vol. 68

constituents of the innate immune system. Although prokaryotes are devoid of any kind of immune response, some of them are also known to produce peptides with antimicrobial potential (Cheigh and Pyun 2005).

Generally, bacterial AMPs provide their host with a selective advantage. This advantage may imply that prokaryotic strains are able to successfully defend their assigned ecological niches against putative competitors (Bhatti et al. 2004).

Regardless of their origin, AMPs are gene-encoded, low-molecular weight proteins, generally consisting of fewer than 100 amino acid residues (Ganz 2005). Strikingly, there is little sequence or structural similarity to perceive among them. Although they exhibit a rather wide range of variance, it is impressive to note that all AMPs seem to operate via the same fundamental mode of action. This involves the permeabilisation of microbial membranes, concomitantly resulting in growth arrest of putative pathogens or rival strains. An underlying prerequisite for the membrane perturbing effect is certainly the cationic net charge of AMPs, which is a common characteristic for this outstanding group of peptides. Furthermore, AMPs generally exhibit an amphipathic configuration (De Smet and Contreras 2005), which is also assumed to be involved in the process of membrane permeabilisation.

The characteristic event of membrane permeation consists of the electrostatic interaction between AMPs and membranes. It has been suggested that positively charged peptides can displace charge-neutralising cations, which were found to localise on membrane surfaces. Upon removal, AMPs can bind to negatively charged membrane constituents, such as lipopolysaccharides. Alternatively, AMPs can neutralise the membrane charge within the affected area, subsequently resulting in the permeabilisation of membranes (Bowdish et al. 2005).

2.1 Function

Antimicrobial peptides form the first line of innate host defence in multicellular organisms. In contrast to the adaptive immune system, which may take days or weeks until it successfully responds to invasive attack, the innate immune system provides a rapid means to combat pathogen infection right from the start (Clark and Kupper 2005).

Presumably, all metazoans have evolved an inborn defence mechanism. Distinguishing features of this innate immunity comprises of pattern recognition receptors (PRR), which usually exhibit a broad range of specificity. PRR are able to recognize many related molecular structures, referred to as pathogen-associated molecular patterns (PAMPs) (Nurnberger et al. 2004; Zipfel and Felix 2005). PAMPs generally show little variance. They typically consist of polysaccharides and polynucleotides, exclusively present in the invading pathogen. No memory of prior exposure to a certain pathogen is required for PRR expression, which explains the rapidity with which the innate immune response is able to respond to invasive attack.

Some organisms, such as plants and lower animals, do not possess an adaptive immune response and are therefore utterly depend on their innate immune system. However, although devoid of an acquired immunity, these organisms are highly successful in protecting themselves against life-threatening invaders. This circumstance clearly speaks in favour of the efficiency of the innate immune system, which has stood the test of time for million of years.

Remarkably, the effective range of AMPs is not exclusively restricted to the innate immune system. These peptides were also found to trigger and to interact with the adaptive immune response (Oppenheim et al. 2003). In metabolic terms, this is a greatly economic means of responding to pathogen attack. The host merely invests energy into the expression of basic defence machinery.

General elicitors, collectively termed now as pathogen-associated molecular patterns (PAMPs; Fig. 1), interact with host receptors, such as pattern recognition receptors (PRRs) in plants or Toll-like receptors (TLRs) in animals. The receptors, in turn, translate the signal into cellular reactions that result in the activation of plant defence reactions (Fliegmann et al. 2004).

The correlation between environmental stimulus and defence gene expression may be diverse. However, AMPs are a potent means of providing their host with a selectional advantage, be it either the protection against pathogen invasion or the defence of an ecological niche against a putative competitor.

Every organism encounters pathogen invasion most of the time, which implies that parasites try to enter their host with the aim to feed and to propagate at its expense. Pathogen invasion can constitute a considerable metabolic burden for the affected organism, significantly reducing its fitness and chances for survival. In order to successfully counteract parasite intrusion, the host mounts an immune response. This includes the production of AMPs, which provides it with a selectional advantage over non-producing strains. The defence of a certain habitat or ecological niche by means of AMP activity particularly holds true for prokaryotes. Both *Lactococcus lactis* subsp. *lactis* and *Listeria monocytogenes* are lactic acid bacteria sharing the same habitats. *L. lactis* produces an AMP referred to as nisin. This peptide has proven to effectively inhibit the growth of *L. monocytogenes* (Bhatti et al. 2004), thereby keeping the putative rival for nutrients successfully at bay.

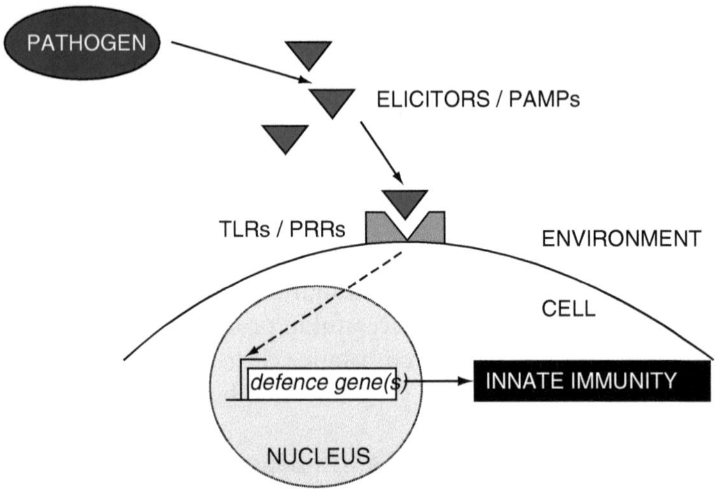

Fig. 1. Schematic drawing outlining the general recognition of elicitors or pathogen-associated molecular patterns (PAMPs). Defined receptors, such as the pattern recognition factors (PRRs) in plants or the Toll-like receptors (TLRs) in animals, perceive these pathogenderived signals, consequently resulting in the transcription of specifically associated defence genes

2.2 Mechanism of action

The molecular activity of AMPs can generally be ascribed to an electrostatic relation between positively charged peptides and negatively charged microbial membranes (Park and Hahm 2005). For many AMPs described so far, this interaction appears to be highly unspecific, since it does not seem to operate via a receptor-mediated mechanism. Nevertheless, the consequences for an invading pathogen under the influence of AMP activity may be dramatic: peptide–membrane interactions can result in the permeabilization of membranes, loss of membrane potential and the discharge of cytosolic metabolites (Tossi and Sandri 2002).

Outstanding characteristics such as size, amino acid composition, amphipathicity and cationic charge facilitate the attachment and insertion of AMPs into membrane bilayers. Three models have been put forward to explain the formation of pores: the barrel-stave, the carpet and the toroidal model. The publications by Park and Hahm (2005) and Zemel (2003) provide detailed information on these model mechanisms. In any event, however helpful the character of these models may be, they provide only scant insight into how peptide damage and killing of microorganisms truly occurs.

For some AMPs, the interaction with specific membrane-associated targets or receptors has recently been described (Thevissen et al. 2004).

It was shown that some antifungal peptides interact with fungal glucosylceramides. Interestingly, the analysed AMPs originated from different eukaryotic kingdoms, which leads to the conclusion that they once must have evolved from a single precursor molecule (Thevissen et al. 2004).

Several other observations suggest that the translocation of AMPs across membranes can influence septum formation in ascomycetes. Furthermore, the synthesis of cell walls, nucleic acids and proteins may also be affected, as well as the activity of certain enzymes (Olmo et al. 2001; Park and Hahm 2005).

2.3 Classification

Due to the vast variety of peptides displaying antimicrobial potential, the classification of AMPs into different categories is not an easy task. Neither a taxonomic nor a functional classification seems to be sufficiently rigorous, therefore some rather broad structural characteristics have been applied for the categorisation of these peptides. A descriptive summary on important AMP subfamilies is given by Marshall and Arenas (http://www.ejbiotechnology. info/content/vol6/issue3/full/1).

In plants, eight distinct classes of AMPs have been identified so far (Garcia-Olmedo et al. 1998; Lay and Anderson 2005). In humans, three major groups of AMPs are characterised (De Smet and Contreras 2005), whereas the innate immune system of insects is said to consist of seven important families of peptides (Royet et al. 2005). Nonetheless, it is worth mentioning that the definite classification of many individual AMPs still remains a controversial issue.

There are presently two extensive AMP databases freely available on the world-wide-web (http://www.bbcm.univ.trieste.it/~tossi/pag1.htm and http://aps.unmc.edu/AP/main.php).

3 Structure and gene regulation of antimicrobial peptides

The great attractiveness of AMPs for medicine or applied biotechnology can be attributed to their biological origin, high sustainability, broad diversity and appealing range of specificity. In order to best exploit these peptides, detailed knowledge has to be gathered concerning their tertiary structure, their mode of action and the regulation of AMP-encoding genes.

X-ray crystal diffraction, solution nuclear magnetic resonance (NMR) and theoretical modelling are techniques by which the three-dimensional structure of proteins and peptides can be resolved. At present, none of these methods seems suitable to provide analyses data in a straightforward manner (http://course.wilkes.edu/bioinformatics/stories/storyReader$130).

Many attempts have been made to gain more insight into the regulation of AMP-encoding genes. Whilst expression patterns of genes from mammals and insects are especially well investigated, research into respective genes in bacteria, fungi and plants is only now catching up. Recent data provide valuable information about the stunning complexity by which organisms encounter stress situations within the scope of pathogen attack.

3.1 Plants

Approximately 10,000 years ago, humans started to practise agriculture. Since then, a steady improvement of crop plants has been achieved in order to meet the growing demand for renewable primary products. Today, crop yields can be pushed to the maximum by various means (Wollenweber et al. 2005); however, a substantial part of the annual harvest falls prey to pathogen attack.

It is estimated that fungi alone destroy approximately 17% crop per annum worldwide, resulting in huge economic losses (http://www.cheminova.com/en/insects_weeds_and_ fungi.htm).

Extremely alarming is the observation that many crop-protecting chemicals, generally applied by way of a precaution, seem to have lost their potency (Sundin 2001; McManus et al. 2002). The steady increase of resistant pathogens imposes extreme pressure on farmers and the chemical industry to rapidly find new, efficient and cheap antibiotics. Ideally, these substances should combine a high degree of activity and a narrow susceptibility range, while still being completely compatible with the environment. Thus, ideal candidates for future crop-protecting agents are represented by AMPs, which still hold high potential for optimisation to demand by genetic engineering.

3.1.1 Antimicrobial peptides

Plants are sessile organisms, and are therefore exceptionally dependent on their immune response. Within this scope, plants produce a whole battery of AMPs and proteins, all intended to protect the organism against pathogen attack. Interestingly, the encounter between plant and pathogen can follow two different ways of progression—the so-called susceptible or the resistant route of interaction.

During the resistant route of interaction, which is also referred to as the hypersensitive response, the plant aims to literally trap the pathogen within necrotic plant tissue and layers of decomposing cells (Greenberg and Yao 2004). The susceptible route of interaction implies that the infection can initially ensue (Lay and Anderson 2005). In this case, the plant has to fall back on its repository of defence-related cell responses, among which the expression of AMPs is included.

As previously stated, the classification of AMPs into distinct groups or families is a source of controversy. Based on amino acid sequence identities and spacing between cysteines residues, it has been suggested to differentiate between the following families of plant AMPs: thionins, plant defensins, lipid-transfer proteins, hevein- and knotting-type peptides, and AMPs from *Macademia integrifolia*, *Impatiens balsamina* and the group of cyclotides (Lay and Anderson 2005). Other sources state that snakines as well as the multiubiquitine binding protein (MBP1) serve to constitute complemental classes of AMPs within this listing of families (Garcia-Olmedo et al. 1998). Apart from these variations, the molecular size of AMPs provides less reason for debate, since it typically ranges between 2 and 9 kDa. The expression of AMPs is described to be either be constitutive or inducible (Garcia-Olmedo et al. 1998). Table 1 gives an overview on the basic characteristics of the individual peptide families summarized above.

3.1.2 Structure

Antimicrobial peptides from plants generally possess four to eight cysteine residues, which result in the formation of two to four disulfide bridges (Table 1). However, regarding the number of disulfide bridges observed in AMPs, there are several exceptions to the rule.

Snakin-1 was identified to be the only peptide to date containing 12 cysteines, ultimately resulting in the arrangement of six disulfide bridges (Segura et al. 1999). Others, such as the chitin-binding protein from *Euonymus europaeus* (Ee-CBP) and the AMPs from *Eucommia ulmoides* (e.g. EAFP2), contain five disulfide bridges (Van den Bergh et al. 2002; Huang et al. 2004). The hevein-type peptide Ac-AMP2, isolated from *Amaranthus caudatus* (el Bouyoussfi et al. 1997), contains three disulfide bridges.

Intramolecular disulfide bridges contribute considerably to the tight folding and characteristic stability of AMPs. This stability is indicated by the extraordinarily high heat and protease resistance of these peptides (Lay and Anderson 2005).

Table 1. Families of plant antimicrobial peptides as suggested by Lay and Anderson (2005)

AMP family	Prominent member	Organism	Number of amino acid residues	Consensus sequence	Reference
Thionins (8-Cys type)	α-Purothionin	*Triticum aestivum*	45	2-CC-7-C-3-C-8-C-3-C-1-C-8-C-6	(Ohtani et al. 1977)
Plant defensins	Rs-AFP	*Raphanus sativus L.*	51	3-C-10-C-5-C-3-C-9-C-8-C-1-C-3-C	(Terras et al. 1992)
Lipid transfer proteins	Ace-AMP	*Allium cepa L.*	93	3-C-9-C-12-CC-18-C-1-C-23-C-15-C-4	(Cammue et al. 1995)
Hevein-type peptides	Ac-AMP2	*Amaranthus caudatus*	30	3-C-4-C-4-CC-5-C-6-C-2	(Broekaert et al. 1992)
Knotting-type peptides	Mj-AMP1	*Mirabilis jalapa L.*	36	1-C-6-C-8-CC-3-C-10-C-3	(Cammue et al. 1992)
Macademia	Mi-AMP1	*Macadamia integrifolia*	76	10-C-9-C-1-C-25-C-14-C-11-C	(McManus et al. 1999)
Impatiens	Ib-AMP1	*Impatiens balsamina*	20	5-CC-8-C-3-C	(Patel et al. 1998)
Cyclotides	Katala B1	*Oldenlandia affinis*	29	1-C-3-C-4-C-4-C-1-C-4-C-6	(Jennings et al. 2001)

3.1.3 Regulation

Both biotic and abiotic stresses can trigger a network of individual signalling pathways, which all seem to be extensively interconnected with each other (Pieterse and Van Loon 2004). Obviously, this cross-talk is very sensible, because it enables the organism to fine-tune and coordinate its defence responses in accordance with the individual stress applied. In general, the reception of species-specific elicitor molecules constitutes the initial step within a cascade of transduction events.

Interestingly, the elicitors are generally recognised by distinct receptors localised on the surface of the affected host cell. The receptors consequently translate the signal into defined events taking place in the plasma membrane, the cytosol and/or the nucleus (Shirasu et al. 1996). Regarding signal recognition and signal transduction, there are numerous highly informative review articles available to date. The interested reader is referred to publications by Dangl and Jones (2001) and Asai et al. (2002).

Most interestingly, each signalling pathway comprises a specific phytohormone that serves to amplify and spread the host immune response immediately upon pathogen recognition. The most prominent phytohormones are salicylic acid (SA), ethylene (ET) and the jasmonates (JA). Depending on which of these hormones constitute the central signalling component, the designation of the corresponding pathway applies accordingly. Lately, a number of excellent review articles have been published that address the complexity of ET, JA and SA signalling in plants. In this regard, the reader is referred to articles from Pieterse and Van Loon (2004) and McGrath (2005).

Presently, the most comprehensive understanding of defence gene regulation is achieved in *Arabidopsis*. Therefore, the following paragraphs will summarize the latest progress acquired for this model system exemplarily. Signalling pathways in *Arabidopsis* have been investigated to a great extent; however, there still are several pieces missing to complete the entire picture of defence gene regulation in this organism. Table 2 summarizes the recently described factors and elements involved in ET, JA and SA network signalling pathways, all being ultimately involved in pathogenesis-related defence gene expression.

Regarding the *Arabidopsis* defensin PDF1.2, its expression was shown to be induced by exogenous treatment with either ethylene or methyl jasmonate. In contrast, the application of salicylic acid had no effect on *pdf1.2* expression (Penninckx et al. 1998).

The coronatine-insensitive 1 (COI1) gene encodes an F-box protein involved in the ubiquitin-proteosome pathway, and is required for

Table 2. Recently described factors and elements involved in ethylene (ET), jasmonate (JA) and salicylic acid (SA) signalling networks of *Arabidopsis*

Factors/ elements	Characteristics and function	Reference
ERF1	Ethylene response factor (ERF) transcription factor. Involved in converging cross-talk between ET and JA signalling pathways	Lorenzo et al. (2003)
NPR1	SA-activated transcription factor. Modulates cross-talk between JA and SA signalling pathways	Pieterse and Van Loon (2004)
WRKY70	SA-activated transcription factor involved in cross-talk between JA and SA signalling pathways	Li et al. (2004)
AtMYC2	Transcription factor involved in JA-dependent wound-response gene induction	Boter et al. (2004)
SCF^{COI1}	The COI1-associated protein complex (SCF^{COI1}) constitutes an intermediary module in JA-mediated ERF1 activation	Devoto et al. (2002)
MPK4	The MAP-kinase 4 regulator is involved in SA-dependent defence gene repression and JA-dependent defence gene activation	Andreasson et al. (2005)
GCC-box	Promoter element involved in the transcriptional induction of JA-dependent defence genes	Brown et al. (2003)

response to jasmonates in *Arabidopsis* (Devoto et al. 2002). Recently, research has identified that COI1 associates with other proteins to form ubiquitin–ligase complexes, designated as SCF^{COI1}. These complexes constitute an intermediary module between JA and ET signalling via the activation of the ethylene response factor 1 (ERF1) (Devoto et al. 2002; Lorenzo et al. 2003).

It appears that ethylene-responsive element binding factors (ERF) play a significant role in defence gene induction on the whole. They are known to consist of a small subfamily of ERF proteins that can act as transcriptional repressors. One such repressor, AtERF4, is capable of modulating ethylene and abscisic acid responses in *Arabidopsis* (Yang et al. 2005).

The connection between SCF^{COI1} and ERF1 explains why concomitant induction of the ET and the JA response pathways are required for transcriptional induction of the *Arabidopsis* defensin-encoding gene *pdf1.2.* In Fig. 2, the interconnection of signalling pathways and the subsequent induction of characterised defence genes is outlined schematically. Arrows indicate positive interaction, while dashed lines represent negative interaction.

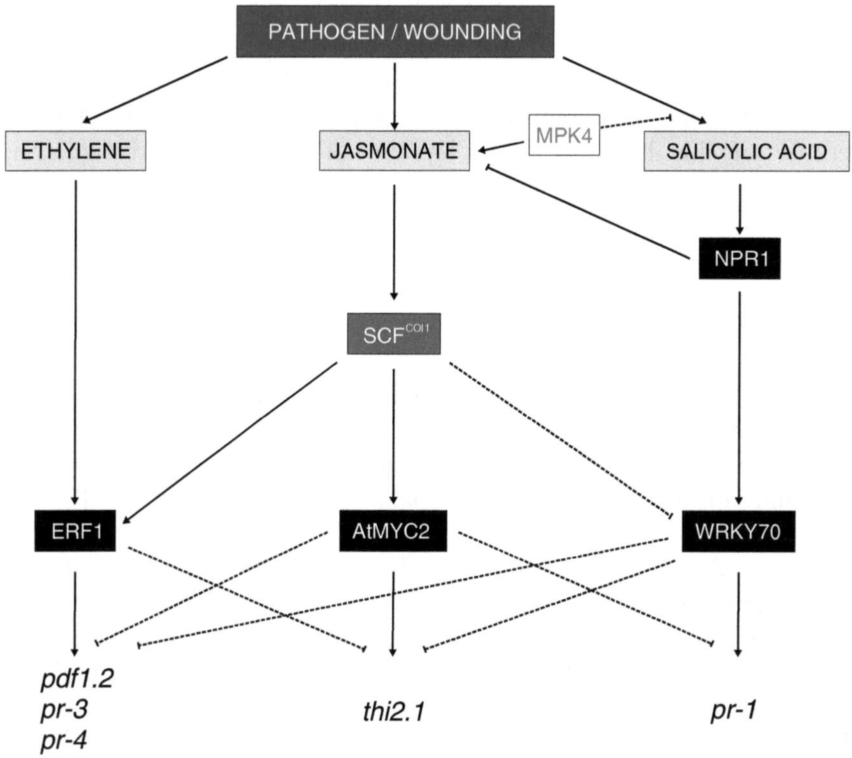

Fig. 2. Model outlining the interaction of key players in *Arabidopsis* defence gene induction

The MAP kinase 4 (MPK4) is regarded as a regulator of pathogen defence responses, because it is required for both repression of SA-dependent resistance and for activation of JA-dependent defence gene expression. The identification of the MPK4 substrate 1 (MKS1) is proposed to contribute to MPK4-regulated defence activation by coupling the kinase to specific WRKY transcription factors (Andreasson et al. 2005).

Regarding the genetic organisation of *pdf1.2*, a so-called GCC-box was identified in the promoter region of this gene. This element was shown to be an essential prerequisite for JA-dependent induction of *pdf1.2* expression (Brown et al. 2003). The pathogenesis-related genes *pr-3* and *pr-4*, encoding for a basic chitinases (PR-3) and hevein-like protein (PR-4), underlie the same signalling induction pattern as *pdf1.2*. This implies that *pr-3*, *pr-4* and *pdf1.2* are repressed by the wound-response inducing transcription factor AtMYC2. While repressing pathogenesis-related response genes, AtMYC2 is concomitantly involved in the up-regulation of wound response genes, such as the plant thionin-encoding gene *thi2.1*. Likewise, the SA-dependent expression of *pr-1*, a gene coding for a small, cationic AMP in *Arabidopsis*, is repressed by AtMYC2 and SCF^{COI1} (see Fig. 2). Because of this highly complex system of interconnected signalling pathways, the plant is able to choose the right set of genes according to the situational demand.

3.2 Insects

Insects are highly successful animals, not least because of their large diversity and the ability to adapt to many different habitats. Equally diverse is the spectrum of AMPs synthesised by this large kingdom of species. The authors Bulet and Stocklin (2005) provide valuable information about AMPs from insects.

The fruit fly *Drosophila melanogaster* is by far the genetically best characterised insect to date. Since it is fully sequenced and has the undisputable advantage of possessing short generation times, *Drosophila* is predestined to serve as a valuable model system. In this insect, seven distinct families of AMPs have been described. These consist of defensin, drosocin, metchnikowin, cecropins, drosomycins, attacins and diptericins (Tzou et al. 2002). An excellent review on innate immunity in *Drosophila* was published by the authors Kim and Kim (2005).

While the families of defensin, drosocin and metchnikowin consist of only single members, the cecropins, attacins and diptericins are composed of four, four and two members, respectively. With seven members, the drosomycins constitute the largest family of AMPs in *Drosophila*.

Two distinct signalling pathways are known to regulate the expression of these families of peptides (Fig. 3). While the Toll pathway is predominantly triggered by fungal and Gram-positive bacterial infection, the immune deficiency (Imd) pathway was shown to be additionally induced by Gram-negative bacterial strains. For a review on sensing and signalling in *Drosophila*, the reader is referred to the recent article by Royet et al. (2005).

In brief, Toll pathway induction leads to proteolytic cleavage of a cytokine named Spaetzle. In the course of signal distribution, the inhibitory protein referred to as Cactus is degraded, concomitantly releasing a transcription factor designated as Dif. In analogy, the Imd pathway results in cleavage and activation of a transcription factor named Relish. Both factors, Dif and Relish, can subsequently activate the expression of AMP encoding genes (Royet et al. 2005).

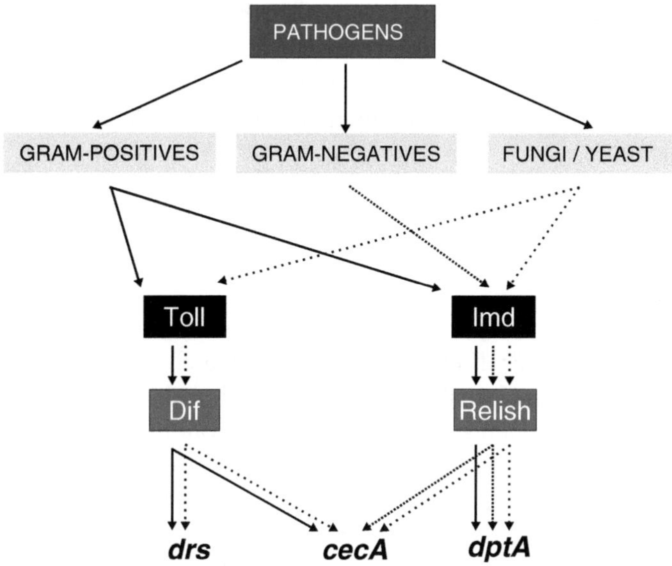

Fig. 3. Simplified model illustrating the network of defence gene induction in *Drosophila*. For further explanation, see text

The expression of genes encoding for the AMPs Cecropin A1 and A2 (CecA) were shown to be activated by Gram-positive and Gram-negative bacteria, as well as by fungi. In analogy, Diptericin A (DptA) gene expression can also be triggered by the same three groups of organisms, while the Drosomycin (Drs) gene is only activated upon attack by Gram-positive bacteria and fungi (Hedengren-Olcott et al. 2004).

Obviously, the Toll and the Imd pathways are stimulated by different pathogen-associated molecular patterns (PAMPs), resulting in differential induction of AMP-encoding genes. The induced gene expression of AMPs in response to pathogen attack relies on the translocation of the so-called Rel transcription factors into the nucleus.

The Rel family of transcription factors includes the proteins Dorsal, Dif and Relish. Characteristically, the Rel homology domain is responsible for DNA binding and regulated nuclear import (Jia et al. 2002).

The nuclear localisation enables Relish, Dif and Dorsal to bind to κB-like motifs in the promoter region of inducible genes.

A further putative promoter element referred to as region 1 (R1) has recently been identified (Uvell and Engstrom 2003).

Although it is generally assumed that the expression of AMP encoding genes in *Drosophila* is exclusively regulated by those two pathways, it has recently been shown that aging, circadian rhythms and mating also seem to influence AMP expression rather dramatically. A male seminal peptide, transferred during copulation, was demonstrated to be the major agent eliciting transcription of the metchnikowin and of other AMP-encoding genes (Peng et al. 2005).

3.3 Mammals

To date, a wide variety of peptides with antimicrobial activity have been isolated from mammals (Braff et al. 2005). Unfortunately, the most aggravating factor in their functional analysis is the observation that mammalian AMPs often appear inactive when assayed under culture conditions (Dorschner et al. 2006). Nevertheless, human AMPs attract much scientific interest, understandable when bearing in mind the enormous demand for new medical strategies in the battle against pathogen infection (Chen et al. 2005).

Three important families of AMPs are differentiated in humans: the group of defensins (α- and β-defensins), the histatins and the cathelicidins, of which the latter consists of a single member only (LL-37) (Fig. 4). Information regarding structure and mode of action of human AMPs was recently published in the review article by De Smet and Contreras (2005).

The histatins and the cathelicidins consist of 24–38 amino acid residues. Since they do not contain any cysteines, these peptides are devoid of stabilising disulfide bridges and appear to exist in a linear structure (De Smet and Contreras 2005).

The general belief is that the sole function of host defence peptides is restricted to innate immunity alone. Obviously, mammalian AMPs do also have a number of immunomodulatory functions, which include altering host gene expression and acting as chemokines and/or inducing chemokine production. Furthermore, they are involved in inhibiting lipopolysaccharide-induced pro-inflammatory cytokine production, promoting wound healing and modulating the responses of dendritic cells (Bowdish et al. 2005).

Innate immune response of insects and mammals bear some striking similarities (http://www.sdbonline.org/fly/torstoll/traf1-1.htm), which suggests that they must have evolved from one ancestral system. In analogy with gene expression in *Drosophila*, the induction of pathogen-induced genes in mammals relies on the translocation of Rel transcription factors into the nucleus. Subsequent binding to κB-like motifs in the promoter region of inducible genes results in AMP expression. For further details, the reader is referred to the review by Girardin (2003).

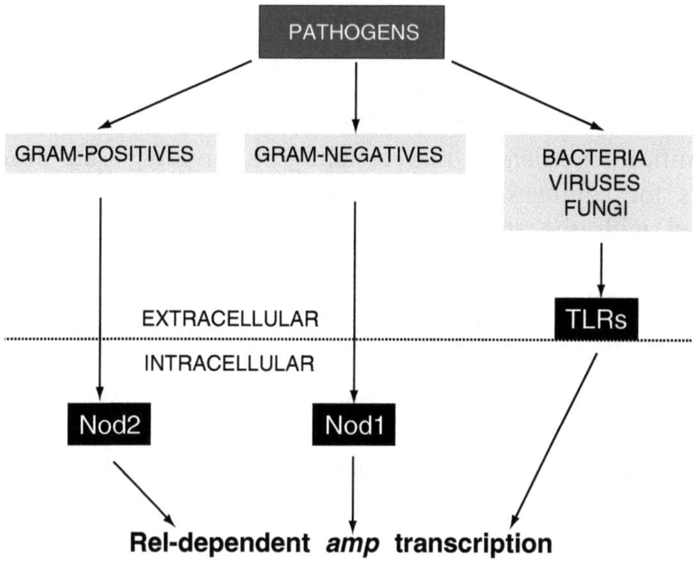

Fig. 4. Simplified model illustrating the network of defence gene induction in mammals. For further explanation, see text

While the nucleotide-binding oligomerisation domain (Nod) proteins enable the intracellular reception of PAMPs, 10 different Toll-like receptors (TLRs) permit extracellular recognition of various different pathogens. Most TLR ligands identified so far are conserved microbial products, which signal the presence of an infection. For more detailed information on the role of Toll-like receptors in pathogen recognition, the reader is referred to the review by Janssens and Beyaert (2003).

As previously mentioned, innate immune responses of mammals and insects share some considerable similarities. Thus, signal transduction and defence gene activation is, in general, initially performed on the insect model first. In subsequent analyses the transferability of data into the mammalian regulatory system is screened.

4 Application of antimicrobial peptides

On the one hand, expectations regarding adequate application of new antibiotics in fields such as agriculture, food industry and medicine are rising steadily. On the other hand, the breakthrough of AMP application in biotechnology still has not been realised yet. Certainly the question arises as to why is this so in times where humans are desperately seeking for new alternatives in the combat against pathogen infection? Somehow, the biotechnological industry is still reserved in fully appreciating the great potential that AMPs do offer (Tossi 2005).

A significant bottleneck may certainly be the large-scale production of AMPs for commercial applications. Further important aspects are concerned with the stabilization of secondary structure elements to improve receptor–ligand recognition. Only lately, a novel application for optimising receptor–ligand interaction was reported, aimed to create new antimicrobial agents using the highly folded thionin from *Pyrularia puberaas* as a template (Vila-Perello et al. 2006).

There are two putative mechanisms through which AMPs may acquire practical application: They could either be applied directly to the place of action; alternatively, AMP-encoding genes could be transferred into heterologous hosts by means of genetic engineering.

4.1 Crop protection

The great advantage of disease resistance strategies is that they can reduce chemical input into the environment. Thus, these alternative methods represent powerful approaches to contribute to sustainable agriculture. Many publications are available that broach the issue of novel plant engineering methodologies, all of them aiming at the receipt of fruitful transgenic plants. Examples include the macadamia antimicrobial peptide 1 (MiAMP1), which was successfully expressed in transgenic lines of *Brassica napus* L. (http://www.regional.org.au/au/gcirc/4/508.htm). Likewise, the expression of the synthetic peptide D4E1, which was designed on the basis of Cecropin B from moths, was shown to have broad-spectrum antimicrobial action against significant plant pathogens (http://www. indsp.org/NRTPAP.php).

Further descriptive information concerning the application of AMPs in crop protection can be found in the review articles by Bostock (2005) and Gurr and Rushton (2005).

Specifically in Europe the reservation against the use of transgenic crops is eminent. Although the use of conventional crop-protecting agents is denounced as harmful for the environment and the consumer, the biological alternative of AMP application in agriculture still has not succeeded yet. However, the advantages of AMP characteristics should be convincing: These highly potent peptides originate from biological sources, thus biodegradability does not constitute an issue, neither for humans nor for the environment. Therefore, AMPs are sustainable resources with comparatively cheap production costs. Ultimately, AMPs appear to be particularly predestined for the protection of crops against pathogen infection.

4.2 Food preservation

Bacteriocins are bacterially produced antimicrobial peptides which exhibit both narrow or broad host ranges. Many of these bioactive compounds are produced by food-grade lactic acid bacteria (Cotter et al. 2005). A prominent representative for the group of bacteriocins is nisin. This antimicrobial molecule is produced by *Lactococcus lactis* and has been used as a food preservative for over 50 years (Reddy et al. 2004b). More precisely, nisin exhibits antibacterial activity against Gram-positive bacteria, thereby extending the shelf-life of dairy products considerably. Consequently, nisin can be regarded as a prime example for the successful application of AMPs in biotechnology.

The recent review from Mierau and Kleerebezem (2005) provides extensive information about the use and the optimisation of *Lactococci* and other Gram-positive bacteria in modern biotechnological applications.

In contrast to chemically derived antibiotics, nisin and other biologically-derived AMPs are biodegradable molecules; thus, they are perfectly suited for practical applications in the food industry.

4.3 Medical application

Their wide spectrum of therapeutic potential suggests that AMPs may provide powerful tools in the treatment of cancer (Tanaka et al. 2001), viral (Chernysh et al. 2002) or parasitic infection (Vizioli and Salzet 2002). In fact, several biotechnological companies are currently attempting to bring AMPs to an applicable stage. For instance, one of these companies is particularly interested to develop the next generation of innate immune modulators and shows particular interest in the human host defence peptide LL-37.

Nisin, which exhibits spermicidal and antimicrobial properties, has now been suggested as a safe vaginal contraceptive for future therapeutic interventions in sexually transmitted infections. For detailed information, the reader is referred to the review article by Yedery and Reddy (2005).

However, despite the euphoric prognoses about their advantages in practical application, only few AMPs are actually in clinical and commercial use, including nisin, polymixin B and gramicidin S (Bradshaw 2003).

5 What more is to come?

Although the demand for new antibiotics is rising continuously, AMPs still have not found their way into the broad sectors of biotechnology. This is rather surprising, since the vast amounts of different antimicrobial molecules seem to be highly competitive alternatives to conventional antibiotics.

A prerequisite for the putative application of peptides with antimicrobial potential is the analysis of their mode of action: The determination of host ranges as well as the identification of specific AMP targets is absolutely required. Furthermore, factors such as the optimisation of production cost, toxicity against eukaryotic cell types and development of allergic reactions against these peptides have to be considered, to name but a few (Bradshaw 2003).

Numerous AMPs have been expressed in transgenic plants, but only few have proven successful in protecting the plant against pathogen attack (http://www.isb.vt.edu/news/2005/news05.Oct.htm). However, in silico modelling is a comparatively new method that provides sophisticated possibilities to optimise AMP characteristics. The application of this method promises the economic production and successful expression of novel drugs customised to the individual need of the consumer. For further leading information regarding the design of new antimicrobial drugs, the reader is referred to the review article by Monk and Harding (2005).

6 Conclusions

The information regarding signalling pathways involved in the transcriptional regulation of AMP-encoding genes is substantial. In contrast, details about the regulation of individual AMPs on the genetic level is comparatively sparse. Although our knowledge about the involvement of transcription factors and regulatory elements in promoter regions or transcriptional units of pathogen-inducible genes has progressed, we still seem distant from understanding the whole complexity by which organisms are capable of reacting towards life-endangering situations.

In order to bring AMPs a step further towards their application in medicine or biotechnology, many more efforts have to be undertaken to scrutinise the molecular events involved in defence gene transcription.

Finally, it should be noted that this large group of bioactive peptides provides valuable templates for the design of novel, highly effective antibiotics.

References

Andreasson E, Jenkins T, Brodersen P, Thorgrimsen S, Petersen NH, Zhu S, Qiu JL, Micheelsen P, Rocher A, Petersen M, Newman MA, Bjorn Nielsen H, Hirt H, Somssich I, Mattsson O, Mundy J (2005) The MAP kinase substrate MKS1 is a regulator of plant defense responses. Embo J 24:2579–2589

Asai T, Tena G, Plotnikova J, Willmann MR, Chiu WL, Gomez-Gomez L, Boller T, Ausubel FM, Sheen J (2002) MAP kinase signalling cascade in *Arabidopsis* innate immunity. Nature 415:977–983

Bhatti M, Veeramachaneni A, Shelef LA (2004) Factors affecting the antilisterial effects of nisin in milk. Int J Food Microbiol 97:215–219

Bostock RM (2005) Signal crosstalk and induced resistance: straddling the line between cost and benefit. Annu Rev Phytopathol 43:545–580

Boter M, Ruiz-Rivero O, Abdeen A, Prat S (2004) Conserved MYC transcription factors play a key role in jasmonate signaling both in tomato and Arabidopsis. Genes Dev 18: 1577–1591

Bowdish DM, Davidson DJ, Hancock RE (2005) A re-evaluation of the role of host defence peptides in mammalian immunity. Curr Protein Pept Sci 6:35–51

Bradshaw J (2003) Cationic antimicrobial peptides : issues for potential clinical use. BioDrugs 17:233–240

Braff MH, Bardan A, Nizet V, Gallo RL (2005) Cutaneous defense mechanisms by antimicrobial peptides. J Invest Dermatol 125:9–13

Broekaert WF, Marien W, Terras FR, De Bolle MF, Proost P, Van Damme J, Dillen L, Claeys M, Rees SB, Vanderleyden J et al. (1992) Antimicrobial peptides from *Amaranthus caudatus* seeds with sequence homology to the cysteine/glycine-rich domain of chitin-binding proteins. Biochemistry 31:4308–4314

Brown RL, Kazan K, McGrath KC, Maclean DJ, Manners JM (2003) A role for the GCC-box in jasmonate-mediated activation of the PDF1.2 gene of *Arabidopsis*. Plant Physiol 132: 1020–1032

Bulet P, Stocklin R (2005) Insect antimicrobial peptides: structures, properties and gene regulation. Protein Pept Lett 12:3–11

Cammue BP, De Bolle MF, Terras FR, Proost P, Van Damme J, Rees SB, Vanderleyden J, Broekaert WF (1992) Isolation and characterization of a novel class of plant antimicrobial peptides form *Mirabilis jalapa* L. seeds. J Biol Chem 267:2228–2233

Cammue BP, Thevissen K, Hendriks M, Eggermont K, Goderis IJ, Proost P, Van Damme J, Osborn RW, Guerbette F, Kader JC et al. (1995) A potent antimicrobial protein from onion seeds showing sequence homology to plant lipid transfer proteins. Plant Physiol 109: 445–455

Cheigh CI, Pyun YR (2005) Nisin biosynthesis and its properties. Biotechnol Lett 27: 1641–1648

Chen H, Xu Z, Peng L, Fang X, Yin X, Xu N, Cen P (2005) Recent advances in the research and development of human defensins. Peptides 27:931–940

Chernysh S, Kim SI, Bekker G, Pleskach VA, Filatova NA, Anikin VB, Platonov VG, Bulet P (2002) Antiviral and antitumor peptides from insects. Proc Natl Acad Sci USA 99: 12628–12632

Clark R, Kupper T (2005) Old meets new: the interaction between innate and adaptive immunity. J Invest Dermatol 125:629–637

Cotter PD, Hill C, Ross RP (2005) Bacteriocins: developing innate immunity for food. Nat Rev Microbiol 3:777–788

Dangl JL, Jones JD (2001) Plant pathogens and integrated defence responses to infection. Nature 411:826–833

De Smet K, Contreras R (2005) Human antimicrobial peptides: defensins, cathelicidins and histatins. Biotechnol Lett 27:1337–1347

Devoto A, Nieto-Rostro M, Xie D, Ellis C, Harmston R, Patrick E, Davis J, Sherratt L, Coleman M, Turner JG (2002) COI1 links jasmonate signalling and fertility to the SCF ubiquitin-ligase complex in *Arabidopsis*. Plant J 32:457–466

Dorschner RA, Lopez-Garcia B, Peschel A, Kraus D, Morikawa K, Nizet V, Gallo RL (2006) The mammalian ionic environment dictates microbial susceptibility to antimicrobial defense peptides. Faseb J 20:35–42

el Bouyoussfi M, Laus G, Verheyden P, Wyns L, Tourwe D, Van Binst G (1997) Location of the three disulfide bonds in an antimicrobial peptide from *Amaranthus caudatus* using mass spectrometry. J Pept Res 49:336–340

Fliegmann J, Mithofer A, Wanner G, Ebel J (2004) An ancient enzyme domain hidden in the putative beta-glucan elicitor receptor of soybean may play an active part in the perception of pathogen-associated molecular patterns during broad host resistance. J Biol Chem 279: 1132–1140

Ganz T (2005) Defensins and other antimicrobial peptides: a historical perspective and an update. Comb Chem High Throughput Screen 8:209–217

Garcia-Olmedo F, Molina A, Alamillo JM, Rodriguez-Palenzuela P (1998) Plant defense peptides. Biopolymers 47:479–491

Girardin SE, Philpott DJ, Lemaitre B (2003) Sensing microbes by diverse hosts. Workshop on pattern recognition proteins and receptors. EMBO Rep 4:932–936

Greenberg JT, Yao N (2004) The role and regulation of programmed cell death in plant-pathogen interactions. Cell Microbiol 6:201–211

Gurr SJ, Rushton PJ (2005) Engineering plants with increased disease resistance: how are we going to express it? Trends Biotechnol 23:283–290

Hedengren-Olcott M, Olcott MC, Mooney DT, Ekengren S, Geller BL, Taylor BJ (2004) Differential activation of the NF-kappaB-like factors Relish and Dif in *Drosophila melanogaster* by fungi and Gram-positive bacteria. J Biol Chem 279:21121–21127

Huang RH, Xiang Y, Tu GZ, Zhang Y, Wang DC (2004) Solution structure of *Eucommia* antifungal peptide: a novel structural model distinct with a five–disulfide motif. Biochemistry 43:6005–6012

Janssens S, Beyaert R (2003) Role of Toll-like receptors in pathogen recognition. Clin Microbiol Rev 16:637–646

Jennings C, West J, Waine C, Craik D, Anderson M (2001) Biosynthesis and insecticidal properties of plant cyclotides: the cyclic knotted proteins from *Oldenlandia affinis*. Proc Natl Acad Sci USA 98:10614–10619

Jia S, Flores-Saaib RD, Courey AJ (2002) The Dorsal Rel homology domain plays an active role in transcriptional regulation. Mol Cell Biol 22:5089–5099

Kim T, Kim YJ (2005) Overview of innate immunity in *Drosophila*. J Biochem Mol Biol 38: 121–127

Lay FT, Anderson MA (2005) Defensins—components of the innate immune system in plants. Curr Protein Pept Sci 6:85–101

Lemaitre B (2004) The road to Toll. Nat Rev Immunol 4:521–527

Li J, Brader G, Palva ET (2004) The WRKY70 transcription factor: a node of convergence for jasmonate-mediated and salicylate-mediated signals in plant defense. Plant Cell 16:319–331

Lorenzo O, Piqueras R, Sanchez-Serrano JJ, Solano R (2003) ETHYLENE RESPONSE FACTOR1 integrates signals from ethylene and jasmonate pathways in plant defense. Plant Cell 15:165–178

McGrath KC, Dombrecht B, Manners JM, Schenk PM, Edgar CI, Maclean DJ, Scheible WR, Udvardi MK, Kazan K (2005) Repressor- and activator-type ethylene response factors functioning in jasmonate signaling and disease resistance identified via a genome-wide screen of *Arabidopsis* transcription factor gene expression. Plant Physiol 139:949–959

McManus AM, Nielsen KJ, Marcus JP, Harrison SJ, Green JL, Manners JM, Craik DJ (1999) MiAMP1, a novel protein from *Macadamia integrifolia* adopts a Greek key beta-barrel fold unique amongst plant antimicrobial proteins. J Mol Biol 293:629–638

McManus PS, Stockwell VO, Sundin GW, Jones AL (2002) Antibiotic use in plant agriculture. Annu Rev Phytopathol 40:443–465

Mierau I, Kleerebezem M (2005) 10 years of the nisin-controlled gene expression system (NICE) in *Lactococcus lactis*. Appl Microbiol Biotechnol 68:705–717

Monk BC, Harding DR (2005) Peptide motifs for cell-surface intervention: application to anti-infective and biopharmaceutical development. BioDrugs 19:261–278

Nimchuk Z, Eulgem T, Holt BF, 3rd, Dangl JL (2003) Recognition and response in the plant immune system. Annu Rev Genet 37:579–609

Nurnberger T, Brunner F, Kemmerling B, Piater L (2004) Innate immunity in plants and animals: striking similarities and obvious differences. Immunol Rev 198:249–266

Ohtani S, Okada T, Yoshizumi H, Kagamiyama H (1977) Complete primary structures of two subunits of purothionin A, a lethal protein for brewer's yeast from wheat flour. J Biochem (Tokyo) 82:753–767

Olmo N, Turnay J, Gonzalez de Buitrago G, Lopez de Silanes I, Gavilanes JG, Lizarbe MA (2001) Cytotoxic mechanism of the ribotoxin alpha-sarcin. Induction of cell death via apoptosis. Eur J Biochem 268:2113–2123

Oppenheim JJ, Biragyn A, Kwak LW, Yang D (2003) Roles of antimicrobial peptides such as defensins in innate and adaptive immunity. Ann Rheum Dis 62 Suppl 2: ii17–21

Park Y, Hahm KS (2005) Antimicrobial peptides (AMPs): peptide structure and mode of action. J Biochem Mol Biol 38:507–516

Patel SU, Osborn R, Rees S, Thornton JM (1998) Structural studies of Impatiens balsamina antimicrobial protein (Ib-AMP1). Biochemistry 37:983–990

Peng J, Zipperlen P, Kubli E (2005) Drosophila sex-peptide stimulates female innate immune system after mating via the Toll and Imd pathways. Curr Biol 15:1690–1694

Penninckx IA, Thomma BP, Buchala A, Metraux JP, Broekaert WF (1998) Concomitant activation of jasmonate and ethylene response pathways is required for induction of a plant defensin gene in *Arabidopsis*. Plant Cell 10:2103–2113

Pieterse CM, Van Loon LC (2004) NPR1: the spider in the web of induced resistance signaling pathways. Curr Opin Plant Biol 7:456–464

Reddy KV, Yedery RD, Aranha C (2004a) Antimicrobial peptides: premises and promises. Int J Antimicrob Agents 24:536–547

Reddy KV, Aranha C, Gupta SM, Yedery RD (2004b) Evaluation of antimicrobial peptide nisin as a safe vaginal contraceptive agent in rabbits: in vitro and in vivo studies. Reproduction 128:117–126

Royet J, Reichhart JM, Hoffmann JA (2005) Sensing and signaling during infection in *Drosophila*. Curr Opin Immunol 17:11–17

Segura A, Moreno M, Madueno F, Molina A, Garcia-Olmedo F (1999) Snakin-1, a peptide from potato that is active against plant pathogens. Mol Plant Microbe Interact 12:16–23

Shirasu K, Dixon RA, Lamb C (1996) Signal transduction in plant immunity. Curr Opin Immunol 8:3–7

Sundin GW (2001) Antibiotic resistance affects plant pathogens. Science 291:2551

Tanaka K, Fujimoto Y, Suzuki M, Suzuki Y, Ohtake T, Saito H, Kohgo Y (2001) PI3–kinase p85alpha is a target molecule of proline-rich antimicrobial peptide to suppress proliferation of ras-transformed cells. Jpn J Cancer Res 92:959–967

Terras FR, Schoofs HM, De Bolle MF, Van Leuven F, Rees SB, Vanderleyden J, Cammue BP, Broekaert WF (1992) Analysis of two novel classes of plant antifungal proteins from radish (*Raphanus sativus* L.) seeds. J Biol Chem 267:15301–15309

Thevissen K, Warnecke DC, Francois IE, Leipelt M, Heinz E, Ott C, Zahringer U, Thomma BP, Ferket KK, Cammue BP (2004) Defensins from insects and plants interact with fungal glucosylceramides. J Biol Chem 279:3900–3905

Tossi A (2005) Host defense peptides: roles and applications. Curr Protein Pept Sci 6:1–3

Tossi A, Sandri L (2002) Molecular diversity in gene-encoded, cationic antimicrobial polypeptides. Curr Pharm Des 8:743–761

Tzou P, Reichhart JM, Lemaitre B (2002) Constitutive expression of a single antimicrobial peptide can restore wild-type resistance to infection in immunodeficient *Drosophila* mutants. Proc Natl Acad Sci USA 99:2152–2157
Uvell H, Engstrom Y (2003) Functional characterization of a novel promoter element required for an innate immune response in *Drosophila*. Mol Cell Biol 23:8272–8281
Van den Bergh KP, Proost P, Van Damme J, Coosemans J, Van Damme EJ, Peumans WJ (2002) Five disulfide bridges stabilize a hevein-type antimicrobial peptide from the bark of spindle tree (*Euonymus europaeus* L.). FEBS Lett 530:181–185
Vila–Perello M, Tognon S, Sanchez-Vallet A, Garcia-Olmedo F, Molina A, Andreu D (2006) A minimalist design approach to antimicrobial agents based on a thionin template. J Med Chem 49:448–451
Vizioli J, Salzet M (2002) Antimicrobial peptides versus parasitic infections? Trends Parasitol 18:475–476
Wang Z, Wang G (2004) APD: the antimicrobial peptide database. Nucleic Acids Res 32: D590–592
Wollenweber B, Porter JR, Lubberstedt T (2005) Need for multidisciplinary research towards a second green revolution. Curr Opin Plant Biol 8:337–341
Yang Z, Tian L, Latoszek-Green M, Brown D, Wu K (2005) *Arabidopsis* ERF4 is a transcriptional repressor capable of modulating ethylene and abscisic acid responses. Plant Mol Biol 58:585–596
Yedery RD, Reddy KV (2005) Antimicrobial peptides as microbicidal contraceptives: prophecies for prophylactics—a mini review. Eur J Contracept Reprod Health Care 10:32–42
Zasloff M (2002) Antimicrobial peptides of multicellular organisms. Nature 415:389–395
Zemel A, Fattal DR, Ben-Shaul A (2003) Energetics and self-assembly of amphipathic peptide pores in lipid membranes. Biophys J 84:2242–2255
Zhao J, Davis LC, Verpoorte R (2005) Elicitor signal transduction leading to production of plant secondary metabolites. Biotechnol Adv 23:283–333
Zipfel C, Felix G (2005) Plants and animals: a different taste for microbes? Curr Opin Plant Biol 8:353–360

Professor Dr. Ulf Stahl,
Silke Hagen
Technische Universität Berlin
Institut für Mikrobiologie und Genetik
D-13355 Berlin
Telefon: + 49-30-314-72271
Fax: + 49-30-314-72922
e-mail: Ulf.Stahl@LB.tu-berlin.de

Function of Genetic Material: Assembly Factors of the Photosynthetic Machinery in Cyanobacteria

Jörg Nickelsen, Marc M. Nowaczyk and Birgit Klinkert

1 Introduction

The development of oxygenic photosynthesis by cyanobacteria, more than 3 billion years ago, dramatically changed life on Earth (Xiong and Bauer 2002). Organic compounds were provided due to the fixation of inorganic carbon and, concomitantly, all today`s atmospheric oxygen was produced as a byproduct. During photosynthesis, sunlight drives the transfer of electrons from water to NADP and, simultaneously, a transmembrane electrochemical gradient is generated which is utilized for the synthesis of ATP, the universal energy carrier in a living cell.

The underlying photosynthetic electron flow takes place across a specialized membrane system, the so-called thylakoids, which can be found in both prokaryotic cyanobacteria and evolutionary related chloroplasts of eukaryotic algae and plants. Despite ultrastructural differences, these phylogenetically diverse thylakoids share a common composition of four major membrane complexes constituting the basic photosynthetic apparatus. They include photosystems I and II (PS I and II) as well as the cytochrome-b_6f complex (cyt b_6f) and the chloroplast ATP synthase (F-ATPase) as delineated in Fig. 1.

PS II serves as a water-plastoquinone oxidoreductase in which the primary charge separation occurs when light energy is captured by a harvesting antenna and channelled to a specialized P680 chlorophyll a pair in the reaction centre. The electron is transferred to the initial acceptor pheophytin and the resulting electron gap at P_{680} is filled, via the D1-Y_Z tyrosine, by an electron from water which has been oxidized in a controlled fashion by the manganese cluster at the lumenal side of PS II. From pheophytin electrons are transferred to plastoquinone A (Q_A), and then they are used to reduce plastoquinone B (Q_B) to plastohydroquinone which diffuses into the membrane and constitutes the plastoquinone pool.

Next, the cyt b_6f complex serving as a plastoquinone-plastocyanin oxidoreductase shuttles the electrons to plastocyanin, a soluble lumenal copper protein. Finally in a second light-driven step, the translocation of

Progress in Botany, Vol. 68

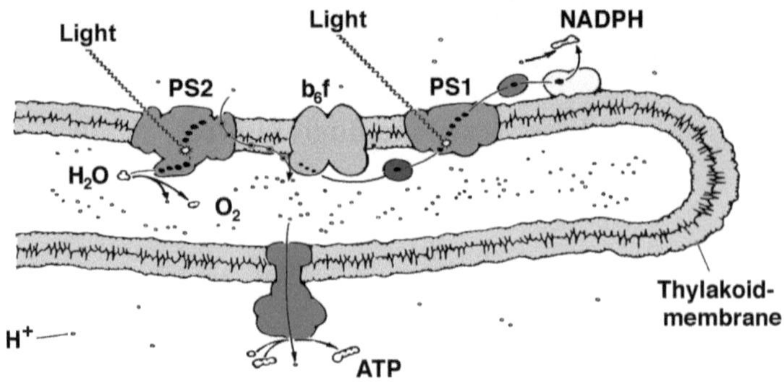

Fig. 1. Schematic model for the major protein complexes of the thylakoid membrane. For explanation, see text

these electrons from plastocyanin to the stromally located ferredoxin is mediated by PS I via several co-factors. Reduced ferredoxin is subsequently used as a reducing reagent in various biosynthetic pathways including NADPH production. The charge separation in PS II and PS I, as well as the electron transfer by the cyt b_6f complex, results in the formation of a proton motive force which drives ATP synthesis by the fourth complex, i.e. the chloroplast F-ATPase.

This overall picture of photosynthesis has recently been depicted in much greater detail by major breakthroughs in the structural analysis of the molecular machinery mediating photosynthetic electron flow (for a recent review, see Nelson and Ben-Shem 2004). For PS II, crystal structures from the cyanobacteria *Thermosynechococcus elongatus* and *Thermosynechococcus vulcanus* have been resolved at high resolution (Zouni et al. 2001; Kamiya and Shen 2003; Ferreira et al. 2004; Loll et al. 2005). In addition, the structural characteristics of cyanobacterial and plant PS I, as well as the cyt b_6f complexes from the cyanobacterium *Mastigocladus laminosus* and the green alga *Chlamydomonas reinhardtii*, have been determined in detail (Jordan et al. 2001; Ben-Shem et al. 2003; Kurisu et al. 2003; Stroebel et al. 2003).

These structural data allowed several burning questions to be answered regarding the structure/function relationships within single thylakoid membrane complexes. Accordingly, a precise picture of how photosynthesis works can be drawn today. However, in contrast, very little information is available on how the photosynthetic apparatus is built up during thylakoid membrane biogenesis. In eukaryotes, this process is complicated by the fact that due to the endosymbiotic origin of chloroplasts, the various subunits of the thylakoid protein complexes are encoded by two different genetic systems, either the nuclear or the plastid genome. Thus, a sophisticated and versatile regulatory network between these two genetic compartments had to be established during evolution guaranteeing coordinated expression of photosynthetic complex subunits (Nickelsen 2003). Obviously, cyanobacteria, as prokaryotes, do not share similar regulatory principles; however, the basic processes of protein insertion into the thylakoid membrane and subsequent subunit assembly are likely to be the same in both chloroplasts and cyanobacterial cells. As such, cyanobacteria in general, and especially the well-characterized species *Synechocystis* sp. PCC 6803 (hereafter *Synechocystis* 6803), serve as a powerful model system to study general aspects of photosynthetic complex formation within thylakoids. Furthermore, a growing list of completely sequenced cyanobacterial genomes is available at Cyanobase (http://www.kazusa.or.jp/cyano) facilitating phylogenetic analyses of the underlying genes.

2 Assembly/stability of thylakoid membrane complexes

Recent work has demonstrated that besides the subunits themselves, several trans-acting factors are involved in the assembly process of photosynthetic complexes. These assembly factors do not form parts of the final functional unit, but interact only transiently with distinct subunits. They can generally be divided into four principle groups.

First, some complex subunits must be post-translationally modified by enzyme activities prior to complete assembly. The most prominent example is represented by the C-terminal processing of the PS II reaction centre subunit D1 which is required for further docking of the luminal oxygen evolving complex (Anbudurai et al. 1994; Shestakov et al. 1994). A second group includes factors that generally assist the insertion of proteins into or across the thylakoid membrane, and usually have a broader spectrum of target subunits like the Sec and Tat translocons (Di Cola et al. 2005), the chloroplast signal recognition particle (Schünemann 2004) or the membrane integrase Alb3 which is homologous to YidC from *Escherichia coli* and Oxa1 from mitochondria (Yi and Dalbey 2005). In particular, the photosynthetic membrane contains an extraordinary high number of various pigments and other co-factors which have to be synthesized and incorporated into their cognate apoproteins. The third group of assembly-assisting factors, therefore, is responsible for the correct attachment of iron-sulphur clusters or the incorporation of chlorophylls and carotenoids.

In *Arabidopsis thaliana*, recent genetic analyses revealed several genes with homologues in cyanobacteria which are involved in iron-sulphur cluster assembly and, consequently, PS I biogenesis (for an overview, see Kessler and

Papenbrock 2005; Pilon et al. 2006). For instance, the universally conserved Hcf101 protein was shown to be involved in [4Fe-4S]-cluster biogenesis in *A. thaliana* (Lezhneva et al. 2004; Stöckel and Oelmüller 2004). The Hcf101 homologue from *Synechocystis* 6803, *slr0067*, appears to be an essential gene, since upon its insertional inactivation, no complete segregation of wild-type and mutant alleles could be obtained (Kessler and Papenbrock 2005). However, despite an apparent conservation between the nif/isc/suf pathways in cyanobacteria and plastids, also factors specific for vascular plants, like the Apo protein family, have been described (Amann et al. 2004). This would suggest distinct differences in iron-sulphur cluster assembly/stabilization between pro- and eukaryotes.

Finally, the last group includes those factors which act in a chaperone-like fashion and appear to mediate a correct subunit assembly by spatial and temporal means. This review is aimed at briefly summarizing current knowledge of the complex assembly process in cyanobacteria by focussing on this latter group of assembly factors. To date, the analysis of photosynthetic mutants has revealed that they are involved in the formation of both functional PS I and PS II.

2.1 Assembly/stability of photosystem I

The crystal structures of both cyanobacterial and plant PS I have recently been released (Jordan et al. 2001; Ben-Shem et al. 2003). Although the core of the PS I reaction centres is highly conserved amongst all organisms performing oxygenic photosynthesis, the general organization of higher order structures is different. In contrast to algae and plants, cyanobacteria contain a trimeric PS I complex whose formation is dependent on the PsaL subunit (Chitnis and Chitnis 1993). Furthermore, under iron stress conditions, cyanobaterial PS I assembles with an unusual antenna consisting of 18 monomers of the CP43-like IsiA protein (Bibby et al. 2001; Boekema et al. 2001).

In cyanobacteria, each photosystem I monomer was shown to comprise 12 subunits which are listed in Table 1 along with some of their molecular characteristics. The two major subunits PsaA and PsaB coordinate most of the co-factors as well as the chlorophyll special pair P700. Their structural organization resembles that of the D1/D2 and CP43/CP47 subunits of PS II suggesting that both photosystems derived from a common ancestor (Baymann et al. 2001). Subunits PsaC, PsaE, and PsaD are involved in binding of Ferredoxin at the stromal side while PsaF mediates docking of plastocyanin from the lumenal side of PS I, at least in algae and plants. Its precise role in cyanobacteria has still to be elucidated like that of other subunits (Table 1).

Table 1. Subunits of PS I

Subunit	MW[a]	TMH[b]	Function
PsaA	82.9	11	Light harvesting, charge separation
PsaB	81.3	11	Light harvesting, charge separation
PsaC	8.8	0	Electron transfer, 2 Fe-S cluster
PsaD	15.6	0	Binding of ferredoxin/flavodixin
PsaE	8.1	0	Binding of ferredoxin/flavodixin
PsaF	18.2	2	Binding of plastocyanin/cytochrome c_6
PsaI	4.4	1	Functional organization of PsaL
PsaJ	4.5	1	Functional organization of PsaF
PsaK	13.7	2	Function unknown
PsaL	16.6	2	Trimer formation in cyanobacteria, binds Ca^{2+}
PsaM	3.4	1	Function unknown
PsaX	4.9	1	Function unknown, not found in all cyanobacteria

[a]Calculated MW of the mature protein (kDa)
[b]Number of TMH (trans membrane helices)

However, the assembly of the cyanobacterial PS I complex appears to depend on several trans-acting facilitators, a few of which have recently been identified (for review, see Schwabe and Kruip 2000; Saenger et al. 2002). These include homologues of conserved chloroplast open reading frames, namely Ycf3, Ycf4 and Ycf37 as well as the proteins RubA and BtpA.

Ycf4: The first factor required for stable accumulation of PS I was discovered by the identification of the *Synechocystis* 6803 open reading frame 184 which was subsequently named *sll0226* (Wilde et al. 1995). It encodes a homologue of the 20 kDa Ycf4 protein which is encoded by the chloroplast genomes of algae and higher plants. Disruption of *sll0226* leads to reduced levels of PS I which are still capable of performing photosynthetic charge separation and, thus, mutant cells grow photoautotrophically at almost wild-type rates (Wilde et al. 1995). In contrast, thylakoid-localized Ycf4 was shown to be essential for PS I accumulation in chloroplasts of both the green alga *C. reinhardtii* and the vascular plant *Nicotiana tabacum* (Boudreau et al. 1997; Ruf et al. 1997). This indicates that despite a conservation of 43 % identical amino acids between the alga and cyanobacterial proteins, the function of Ycf4 has apparently shifted during evolution towards a more essential role for PS I biogenesis in chloroplasts (Rochaix et al. 2004). The precise function of Ycf4, however, has still to be elucidated.

Ycf3: Ycf3 represents another PS I-related open reading frame which is, with the exception of *Euglena gracilis* (Martin et al. 2002), usually encoded in the chloroplast genomes of eukaryotes, and to date, represents one of the best-characterized, transiently acting PS I assembly factors. Ycf3 from *Synechocystis* 6803 (*slr0823*) shares a high degree of conservation with the chloroplastic homologues from algae and land plants. Consistently, all Ycf3 proteins contain three so-called TPR (tetratrico peptide repeats) units.

The TPR domain consists of multiple units (3–16) of a degenerate motif comprising 34 amino acids which forms two amphipathic α-helices. The crystal structure of TPR domains revealed that they can form right-handed superhelices which then provide the surface for mediating specific protein-protein interactions (for review, see Blatch and Lässle 1999; D'Andrea and Regan 2003). TPR proteins are ubiquitously distributed throughout all kingdoms of life and participate in a wide range of different molecular functions, such as cell division, DNA replication, RNA maturation and protein transport. Interestingly, several TPR proteins have recently been shown to be involved in chloroplast biogenesis in eukaryotes by affecting chloroplast gene expression at the posttranscriptional level (Boudreau et al. 2000; Vaistij et al. 2000; Felder et al. 2001; Sane et al. 2005) or chlorophyll biosynthesis (Meskauskiene et al. 2001). Cyanobacteria contain a relatively high number of putative TPR proteins as compared to other Gram negative bacteria. For instance, the bioinformatical inspection of the *Anabena* 7120 and *Synechocystis* 6803 genomes revealed 65 and 23 TPR proteins, respectively, whereas only 11 candidates were identified in *E. coli* (J. Nickelsen, unpublished data). To date, the function of most of the cyanobacterial TPR domain-containing polypeptides is unknown. However, as mentioned later on, the systematic inactivation of the respective open reading frames has shown that, in addition to Ycf3, at least two more of them are involved in the assembly of photosynthetic complexes.

Ycf3 knockout strains have been obtained for *C. reinhardtii* and *N. tabacum* and their corresponding phenotypes indicated a strict requirement of Ycf3 for the activity of PS I in both algae and higher plants on a post-translational level (Boudreau et al. 1997; Ruf et al. 1997). The Ycf3 protein was localized to thylakoid membranes, but it was not stably associated with PS I. A temperature-sensitive *ycf3* mutant was constructed in *C. reinhardtii* and used in temperature-shift experiments, thereby demonstrating that Ycf3 is indeed required for the assembly but not the stability of PS I (Naver et al. 2001). Furthermore, the PS I subunits PsaA and PsaD were identified as direct interaction partners for Ycf3 by applying immunoprecipitation and immunoblot experiments. In cyanobacteria, probably homo-oligomeric

Ycf3 is essential for PS I activity, too (Schwabe 2003; Schwabe et al. 2003). However, in contrast to its chloroplast counterparts, it appears to be mainly localized to the plasma membrane but not the thylakoids (Zak et al. 2001) pointing to a role for Ycf3 during the early steps of PS I biogenesis (see section 3).

Ycf37: Similar to Ycf3, also the Ycf37 protein contains three TPR units and was suggested to be involved in the biogenesis and/or stability of PS I. It is encoded by the open reading frame *slr0171* in *Synechocystis* 6803 and homologues have been identified in the chloroplast genomes of the non-green algae *Cyanophora paradoxa*, *Cyanidium caldarium*, *Porphyra purpurea* and *Guillardia theta* (Wilde et al. 2001). In green algae and higher plants, the *ycf37* gene is located in the respective nuclear genomes. In contrast to *ycf3*, deletion of cyanobacterial *ycf37* had only a moderate effect on PS I activity resulting in a decrease of 30 % of trimeric PS I as compared to the wild-type (Dühring et al. 2005). Blue-native PAGE revealed the existence of a novel, high-light dependent, monomeric PS I complex which is absent in the *ycf37* mutant suggesting that Ycf37 is involved in stabilization of this intermediate form of PS I (Dühring et al. 2005).

RubA: In *Synechococcus* sp. PCC 7002, the *rubA* gene encodes an unusual rubredoxin which is strictly required for PS I activity (Shen et al. 2002a). Detailed molecular analysis revealed that RubA protein is localized to thylakoid but not plasma membranes of cyanobacteria as well as chloroplast thylakoids from algae and vascular plants. It might transiently associate with monomeric PS I but no evidence for an association with either trimeric PS I or PS II was obtained. In a *rubA* mutant, PS I content was reduced to 40% of the wild-type level and residual complexes completely lacked the stromal subunits PsaC, PsaD and PsaE (Shen et al. 2002a). It was proposed that RubA is specifically required for the assembly of the F(X) iron-sulphur cluster (Shen et al. 2002b).

BtpA: Another factor involved in PS I accumulation is represented by the so-called BtpA (biogenesis of thylakoid protein) protein which is encoded by the open reading frame *sll0634* in *Synechocystis* 6803. In a *btpA* mutant, both the PsaA and PsaB PS I reaction centre subunits are significantly reduced while the level and activity of PS II is unaffected (Bartsevich and Pakrasi 1997). The 30 kDa BtpA polypeptide belongs to the structural class of TIM-barrel enzymes (Wierenga 2001) and was shown to be an extrinsic membrane protein which is localized to cytoplasmic face of the thylakoids (Zak et al. 1999). More recent studies suggest that BtpA is involved in PS I stabilization under low temperature conditions (Zak and Pakrasi 2000). However, the precise role of BtpA remains to be elucidated, especially, in view of the fact that close relatives can be identified in various non-photosynthetic organisms whereas no homologues exist in higher plants (Bartsevich and Pakrasi 1997).

2.2 Assembly/stability of photosystem II

PS II comprises at least 19 protein subunits (Table 2), 36 chlorophyll a molecules, seven ß-carotenes, two pheophytines, two plastoquinones, one heme b, one heme c, a none heme Fe and one OEC per monomer in cyanobacteria (Ferreira et al. 2004). The monomeric PS II core complex represents the smallest functional unit. It is composed of several membrane integral subunits (PsbA, PsbB, PsbC, PsbD, PsbE, PsbF, PsbI) and the extrinsic proteins PsbO, PsbU, PsbV which shield the oxygen evolving complex. The native complex is mainly present as a dimer in the thylakoid membrane (Rögner et al. 1996), due to structural constraints for optimized energy transfer within the complex (Mustardy et al. 1992). On the other hand, optimized energy transfer within the thylakoid membrane (e.g. state transitions) is influenced by the monomeric/dimeric state PS II (Kruip 1994). The core centre proteins D1 and D2 each contain five transmembrane helices and bind most of the redox centers of the intrinsic electron transfer chain, i.e. six chlorophyll a molecules, two phaeophytins, two quinones and a none-heme iron.

Most of the chlorophyll a molecules in the complex (30 of 36) are bound by the intrinsic antenna proteins CP47 and CP43, which form six transmembrane helices each. The arrangement of the transmembrane helices is similar to that of the PS I subunits PsaA and PsaB, but there are slight differences in the orientation of the chlorophyll molecules which might explain the slower energy transfer compared to PS I (de Weerd et al. 2002). The large number of small subunits, often build by only one transmembrane helix, is remarkable compared with other membrane protein complexes, although the function for most of these subunits is as yet unknown (for review, see Shi and Schröder 2004).

Due to the large complexity of PS II, its assembly is very likely to be a multistep process guided by several chaperones and other factors. The first step includes the integration of the transmembrane helices into the lipid phase followed by the formation of an initial PS II precomplex. It was shown by applying pulse labelling experiments and mutational studies that, at very early stages of PS II assembly, a D1/D2 heterodimer is build which associates sequentially with Cyt-b559 and PsbI (Aro et al. 2004; Komenda et al. 2004). The integration of the other small subunits is less clear. Some are present at this early stage (e.g. PsbL, PsbJ), whereas others might be attached after the incorporation of the intrinsic antenna proteins CP43 and CP47 (Swiatek et al. 2003;

Table 2. Subunits of PS II

Gene	Synonyms	MW[a]	TMH[b]	Function	Prokar-yotes	Eukar-yotes
PsbA	D1	38	5	Primary reaction	+	+
PsbB	CP47	56.3	6	Light harvesting	+	+
PsbC	CP43	50.1	6	Light harvesting	+	+
PsbD	D2	39.2	5	Primary reaction	+	+
PsbE	Cyt b559[a]	9.2	1	Photoprotection and assembly	+	+
PsbF	Cyt b559[b]	4.4	1	Photoprotection and assembly	+	+
PsbG				Not PS II	–	–
PsbH		7.7	1	Q_A to Q_B transfer, PS II repair	+	+
PsbI		4.2	1	Dimerization and photoprotection	+	+
PsbJ		4.1	1	Involved in assembly	+	+
PsbK		4.3	1	Stabilization and pigment binding	+	+
PsbL		4.4	1	Stabilization	+	+
PsbM		3.7	1	Stabilization of dimmer	+	+
PsbN		4.7	1	Unknown, probably not PS II	(+)	(+)
PsbO	33 kDa	33	0	Stabilization of OEC	+	+
PsbP	23 kDa	23	0	Binding of Ca^{2+} and Cl^-	(+)[c]	+
PsbQ	16 kDa	16.5	0	Binding of Ca^{2+} and Cl^-	(+)[c]	+
PsbR		15.1	1	Stabilization of OEC	–	+
PsbS		26.1	4	Photoprotection	–	+
PsbTc		4.0	1	Dimerization	+	+
PsbTn		3.0	0	Unknown	–	+
PsbU	12 kDa	12	0	Stabilization of OEC	+	–
PsbV	Cyt c550	17.8	0	Stabilization of OEC	+	–
PsbW		12.1	1	Dimerization	–	+
PsbX		4.1	1	Stabilization	+	+
PsbY		4	1	Unknown	+	+
PsbZ	Ycf9	6.7	2	Stabilization of supercomplex	+	+

[a]Calculated MW of the mature protein (kDa)
[b]Number of TMH (trans membrane helices)
[c]These proteins seem to be more transiently associated with PS II in cyanobacteria and might have a different function compared to green algae/vascular plants

Komenda et al. 2004; Rokka et al. 2005). These early processes could take place spontaneously or guided by some yet unknown factors.

HCF136: One potential candidate for an essential factor mediating the early PS II assembly steps is the luminal protein HCF136 (ycf48). It was shown that an *A. thaliana* T-DNA insertion mutant of the *HCF136* gene was devoid of any PS II activity and none of the nuclear- and plastome-encoded PS II subunits accumulated to significant levels (Meurer et al. 1998). HCF136 associates with a PS II precomplex in the thylakoid membrane containing at least D2 and cyt-b559 further supporting the idea of a factor acting during early assembly processes (Plücken et al. 2002). Potential homologues of this protein could be identified on the genome level in several cyanobacteria including the reading frame *slr2034* in *Synechocystis* 6803. A *slr2034* homologue from *Synechococcus* 7002 was, however, found to be dispensable for PS II activity (Shen et al. 2002b). Hence, additional work is required to elucidate the precise function of HCF136 during the biogenesis of both pro- and eukaryotic PS II.

PratA: Once the D1 precursor protein pD1 has been inserted into the membrane another early assembly step is represented by the C-terminal maturation of pD1 prior to the formation of the oxygen evolving complex. In all organisms performing oxygenic photosynthesis except the eukaryotic alga *Euglena gracilis* the D1 protein is synthesized with an extension of 8-16 amino acids (Diner et al. 1988; Svensson et al. 1991). This extension is usually cleaved after its translocation into the luminal space by the endoprotease CtpA (Anbudurai et al. 1994; Shestakov et al. 1994). The removal of the extension is strictly required for the proper subsequent assembly of the oxygen evolving complex at the luminal side of PS II (Roose and Pakrasi 2004). However, the absence of the extension by itself has only a moderate effect on PS II activity leaving the question open why it has been retained during evolution (Ivleva et al. 2000; Kuvikova et al. 2005).

Recently, it was shown that in *Synechocystis* 6803, a periplasmic factor, named PratA, is involved in the processing of the D1 subunit (Klinkert et al. 2004). Interestingly and similar to *ycf3* and *ycf37*, the *PratA* gene (*slr2048*) encodes a TPR protein containing nine TPR units. Targeted inactivation of *PratA* resulted in a drastically reduced PS II content and accumulation of unprocessed pD1, suggesting that this factor facilitates CtpA function (Klinkert et al. 2004). In agreement with this, a similar D1 maturation-assisting activity was also postulated for spinach, based on in vitro binding studies (Yamamoto et al. 2001). Soluble PratA binds to the D1 C-terminus and, most intriguingly, it was shown to be localized to the periplasmic cell compartment supporting the idea that the early PS II assembly steps take place at the plasma membrane in *Synechocystis* 6803 (Fulda et al. 2000; Klinkert et al. 2004; see section 3).

Psb27: The early events of PS II biogenesis might also depend on another protein called Psb27. This factor has been detected in PS II complexes accumulating in a CtpA deletion mutant of *Synechocystis* 6803 (Roose and Pakrasi 2004). The 11-kDa Psb27 protein was shown to contain an N-terminal lipid modification typical for bacterial lipoproteins (Nowaczyk, unpublished results). It is tightly associated with a preassembled PS II subfraction missing the extrinsic proteins PsbO, PsbU, PsbV and therefore devoid of any oxygen evolving activity (Nowaczyk, unpublished results). In conclusion, available data suggest that it might be involved in the processing of pD1 and/or, more generally, in the assembly of the PS II donor site.

PsbQ/PsbP: Furthermore, two additional potential subunits at the donor side are necessary for the biogenesis of fully active PS II complexes in cyanobacteria. To date, the accepted view was that green algae and higher plants possess the extrinsic proteins PsbP, PsbQ and PsbO, whereas cyanobacteria, instead, contain PsbO, PsbU and PsbV subunits. However, Thornton et al. (2004) were able to identify homologues of the higher plant proteins PsbP and PsbQ in *Synechocystis* 6803. They showed that both proteins are associated with the PS II complex and necessary for its optimal function (Thornton et al. 2004; Summerfield et al. 2005a,b). PsbQ seems to be present in equal amounts compared to other PS II subunits whereas PsbP is clearly substochiometric in *Synechocystis* 6803. The recent PS II crystal structures from the thermophile cyanobacteria *T. elongatus* or *T. vulcanus* did not indicate the presence of any of these additional luminal subunits. Thus, it could be that they are loosely attached to the complex and lost during preparation or, alternatively, they only accumulate under specific conditions in cyanobacteria. This last option would indicate a modified function compared to their homologues in higher plants or green algae. In conclusion, the role which PsbQ and PsbP play in cyanobacteria still remains to be elucidated.

Psb29: An additional protein, which was detected in PS II preparations in substoichiometric amounts, is Psb29 (Kashino et al. 2002). Deletion of the *Psb29* gene (*sll1414*) in *Synechocystis* 6803 led to slower growth rates under high light conditions, increased light sensitivity, and lower PS II efficiency of mutant cells. A similar phenotype was observed for a mutant in which the *Psb29* homologue from *A. thaliana* had been inactivated by a T-DNA insertion (Keren et al. 2005). Both cyanobacterial and plant mutants exhibited a light-dependent increase in the proportion of uncoupled proximal antennae in PS II. Hence, it was postulated that the evolutionary conserved Psb29 protein plays a critical role during biogenesis of PS II (Keren et al. 2005).

Psb28: At the final stage of the PS II assembly process, dimerization of the monomeric complex and further supramolecular organization of the dimeric PS II complexes occurs in the thylakoid membrane. One factor involved in these higher order structures of PS II might be Psb28. Upon inactivation of this protein in *Synechocystis* 6803, PS II becomes mobile, suggesting that Psb28 serves as an anchor which prevents lateral diffusion in the thylakoid membrane (S. Purton and C. Mullineaux, personal communication). Nevertheless, in wild-type cells, red light-controlled mobilization of PS II is required for efficient recovery after photoinactivation (Sarcina et al. 2005). This suggests that Psb28 acts as a structural organizer and, additionally, regulates the dynamics of PS II mobility.

2.3 Repair of photosystem II

Photosynthetic water splitting is inevitable, coupled with the formation of reactive oxygen species. Damage to PS II is mainly caused by singlet oxygen generated from triplet chlorophyll species. In living cells, many detoxification systems are present that scavenge radicals and active oxygen species (Dietz et al. 2002). Nevertheless, even at low light intensities, PS II is subject to inherent damage underlining the requirement for an efficient repair system (Anderson et al. 1997). Under light stress conditions, however, this repair system is not able to cope with damage anymore and, consequently, PS II is irreversibly inactivated (Aro et al. 1993). It is known for a long time that the D1 Protein of PS II is the major target for photodamage. Consequently, it has to be frequently replaced by a newly synthesized one, and thus exhibits the highest turn-over rate of all photosynthetic subunits (Aro et al. 1993). More recently, the reaction centre protein D2 and the small subunit PsbH have also been shown to be replaced during the photoinhibition repair cycle of PS II (Bergantino et al. 2003).

The various steps of the PS II repair cycle include:

1) Photodamage to PS II, electron transport is impaired
2) Conformational change, signal to remove the damaged subunits
3) Monomerization of PS II and partial disassembly (e.g., removal of CP43)
4) Degradation of the damaged subunits and replacement
5) Reassembly of the other intrinsic subunits
6) D1-processing, binding of extrinsic proteins and photoactivation
7) Dimerization of the active monomer

At the moment it remains elusive which factors are assisting this process. Steps 5–7 are comparable to the de novo assembly of PS II complexes whereas steps 1–4 represent unique features of the repair cycle and, therefore, specific factors are likely to be involved in this part of the cycle. For instance, much work has been invested to identify the proteases that are involved in the degradation of the D1 subunit. The current model for D1 degradation predicts that in a first step DegP2, a member of the DegP/HtrA protein family, performs the primary cleavage event in the stromal loop connecting transmembrane helices D and E (Haussuhl et al. 2001), whereas a protease of the FtsH family degrades the resulting fragments of this primary cleavage event (Lindahl et al. 2000). In *Synechocystis* 6803, the DegP2 protease seems to be dispensable for rapid D1-turnover and, therefore the FtsH protease might play a more general role for the PS II repair cycle in cyanobacteria (Nixon et al. 2005; Kamata et al. 2005).

3 Subcompartementalization of thylakoid membrane biogenesis

Besides the question of how the multisubunit complexes of the thylakoid membrane are assembled, another intriguing and still unresolved aspect concerns the spatial organization of the underlying processes. Recent years have seen the development of a hypothesis which predicts a vesicular lipid transport system in chloroplasts (Vothknecht and Soll 2002). This system would transfer galactolipids from the inner envelope membrane to the thylakoids. Several lines of evidence support the idea of intraorganellar vesicle transport in plastids resembling the one which is usually found in the cytoplasm.

For instance, in *C. reinhardtii* chloroplasts, a membranous subfraction, the so-called "low density membrane" (LDM) system has been identified during cell fractionation experiments (Zerges and Rochaix 1998). LDMs resemble the inner envelope with regard to their acyl lipid composition and are partially associated with thylakoids in a Mg^{2+} dependent fashion. Thus, they were postulated to represent an intermediate state of thylakoid membrane formation which originates from the inner chloroplast envelope (Zerges 2000). Moreover, a dynamin-like protein (ADL-1) from *A. thaliana* was found to be required for thylakoid membrane formation (Park et al. 1998) and the VIPP1 protein was implicated in budding of small vesicles from the inner envelope membrane, a process that can be followed in wild-type plants after exposure to low temperature conditions of 4°C (Kroll et al. 2001).

In *Synechocystis* 6803, the inactivation of the VIPP1 homologue (*sll0617*) also led to impaired thylakoid membrane formation, suggesting an evolutionary conserved function of this protein (Westphal et al. 2001). VIPP1 shares homologies with PspA, a bacterial protein which is induced under various stress conditions including phage invasion, inhibition of lipid biosynthesis and secretion defects. Nevertheless, VIPP1 coexists with PspA in cyanobacteria and contains an extra C-terminal extension which was proposed to specify its function during thylakoid membrane biogenesis.

Though the precise roles both proteins play, to date, remain elusive, they have been shown to form homomultimers with ring-shaped morphology (Aseeva et al. 2004; Hankamer et al. 2004).

To date, it is still an open question whether vesicle transport from the plasma membrane to the thylakoids occurs in cyanobacteria. Alternatively, lateral fusions of the two membranes types can be envisaged based on electron micrographical data (Spence et al. 2003). Independent of the precise molecular details, a transfer mechanism for lipids and, more important in the context of this reviews topic, also protein complexes must be considered. The recent observation that the initial steps of photosystem assembly take place at the plasma membrane but not the thylakoids strongly support this idea (Zak et al. 2001).

Based on an efficient two-dimensional separation technique, plasma and thylakoid membranes were separated and subjected to immunodetection of both PS II and PS I subunits. The data revealed that the D1, D2 and cytochrome b559 subunits of PS II, as well as the PsaA and PsaB reaction centre subunits of PS I, are present not only in thylakoids, but also in plasma membranes though to a substantially lesser extent. In contrast, PsbB and PsbC as well as other subunits of PS II and PS I were not present in the plasma membrane fraction. Furthermore, the C-terminal D1 processing

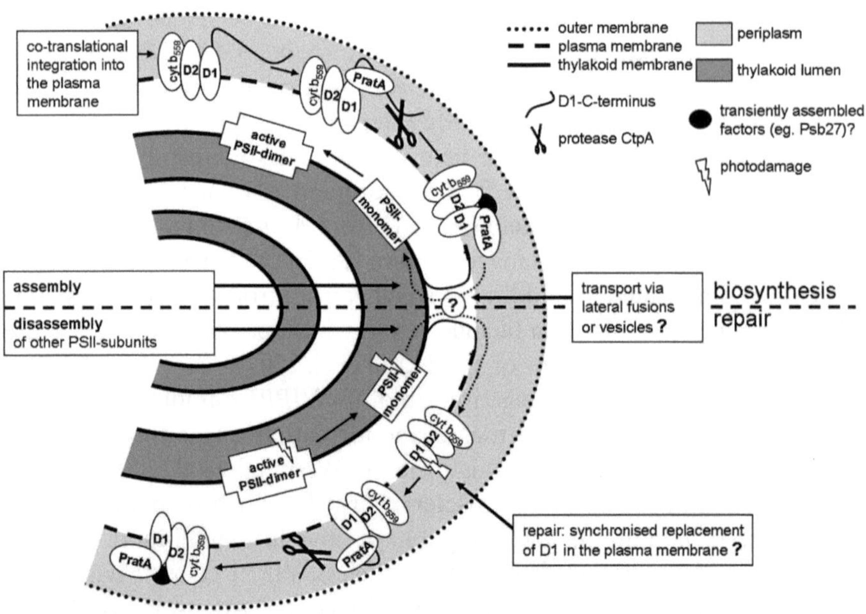

Fig. 2. Model for PS II biogenesis and repair in cyanobacteria. For explanation, see text

peptidase CtpA as well as the abovementioned PS I assembly factors Ycf3 and Ycf4 were almost exclusively detected in the plasma membrane and not the thylakoids. Finally, the plasma membrane-localized core-complexes contained chlorophyll molecules and were able to undergo light-induced charge separation (Zak et al. 2001; Keren et al. 2005).

A second, more genetic piece of evidence for a spatially separated photosystem assembly process was obtained with the identification of the PratA factor mentioned before (Klinkert et al. 2004; see also section 2.2). This soluble D1-associated protein was localized to the periplasm of *Synechocystis* 6803 cells, further supporting the idea of a plasma membrane-localized D1 maturation system containing CtpA (Fulda et al. 2000; Klinkert et al. 2004). The structure of the PratA signal sequence suggests that it is transported via the Sec pathway across the plasma membrane of *Synechocystis* cells and, consistently, PratA was detected in a proteomic approach aimed at mapping periplasmic proteins from *Synechocystis* 6803 (Fulda et al. 2000). Moreover, cell fractionation experiments indicated that the vast majority of the protein is indeed located in the periplasm and not the thylakoid lumen (Klinkert et al. 2004). This adds to the notion that a strict protein sorting system exist in cyanobacteria which targets proteins to either the periplasm or the thylakoid lumen (Spence et al. 2003).

Based on the available evidence, the actual picture arising for the spatial and temporal regulation of PS II biogenesis including some of the involved factors is depicted in Fig. 2. During protein synthesis, both the D2 and the precursor D1 (pD1) subunits are co-translationally inserted into the plasma membrane where they form together with cytochrome b599 the PS II core complex (Komenda et al. 2004; Keren et al. 2005). This plasma membrane-localized core complex exhibits chlorophyll fluorescence characteristics similar to isolated core complexes from thylakoids indicating proper integration of pigment molecules (Keren et al. 2005). As a next step, the precursor pD1 exposing its C-terminus to the periplasm is processed by the plasma membrane associated CtpA protease, a process which is assisted by the soluble PratA factor. At this stage, Psb27 might also already be associated with the core complex from the periplasmic site because it accumulates in PS II complexes from a *CtpA* deletion mutant containing only non-processed D1. Furthermore, D1 appears to interact directly with Psb27 (J. Nickelsen, unpublished results). Next, this core complex travels via lateral fusions, vesicles or an yet unidentified pathway to the thylakoid membrane where the assembly of monomeric and finally dimeric PS II continues with docking of CP47 and CP43 as well as the remaining subunits (Komenda et al. 2004; Keren et al. 2005). Intriguingly, these latter subunits, especially CP43, were exclusively found in thylakoid membranes but not in the plasma membrane

(Zak et al. 2001). Moreover, structural studies of PS II have revealed that CP43 contributes ligands to the manganese cluster responsible for water oxidation. Thus, active PS II can only be formed in thylakoids due to an adequate separation of assembly steps and activity. This separation is likely to protect the photosynthetic machinery from partial electron transfer reactions creating potentially harmful radical species.

While this model appears to apply for explaining the de novo formation of PS II complexes, a still unresolved matter concerns the processes which are connected to the PS II repair cycle. In chloroplasts, the replacement of photodamaged D1 protein takes place at stromal thylakoid membranes and repaired PS II then moves back to the grana regions of thylakoids (Adir et al. 1990). Therefore, it was assumed that in cyanobacteria, PS II is also repaired at thylakoids. However, in view of the above-mentioned model, one might also speculate whether photodamaged PS II complexes move back to the plasma membrane where the insertion of new D1 protein is achieved (Keren et al. 2005). As mentioned above, the Psb28 dependent mobilization of PS II has been reported to be required for the rapid initiation of recovery from photoinhibition (Sarcina et al. 2005). Furthermore, two other lines of evidence would support the idea of a plasma membrane-localized D1 exchange. First, the C-terminal D1 processing protease CtpA, which is also required during D1 replacement, was only localized to the plasma membrane but not the thylakoid membrane in *Synechocystis* 6803 (Zak et al. 2001). Second, the recently discovered Hho protease from *Synechocystis* 6803 has been shown to be involved in the degradation of photodamaged D1 (I. Adamska, personal communication), and, similar to PratA, this enzyme was detected in the periplasm (Fulda et al. 2000).

4 Conclusions and perspectives

The biogenesis of the photosynthetic apparatus in cyanobacteria is a stepwise process which requires various trans-acting factors assisting the assembly of single complexes in a chaperone-like fashion. To date, probably only a handful of these factors have been identified and, for only a few of these, a precise function can be assigned (Table 3). Thus, current genetic and biochemical work focuses on the identification of additional assembly factors for all the major complexes of the thylakoid membrane, i.e. PS II, PS I, cyt b_6f and the F-ATPase. One principle that becomes evident is that apparently TPR proteins like Ycf3, Ycf37 and PratA play a crucial role in this context, reminiscent to the situation of regulators for RNA metabolism in the chloroplast of eukaryotes (Nickelsen 2003). Therefore, systematic genetic approaches aimed at inactivation of all 23 TPR domain-containing genes of

Table 3. Transient components of cyanobacterial PSI and PS II

Protein (reading frame)	Homologue in chloroplasts	Protein features	Mutant phenotype
Ycf4 (*sll0226*)	+	Plasma membrane protein	Reduced PSI level
Ycf3 (*slr0823*)	+	Plasma membrane associated TPR-protein	Essential for PSI activity
Ycf37 (*slr0171*)	+	TPR-protein	Reduced PSI level
RubA (*slr2033*)	+	Unusual rubredoxin associated with thylakoid membranes	Reduced PSI level
BtpA (*sll0634*)	–	TIM-barrel protein associated with thylakoid membranes	Reduced PSI level
HCF136 (*slr2034*)	+	Luminal protein associated with preassembled PS II	None
CtpA (*slr0008*)	+	D1 C-terminal processing protease	Essential for PS II activity
PratA (*slr2048*)	–	Periplasmic TPR protein interacting with D1	Reduced PS II level
PsbP (*sll1418*)	+	Potential lipoprotein at the donor site of PS II	Reduced PS II activity
PsbQ (*sll1638*)	+	Potential lipoprotein at the donor site of PS II	Reduced PS II activity
Psb27 (*slr1675*)	+	Potential lipoprotein at the donor site of PS II	Not reported
Psb28 (*sll1398 slr1739*)	+	Acceptor site factor of PS II	Not reported
Psb29 (*sll1414*)	+	Acceptor site factor of PS II	Lower PS II efficiency

Synechocystis 6803 have been initiated. Preliminary data suggest that at least two more TPR proteins are required for efficient photosynthetic activity (J. Nickelsen, unpublished results). With the availability of the respective mutant strains, their biochemical characterization, for instance by combined protein pulse-labelling/2D BN-PAGE analyses, will help to precisely determine the affected complex assembly steps (Komenda et al. 2004). Furthermore, the isolation of photosynthetic subcomplexes via His-tagged subunits and their subsequent mass spectrometric analysis represents a complementary approach for the identification of factors which transiently associate with intermediate forms or partially assembled photosystems (Kashino et al. 2002; Nowaczyk 2005).

As a next step, the precise interaction partners of the regulatory factors have to be identified by biochemical means or, alternatively, by using the yeast split-ubiquitin system which allows one to monitor the interaction between membrane proteins in a living cell (Stagljar et al. 1998; Pasch et al. 2005). With regard to protein/protein interactions, an intriguing question concerns the association of the assembly factors not only with distinct complex subunits but also between the various trans-acting factors themselves. If such interactions exist, they might form a kind of assembly scaffold organizing photosystem biogenesis.

Finally, an aspect that has not been addressed in this review concerns the incorporation of the various low molecular weight co-factors such as iron-sulphur clusters, chlorophylls and carotenoids into the respective apoproteins. This process probably interferes with assembly of the protein subunits and, thus, adds another level of complexity to the generation of functional photosynthetic machinery.

References

Adir N, Shochat S, Ohad I (1990) Light-dependent D1 protein synthesis and translocation is regulated by reaction center II. Reaction center II serves as an acceptor for the D1 precursor. J Biol Chem 265:12563–12568

Amann K, Lezhneva L, Wanner G, Herrmann RG, Meurer J (2004) Accumulation of photosystem 1, a member of a novel gene family, is required for accumulation of [4Fe-4S] cluster-containing chloroplast complexes and antenna proteins. Plant Cell 16:3084–3097

Anbudurai PR, Mor TS, Ohad I, Shestakov SV, Pakrasi HB (1994) The *ctpA* gene encodes the C–terminal processing protease for the D1 protein of the photosystem II reaction center complex. Proc Natl Acad Sci USA 91:8082–8086

Anderson JM, Park YI, Chow WS (1997) Photoinactivation and photoprotection of photosystem II in nature. Physiologia Plantarum 100:214–223

Aro EM, McCaffery S, Anderson JM (1993) Photoinhibition and D1 protein degradation in peas acclimated to different growth irradiances. Plant Physiol 103:835–843

Aro EM, Suorsa M, Rokka A, Allahverdiyeva Y, Paakkarinen V, Saleem A, Battchikova N, Rintamaki E (2004) Dynamics of photosystem II: a proteomic approach to thylakoid protein complexes. J Exp Bot 56:347–356

Aseeva E, Ossenbuhl F, Eichacker LA, Wanner G, Soll J, Vothknecht UC (2004) Complex formation of Vipp1 depends on its alpha-helical PspA-like domain. J Biol Chem 279:35535–35541

Bartsevich VV, Pakrasi HB (1997) Molecular identification of a novel protein that regulates biogenesis of photosystem I, a membrane protein complex. J Biol Chem 272:6382–6387

Baymann F, Brugna M, Muhlenhoff U, Nitschke W (2001) Daddy, where did (PS)I come from? Biochim Biophys Acta 1507:291–310

Ben–Shem A, Frolow F, Nelson N (2003) Crystal structure of plant photosystem I. Nature 426:630–635

Bergantino E, Brunetta A, Touloupakis E, Segalla A, Szabo I, Giacometti GM (2003) Role of the PSII-H subunit in photoprotection:novel aspects of D1 turnover in *Synechocystis* 6803. J Biol Chem 278:41820–41829

Bibby TS, Nield J, Partensky F, Barber J (2001) Oxyphotobacteria. Antenna ring around photosystem I. Nature 413:590

Blatch GL, Lässle M (1999) The tetratricopeptide repeat:a structural motif mediating protein-protein interactions. Bioessays 21:932–939

Boekema EJ, Hifney A, Yakushevska AE, Piotrowski M, Keegstra W, Berry S, Michel KP, Pistorius EK, Kruip J (2001) A giant chlorophyll-protein complex induced by iron deficiency in cyanobacteria. Nature 412:745–748

Boudreau E, Takahashi Y, Lemieux C, Turmel M, Rochaix JD (1997) The chloroplast *ycf3* and *ycf4* open reading frames of *Chlamydomonas reinhardtii* are required for the accumulation of the photosystem I complex. EMBO J 16:6095–6104

Boudreau E, Nickelsen J, Lemaire SD, Ossenbühl F, Rochaix JD (2000) The *Nac2* gene of *Chlamydomonas* encodes a chloroplast TPR-like protein involved in *psbD* mRNA stability. EMBO J 19:3366–3376

Chitnis VP, Chitnis PR (1993) PsaL subunit is required for the formation of photosystem I trimers in the cyanobacterium *Synechocystis* sp. PCC 6803. FEBS Lett 336:330–334

D'Andrea LD, Regan L (2003) TPR proteins:the versatile helix. Trends Biochem Sci 28:655–662

de Weerd FL, van Stokkum IH, van Amerongen H, Dekker JP, van Grondelle R (2002) Pathways for energy transfer in the core light-harvesting complexes CP43 and CP47 of photosystem II. Biophys J 82:1586–1597

Di Cola A, Klostermann E, Robinson C (2005) The complexity of pathways for protein import into thylakoids:it's not easy being green. Biochem Soc Trans 33:1024–1027

Dietz KJ, Horling F, Konig J, Baier M (2002) The function of the chloroplast 2–cysteine peroxiredoxin in peroxide detoxification and its regulation. J Exp Bot 53:1321–1329

Diner BA, Ries DF, Cohen BN, Metz JG (1988) COOH-terminal processing of polypeptide D1 of the photosystem II reaction center of *Scenedesmus obliquus* is necessary for the assembly of the oxygen-evolving complex. J Biol Chem 263:8972–8980

Dühring U, Irrgang KD, Lunser K, Kehr J, Wilde A (2006) Analysis of photosynthetic complexes from a cyanobacterial *ycf37* mutant. Biochim Biophys Acta 1757:3–11

Felder S, Meierhoff K, Sane AP, Meurer J, Driemel C, Plucken H, Klaff P, Stein B, Bechtold N, Westhoff P (2001) The nucleus-encoded *HCF107* gene of *Arabidopsis* provides a link between intercistronic RNA processing and the accumulation of translation–competent *psbH* transcripts in chloroplasts. Plant Cell 13:2127–2141

Ferreira KN, Iverson TM, Maghlaoui K, Barber J, Iwata S (2004) Architecture of the photosynthetic oxygen–evolving center. Science 303:1831–1838

Fulda S, Huang F, Nilsson F, Hagemann M, Norling B (2000) Proteomics of *Synechocystis* sp. strain PCC 6803. Identification of periplasmic proteins in cells grown at low and high salt concentrations. Eur J Biochem 267:5900–5907

Hankamer BD, Elderkin SL, Buck M, Nield J (2004) Organization of the AAA(+) adaptor protein PspA is an oligomeric ring. J Biol Chem 279:8862–8866

Haussuhl K, Andersson B, Adamska I (2001) A chloroplast DegP2 protease performs the primary cleavage of the photodamaged D1 protein in plant photosystem II. EMBO J 20:713–722

Ivleva NB, Shestakov SV, Pakrasi HB (2000) The carboxyl-terminal extension of the precursor D1 protein of photosystem II is required for optimal photosynthetic performance of the cyanobacterium *Synechocystis* sp. PCC 6803. Plant Physiol 124:1403–1412

Jordan P, Fromme P, Witt HT, Klukas O, Saenger W, Krauss N (2001) Three-dimensional structure of cyanobacterial photosystem I at 2.5 A resolution. Nature 411:909–917

Kamata T, Hiramoto H, Morita N, Shen JR, Mann NH, Yamamoto Y (2005) Quality control of photosystem II:an FtsH protease plays an essential role in the turn-over of the reaction centre D1 protein in *Synechocystis* PCC 6803 under heat stress as well as light stress conditions. Photochem Photobiol Sci 4:983–990

Kamiya N, Shen JR (2003) Crystal structure of oxygen-evolving photosystem II from *Thermosynechococcus vulcanus* at 3.7-A resolution. Proc Natl Acad Sci USA 100:98–103

Kashino Y, Lauber WM, Carroll JA, Wang Q, Whitmarsh J, Satoh K, Pakrasi HB (2002) Proteomic analysis of a highly active photosystem II preparation from the cyanobacterium *Synechocystis* sp. PCC 6803 reveals the presence of novel polypeptides. Biochemistry 41:8004–8012

Keren N, Liberton M, Pakrasi HB (2005) Photochemical competence of assembled photosystem II core complex in cyanobacterial plasma membrane. J Biol Chem 280:6548–6553

Kessler D, Papenbrock J (2005) Iron–sulfur cluster biosynthesis in photosynthetic organisms. Photosynth Res 86:391–407

Klinkert B, Ossenbühl F, Sikorski M, Berry S, Eichacker L, Nickelsen J (2004) PratA, a periplasmic tetratricopeptide repeat protein involved in biogenesis of photosystem II in *Synechocystis* sp. PCC 6803. J Biol Chem 279:44639–44644

Komenda J, Reisinger V, Müller BC, Dobáková M, Eichacker LA (2004) Accumulation of the D2 Protein is a key regulatory step for assembly of the photosystem II reaction center complex in *Synechocystis* PCC 6803. J Biol Chem 279:48620–48629

Kroll D, Meierhoff K, Bechtold N, Kinoshita M, Westphal S, Vothknecht UC, Soll J, Westhoff P (2001) *VIPP1*, a nuclear gene of *Arabidopsis thaliana* essential for thylakoid membrane formation. Proc Natl Acad Sci USA 98:4238–4242

Kruip J, Bald D, Boekema EJ, Rögner M (1994) Evidence for the existence of trimeric and monomeric Photosystem I complexes in thylakoid membranes. Photosynth Res 40:279–286

Kurisu G, Zhang H, Smith JL, Cramer WA (2003) Structure of the cytochrome b6f complex of oxygenic photosynthesis: tuning the cavity. Science 302:1009–1014

Kuvikova S, Tichy M, Komenda J (2005) A role of the C-terminal extension of the photosystem II D1 protein in sensitivity of the cyanobacterium *Synechocystis* PCC 6803 to photoinhibition. Photochem Photobiol Sci 4:1044–1048

Lezhneva L, Amann K, Meurer J (2004) The universally conserved HCF101 protein is involved in assembly of [4Fe-4S]-cluster-containing complexes in *Arabidopsis thaliana* chloroplasts. Plant J 37:174–185

Lindahl M, Spetea C, Hundal T, Oppenheim AB, Adam Z, Andersson B (2000) The thylakoid FtsH protease plays a role in the light-induced turnover of the photosystem II D1 protein. Plant Cell 12:419–431

Loll B, Kern J, Saenger W, Zouni A, Biesiadka J (2005) Towards complete co-factor arrangement in the 3.0 A resolution structure of photosystem II. Nature 438:1040–1044

Martin W, Rujan T, Richly E, Hansen A, Cornelsen S, Lins T, Leister D, Stoebe B, Hasegawa M, Penny D (2002) Evolutionary analysis of *Arabidopsis*, cyanobacterial, and chloroplast genomes reveals plastid phylogeny and thousands of cyanobacterial genes in the nucleus. Proc Natl Acad Sci USA 99:12246–12251

Meskauskiene R, Nater M, Goslings D, Kessler F, op den Camp R, Apel K (2001) FLU:a negative regulator of chlorophyll biosynthesis in *Arabidopsis thaliana*. Proc Natl Acad Sci USA 98:12826–12831

Meurer J, Plucken H, Kowallik KV, Westhoff P (1998) A nuclear–encoded protein of prokaryotic origin is essential for the stability of photosystem II in *Arabidopsis thaliana*. EMBO J 17:5286–5297

Mustardy L, Cunningham FX, Jr., Gantt E (1992) Photosynthetic membrane topography: quantitative in situ localization of photosystems I and II. Proc Natl Acad Sci USA 89:10021–10025

Naver H, Boudreau E, Rochaix JD (2001) Functional studies of Ycf3:its role in assembly of photosystem I and interactions with some of its subunits. Plant Cell 13:2731–2745

Nelson N, Ben-Shem A (2004) The complex architecture of oxygenic photosynthesis. Nat Rev Mol Cell Biol 5:971–982

Nickelsen J (2003) Molecular genetics of chloroplast biogenesis In:Esser K, Ulrich L (eds) Progress in botany genetics—physiology—systematics—ecology. Springer–Verlag, Berlin Heidelberg, pp 53–73

Nixon PJ, Barker M, Boehm M, de Vries R, Komenda J (2005) FtsH–mediated repair of the photosystem II complex in response to light stress. J Exp Bot 56:357–363
Nowaczyk M (2005) Untersuchungen zur Struktur, Funktion und Dynamik von Photosystem II aus dem thermophilen Cyanobakterium *Thermosynechococcus elongatus*, Dissertation, Ruhr-Universität Bochum
Park JM, Cho JH, Kang SG, Jang HJ, Pih KT, Piao HL, Cho MJ, Hwang I (1998) A dynamin–like protein in *Arabidopsis thaliana* is involved in biogenesis of thylakoid membranes. EMBO J 17:859–867
Pasch JC, Nickelsen J, Schunemann D (2005) The yeast split-ubiquitin system to study chloroplast membrane protein interactions. Appl Microbiol Biotechnol 69:440–447
Pilon M, Abdel-Ghany SE, Van Hoewyk D, Ye H, Pilon-Smits EA (2006) Biogenesis of iron-sulfur cluster proteins in plastids. Genet Eng (NY) 27:101–117
Plücken H, Muller B, Grohmann D, Westhoff P, Eichacker LA (2002) The HCF136 protein is essential for assembly of the photosystem II reaction center in *Arabidopsis thaliana.* FEBS Lett 532:85–90
Rochaix JD, Perron K, Dauvillee D, Laroche F, Takahashi Y, Goldschmidt-Clermont M (2004) Post–transcriptional steps involved in the assembly of photosystem I in *Chlamydomonas.* Biochem Soc Trans 32:567–570
Rögner M, Boekema EJ, Barber J (1996) How does photosystem 2 split water? The structural basis of efficient energy conversion. Trends Biochem Sci 21:44–49
Rokka A, Suorsa M, Saleem A, Battchikova N, Aro EM (2005) Synthesis and assembly of thylakoid protein complexes. Multiple assembly steps of photosystem II. Biochem J 15:159–168
Roose JL, Pakrasi HB (2004) Evidence that D1 processing is required for manganese binding and extrinsic protein assembly into photosystem II. J Biol Chem 279:45417–45422
Ruf S, Kossel H, Bock R (1997) Targeted inactivation of a tobacco intron–containing open reading frame reveals a novel chloroplast-encoded photosystem I-related gene. J Cell Biol 139:95–102
Saenger W, Jordan P, Krauss N (2002) The assembly of protein subunits and cofactors in photosystem I. Curr Opin Struct Biol 12:244–254
Sane AP, Stein B, Westhoff P (2005) The nuclear gene *HCF107* encodes a membrane-associated R-TPR (RNA tetratricopeptide repeat)-containing protein involved in expression of the plastidial *psbH* gene in *Arabidopsis.* Plant J 42:720–730
Sarcina M, Bouzovitis N, Mullineaux CW (2005) Mobilization of photosystem II induced by intense red light in the cyanobacterium *Synechococcus* sp PCC7942. Plant Cell 18:457–464
Schünemann D (2004) Structure and function of the chloroplast signal recognition particle. Curr Genet 44:295–304
Schwabe TM (2003) Analysis of the biogenesis and dynamics of the membrane protein complex photosystem I in cyanobacteria., Dissertation, Ruhr-Universität Bochum
Schwabe TM, Kruip J (2000) Biogenesis and assembly of photosystem I. Indian J Biochem Biophys 37:351–359
Schwabe TM, Gloddek K, Schluesener D, Kruip J (2003) Purification of recombinant BtpA and Ycf3, proteins involved in membrane protein biogenesis in *Synechocystis* PCC 6803. J Chromatogr B Analyt Technol Biomed Life Sci 786:45–59
Shen G, Antonkine ML, van der Est A, Vassiliev IR, Brettel K, Bittl R, Zech SG, Zhao J, Stehlik D, Bryant DA, Golbeck JH (2002a) Assembly of photosystem I. II. Rubredoxin is required for the in vivo assembly of F(X) in *Synechococcus* sp. PCC 7002 as shown by optical and EPR spectroscopy. J Biol Chem 277:20355–20366
Shen G, Zhao J, Reimer SK, Antonkine ML, Cai Q, Weiland SM, Golbeck JH, Bryant DA (2002b) Assembly of photosystem I. I. Inactivation of the rubA gene encoding a membrane–associated rubredoxin in the cyanobacterium Synechococcus sp. PCC 7002 causes a loss of photosystem I activity. J Biol Chem 277:20343–20354
Shestakov SV, Anbudurai PR, Stanbekova GE, Gadzhiev A, Lind LK, Pakrasi HB (1994) Molecular cloning and characterization of the *ctpA* gene encoding a carboxyl–terminal

processing protease. Analysis of a spontaneous photosystem II–deficient mutant strain of the cyanobacterium *Synechocystis* sp. PCC 6803. J Biol Chem 269:19354–19359

Shi LX, Schröder WP (2004) The low molecular mass subunits of the photosynthetic supracomplex, photosystem II. Biochim Biophys Acta 1608:75–96

Spence E, Sarcina M, Ray N, Moller SG, Mullineaux CW, Robinson C (2003) Membrane-specific targeting of green fluorescent protein by the Tat pathway in the cyanobacterium *Synechocystis* PCC6803. Mol Microbiol 48:1481–1489

Stagljar I, Korostensky C, Johnsson N, te Heesen S (1998) A genetic system based on split–ubiquitin for the analysis of interactions between membrane proteins in vivo. Proc Natl Acad Sci USA 95:5187–5192

Stöckel J, Oelmüller R (2004) A novel protein for photosystem I biogenesis. J Biol Chem 279:10243–10251

Stroebel D, Choquet Y, Popot JL, Picot D (2003) An atypical haem in the cytochrome b(6)f complex. Nature 426:413–418

Summerfield TC, Shand JA, Bentley FK, Eaton-Rye JJ (2005a) PsbQ (Sll1638) in *Synechocystis* sp. PCC 6803 is required for photosystem II activity in specific mutants and in nutrient-limiting conditions. Biochemistry 44:805–815

Summerfield TC, Winter RT, Eaton-Rye JJ (2005b) Investigation of a requirement for the PsbP-like protein in *Synechocystis* sp. PCC 6803. Photosynth Res 84:263–268

Svensson B, Vass I, Styring S (1991) Sequence analysis of the D1 and D2 reaction center proteins of photosystem II. Z Naturforsch [C] 46:765–776

Swiatek M, Regel RE, Meurer J, Wanner G, Pakrasi HB, Ohad I, Herrmann RG (2003) Effects of selective inactivation of individual genes for low–molecular–mass subunits on the assembly of photosystem II, as revealed by chloroplast transformation:the *psbEFLJ*operon in *Nicotiana tabacum*. Mol Genet Genom 268:699–710

Thornton LE, Ohkawa H, Roose JL, Kashino Y, Keren N, Pakrasi HB (2004) Homologs of plant PsbP and PsbQ proteins are necessary for regulation of photosystem II activity in the cyanobacterium *Synechocystis* 6803. Plant Cell 16:2164–2175

Vaistij FE, Boudreau E, Lemaire SD, Goldschmidt-Clermont M, Rochaix JD (2000) Characterization of Mbb1, a nucleus-encoded tetratricopeptide-like repeat protein required for expression of the chloroplast *psbB/psbT/psbH* gene cluster in *Chlamydomonas reinhardtii*. Proc Natl Acad Sci USA 97:14813–14818

Vothknecht UC, Soll J (2002) Chloroplast quest:a journey from the cytosol into the chloroplast and beyond. Rev Physiol Biochem Pharmacol 145:181–222

Westphal S, Heins L, Soll J, Vothknecht UC (2001) *Vipp1* deletion mutant of *Synechocystis*: a connection between bacterial phage shock and thylakoid biogenesis? Proc Natl Acad Sci USA 98:4243–4248

Wierenga RK (2001) The TIM-barrel fold:a versatile framework for efficient enzymes. FEBS Lett 492:193–198

Wilde A, Hartel H, Hubschmann T, Hoffmann P, Shestakov SV, Börner T (1995) Inactivation of a *Synechocystis* sp strain PCC 6803 gene with homology to conserved chloroplast open reading frame 184 increases the photosystem II-to-photosystem I ratio. Plant Cell 7:649–658

Wilde A, Lunser K, Ossenbühl F, Nickelsen J, Börner T (2001) Characterization of the cyanobacterial *ycf37*:mutation decreases the photosystem I content. Biochem J 357:211–216

Xiong J, Bauer CE (2002) Complex evolution of photosynthesis. Annu Rev Plant Biol 53:503–521

Yamamoto Y, Inagaki N, Satoh K (2001) Overexpression and characterization of carboxyl-terminal processing protease for precursor D1 protein:regulation of enzyme-substrate interaction by molecular environments. J Biol Chem 276:7518–7525

Yi L, Dalbey RE (2005) Oxa1/Alb3/YidC system for insertion of membrane proteins in mitochondria, chloroplasts and bacteria (review). Mol Membr Biol 22:101–111

Zak E, Pakrasi HB (2000) The BtpA protein stabilizes the reaction center proteins of photosystem I in the cyanobacterium *Synechocystis* sp. PCC 6803 at low temperature. Plant Physiol 123:215–222

Zak E, Norling B, Andersson B, Pakrasi HB (1999) Subcellular localization of the BtpA protein in the cyanobacterium *Synechocystis* sp. PCC 6803. Eur J Biochem 261:311–316

Zak E, Norling B, Maitra R, Huang F, Andersson B, Pakrasi HB (2001) The initial steps of biogenesis of cyanobacterial photosystems occur in plasma membranes. Proc Natl Acad Sci USA 98:13443–13448

Zerges W (2000) Translation in chloroplasts. Biochimie 82:583–601

Zerges W, Rochaix JD (1998) Low density membranes are associated with RNA-binding proteins and thylakoids in the chloroplast of *Chlamydomonas reinhardtii.* J Cell Biol 140:101–110

Zouni A, Witt HT, Kern J, Fromme P, Krauss N, Saenger W, Orth P (2001) Crystal structure of photosystem II from *Synechococcus elongatus* at 3.8 A resolution. Nature 409:739–743

Jörg Nickelsen,
Marc M. Nowaczyk*,
Birgit Klinkert
Lehrstuhl für Allgemeine und Molekulare Botanik and
*Lehrstuhl für Biochemie der Pflanzen
Ruhr-Universität Bochum
Universitätsstr. 150
44780 Bochum
Germany
e-mail: joerg.nickelsen@rub.de

Extranuclear Inheritance: Genetics and Biogenesis of Mitochondria

Georg Michaelis and Elke Pratje

1 Introduction

In the period since our last review of this topic in *Progress in Botany* (Esser et al. 2005), the number of mitochondrial genome sequences available in GOBASE has increased dramatically (see below). This is a striking reflection of the vitality of the field. In the following, we first provide a brief overview of relevant publications that have appeared during the past 2 years, before focusing on a selection of notable contributions of particular interest.

The 6th International Congress on Plant Mitochondria took place in Perth, Western Australia, in 2002. The proceedings of this meeting were published in 2004 by Kluwer Academic in a volume entitled *Plant Mitochondria: From Genome to Function*, edited by three members of the Organizing Committee (D.A. Day, A.H. Miller and J. Whelan).

The next congress in the series was organized by J.-M. Grienenberger and colleagues from the IBMP in Strasbourg, and took place in Obernai, France, in 2005. Here, Arnold J. Bendich summarized his remarkable findings on the decline in levels of chloroplast DNA during leaf development. As leaves mature, the amounts of chloroplast DNA decline and the percentage of chloroplasts without detectable DNA increases. This is true for *Medicago*, pea, and especially for maize. In the case of tobacco, which can readily be regenerated from somatic cells, most chloroplasts retain some of their DNA (Oldenburg and Bendich 2004; Rowan et al. 2004; Shaver et al. 2005). Thus, degradation of chloroplast DNA may limit the potential for plant regeneration from a leaf cell. In view of the wide-ranging implications of these findings, Li et al. (2005) have reinvestigated the age dependency of genome copy numbers by analysing chloroplast and mitochondrial DNA amounts in *Arabidopsis* and tobacco plants. They reported that copy numbers of organelle DNA remain remarkably constant during leaf development in these species. Unfortunately, these authors did not include maize in their study.

A meeting on "Cross-talk between Nucleus and Organelles", organized by L. Del Giudice, D.R. Massardo and K. Wolf, took place in September 2004 in

Progress in Botany, Vol. 68

Naples, Italy. The contributions, covering chloroplasts as well as mitochondria, were published in a special issue of the journal *Gene* (volume 354, July 2005, edited by R.A. Butow, L. Del Giudice and K. Wolf), which includes not only research articles but also reviews on retrograde regulation of multidrug resistance, biogenesis of cytochrome oxidase, RNA maturation in mitochondria, chloroplast membrane transport, cross-talk between chloroplasts and the nucleus and mitochondria in *Arabidopsis*, mutations in human DNA polymerase γ, and mitochondrial diseases.

Volume 8 of *Topics in Current Genetics* (Springer Verlag), edited by E.M. Koehler and M.F. Bauer, which appeared in 2004, is dedicated to mitochondria. This collection, entitled "Mitochondrial Function and Biogenesis", contains articles on the origin of mitochondria, protein import, tRNA editing, protein quality control, cytochrome oxidase assembly, oxidative phosphorylation, mitochondrial DNA maintenance, mitochondrial fission and fusion, voltage-dependent anion channels (VDAC), mammalian mitochondria, synthesis and transport of lipids, and steroid synthesis in the mitochondrion.

Finally, the reader is referred to a number of reviews on evolution of mitochondrial DNA (Bullerwell and Gray 2004; Knoop 2004; Bullerwell and Lang 2005), inheritance and recombination of mitochondrial genomes (Barr et al. 2005), group I introns (Haugen et al. 2005), nuclear insertions of organelle DNA (Leister 2005), organelle nuclei (Sakai et al. 2004), CMS (Hanson and Bentolila 2004), mitochondria of protists (Gray et al. 2004), plant organelle positioning (Wada and Suetsugu 2004), mitochondrial carriers in *Arabidopsis* (Picault et al. 2004), uncoupling proteins (Hourton-Cabassa et al. 2004), iron-sulphur proteins (Balk and Lobreaux 2005; Barras et al. 2005), NAD(P)H dehydrogenases (Rasmusson et al. 2004), respiratory metabolism (Fernie et al. 2004), protein oxidation in plant mitochondria as a stress indicator (Moller and Kristensen 2004), and programmed plant cell death (van Doorn 2005).

2 Mitochondrial genomes and proteomes

2.1 Mitochondrial genomes

The number of complete mitochondrial genome sequences in the organelle genome database GOBASE has now reached 3043 (release 14 of December 2005), compared with the 408 listed in release 6 of 2003. The number of organisms for which complete mitochondrial sequences are available is 808. This sample, however, includes relatively few higher plants, because their mitochondrial genomes are particularly large and complex. In addition to the previously published mitochondrial genomes of *Arabidopsis thaliana*, *Beta vulgaris*, *Oryza sativa* and *Brassica napus*, recent reports have added the complete nucleotide sequences of the tobacco (*Nicotiana tabacum*) mitochondrial genome (430 kb; Sugiyama et al. 2005), as well as those of maize (*Zea mays*) (569 kb; Clifton et al. 2004) and wheat (*Triticum aestivum*) (452 kb; Ogihara et al. 2005). In addition, Satoh et al. (2004) have determined the complete nucleotide sequence of the mitochondrial genome for cytoplasmic male-sterile (CMS) sugar beet, permitting its comparison with the previously published sequence for the male-fertile sugar beet. The CMS genome contains four transcribed ORFs which are missing from the normal genome and are therefore potential candidates for the cause of CMS. The same group also investigated the protein profile of sugar beet mitochondria and found a variant atp6 fusion protein that is expressed in Owen CMS plants but not in normal sugar beet (Yamamoto et al. 2005).

2.2 Inheritance of mitochondrial DNA

Plant mitochondrial DNA shows a high frequency of homologous recombination following protoplast fusion. This observation implies the occurrence of mitochondrial fusion. In the yeast *Saccharomyces cerevisiae* and in animal systems, many genes have been identified that are involved in fusion and fission of the organelle (reviewed in Esser et al. 2005; Okamoto and Shaw 2005). In contrast, while some fission-related genes have been described in a red alga and in some higher plants (Nishida et al. 2003; Arimura et al. 2004; Logan et al. 2004), fusion-related genes homologous to those from yeast and animals have not been detected in plant sequence databases. Arimura et al. (2004) have recently demonstrated mitochondrial fusion in onion cells using a photoconvertible fluorescent protein. Frequent fusions and fissions were observed. Fission occasionally results in an unequal distribution of mitochondrial nucleoids, and fusion would provide a mechanism for mixing the heterogeneous mitochondrial genomes.

Sheahan et al. (2004) have observed a phase of elongation in tobacco mesophyll protoplasts before the protoplasts divide. This correlates with extensive fusions of mitochondria which ensure mixing and even distribution of mitochondrial nucleoids (Sheahan et al. 2005). The fusion of mitochondria requires cytoplasmic protein synthesis, an electrochemical membrane potential, intact microtubules and kinesin, but ATP seems not to be necessary.

Mitochondrial plasmids are found in several species of higher plants and fungi. The majority of them are linear, double-stranded DNA molecules with terminal inverted repeats. These plasmids are generally inherited maternally, like the main mitochondrial genome. One exception is the mitochondrial plasmid mF of the protist *Physarum polycephalum*, which circumvents uniparental inheritance by promoting mitochondrial fusions (Sakurai et al. 2004). Thus, the mF plasmid is transmitted to all progeny in crosses between strains with (mF+) and without (mF–) this plasmid. This is the case even when the mtDNA is transmitted from the strain that lacks the plasmid. Uniparental inheritance of mitochondrial DNA is mediated by selective degradation of mtDNA from one (usually the paternal) parent (Moriyama and Kawano 2003). The mF plasmid from the paternal parent must avoid this degradation. After crossing of mF– with mF+ cells, rapid fusion of the mitochondria from both parents is observed. In the resulting giant mitochondria, integration of the mF ensues in up to 80% of the maternal mtDNA molecules. Both integrated and free plasmids escape selective digestion, whereas the mtDNA of the paternal parent is degraded.

2.3 Mitochondrial proteomes

Three strategies are currently being pursued to identify all the proteins present in mitochondria of a given species (the mitochondrial proteome):

(i) *Bioinformatic evaluation of genomic data obtained from completely sequenced organisms.* So far, the few proteins encoded by mitochondrial DNA all seem to remain in the organelle. By far the majority of mitochondrial proteins are encoded in the nucleus, and carry N-terminal signal peptides that contain targeting and processing information. In the bioinformatic approach, characteristic features of these N-terminal signal peptides are used to predict mitochondrial localization. One problem is that proteins of the outer mitochondrial membrane, as well as small proteins, are not proteolytically processed and do not contain detectable signal peptides. Furthermore, the various prediction algorithms used can give conflicting results. Therefore, the bioinformatic approach gives only rather crude estimates, indicating, for example, that 5–10% of all *Arabidopsis thaliana* proteins (1500–3000) are targeted to mitochondria. This approach can be extended by comparing sequence similarity and experimental evidence of mitochondrial localization between diverse species.

(ii) *Experimental localization studies on individual proteins.* This strategy is straightforward, but the number of functionally identified and localized proteins is not high. In yeast, the GFP (green fluorescence protein) fusion technology is often used for intracellular localization studies, but mislocalization can occur when small proteins are fused to the larger GFP protein.

(iii) *High-resolution separation of proteins by two-dimensional gel electrophoresis combined with subsequent identification by mass spectrometry.* General problems of this approach are difficulties encountered in obtaining pure mitochondria, solubilizing membrane proteins, and detecting proteins that occur in only a few copies per cell. Extensive studies of mitochondrial proteomes have recently been reported for yeast, mouse, human and higher plants. In the nuclear genome of the simple eukaryote *S. cerevisiae* (12,156.6 kb, with some 6600 predicted ORFs), genes for 750 mitochondrial proteins have been identified, accounting for 90% of all putative mitochondrial proteins (Sickmann et al. 2003).

In higher plants, as many as 2000–3000 proteins are predicted to be destined for mitochondria (for reviews, see Braun and Millar 2004; Millar et al. 2005). More than 400 mitochondrial proteins have been identified. This approach is complemented by the analysis of individual respiratory complexes, supercomplexes (Eubel et al. 2004; Dudkina et al. 2005) and submitochondrial compartments such as the outer mitochondrial membrane (Pfanner et al. 2004). It was shown for example that complex I of the respiratory chain includes carbonic anhydrase (Sunderhaus et al. 2006). More than 70 of the identified proteins of *Arabidopsis thaliana* mitochondria lack similarity to any protein of known function. These results demonstrate the power of proteomics to discover novel or plant-specific mitochondrial proteins and provide a basis for elucidating their functions. Another problem is the regulation of the composition of the mitochondrial proteome, which is not static and changes in response to developmental stage, environmental influences, and nuclear background or mutations. One outstanding challenge will also be to understand the post-translational modifications of the plant mitochondrial proteome in which redox state, oxidative stress, and protein phosphorylation seem to play major roles.

3 RNA import and stability

3.1 tRNA import

In most organisms, the mitochondrial DNA does not encode the complete set of tRNA genes required for the translation of mitochondrial mRNAs. The missing tRNAs are encoded in the nucleus and must therefore be imported from the cytosol.

It was previously thought that mitochondria of the yeast *S. cerevisiae* import only a single tRNA, a lys-tRNA (Tarassov et al. 1995). This lys-tRNA is aminoacylated by the cytosolic lysyl-tRNA synthetase and imported via the mitochondrial protein import system in a complex with the precursor of the mitochondrial lysyl-tRNA synthetase.

Recently, Rinehart et al. (2005) described the import of two isoacceptor gln-tRNAs, the cytosolic gln-tRNA(UUG) and gln-tRNA(CUG). In-vitro import of the gln-tRNAs into isolated yeast mitochondria was found to be independent of exogenous cytosolic factors, in contrast to the uptake of the lysine tRNA. The gln-tRNAs are aminoacylated by the cytoplasmic glutaminyl-tRNA synthetase which is also present in mitochondria, as indicated by the finding that a gln-tRNA synthetase-GFP fusion protein was found in the cytoplasm and the mitochondria. It has been suggested that the mitochondrially encoded gln-tRNA(UUG) decodes CAA codons, but not CAG codons, due to the presence of a 5-carboxymethylaminomethyl-2-thio-uridine modification at the first anticodon nucleotide (Nakai et al. 2004). The imported, nucleus-encoded gln-tRNA(CUG) may therefore serve to decode the CAG codons. A function for the imported gln-tRNA(UUG) is less evident.

In human cells, all mitochondrial tRNAs are encoded in the mitochondrial DNA and thus tRNAs do not normally need to be imported. Nevertheless, isolated human mitochondria can import the nucleus-encoded lys-tRNA from yeast, indicating that the human organelles possess a tRNA import mechanism that is flexible enough to permit the uptake of tRNAs originating from other species. Indeed, a mutation in the mitochondrial lys-tRNA gene which causes the MERRF (myoclonic epilepsy with ragged-red fibers) syndrome in humans can be rescued by the imported, nucleus-encoded yeast lys-tRNA in cell culture (Kolesnikova et al. 2004). Mahata et al. (2005) have also shown that a protein complex from *Leishmania* induces the import of the human cytoplasmic lys-tRNA into human mitochondria, and that this imported lys-tRNA also restores the mitochondrial translation system in organelles from patients with MERRF or the Kearn–Sayre syndrome.

In contrast to yeast and human which import few if any tRNAs, trypanosomatids and apicomplexans must import all tRNAs (at least 24), as the mitochondrial genomes of these two groups of parasitic protozoa completely lack tRNA genes.

With the exception of the initiator met-tRNA, which is cytosol-specific, all other trypanosomal tRNAs function both in the cytosol and in mitochondria. Between 1% and 8% of each tRNA is found in the organelle. Esseiva et al. (2004a) identified the nucleotides responsible for the different targeting of the closely related initiator and elongator met-tRNAs. They showed that the T-stem specifies the cytosol-specific localization of the initiator met-tRNA. Since the elongator met-tRNA represents the only met tRNA in the mitochondria it has to function as both elongator (Met) and initiator (fMet). After import a fraction of the charged tRNA is partially formylated, bound by initiation factor 2 and brought to the ribosomes (Charrière et al. 2005).

A similar situation has been described for the apicomplexan parasite *Toxoplasma gondii*. In this case too, the complete set of mitochondrial tRNAs, with the exception of the cytosol-specific initiator met-tRNA, must be imported (Esseiva et al. 2004b).

In *Leishmania tarentolae* the sorting of the tRNAs is regulated by a modification of the wobble position (Kaneko et al. 2003; see Esser et al. 2005). The cytosolic glu-tRNA(UUC) and gln-tRNA(UUG) have 5-methoxycarbonylmethyl-2-thio-uridine at this position, which prevents import into the organelle. The corresponding mitochondrial tRNAs lack the thio modification. However, such a modification could not be detected in the cytosolic gln-tRNA of *T. gondii* (Esseiva et al. 2004b).

tRNA import into the mitochondria of trypanosomatids is independent of the protein import machinery, but requires a receptor and ATP (Bhattacharyya and Adha 2004). The tRNAs are imported as mature molecules, at least in *Leishmania*. The D-domain appears to contain an import signal which is recognized by the RNA receptor.

A similar situation is found in higher plants. The mitochondria of higher plants must import about 30–50% of their tRNAs, depending on the organism considered. The process requires a mitochondrial receptor, ATP and a membrane potential, but proceeds in the absence of cytosolic proteins. The D-domain, the anticodon and the T-domain of val-tRNA all seem critical for import into the mitochondria of tobacco cells (Laforest et al. 2005). It has therefore been postulated that the conformation of the tRNA, rather than a specific import sequence, is recognized. The essential roles of the anticodon and the D-domain have also been shown for gly-tRNAs import. Salinas et al. (2005) exchanged sequences between the three isoacceptor gly-tRNAs of tobacco: gly-tRNA(UCC), gly-tRNA(CCC), both of which are present in the cytosol and mitochondria, and the cytosol-specific gly-tRNA(GCC). They demonstrated that the anticodon and the D-domain are required, but not sufficient for promoting import into tobacco mitochondria.

3.2 RNA stability

RNA stability plays an important role in mitochondrial gene expression but the underlying mechanisms that regulate mRNA half-life differ between organisms. Polyadenylation can act either to stabilize mitochondrial mRNAs, as in human (Temperley et al. 2003), or as a signal for destabilization, as in plant mitochondria (Gagliardi et al. 2001; Kuhn et al. 2001; Binder and Brennicke 2003).

Mammalian mitochondrial mRNAs have short poly(A) tails that are post-transcriptionally added. One function of this polyadenylation is to generate a stop codon. Mitochondrial genomes are often so compact that in some genes stop codons are not encoded in the mtDNA but rather created by the addition of poly(A) tails to produce UAA stop codons. Nagaike et al. (2005) demonstrated that human mitochondrial mRNAs are stabilized by polyadenylation. The combined action of mitochondrial-specific poly(A) polymerase and polynucleotide phosphorylase regulate this polyadenylation. Reducing the level of poly(A) polymerase by RNA interference reduces the levels of mitochondrial mRNAs and their translation products.

Polyadenylation also plays a role in RNA stability in trypanosomal mitochondria. In vitro RNA turnover studies revealed that a poly(A) tail protects fully or partially edited mRNA from degradation (Kao and Read 2005), but unedited polyadenylated mRNAs are rapidly degraded (Ryan et al. 2003). Recently Ryan and Read (2005) showed that degradation of polyadenylated mRNAs is UTP dependent and requires RET1, the RNA editing terminal uridylyl transferase.

All mitochondrial mRNAs in *S. cerevisiae* appear to lack poly(A) tails. The enzymes involved in mitochondrial RNA degradation were first identified in this organism. The mitochondrial RNA degradosome of yeast contains two subunits, an NTP-dependent 3′→5′ exoribonuclease (DSS1) and an RNA helicase (SUV3), which are responsible for substrate recognition and RNA unwinding, respectively (Dziembowski et al. 2003). DSS1 homologues seem to be absent in human and plant mitochondria, whereas SUV3 homologs have been identified in both organisms (Gagliardi et al. 2004).

Trypanosoma brucei also possesses an SUV3 homologue (cited in Ryan and Read 2005) and a DSS1 exoribonuclease that affects degradation of edited and unedited mRNAs and guide RNAs (Penschow et al. 2004). In trypanosomes, extensive nucleolytic cleavage and trimming are required to produce mature RNAs. This is due to the fact that the maxi- and minicircle genomes are transcribed as polycistronic precursor RNAs. Improperly processed or edited RNAs need to be rapidly removed (Table 1).

Table 1. Functions of mitochondrial mRNA polyadenylation

Organisms	Polyadenylation of *mitochondrial mRNA*	*Function*
Yeast	None	
Human	All	Stability
Higher plants	Few	Degradation

4 Protein import into plant mitochondria

4.1 Import receptor and import sequences

Mitochondrial protein import pathways have been analysed in great detail in the yeast *S. cerevisiae* (for reviews, see Tokatlidis et al. 2000; Rehling et al. 2001; Taylor and Pfanner 2004). Subsequent analysis of mitochondrial protein import in plants revealed that the plant import apparatus differs considerably from that of yeast (Lister et al. 2005). The *Arabidopsis* genome sequence has been searched for components of the mitochondrial protein import apparatus. The results are available at http://millar3.biochem.uwa.edu.au/~lister/index.html (Lister et al. 2003). In animals and fungi, Tom20 and Tom70 are the receptor components of the outer mitochondrial membrane. Proteins related to these two receptor proteins have not been found in plants. Instead unrelated 20 kDa and 64 kDa proteins seem to act as presequence receptors in plants (Heins and Schmitz 1996; Chew et al. 2004). Interestingly, the 64 kDa protein shows homology to the plastid protein Toc64. These results indicate that the receptors evolved after the divergence of the animal-fungal lineage from the line leading to higher plants. Overall, some of the components of the mitochondrial protein import apparatus seem to derive from the eubacterial endosymbiont, some arise "de novo" and a third class appears lineage specific (Lister et al. 2005). In the following, we focus on recent results concerning the sequences that mediate protein import into plant mitochondria.

Gene transfer from the mitochondrion to the nucleus is an ongoing process in plants (Adams and Palmer 2003). After gene transfer, the nucleus-encoded protein product has to be targeted back to the organelle. The minimal requirements for protein import into mitochondria were analysed by Murcha et al. (2005b), using as a model the S10 protein of the small ribosomal subunit. The S10 protein is encoded by the nuclear genome in *Arabidopsis*, carrot, lettuce, spinach and maize, but is mitochondrially encoded in soybean. In carrots and lettuce, the S10 protein has acquired N-terminal signal sequences, whereas in maize the S10 protein is imported without an extension. Deletion analysis indicated that the first 20 amino acid residues of the maize protein function as a mitochondrial targeting region. The authors also introduced changes in the first 20 amino acid residues of the mitochondrially encoded S10 protein of soybean in an attempt to induce the in-vitro import of this protein into mitochondria. The amino acid substitutions were chosen so as to correspond to the respective sequence of the nucleus-encoded maize protein. Substitutions that altered overall charge and hydrophobicity and created an amphipathic structure and a binding site for Tom20, the import receptor of the outer mitochondrial membrane permitted uptake of the altered soybean protein. These experiments demonstrate that retargeting to mitochondria can be achieved by two mechanisms: the protein can either acquire an N-terminal signal sequence, or the protein sequence is changed in such a way that a non-cleavable mitochondrial targeting sequence is created.

However, a conserved Tom20 binding site is not present in all proteins imported into mitochondria, implying the existence of additional import receptors or the involvement of different members of the Tom20 gene family (Werhahn et al. 2001; Chew et al. 2004). In *Arabidopsis*, the Tom20 gene family consists of four members.

Most mitochondrial carrier proteins in *S. cerevisiae* are synthesized without transient N-terminal presequences. In contrast, a subset of carrier proteins from plants contains cleavable signal sequences. Murcha et al. (2005a) studied the function of the cleavable extension in a reconstituted import system. Removal of the extensions in the phosphate and adenine nucleotide carriers still allows import into mitochondria, but the extensions significantly stimulate insertion into the inner mitochondrial membrane. Furthermore, fusions of the extensions to carrier proteins which do not contain cleavable extensions also enhance their membrane insertion.

After import into mitochondria, the signal peptides are removed by signal peptidases present in the mitochondrial matrix or the inner mitochondrial membrane (see Michaelis et al. 2005). Proteolytic processing occurs either in one or in two successive steps. Recently, two related proteases (AtPrePI and AtPrePII) that degrade cleaved signal peptides were isolated from *Arabidopsis thaliana* (Bhushan et al. 2005; Stahl et al. 2005). Both proteases are targeted to mitochondria and plastids. They belong to the pitrilysin subfamily with an inverted metal-binding motif. Unfolded peptides of 10–65 amino acid residues serve as the substrates. AtPrePI shows a preference for the N-terminal amphiphilic α-helix and positively charged amino acid residues, whereas AtPrePII did not show any positional preference.

4.2 Dual targeting to mitochondria and plastids

In plants, the coexistence of mitochondria and plastids in the cytosol requires import systems that can discriminate between mitochondrial and plastid proteins. The amino acid compositions of the presequences used by the two systems are similar, although the primary sequences are not conserved. They are characterized by hydrophobic, hydroxylated and positively charged residues. Mitochondrial presequences can form amphiphilic

α-helices, whereas the hydroxylated residues in the plastid presequences may function as phosphorylation sites to initiate import.

Several nucleus-encoded proteins are targeted to both mitochondria and plastids. These proteins are involved in general functions such as DNA replication, transcription, translation, and defence against damage by reactive oxygen species (Mackenzie 2005).

The signal for dual targeting of pea glutathione reductase has been localized to the N-terminal 60 amino acid residues of the preprotein. In-vitro mutagenesis of the presequence and protein import into isolated organelles indicate the presence of overlapping import signals (Chew et al. 2003).

Two in-frame AUG codons for translation initiation are found in the RNA polymerase transcript in the moss *Physcomitrella patens* (Kabeya and Sato 2005). Use of the second AUG codon determines mitochondrial localization of the product, whereas the first AUG codon specifies dual targeting in vitro. Whether the first AUG is used in vivo remains to be verified (Kabeya et al. 2002; Richter et al. 2002; Kabeya and Sato 2005).

A DNA polymerase of *Arabidopsis thaliana* shows dual targeting when the 5′ untranslated leader region (UTR) is present, but deletion of the 5′UTR results in targeting to the plastid (Christensen et al. 2005). This indicates that translation of the mitochondrial enzyme starts in the 5′UTR. However, an AUG start codon is lacking in this region. A CUG codon was identified that is likely to account for initiation of translation. Thus to achieve dual targeting of the DNA polymerase the control of translation initiation is relaxed.

To summarize, dual targeting to mitochondria and plastids is achieved by one presequence with two overlapping recognition sites, by two in-frame initiation codons, or by relaxation of the control of translation, permitting initiation at a non-AUG codon.

5 Autophagic degradation of mitochondria

Protein import into mitochondria and the biogenesis of this organelle are intensely studied topics of research. However, much less is known about the turnover and degradation of this organelle. The degradation pathways permit the cell to eliminate unwanted or damaged organelles and recycle the components.

In plants and fungi, degradation of organelles takes place in the lytic and acidic vacuoles that have remarkably diverse functions (e.g. regulation of turgor pressure, storage, or degradation) and are related to the lysosomes of animal cells. During the past few years, autophagy has emerged as a general process that delivers portions of cytoplasm or entire organelles to the vacuole for degradation and recycling (for reviews, see Klionsky and Emr 2000; Abeliovich and Klionsky 2001). Autophagy seems to begin with the enclosure of an organelle by a membrane, forming an autophagosome that fuses with the vacuole (Fig. 1). Although autophagy is thought to operate randomly and to recycle components such as amino acids to the starved cells, recent evidence suggests that it might also be required for the selective elimination of organelles like mitochondria especially when they are damaged (Lemasters 2005; Priault et al. 2005). Thus, autophagy may play a role in apoptosis (Boya et al. 2005; Gonzalez-Polo et al. 2005), senescence, removal of mutated mitochondrial DNA, uniparental inheritance, and the escape of mitochondrial RNA, DNA (Campbell and Thorsness 1998), or protein into the cytoplasm or the nucleus. The maternally inherited surface antigen of mouse cells is a fragment of the mitochondrially encoded ND1 protein which has to escape from the organelle or the vacuole (Fischer-Lindahl et al. 1991).

Molecular details are emerging from studies on *S. cerevisiae*. In this yeast, autophagy can be induced by the drug rapamycin or by nitrogen starvation. Mutant strains have been characterized in which the degradation of mitochondrial proteins, as monitored by Western blot analysis, is delayed or does not occur. Such strains carry null mutations in the genes for the vacuolar protease Pep4 or the autophagy-specific Apg/Atg proteins, for example Atg5 and Atg8 (Kim et al. 2001; Kuma et al. 2002). Recently, a specific mitochondrial outer membrane protein was described and designated Uth1 (Camougrand et al. 2004; Kissova et al. 2004). In the absence of Uth1, resistance to rapamycin treatment and starvation was observed, two conditions that induce the process of autophagy in yeast. Although the autophagic machinery is functional in the absence of Uth1, this protein is required for the specific degradation of mitochondria, a process which is also referred to as mitophagy to emphasize the non-random nature of the process.

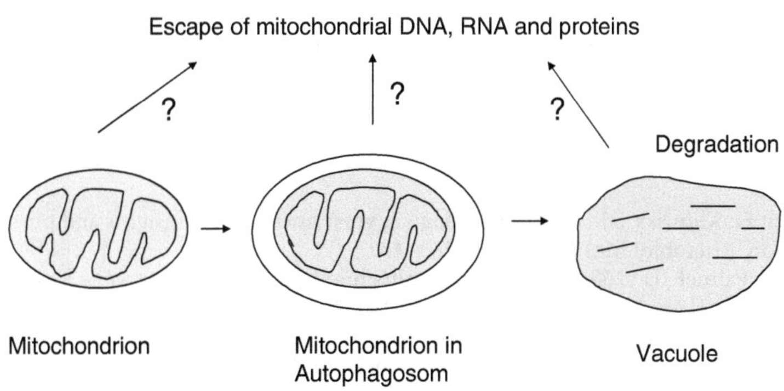

Fig. 1. Autophagic degradation of mitochondria

6 Summary and perspectives

Only a limited number of topics of current research could be presented here. Work on the morphology of mitochondria, their fission and fusion processes, was covered in our previous article (Esser et al. 2005). Here, we chose to concentrate on themes that dominate studies on the biogenesis of mitochondria: aspects of the protein import machinery of higher plants, proteins with dual localization in the cell (in mitochondria and chloroplasts) and RNA import. The systematic definition of the mitochondrial proteome, and especially studies designed to identify the changes that occur in association with cell differentiation and development, will undoubtedly contribute greatly to our understanding of these questions in the near future.

Compared with our knowledge of biosynthetic processes, much less is known about the turnover and degradation of mitochondrial macromolecules (DNA, RNA and proteins) and of the organelle itself. We mentioned above the functions of poly(A) tails in determining the stability of mitochondrial mRNAs and the discovery of the *UTH1* gene required for specific autophagic degradation of mitochondria in yeast. A molecular dissection of this degradation process might give new insights into a variety of phenomena, such as gene transfer from the organelle to the nucleus, elimination of damaged organelle DNA, apoptosis, aging and even cases of uniparental inheritance. Work in this area may also have a bearing on the controversial discussion on the loss of organelle DNA observed during leaf development in some species of higher plants, discussed in the Introduction. These striking findings demonstrate that plant development still poses many open problems, and further surprises are undoubtedly in store.

Acknowledgements

We thank Dr. P. Hardy for critical reading of the manuscript.

References

Abeliovich H, Klionsky DJ (2001) Autophagy in yeast: mechanistic insights and physiological function. Microbiol Mol Biol Rev 65:463–479

Adams KL, Palmer JD (2003) Evolution of mitochondrial gene content: gene loss and transfer to the nucleus. Mol Phylogenet Evol 29:380–395

Arimura S, Yamamoto J, Aida GP, Nakazono M, Tsutsumi N (2004) Frequent fusion and fission of plant mitochondria with unequal nucleoid distribution. Proc Natl Acad Sci USA 101:7805–7808

Balk J, Lobreaux S (2005) Biogenesis of iron-sulfur proteins in plants. Trends Plant Sci 10:324–331

Barr CM, Neiman M, Taylor DR (2005) Inheritance and recombination of mitochondrial genomes in plants, fungi and animals. New Phytol 168:39–50

Barras F, Loiseau L, Py B (2005) How *Escherichia coli* and *Saccharomyces cerevisiae* build Fe/S proteins. Adv Microb Physiol 50:41–101

Bhattacharyya SN, Adhya S (2004) tRNA–triggered ATP hydrolysis and generation of membrane potential by the *Leishmania* mitochondrial tRNA import complex. J Biol Chem 279:11259–11263

Bhushan S, Stahl A, Nilsson S, Lefebvre B, Seki M, Roth C, McWilliam D, Wright SJ, Liberles DA, Shinozaki K, Bruce BD, Boutry M, Glaser E (2005) Catalysis, subcellular localization, expression and evolution of the targeting peptides degrading protease, AtPreP2. Plant Cell Physiol 46:985–996

Binder S, Brennicke A (2003) Gene expression in plant mitochondria: transcriptional and post–transcriptional control. Philos Trans R Soc Lond B Biol Sci 358:181–188

Boya P, Gonzalez–Polo RA, Casares N, Perfettini JL, Dessen P, Larochette N, Metivier D, Meley D, Souquere S, Yoshimori T, Pierron G, Codogno P, Kroemer G (2005) Inhibition of macroautophagy triggers apoptosis. Mol Cell Biol 25:1025–1040

Braun H-P, Millar AH (2004) Proteome analyses for characterization of plant mitochondria. In: Day DA, Millar AH and Whelan J (eds) Plant mitochondria: from genome to function. Kluwer, Dordrecht, pp 143–162

Bullerwell CE, Gray MW (2004) Evolution of the mitochondrial genome: protist connections to animals, fungi and plants. Curr Opin Microbiol 7:528–534

Bullerwell CE, Lang BF (2005) Fungal evolution: the case of the vanishing mitochondrion. Curr Opin Microbiol 8:362–369

Camougrand N, Kissova I, Velours G, Manon S (2004) Uth1p: a yeast mitochondrial protein at the crossroads of stress, degradation and cell death. FEMS Yeast Res 5:133–140

Campbell CL, Thorsness PE (1998) Escape of mitochondrial DNA to the nucleus in *yme1* yeast is mediated by vacuolar-dependent turnover of abnormal mitochondrial compartments. J Cell Sci 111:2455–2464

Charriere F, Tan TH, Schneider A (2005) Mitochondrial initiation factor 2 of *Trypanosoma brucei* binds imported formylated elongator-type tRNA(Met). J Biol Chem 280:15659–15665

Chew O, Whelan J, Millar AH (2003) Molecular definition of the ascorbate-glutathione cycle in *Arabidopsis* mitochondria reveals dual targeting of antioxidant defenses in plants. J Biol Chem 278:46869–46877

Chew O, Lister R, Qbadou S, Heazlewood JL, Soll J, Schleiff E, Millar AH, Whelan J (2004) A plant outer mitochondrial membrane protein with high amino acid sequence identity to a chloroplast protein import receptor. FEBS Lett 557:109–114

Christensen AC, Lyznik A, Mohammed S, Elowsky CG, Elo A, Yule R, Mackenzie SA (2005) Dual-domain, dual-targeting organellar protein presequences in *Arabidopsis* can use non-AUG start codons. Plant Cell 17:2805–2816

Clifton SW, Minx P, Fauron CM, Gibson M, Allen JO, Sun H, Thompson M, Barbazuk WB, Kanuganti S, Taylor C, Meyer L, Wilson RK, Newton KJ (2004) Sequence and comparative analysis of the maize NB mitochondrial genome. Plant Physiol 136:3486–3503

Dudkina NV, Eubel H, Keegstra W, Boekema EG, Braun H-P (2005) Structure of a mitochondrial supercomplex formed by respiratory-chain complexes I and III. Proc Natl Acad Sci USA 102:3225–3229

Dziembowski A, Piwowarski J, Hoser R, Minczuk M, Dmochowska A, Siep M, van der Spek H, Grivell L, Stepien PP (2003) The yeast mitochondrial degradosome. Its composition, interplay between RNA helicase and RNase activities and the role in mitochondrial RNA metabolism. J Biol Chem 278:1603–1611

Esseiva AC, Naguleswaran A, Hemphill A, Schneider A (2004a) Mitochondrial tRNA import in *Toxoplasma gondii*. J Biol Chem 279:42363–42368

Esseiva AC, Marechal-Drouard L, Cosset A, Schneider A (2004b) The T-stem determines the cytosolic or mitochondrial localization of trypanosomal tRNAsMet. Mol Biol Cell 15: 2750–2757

Esser K, Michaelis G, Pratje E (2005) Extranuclear inheritance: mitochondrial genetics and biogenesis. Prog Bot 66:90–111

Eubel H, Heinemeyer J, Sunderhaus S, Braun HP (2004) Respiratory chain supercomplexes in plant mitochondria. Plant Physiol Biochem 42:937–942

Fernie AR, Carrari F, Sweetlove LJ (2004) Respiratory metabolism: glycolysis, the TCA cycle and mitochondrial electron transport. Curr Opin Plant Biol 7:254–261

Fischer-Lindahl K, Hermel E, Loveland BE, Wang CR (1991) Maternally transmitted antigen of mice: a model transplantation antigen. Annu Rev Immunol 9:351–372

Gagliardi D, Perrin R, Marechal-Drouard L, Grienenberger JM, Leaver CJ (2001) Plant mitochondrial polyadenylated mRNAs are degraded by a 3′- to 5′-exoribonuclease activity, which proceeds unimpeded by stable secondary structures. J Biol Chem 276:43541–43547

Gagliardi D, Stepien PP, Temperley RJ, Lightowlers RN, Chrzanowska-Lightowlers ZM (2004) Messenger RNA stability in mitochondria: different means to an end. Trends Genet. 20:260–267. Erratum in: Trends Genet (2005) 21:36

Gonzalez-Polo RA, Boya P, Pauleau AL, Jalil A, Larochette N, Souquere S, Eskelinen EL, Pierron G, Saftig P, Kroemer G (2005) The apoptosis/autophagy paradox: autophagic vacuolization before apoptotic death. J Cell Sci 118:3091–3102

Gray MW, Lang BF, Burger G (2004) Mitochondria of protists. Annu Rev Genet 38:477–524

Hanson MR, Bentolila S (2004) Interactions of mitochondrial and nuclear genes that affect male gametophyte development. Plant Cell 16:S154–169

Haugen P, Simon DM, Bhattacharya D (2005) The natural history of group I introns. Trends Genet 21:111–119

Heins L, Schmitz UK (1996) A receptor for protein import into potato mitochondria. Plant J 9:829–839

Hourton-Cabassa C, Rita Matos A, Zachowski A, Moreau F (2004) The plant uncoupling protein homologues: a new family of energy-dissipating proteins in plant mitochondria. Plant Physiol Biochem 42:283–290

Kabeya Y, Sato N (2005) Unique translation initiation at the second AUG codon determines mitochondrial localization of the phage-type RNA polymerases in the moss *Physcomitrella patens*. Plant Physiol 138:369–382

Kabeya Y, Hashimoto K, Sato N (2002) Identification and characterization of two phage-type RNA polymerase cDNAs in the moss *Physcomitrella patens*: implication of recent evolution of nuclear-encoded RNA polymerase of plastids in plants. Plant Cell Physiol 43:245–255

Kaneko T, Suzuki T, Kapushoc ST, Rubio MA, Ghazvini J, Watanabe K, Simpson L, Suzuki T (2003) Wobble modification differences and subcellular localization of tRNAs in *Leishmania tarentolae*: implication for tRNA sorting mechanism. EMBO J 22:657–667

Kao CY, Read LK (2005) Opposing effects of polyadenylation on the stability of edited and unedited mitochondrial RNAs in *Trypanosoma brucei*. Mol Cell Biol 25:1634–1644

Kim J, Huang WP, Klionsky DJ (2001) Membrane recruitment of Aut7p in the autophagy and cytoplasm to vacuole targeting pathways requires Aut1p, Aut2p, and the autophagy conjugation complex. J Cell Biol 152:51–64

Kissova I, Deffieu M, Manon S, Camougrand (2004) Uth1p is involved in the autophagic degradation of mitochondria. J Biol Chem 279:39068–39074

Klionsky DJ, Emr SD (2000) Autophagy as a regulated pathway of cellular degradation. Science 290:1717–1721

Knoop V (2004) The mitochondrial DNA of land plants: peculiarities in phylogenetic perspective. Curr Genet 46:123–139

Kolesnikova OA, Entelis NS, Jacquin–Becker C, Goltzene F, Chrzanowska–Lightowlers ZM, Lightowlers RN, Martin RP, Tarassov I (2004) Nuclear DNA-encoded tRNAs targeted into mitochondria can rescue a mitochondrial DNA mutation associated with the MERRF syndrome in cultured human cells. Hum Mol Genet 13:2519–2534

Kuhn J, Tengler U, Binder S (2001) Transcript lifetime is balanced between stabilizing stem-loop structures and degradation-promoting polyadenylation in plant mitochondria. Mol Cell Biol 21:731–742

Kuma A, Mizushima N, Ishihara N, Ohsumi Y (2002) Formation of the approximately 350-kDa Apg12-Apg5.Apg16 multimeric complex, mediated by Apg16 oligomerization, is essential for autophagy in yeast. J Biol Chem 277:18619–18625
Laforest MJ, Delage L, Marechal–Drouard L (2005) The T-domain of cytosolic tRNAVal, an essential determinant for mitochondrial import. FEBS Lett 579:1072–1078
Leister D (2005) Origin, evolution and genetic effects of nuclear insertions of organelle DNA. Trends Genet 21:655–663
Lemasters JJ (2005) Selective mitochondrial autophagy, or mitophagy, as a targeted defense against oxidative stress, mitochondrial dysfunction, and aging. Rejuvenation Res 8:3–5
Li W, Ruf S, Bock R (2005) Constancy of organellar genome copy numbers during leaf development and senescence in higher plants. Mol Gen Genomics 275:185–192
Lister R, Murcha MW, Whelan J (2003) The mitochondrial protein import machinery of plants (MPIMP) database. Nucleic Acids Res 31:325–327
Lister R, Hulett JM, Lithgow T, Whelan J (2005) Protein import into mitochondria: origins and functions today. Mol Membr Biol 22:87–100
Logan DC, Scott I, Tobin AK (2004) ADL2a, like ADL2b, is involved in the control of higher plant mitochondrial morphology. J Exp Bot 55:783–785
Mackenzie SA (2005) Plant organellar protein targeting: a traffic plan still under construction. Trends Cell Biol 15:548–554
Mahata B, Bhattacharyya SN, Mukherjee S, Adhya S (2005) Correction of translational defects in patient-derived mutant mitochondria by complex-mediated import of a cytoplasmic tRNA. J Biol Chem 280:5141–5144
Michaelis G, Esser K, Tursun B, Stohn JP, Hanson S, Pratje E (2005) Mitochondrial signal peptidases of yeast: the rhomboid peptidase Pcp1 and its substrate cytochrome c peroxidase. Gene 354:58–63
Millar AH, Heazlewood JL, Kristensen BK, Braun H-P, Moller IM (2005) The plant mitochondrial proteome. Trends Plant Sci 10:36–43
Moller IM, Kristensen BK (2004) Protein oxidation in plant mitochondria as a stress indicator. Photochem Photobiol Sci 3:730–735
Moriyama Y, Kawano S (2003) Rapid, selective digestion of mitochondrial DNA in accordance with the *matA* hierarchy of multiallelic mating types in the mitochondrial inheritance of *Physarum polycephalum*. Genetics 164:963–975
Murcha MW, Millar AH, Whelan J (2005a) The N-terminal cleavable extension of plant carrier proteins is responsible for efficient insertion into the inner mitochondrial membrane. Mol Biol 351:16–25
Murcha MW, Rudhe C, Elhafez D, Adams KL, Daley DO, Whelan J (2005b) Adaptations required for mitochondrial import following mitochondrial to nucleus gene transfer of ribosomal protein S10. Plant Physiol 138:2134–2144
Nagaike T, Suzuki T, Katoh T, Ueda T (2005) Human mitochondrial mRNAs are stabilized with polyadenylation regulated by mitochondria-specific poly(A) polymerase and polynucleotide phosphorylase. J Biol Chem 280:19721–19727
Nakai Y, Umeda N, Suzuki T, Nakai M, Hayashi H, Watanabe K, Kagamiyama H (2004) Yeast Nfs1p is involved in thio-modification of both mitochondrial and cytoplasmic tRNAs. J Biol Chem 279:12363–12368
Nishida K, Takahara M, Miyagishima SY, Kuroiwa H, Matsuzaki M, Kuroiwa T (2003) Dynamic recruitment of dynamin for final mitochondrial severance in a primitive red alga. Proc Natl Acad Sci USA 100:2146–2151
Ogihara Y, Yamazaki Y, Murai K, Kanno A, Terachi T, Shiina T, Miyashita N, Nasuda S, Nakamura C, Mori N, Takumi S, Murata M, Futo S, Tsunewaki K (2005) Structural dynamics of cereal mitochondrial genomes as revealed by complete nucleotide sequencing of the wheat mitochondrial genome. Nucleic Acids Res 33:6235–6250
Okamoto K, Shaw JM (2005) Mitochondrial morphology and dynamics in yeast and multicellular eukaryotes. Annu Rev Genet 39:503–536

Oldenburg DJ, Bendich AJ (2004) Changes in the structure of DNA molecules and the amount of DNA per plastid during chloroplast development in maize. J Mol Biol 344:1311–1330

Penschow JL, Sleve DA, Ryan CM, Read LK (2004) TbDSS-1, an essential *Trypanosoma brucei* exoribonuclease homolog that has pleiotropic effects on mitochondrial RNA metabolism. Eukaryot Cell 3:1206–1216

Pfanner N, Wiedemann N, Meisinger C, Lithgow T (2004) Assembling the mitochondrial outer membrane. Nat Struct Mol Biol 11:1044–1048

Picault N, Hodges M, Palmieri L, Palmieri F (2004) The growing family of mitochondrial carriers in *Arabidopsis*. Trends Plant Sci 9:138–146

Priault M, Salin B, Schaeffer J, Vallette FM, di Rago JP, Martinou JC (2005) Impairing the bioenergetic status and the biogenesis of mitochondria triggers mitophagy in yeast. Cell Death Differ 12:1613–1621

Rasmusson AG, Soole KL, Elthon TE (2004) Alternative NAD(P)H dehydrogenases of plant mitochondria. Annu Rev Plant Biol 55:23–39

Rehling P, Wiedemann N, Pfanner N, Truscott KN (2001) The mitochondrial import machinery for preproteins. Crit Rev Biochem Mol Biol 36:291–336

Richter U, Kiessling J, Hedtke B, Decker E, Reski R, Borner T, Weihe A (2002) Two RpoT genes of *Physcomitrella patens* encode phage-type RNA polymerases with dual targeting to mitochondria and plastids. Gene 290:95–105

Rinehart J, Krett B, Rubio MA, Alfonzo JD, Soll D (2005) *Saccharomyces cerevisiae* imports the cytosolic pathway for Gln-tRNA synthesis into the mitochondrion. Genes Dev 19:583–592

Rowan BA, Oldenburg DJ, Bendich AJ (2004) The demise of chloroplast DNA in *Arabidopsis*. Curr Genet 46:176–181

Ryan CM, Read LK (2005) UTP-dependent turnover of *Trypanosoma brucei* mitochondrial mRNA requires UTP polymerization and involves the RET1 TUTase. RNA 11:763–773

Ryan CM, Militello KT, Read LK (2003) Polyadenylation regulates the stability of *Trypanosoma brucei* mitochondrial RNAs. J Biol Chem 278:32753–32762

Sakai A, Takano H, Kuroiwa T (2004) Organelle nuclei in higher plants: structure, composition, function, and evolution. Int Rev Cytol 238:59–118

Sakurai R, Nomura H, Moriyam Y, Kawano S (2004) The mitochondrial plasmid of the true slime mold *Physarum polycephalum* bypasses uniparental inheritance by promoting mitochondrial fusion. Curr Genet 46:103–114

Salinas T, Schaeffer C, Marechal-Drouard L, Duchene AM (2005) Sequence dependence of tRNA(Gly) import into tobacco mitochondria. Biochimie 87:863–872

Satoh M, Kubo T, Nishizawa S, Estiati A, Itchoda N, Mikami T (2004) The cytoplasmic male-sterile type and normal type mitochondrial genomes of sugar beet share the same complement of genes of known function but differ in the content of expressed ORFs. Mol Genet Genomics 272:247–256

Shaver JM, Oldenburg DJ, Bendich AJ (2005) Changes in chloroplast DNA during development in tobacco, *Medicago truncatula*, pea, and maize. Planta 224:72–82

Sheahan MB, Rose RJ, McCurdy DW (2004) Organelle inheritance in plant cell division: the actin cytoskeleton is required for unbiased inheritance of chloroplasts, mitochondria and endoplasmic reticulum in dividing protoplasts. Plant J 37:379–390

Sheahan MB, McCurdy DW, Rose RJ (2005) Mitochondria as a connected population: ensuring continuity of the mitochondrial genome during plant cell dedifferentiation through massive mitochondrial fusion. Plant J 44:744–755

Sickmann A, Reinders J, Wagner Y, Yoppich C, Zahedi R, Meyer HE, Schonfisch B, Perschil I, Chacinska A, Guiard B, Rehling P, Pfanner N, Meisinger C (2003) The *Saccharomyces cerevisiae* mitochondrial proteome. Proc Natl Acad Sci USA 100:13207–13212

Stahl A, Nilsson S, Lundberg P, Bhushan S, Biverstahl H, Moberg P, Morisset M, Vener A, Maler L, Langel U, Glaser E (2005) Two novel targeting peptide degrading proteases, PrePs, in mitochondria and chloroplasts, so similar and still different. J Mol Biol 349:847–860

Sugiyama Y, Watase Y, Nagase M, Makita N, Yagura S, Hirai A, Sugiura M (2005) The complete nucleotide sequence and multipartite organization of the tobacco mitochondrial genome: comparative analysis of mitochondrial genomes in higher plants. Mol Genet Genomics 272:603–615

Sunderhaus S, Dudkina N, Jansch L, Klodmann J, Heinemeyer J, Perales M, Zabaleta E, Boekema E, Braun HP (2006) Carbonic anhydrase subunits form a matrix-exposed domain attached to the membrane arm of mitochondrial complex I in plants. J Biol Chem 281:6482–6488

Tarassov I, Entelis N, Martin RP (1995) Mitochondrial import of a cytoplasmic lysine-tRNA in yeast is mediated by cooperation of cytoplasmic and mitochondrial lysyl-tRNA synthetases. EMBO J 14:3461–3471

Taylor RD, Pfanner N (2004) The protein import and assembly machinery of the mitochondrial outer membrane. Biochim Biophys Acta 1658:37–43

Temperley RJ et al. (2003) Investigation of a pathogenic mtDNA microdeletion reveals a translation-dependent deadenylation decay pathway in human mitochondria. Hum Mol Genet 12:2341–2348

Tokatlidis K, Vial S, Luciano P, Vergnolle M, Clemence S (2000) Membrane protein import in yeast mitochondria. Biochem Soc Trans 28:495–499

Van Doorn WG (2005) Plant programmed cell death and the point of no return. Trends Plant Sci 10:478–483

Wada M, Suetsugu N (2004) Plant organelle positioning. Curr Opin Plant Biol 7:626–631

Werhahn W, Niemeyer A, Jansch L, Kruft V, Schmitz UK, Braun H-P (2001) Purification and characterization of the preprotein translocase of the outer mitochondrial membrane from *Arabidopsis*. Identification of multiple forms of TOM20. Plant Physiol 125:943–954

Yamamoto MP, Kubo T, Mikami T (2005) The 5′-leader sequence of sugar beet mitochondrial atp6 encodes a novel polypeptide that is characteristic of Owen cytoplasmic male sterility. Mol Genet Genomies 273:342–349

Elke Pratje,
Georg Michaelis
Biozentrum Klein Flottbek
Ohnhorststrasse 18
D-22609 Hamburg
Germany
Tel.: +49-40-42816288;
Fax: +49-40-42816254;
e-mail: pratje@botanik. uni-hamburg.de

Extranuclear Inheritance: Virus-Like DNA-Elements in Yeast

Stefanie Jeske, Friedhelm Meinhardt and Roland Klassen

Abstract

Linear extrachromosomal DNA elements replicating independently of the cellular genome are widespread. Aside from linearity, they all have in common terminal inverted repeat sequences and 5′-covalently linked proteins, which reminds one of linear viral genomes. In eukaryotes they routinely reside in organelles; however, a number of yeasts comprising different genera harbour such elements in the cytoplasm. They exist singly, in pairs or triplets, some conferring a killer and immunity phenotype to their host cells. The killing and immunity character is routinely encoded by the smallest element. Each of the four killer elements known to date encodes a heteromeric toxin, consisting of a chitinase and a hydrophobic domain which corporately facilitate uptake of the toxic subunit into target cells. Regarding the toxin mode of action, such small killer elements must be arranged into two categories. The G1-arresting *Kluyveromyces lactis* toxin, as well as that of *Pichia inositovora*, strictly depend on a protein complex termed Elongator, whereas toxins from *Pichia acaciae* and *Debaryomyces (Wingea) robertsiae* arrest the cell cycle in the S-phase independent of Elongator. Whilst small killer elements rely on the larger plasmids in terms of cytoplasmic transcription, transcript modification, and factors involved in replication initiation, the large plasmids are autonomous elements. Recent work on the *K. lactis* killer plasmid pair, which still represents the paradigm of such elements, and other linear yeast plasmids has strengthened evidence for viral ancestry, because general principles for replication and transcription are realized in these cytoplasmic elements.

1 Introduction

In addition to mitochondrial DNA, yeast cells can harbour two different types of cytoplasmically inherited extranuclear genetic information: cytoplasmic dsRNA viruses of the *Totiviridae* family and virus like dsDNA

Progress in Botany, Vol. 68

elements often referred to as linear plasmids. Since the discovery of the latter, a considerable amount of information concerning their structure and function has accumulated, clearly providing evidence of a viral ancestry. Neither dsRNA nor yeast cytoplasmic dsDNA elements cause any discernible disadvantage to host cells; rather, a killer and killer toxin immunity phenotype is correlated in several instances. Such toxins ensure stable inheritance of the encoding elements by autoselection and concomitantly promote the selective advantage to dominate in environmental scenarios where killer toxin sensitive yeasts compete for limited resources.

Reviews addressing yeast linear plasmids, killer toxins and dsRNA viruses have been published over the years (Stark et al. 1990; Fukuhara 1995; Gunge 1995; Meinhardt and Schaffrath 2001; Schmitt and Breinig 2002; Gunge and Tokunaga 2004; Schaffrath and Meinhardt 2004). Considerable progress concerning yeast linear plasmids and encoded proteins was only achieved quite recently. Here, we review the current state of the art of research on cytoplasmic dsDNA elements from yeast with emphasis on associated killer toxins as well as their presumable viral evolutionary origin.

2 Occurrence

Linear plasmids were firstly discovered in maize (Pring et al. 1977). Subsequently they were detected in other plants and a great number of basidiomycetes and ascomycetes, including yeasts (Meinhardt et al. 1990). Subcellular localization of these elements is in the mitochondria (reviewed in Meinhardt et al. 1990; Kempken et al. 1992; Meinhardt and Rohe 1993). However, yeast linear plasmids are routinely localized in the cytoplasm, with only very few exceptions (see below) (Blaisonneau et al. 1999).

Linear eukaryotic elements, either cytoplasmic or mitochondrial, necessitate specific end-structures to avoid shortening during replicative cycles. This problem is circumvented in bacterial linear plasmids due to closed hairpins. Eukaryotic linear plasmids in general have terminal proteins covalently bound to the 5′ ends (Kikuchi et al. 1984; Stam et al. 1986; Takeda et al. 1996; Kim et al. 2000). They not only protect the ends from exonucleolytic damage, but also serve as primers for replication. The second peculiar structural feature characteristic for both the mitochondrial and cytoplasmic linear plasmids is terminal inverted repetitions, which allow the displaced strand to form a replication intermediate.

The first linear plasmids of a yeast species were discovered in *Kluyveromyces lactis*. Two elements termed pGKL1 and pGKL2 confer a killer phenotype (Gunge et al. 1981). Structurally similar elements were subsequently detected in a number of other genera, including *Saccharomyces*, *Pichia*, *Candida*, *Debaryomyces*, *Botryoascus*, *Trichosporon*, *Saccharomycopsis* and *Schwanniomyces* (see Table 1). Frequently, they form a system of two to three differently sized elements (Table 1), though single plasmids occur also. A systematic screening involving 1805 strains deposited in the CBS yeast culture collection discovered linear plasmids in 32 different strains (Cong et al. 1994; Fukuhara 1995). Based on estimated size and restriction fragment comparisons, some are likely to be identical elements (Cong et al. 1994; Fukuhara 1995; Blaisonneau et al. 1999). The calculated frequency of ~1.8% makes evident the wide distribution among several genera. Only three strains were found to harbour the plasmids in the mitochondria (Blaisonneau et al. 1999), clearly indicating, that such localization is rare in yeasts as opposite to filamentous fungi and plants, in which they are generally localized in mitochondria (Meinhardt et al. 1990).

For the sake of completeness, two exceptions from this rule should be mentioned, i.e. linear plasmids from *Alternaria alternata* (Shepherd 1992), the localization of which is unknown, but apparently not in the mitochondria, and plasmids from the green algae *Chlamydomonas moewusii*, which are localized in chloroplasts (Turmel et al. 1996).

The killer plasmid pair pGKL1 (8.9 kb) and pGKL2 (13.5) of *Kluyveromyces lactis* became the most intensively studied, and thus serve as the paradigm. The smaller pGKL1 encodes the killer toxin, called zymocin, a secreted heterotrimeric protein, that kills other sensitive yeasts, including plasmid free derivatives of the host strain (Gunge et al. 1981; Stark and Boyd 1986; Stark et al. 1990). In addition, the element encodes a toxin-immunity function (Tokunaga et al. 1997), which not only protects the host cell, but also assures stable inheritance of the plasmid system, because occasional loss of pGKL1 is counterselected. The zymocin encoding element pGKL1 is non-autonomous. It depends on the larger pGKL2 in terms of cytoplasmic transcription and replication functions encoded by the latter. pGKL2 is autonomous and may occur also in the absence of pGKL1. Such (partially cured) strains can be found under laboratory conditions. This scenario is rather unlikely in the natural environment due to the autoselective toxin/immunity system encoded by pGKL1.

3 Autonomous elements

Sequence analyses of additional plasmids have disclosed that most of yeast cytoplasmic linear plasmid systems consist of an autonomous (pGKL2-like) element, accompanied by non-autonomous plasmids, which in some

Table 1. Compilation of yeast strains with cytoplasmic linear plasmids

Strain	Plasmids	Size	Reference
Mitochondrial linear plasmids, cryptic			
Pichia hedii, CBS 6931	pPH1	7.1	Blaisonneau et al. 1999
Pichia kluyveri, CBS 7274	pPK1	7.2	Blaisonneau et al. 1999
Pichia kluyveri, CBS 7907	pPK2	7.2	Blaisonneau et al. 1999
Cytoplasmic linear plasmids, cryptic			
Botyroascus cladosporoides, CBS 7434	pBC1A, pBC1B	7.0, 14.0	Fukuhara 1995
Candida tartarivorans, CBS 7955	pCX1A, pCX1B	8, 13	Blaisonneau et al. 1999 (updated from CBS)
Debaryomyces (Pichia) etchellsii, CBS 2011; CBS 2012; CBS5603; CBS 6823	pPE1A, pPE1B	6.7, 12.8	Cong et al. 1994
Debaryomyces hansenii, TK (CBS 7848; ATCC 90624)	pDHL1, pDHL2, pDHL3	8.4, 9.2, 15	Gunge et al. 1993
Debaryomyces hansenii, CBS 770	pDH1A, pDH1B	8.2, 14.4	Cong et al. 1994
Debaryomyces hansenii, CBS 1780	pDH2A, pDH2B	6.5, 15	Cong et al. 1994
Debaryomyces hansenii, CBS 4890	pDH3A, pDH3B	4.7, 13	Cong et al. 1994
Debaryomyces polymorphus, CBS 6453	pDP1	18	Fukuhara 1995
Debaryomyces sp. CBS 7807	pDX1A, pDX1B	8.0, 15	Fukuhara 1995
Pichia nakazawae, CBS 6701	pPN1A, pPN1B, pPN1C	7.0, 8.0, 15	Fukuhara 1995
Pichia pastoris, CBS 7435	pPP1A, pPP1B	9.5, 13	Fukuhara 1995

(Continued)

Table 1. *Continued*

Strain	Plasmids	Size	Reference
Pichia pastoris, NRRL-Y-4290	pPpa1-2, pPpa1-1	8, 11	Banerjee et al. 1998
Saccharomyces kluyveri, IFO 1685	pSKL	14	Kitada and Hishinuma 1987
Saccharomycopsis cratagensis, NRRL-Y-5902	pScr1-3, pScr1-2, pScr1-1	5.0, 7.0, 15	Bolen et al. 1992
Saccharomycopsis malanga, CBS 6531	pSM2A, pSM2B	8.7, 15	Fukuhara 1995
Schwanniomyces occidentalis ATCC 44252	pSoc1-2, pSoc1-1	8.1, 13.4	Chen et al. 2000
Trichosporon pullulans, CBS 2532	pTP1	16	Fukuhara 1995
Cytoplasmic linear plasmids, associated with killer phenotype			
Debaryomyces (Wingea) robertsiae CBS 6693	pWR1A, pWR1B	8.1, 14.6	Cong et al. 1994; Klassen and Meinhardt 2002
Debaryomyces (Wingea) robertsiae CBS 5637	pWR2A, pWR2B	9.0, 15	Cong et al. 1994; Klassen and Meinhardt, unpublished
Kluyveromyces lactis, IFO 1267; CBS 1065; CBS 5618; CBS 8043; NRRL Y-1114	pGKL1, pGKL2	8.9, 13.5	Gunge et al. 1981; Cong et al. 1994
Pichia acaciae, RRL-Y-18665	pPac1-2, pPac1-1	6.8, 13.6	Worsham and Bolen 1990
Pichia inositovora, NRRL-Y-12698	pPin1-3, pPin1-2, pPin1-1	9.7, 13, 18	Ligon et al. 1989; Hayman and Bolen 1991

instances confer a killer phenotype (Hishinuma and Hirai 1991; Bolen et al. 1994; Cong et al. 1994; Klassen et al. 2001, 2002; Klassen and Meinhardt 2002, 2003).

In addition to pGKL2 (Tommasino et al. 1988), the complete nucleotide sequence of autonomous elements was resolved for *Saccharomyces kluyveri* pSKL, *Debaryomyces (Pichia) etchellsii* pPE1B and *Pichia acaciae* pPac1-1 (Hishinuma and Hirai 1991; Klassen et al. 2001; Jeske and Meinhardt 2006). Except for the non-essential ORF1 of pGKL2, all elements are almost identical in genome organization and information content as individual genes display a high degree of similarity (Fig. 1). This suggests a common organization of autonomous elements in different yeast linear plasmid systems. Functional analysis has exclusively been performed with *K. lactis* pGKL2 open reading frames (Tommasino et al. 1988; Wilson and Mecock 1988; McNeel and Tamanoi 1991; Tommasino 1991; Schickel et al. 1996; Schaffrath et al. 1996, 1997, 2000; Schaffrath and Meacock 1996, 2001; Takeda et al. 1996; Larsen and Meinhardt 2000; Tiggemann et al. 2001). Due to the high degree of homology these results may be generalized. As outlined in Fig. 1,

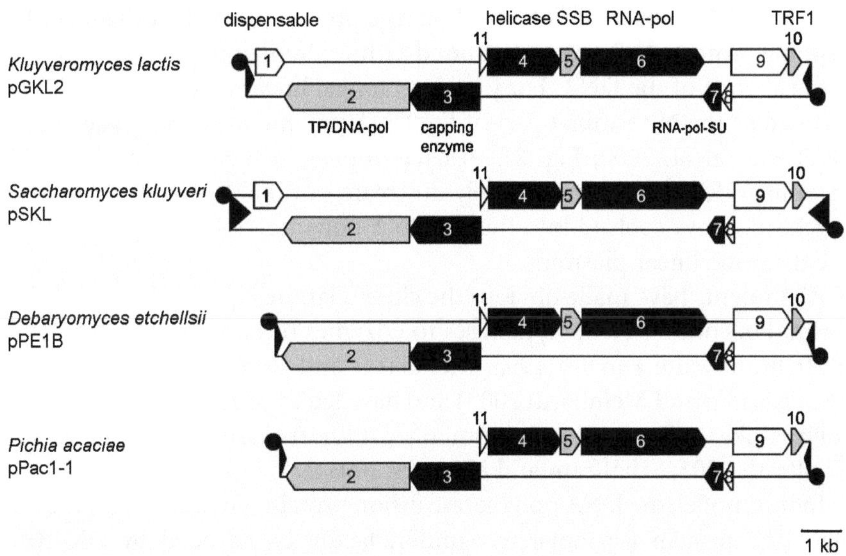

Fig. 1. Schematic representation of completely sequenced yeast autonomous cytoplasmic linear plasmids. Open reading frames (ORFs) are depicted as *arrows*, the orientation of which indicates transcriptional direction. Gene products encoded by *grey shaded* ORFs are involved in cytoplasmic replication, those encoded by *filled* ORFs are instrumental in cytoplasmic transcription; terminal inverted repeats are shown as *black triangles* and the 5′ covalently bound proteins are represented by *black circles*. Numbering of ORFs according to Tommasino et al. (1988) and Larsen et al. (1998)

functions encoded by yeast autonomous elements can be categorized as instrumental in (i) cytoplasmic replication, (ii) transcription and (iii) additional unknown factors.

3.1 Nucleus-independent replication

The crucial enzyme for cytoplasmic replication is the DNA polymerase encoded by ORF2 (Tommasino et al. 1988). In the case of the pGKL system, another (plasmid specific) DNA polymerase is encoded by the accompanying non-autonomous element (Fukuhara 1987; Jung et al. 1987). Both DNA polymerases replicate selectively their "own" plasmid (Kikuchi et al. 1985; Kitada and Gunge 1988). This is different in the case of pPE1 and pPac1 systems, in which only a single DNA polymerase encoding gene (on the autonomous element) is present, the product of which apparently replicates both, the autonomous and the non-autonomous elements (Klassen et al. 2002; Jeske and Meinhardt 2006).

DNA polymerases of eukaryotic linear plasmids belong to a subgroup of the B-type family, which consists of enzymes from bacteriophages and adenoviruses that initiate DNA replication by a protein primed mechanism. The terminal proteins (TP) are the required primers, remaining covalently linked to the 5′ ends of the DNA. For yeast and fungal linear plasmids, the TPs are encoded as the N-terminal part of the DNA polymerase gene (Takeda et al. 1996; Kim et al. 2000; Fig. 2). In bacteriophages, there is a separate TP-encoding ORF located immediately upstream of the DNA polymerase gene, which may have evolved into the TP-DNA polymerase fusion-gene realized in eukaryotic linear plasmids.

Alignments have made obvious the close relationship of eukaryotic linear plasmid encoded DNA polymerases to enzymes of viral origin (Rohe et al. 1991; Braithwaite and Ito 1993; Meinhardt and Rohe 1993; Dufour et al. 2003; Klassen and Meinhardt 2003) and have led to propose a phi29- or adenovirus-like replication mechanism in yeast linear plasmids (Meinhardt and Schaffrath 2001; Schaffrath and Meacock 2001).

In this model, the DNA polymerase fusion protein binds as a complex with the TRF1 protein (terminal recognition factor 1), encoded by ORF10 of pGKL2 (McNeel and Tamanoi 1991; Tommasino 1991) to the terminal inverted repeats on both ends of the plasmid and initiates replication. Concomitantly, the parental strand is displaced and simultaneously covered by single stranded binding proteins (SSB), encoded by ORF5 (Schaffrath and Meacock 2001). The displaced strand might form a so-called panhandle structure via hybridization of the TIRs, which can initiate a novel round of

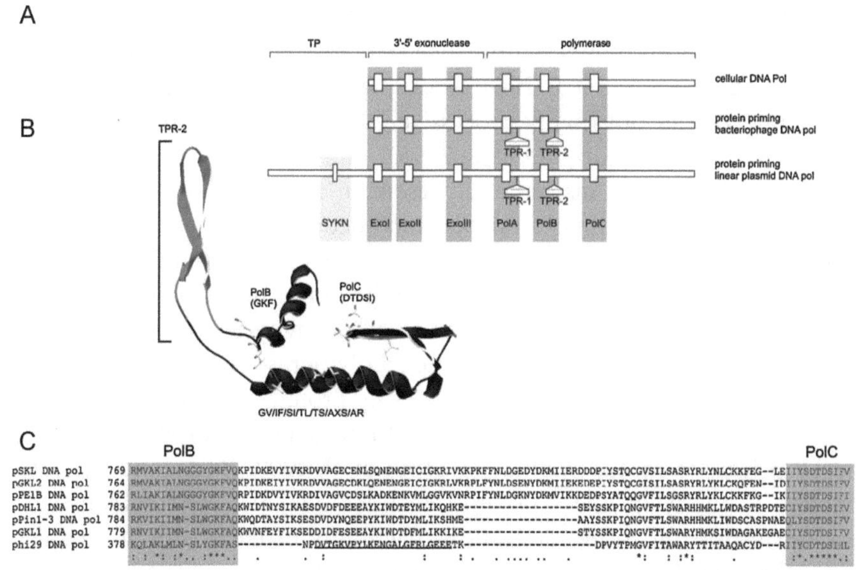

Fig. 2. Domain architecture of DNA polymerases using a protein-primer. **A** Schematic representation of domains in cellular eukaryotic DNA polymerases, protein priming DNA polymerases of phi29-like bacteriophages, and eukaryotic linear plasmids. *TP* terminal protein; *SYKN* moderately conserved TP-motif; *ExoI-III* exonuclease I-III domains; *PolA-C* polymerase domains A-C; *TPR-1, TPR-2* two insertions in the polymerase domains in protein priming DNA polymerases of bacteriophages and linear plasmids. This figure is based on previously presented data (Dufour et al. 2000, 2003) and complemented with linear plasmid encoded DNA polymerases. **B** Structure of the TPR-2 insertion in the phi29 DNA polymerase, based on the PDB file 1X1A (Kamtekar et al. 2004), and visualized with Deep View/Swiss PDB viewer (www.expasy.org/spdbv/). Residues conserved in PolB and PolC regions are highlighted. **C** Alignment of the TPR-2 region of DNA polymerases from yeast linear plasmids and phi29 DNA polymerase. Residues are marked with an *, : or . depending on the degree of conservation. The TPR-2 representing region in the phi29-DNA polymerase (Rodriguez et al. 2005) is underlined. File sources from NCBI (www.ncbi.nlm.nih.gov): pSKL (CAA38621.1) pGKL2 (CAA25568.1) pPE1B (CAC08221.2) pDHL1 (CAA09497.1) pPin1–3 (CAD91889.1) pGKL1 (CAA30603.1) phi29 (P03680). Detailed alignments covering the TPR-1 insertion in linear plasmid encoded DNA polymerases are given in Dufour et al. (2000, 2001)

replication. The mode and time for splitting the covalent TP-DNA polymerase fusion remains obscure; finally, however, the TP remains bound to the 5′ end of the newly synthesized DNA and the polymerase becomes displaced.

The phi29 DNA polymerase has been shown to be capable of replicating the entire phi29 genome without the need for a DNA-helicase or processivity factor (Blanco et al. 1989). A key element for these unique features is an insertion between the polymerase domains B and C, called TPR-2 which apparently mediates the separation of the template from the non-template strand before reaching the polymerase active site. Additionally, this insertion facilitates formation of a doughnut-like structure surrounding the parental DNA duplex (Kamtekar et al. 2004; Rodriguez et al. 2005). Thus, phi29 DNA polymerase carries out highly processive DNA synthesis without the need for a sliding clamp (Rodriguez et al. 2005).

As such, TPR-2 like insertion is present in all DNA polymerases using protein priming, including linear plasmid encoded enzymes (Dufour et al. 2000, 2003; Rodriguez et al. 2005; Fig. 2), it appears likely that linear plasmid encoded DNA polymerases also replicate the entire linear plasmid without the need for a separate DNA helicase or processivity factor. An additional insertion called TPR-1, located between polymerase domains A and B is present in all B-type DNA polymerases using protein primers (Dufour et al. 2003). These domains have been shown to be required for stabilization of the TP-primer terminus in phi29 (Dufour et al. 2000, 2003), and thus may function similarly in linear plasmid encoded enzymes. The evident close relationship of the entire replication machinery of eukaryotic linear plasmids, phi29 related viruses, and adenoviruses strongly suggests a common ancestry (Meinhardt and Rohe 1993).

3.2 Nucleus-independent transcription

Cytoplasmic localization of linear plasmids precludes them from accessing nuclear factors, which are essential for transcription. Due to the active transport of the cellular RNA polymerase into the nucleus the amount of cytoplasmically located enzyme is infinitesimal. Since linear yeast elements possess their own transcription machinery, consisting of both, *cis* and *trans* acting factors, this problem is circumvented.

In-vivo recombination using a nuclearly expressible marker gene (e.g. *S. cerevisisae LEU2*) failed to establish recombinant cytoplasmic plasmids and resulted in hybrid plasmids replicating in the nucleus. The high A+T content (75% average) of the linear plasmids provided artificial ARS-like sequences, which are able to act as nuclear replication origins. Such relocation was already used to study attachment of telomeres to linear hybrid plasmids carrying nuclear marker genes (Kämper et al. 1989a,b, 1991; Gunge et al. 2003). Vice versa, the expression of toxin ORFs on nuclear replicating vectors in plasmid carrying strains failed (Romanos and Boyd 1988; Stark et al. 1990).

In-vivo recombination using marker genes with an in-phase fused conserved sequence element (UCS, upstream conserved sequence), which precedes each linear plasmid ORF, led to the establishment of linear hybrid elements replicating autonomously in the yeast cytoplasm. Stable expression was achieved for the xylose-isomerase gene and the UDP-glucose dehydrogenase, both fused with pGKL2/UCS5, and transferred into pGKL1/ORF2 (Schründer et al. 1996). *Cis* acting elements (UCSs) are highly conserved among the large autonomous and the smaller non-autonomous elements. They span a 6-nucleotide long highly conserved sequence (consensus ATNTGA), which is sufficient for driving transcription of the ORFs located downstream.

In contrast to the highly conserved nucleotide sequence, the distance between a UCS and the respective start codon is more variable ranging from only 18 nt for ORF4 of pSKL to more than 207 nt for ORF9 of pPac1-1, with an average of 20–40 nt (Romanos and Boyd 1988; Stark et al. 1990; Hishinuma and Hirai 1991; Schaffrath et al. 1996; Schickel et al. 1996; Klassen et al. 2001; Jeske and Meinhardt 2006).

Although expression of linear plasmid ORFs is rather weak, determination of transcriptional activity was performed for several linear plasmid ORFs, such as UCS2, 4, 5, 6, 8, 10 and 11 of pGKL2 and UCS1 and 2 of pGKL1 (Schründer and Meinhardt 1995; Schaffrath et al. 1996, 1997; Schickel et al. 1996; Larsen and Meinhardt 2000). Fusion of pPE1B ORF6 with a *gdh*A-reporter gene and in-vivo recombination into ORF2 of pGKL2 did not only show functionality but even interchangeability of UCS elements between different linear plasmid systems (Klassen et al. 2001).

The strength of cytoplasmic promoter activity was determined for the above mentioned UCS2, 4–6, 8 and 10 of pGKL2 and UCS1 and 2 of pGKL1; the most active element drives transcription of the RNA polymerase, followed by the DNA polymerase UCS1 of pGKL1 and the pGKL2/UCS4 driving transcription of the helicase-encoding ORF. The relative activity of UCS elements seems to correlate with the requirements for the linear plasmid system, in which the RNA polymerase plays a key role (Schründer and Meinhardt 1995; Schickel et al. 1996).

3.2.1 Analysis of linear plasmid derived transcripts

Transcription initiation analyses were carried out on pGKL1/ORFs 1, 2, 3 and 4, as well as pGKL2/ORF9. Multiple initiation sites downstream of each UCS element were identified, with a preference for one site. The first site is routinely spaced 8–16 nt from the respective UCS, whereas the distance from the 5′-end of the transcripts to the translational start may vary significantly (Romanos and Boyd 1988; Stark et. al. 1990; Jeske et al. 2006).

Northern analyses of linear plasmid derived transcripts is clearly not suited to distinguish transcripts with slight differences in length due to multiple initiation sites, it is, however, sufficient for comparing electrophoretic mobilities of transcripts to the estimated size of the coding region. Such transcript analyses carried out for pGKL2/ORF5 and 6, pGKL2/ORF2 and 3, and pGKL2/ORF9, suggests monocistronic transcription. Since linear plasmids are tightly packed with genetic information (with partially overlapping ORFs) results obtained suggest transcription termination in a hitherto unknown manner (Schaffrath et al. 1995; Jeske et al. 2006). Up to the present, it remains also totally obscure whether linear plasmid derived transcripts have modifications at their 3′-ends. Computer-aided searches for polyadenylation signals failed to identify such structures, which play an important role in either the addition of a polyA-tail or termination of transcription (data not shown).

In eukaryotes, polyA tails at the mRNA 3′-ends are required for efficient translation initiation and ribosome recycling by forming a loop structure between 5′- and 3′-mRNA ends, mediated by translation initiation factors which are part of the ribosome assembly complex (for review, see Chou 2003, Kapp and Lorsch 2004).

3.2.2 Cytoplasmic transcription machinery

The major players responsible for cytoplasmic transcription are the ORF6 encoded RNA polymerase, and its subunit ORF7p, acting jointly with the cytoplasmic capping enzyme (ORF3p) and the helicase (ORF4p) (Fig. 3). RNA polymerase and its subunit are exclusively encoded by the large, autonomous linear elements and are instrumental in transcription of autonomous plasmid ORFs as well as non-autonomous ones (Wilson and Meacock 1988, Schaffrath et al. 1995, 1997).

ORF6p is a unique RNA polymerase: Comparison of the predicted polypeptides with RNA polymerases of various sources, as well as a conserved domain database identified both β- and β'-domains being located on one polypeptide chain, whereas the *rpo*C domain, regularly located at the C terminius of the β'-domain, is lacking. However, this domain was identified in the predicted ORF7p (Stark et al. 1990; Thuriaux and Sentenac 1992; Schaffrath et al. 1995, 1997). Thus, the linear plasmid encoded RNA polymerase differs in structure from prokaryotic as well as eukaryotic enzymes (Wilson and Meacock 1988). However, when ORF7p is included, which is encoded on the opposite DNA-strand and in the opposite direction arranged without intergenic spacing to ORF6, it resembles the situation in the archaeon *Halobacterium halobium* (Leffers et al. 1989). Despite this, amino acid sequence comparisons carried out exclusively on highly conserved regions revealed similarities with cellular enzymes, rather than viral or prokaryotic polypeptides.

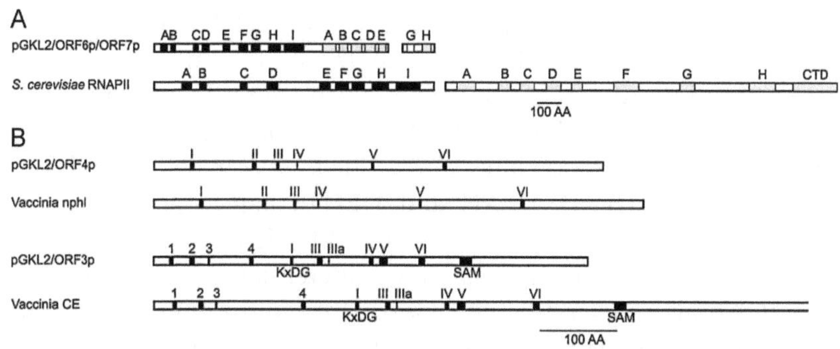

Fig. 3. Transcriptional main players of linear plasmids. pGKL2-encoded proteins involved in cytoplasmic transcription comprise the ORF6 encoded RNA polymerase (ORF6p; CAA30607) and an ORF7 encoded subunit of the latter (ORF7p; CAA30608), a helicase (ORF4p; CAA30605.1) and the cytoplasmic capping enzyme (ORF3p; CAA30604). Arrangement and positions of conserved sequence motifs are given schematically as filled rectangles (*black* or *light grey*). **A** The linear plasmid encoded ORF6p and ORF7p are arranged according to the occurrence of conserved motifs, and compared to the large (subunit 2; P08518) and the largest (subunit 1; P04050) polypeptide of *S. cerevisiae* RNA polymerase II (RNAPII). **B** pGKL2-encoded ORF4p and ORF3p, vaccinia virus nphI (P05807) and the (D1R-encoded) capping enzyme large subunit CE (P04298) are aligned, respectively. NCBI accession numbers are given. Capping enzymes: conserved sequence motifs of the RNA-triphosphatase-domain are given by numbers 1–4, numbers I–VI indicate conserved motifs of the guanylyltransferase; *SAM* S-adenosylmethionine binding domain of the N7-methyltransferase-domain; *KxDG* amino acid motif (single letter code), *x* any amino acid. Helicases: conserved sequence motifs are indicated by numbers I–VI. RNA polymerases: *S. cerevisiae* RNAPII comprises subunit 2 (*left side, black shaded boxes*) and subunit 1 (*right side, grey shaded boxes*), conserved sequence motifs of which are designated A to I (subunit 2) and A to H and CTD (subunit 1). *CTD* C-terminal domain. pGKL2/ORF6 carries conserved motifs, indicated as described for RNAPII, except for domain G and H (pGKL2/ORF7); CTD is lacking

3.2.3 The linear plasmid encoded helicase

ORF4p of pGKL2 displays striking similarities to viral transcriptionally involved helicases and is regarded to function in unwinding secondary structures of the growing RNA strand eventually ensuring fidelity of the RNA polymerase (Tommasino et al. 1988; Stark et al. 1990).

Commonly, helicases are involved in a variety of cellular processes, such as RNA-metabolism or replication, and can be divided into three different superfamilies, based on sequence analyses (Koonin 1991; Kadare and Haenni 1997; de la Cruz et al. 1999). Both superfamilies 1 and 2 possess seven typical and conserved sequence motifs, but with different spacings; superfamily 3 helicases possess only three conserved domains, which differ from those of 1 and 2. Typical for helicases is the purine nucleotide binding motif firstly identified by Walker et al. (1982), consisting of the NTP β- and γ-phosphate binding motif (motif I), with GxGKT as the minimum core amino acid sequence. The second part of the Walker-motif (motif II), contains a D for Asp chelating the Mg^{2+} ion of the Mg-NTP complex (Gorbalenya and Koonin 1993). Furthermore, superfamily 2 helicases can belong to three different subgroups (DEAD, DEAH and DExD), named according to the sequence of their ATP-hydrolysis motif (motif II). In addition, the five additional conserved sequence motifs allow differentiation of the three subgroups (Schmidt and Linder 1992; Gorbalenya and Koonin 1993; Kadare and Haenni 1997; de la Cruz et al. 1999) functionally corresponding with different categories.

Available data on helicases belonging to the three subfamilies was compiled and suggests that DEAD-motif helicases are related to transcription. DEAH-family members play a role in RNA-splicing, and DExD helicases are involved in a variety of processes affecting nucleic acids, such as RNA export, degradation or translation and DNA replication (Jankowsky and Jankowsky 2000).

The linear plasmid encoded ORF4p not only possesses a conserved motif I, but also a conserved DEVH motif, which, in combination with other conserved sequences, defines ORF4p belonging to the DExH subgroup.

The genome of the vaccinia virus encodes multiple helicases functioning in unwinding DNA or resolving RNA-secondary structures, but only the energy coupling factor nphI, which belongs to the DExH subgroup is closely related to ORF4p. Whilst it does not possess an unwinding function for nucleic acids, this factor binds to single stranded DNA and hydrolyses ATP supplying the progressing RNA polymerase with energy in a hitherto unknown manner (Paoletti et al. 1974; Rodriguez et al. 1986; Broyles and Moss 1987; Deng and Schuman 1998).

3.2.4 The capping enzyme and RNA capping

The linear plasmid ORF3p possesses conserved amino acid sequence motifs of the nucleotidyltransferase superfamily, precisely of mRNA capping enzymes (Larsen et al. 1998).

An RNA cap structure is required for efficient translation initiation. It consists of a N7-methylated guanylyl residue, which is covalently linked to the RNA 5′-end by an unusual 5′-5′ triphoshate bridge. Three enzymatic functions are required for cap formation: RNA-triphosphatase, guanylyltransferase and N7-cap-methyltransferase (for review, see Bisaillon and

Lemay 1997). The newly synthesized transcript possesses a 5′-triphosphate end, which is shortened by the hydrolysis of the γ-phosphate through the action of the RNA-triphosphatase, resulting in a RNA-diphosphate end. A covalent guanylyltransferase-GMP intermediate is formed by a nucleophilic attack on the γ-phosphate of a GTP, which results in a phosphoamide-bridge between the guanylyl residue and the ε-amino group of the essential active site lysine residue, which is located in the typical KxDG amino acid motif. The GMP-residue is then transferred to the RNA-diphosphate end, resulting in a 5′-5′-triphosphate bridge (Shuman and Hurwitz 1981; Cong and Shuman 1993; Niles and Christen 1993; Shuman and Schwer 1995). The capping reaction is made irreversible by the addition of a methyl-group to the N7 position of the guanylyl-residue, resulting in the so-called cap0 structure. Addition of methyl groups to the 2′-OH group of first or second riboses of the RNA-strand is possible, too (cap1, cap2, respectively). Whether further riboses are 2′-methylated or not is not yet clear (Shuman 1995, Bisaillon and Lemay 1997).

Experimental evidence for two of three main reactions required for cap addition has been obtained for the predicted cytoplasmatic capping enzyme ORF3p, i.e. the RNA-triphosphatase and the guanylyltransferase activity. Both functions are located on the same polypeptide, and, not surprisingly, a binding site for S-adenosylmethionine providing the methyl group required for cap methylation is located on the same polypeptide too (Larsen et al. 1998; Tiggemann et al. 2001). The linear plasmid encoded capping enzymes resemble enzymes of *pox*- and *iridoviridae* (see Fig. 4) regarding the arrangement of conserved motifs of the RNA-triphosphatase, which are located N-terminal, the guanylyltransferase (located centrally), and the SAM-binding site, which is located in the C-terminus (Shuman 1995; Larsen et al. 1998; Klassen et al. 2001).

The most thoroughly investigated cytoplasmic capping system is the vaccinia enzyme, consisting of two subunits, encoded by D1 and D12, respectively; the large subunit (D1) compares to ORF3p.

The methyltransferase activity of the large polypeptide D1 is stimulated by interaction with the small subunit D12 (Martin et al. 1975; Schuman et al. 1980; Niles et al. 1986, 1989; Shuman 1989; Shuman and Morham 1990; Cong and Shuman 1992; Higman et al. 1992; Larsen et al. 1998; Higman et al. 1994; Mao and Shuman 1994).

For ORF3p of linear plasmids it remains to be elucidated, whether an additional second subunit is required for efficient methyltransferase-activity; proteins encoded by ORFs 8, 9 or 11 with hitherto unknown functions may be candidates.

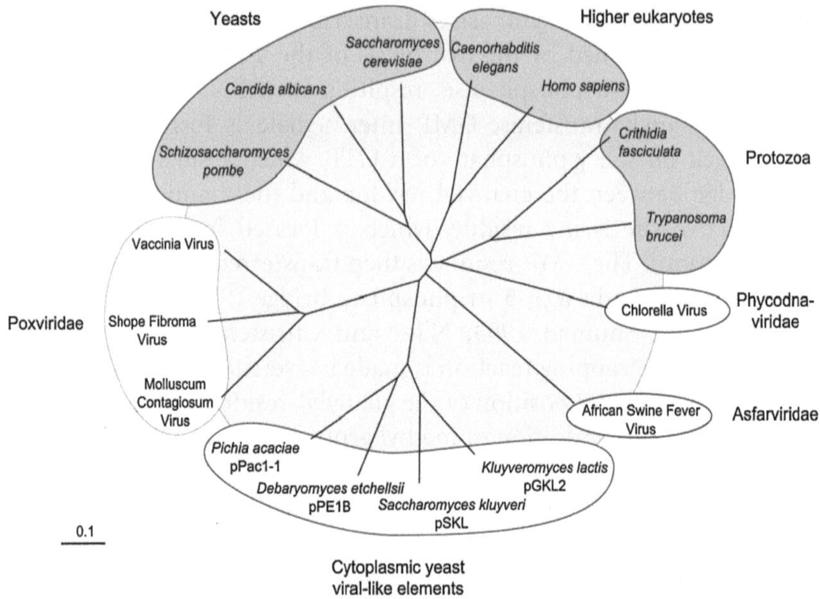

Fig. 4. Phylogenetic relationships of guanylyltransferase core regions. Calculation was done using the program ClustalW (EMBL HUSAR genius net service program package) with standard parameters. Concerning capping of mRNA the yeast cytoplasmic elements have their nearest relatives within the Poxviridae, whereas nuclear capping enzymes of yeasts are more closely related to the nuclear capping apparatus of other eukaryotes. Viral elements in *blank areas*, nuclear capping enzymes in *shaded areas*. Core regions spanning guanylyltransferase motifs I–VI were used for calculations. NCBI accession numbers of capping enzymes: *Kluyveromyces lactis* pGKL2 (CAA306049), *Saccharomyces kluyveri* pSKL (CAA38622), *Pichia etchellsii* pPE1B (CAC08222), *Pichia acaciae* pPac1-1 (CAJ57276), Molluscum contagiosum virus (NP_044041), Shope fibroma virus (P25950), Vaccinia virus (NP_063758), African swine fever virus (CAA50806), *Paramecium bursaria* Chlorella virus 1 (Q84424), *Schizosaccharomyces pombe* (P40997), *Candida albicans* (P78587), *Saccharomyces cerevisiae* (Q01159), *Caenorhabditis elegans* (Q17607), *Homo sapiens* (O60942), *Crithidia fasciculata* (AAC27927), *Trypanosoma brucei* (AAC27926)

4 Non-autonomous elements

In addition to the invariably present autonomous elements consisting of at least 12.6 kb (*P. acaciae* pPac1-1), smaller, non-autonomous linear plasmids can be seen in most of the strains carrying linear cytoplasmic plasmids (Table 1). Their sizes vary considerably, ranging from 6.7 kb (*D. etchellsii* pPE1A) to 9.8 kb (*Pichia inositovora* pPin1-3). The dependency of these elements on the larger plasmid can be inferred from the fact that the gene content must be lower compared to pPac1-1, which presumably represents the smallest possible autonomous element. Thus, simply by size, such small elements cannot encode all the functions required for cytoplasmic inheritance. In fact, partially cured strains were recovered in several instances, carrying only the large (autonomous) element. The opposite, a small (i.e. <12.8 kb) cytoplasmic linear plasmid existing without the larger one has never been observed (Niwa et al. 1981; Worsham and Bolen 1990; Hayman and Bolen 1991; Gunge et al. 1993; Bolen et al. 1994). The non-autonomous elements can functionally be divided into two groups: the cryptic elements and the killer plasmids which confer a killer- and immunity phenotype to the host cell.

4.1 Elements associated with killer phenotypes

Today, there are five known DNA-killer plasmids, i.e. pGKL1 from *K. lactis*, pPac1-2 from *P. acaciae*, pPin1-3 from *P. inositovora* and two possibly closely related plasmids, pWR1A and pWR2A from two different *Debaryomyces robertsiae* strains (Gunge et al. 1981; Worsham and Bolen 1990; Hayman and Bolen 1991; Klassen and Meinhardt 2002; Klassen and Meinhardt, unpublished). Except for pWR2A, nucleotide sequences are available for all of them (Hishinuma et al. 1984; Stark et al. 1984; Sor and Fukuhara 1985; Klassen and Meinhardt 2003; Klassen et al. 2004). Based on genome organization, gene content and toxin function they are grouped into two subtypes, called here type I and type II toxin encoding elements (Fig. 5).

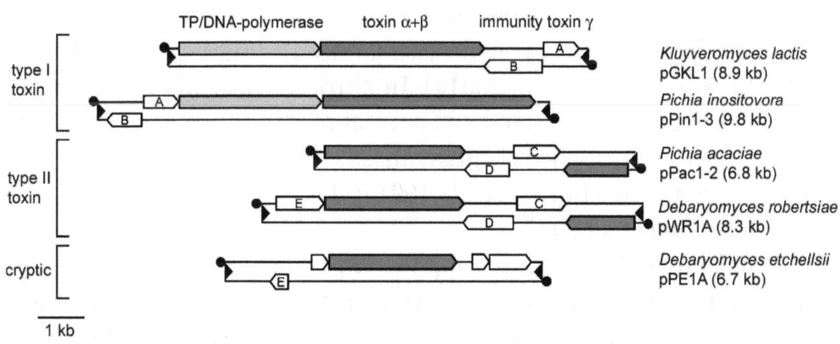

Fig. 5. Non-autonomous yeast linear plasmids. Similar ORFs are indicated by similar gray shading or letters (A–E). Based on genome organization and functional analysis of encoded toxins, elements are grouped into cryptic, type I and type II toxin encoding elements. Functions, if known, are indicated on top. *P. acaciae* and *D. robertsiae* ORFs marked with D encode cell cycle arresting subunits functionally distinct from zymocin γ. Source of data: pGKL1 (Hishinuma et al. 1994; Stark et al. 1994; Sor and Fukuhara 1985), pPin1-3 (Klassen and Meinhardt 2003), pPac1-2, pWR1A (Klassen et al. 2004), pPE1A (Klassen et al. 2002)

4.1.1 Killer plasmids encoding type I toxins

The prototype of a type I toxin encoding element is represented by pGKL1 from *K. lactis*. The element carries four ORFs, encoding a plasmid specific TP/DNA polymerase fusion protein (ORF1), the immunity function (ORF3) and the three subunits (α, β, γ with molecular masses of 99, 30, and 28 kDa, respectively) of the killer toxin zymocin (ORF2 and 4) (Hishinuma et al. 1984; Stark et al. 1984, 1990; Sor and Fukuhara 1985; Stark and Boyd 1986; Tokunaga et al. 1987).

The α- and β-subunits of zymocin arise from the post-translationally processed ORF2 product, while the γ-subunit is separately encoded by ORF4. As typical for secreted proteins, Orf2p and Orf4p are first translocated to the endoplasmic reticulum (ER) where the residing signal peptidase cleaves the signal peptides of both precursors (Tokunaga et al. 1990). During secretion, these processed precursors reach the Golgi network, where the Kex1 (Kex2 in *S. cerevisiae*) protease cleaves Orf2p at position KR_{29} and KR_{894} to yield mature α- and β-subunits (Stark and Boyd 1986; Stark et al. 1990).

All three subunits are assembled into holotoxin (zymocin), in which β- and γ-subunits are covalently connected via a disulfide bridge and the α-subunit contains at least one internal disulfide bridge (Stark and Boyd 1986; Stark et al. 1990).

The first known interaction of zymocin with the target cell is binding of the holotoxin to chitin in the cell wall. This specificity can be attributed to the α-subunit, since it contains a cysteine rich chitin-binding domain typical for plant chitin binding proteins and chitinases as well as a chitinase domain belonging to glycosyl hydrolase family 18 (Stark et al. 1990; Butler et al. 1991a; Jablonowski et al. 2001a). In vitro, chitin binding and chitinase activity have been demonstrated and several mutations in chitin biosynthesis genes, affecting the major chitinsynthase (Chs3) of *S. cerevisiae* render it resistant to zymocin (Butler et al. 1991a; Takita and Castilho-Valavivius 1993; Jablonowski et al. 2001a).

The crucial toxin factor, mediating the cell cycle arrest of the target cell in G1 is the γ-subunit, which acts intracellularly and is translocated into the target cell subsequent to chitin binding via the α-subunit (Butler et al. 1991a,b; Stark et al. 1990). This has been directly demonstrated (Mehlgarten and Schaffrath 2004) and also indirectly by the finding, that intracellular expression of the γ-subunit from the tightly regulated *GAL1* promoter mimics the G1 arrest induced by exo-zymocin (Tokunaga et al. 1989; Butler et al. 1991c). Although the mechanism of transmembrane passage is not entirely understood, it is assumed that the strongly hydrophobic β-subunit is involved in this process (Stark et al. 1990; Jablonowski et al. 2001a; Mehlgarten and Schaffrath 2004; Zink et al. 2005). Transmembrane passage of γ, but not docking of holotoxin, requires the major plasma membrane sphingolipid, mannosyl-diinositolphospho-ceramide, which is synthesized by the Kti6/Ipt1 protein (Zink et al. 2005). The γ-subunit, once taken up by the cell, has been shown to rely on intracellular activation mediated by the plasma membrane ATPase Pma1, whose chronological engagement in toxin action follows the mannosyl-diinositolphospho-ceramide dependent step (Mehlgarten and Schaffrath 2004; Zink et al. 2005). Future tasks will be to elucidate whether toxicity of γ involves a breakdown of the disulphide bridge, that connects β and γ in holotoxin and to define the functional roles of factors involved in early toxin response such as mannosyl-diinositolphospho-ceramide and Pma1.

The intracellular action of the imported γ-subunit strictly depends on a six-subunit (Elp1-6) protein complex termed Elongator, which was originally found to be involved in the elongation step of RNA polymerase II transcription by virtue of its histone acetyltransferase activity, residing in the Elp3 subunit (Otero et al. 1999; Wittschieben et al. 1999; Frohloff et al. 2001; Winkler et al. 2001). Recently, Elongator was also found to be implied in such diverse processes such as tRNA modification and exocytosis (Huang et al. 2005; Rahl et al. 2005). In addition to the requirement of Elongator subunits Elp1-6 for zymocin action, removal of several Elongator interacting proteins, i.e. Kti11-13, Sit4, Sap185, Sap190 as well as inactivation of the histone acetyl transferase activity by site directed mutagenesis all confer immunity towards zymocin (Frohloff et al. 2001, 2003; Jablonowski et al. 2001b,c, 2004; Fichtner et al. 2002a,b, 2003; Fichtner and Schaffrath 2002; Mehlgarten and Schaffrath 2003).

Though there are several lines of evidence for an interference of zymocin with RNA polymerase II transcription (Jablonowski et al. 2001b; Jablonowski and Schaffrath 2002; Kitamoto et al. 2002), the effective zymocin γ-subunit was recently shown to display relevant similarities to bacterial nucleases; it specifically cleave three tRNA species ($tRNA^{Glu}_{UUC}$; $tRNA^{Lys}_{UUU}$ and $tRNA^{Gln}_{UUG}$) at the 3′ side of the wobble nucleoside (Lu et al. 2005; Jablonowski et al. 2006). The wobble nucleoside uridine, which is apparently recognized by zymocin, carries in all three target tRNAs a modification, i.e. a 5-methoxy-carbonyl-methyl group (mcm^5) being synthesized in an Elongator dependent manner (Huang et al. 2005). Another zymocin relevant tRNA modifying enzyme is Trm9/Kti1, which is a methyltransferase that probably acts in concert with Elongator to carry out wobble nucleoside modification (Kalhor and Clarke 2003; Jablonowski et al. 2006). While loss of Elongator and of Elongator interacting proteins Kti11-13 suppresses the synthesis of the entire mcm[5] group, *trm9/kti1* mutants only lack the methyl group of mcm[5] (Huang et al. 2005). Thus, resistance of Elongator, *KTI11-13* and *TRM9* deficient mutants is apparently due to inefficient recognition and cleavage of target tRNAs by zymocin due to non-(*elp1-6; kti11-13*) or incompletely (*trm9*) modified tRNA wobble uridine (Kalhor and Clarke 2003; Lu et al. 2005). In line with the assumption, that tRNA is the target of zymocin, is the finding, that overexpression of target $tRNA^{Glu}$ confers resistance to zymocin (Butler et al. 1994; Jablonowski et al. 2006). The latter is abolished when the methyltransferase Trm9 is simultaneously overexpressed, suggesting that overexpressed $tRNA^{Glu}$ confers resistance because it remains hypomethylated, and is thus not recognized by zymocin γ (Lu et al. 2005; Jablonowski et al. 2006).

Compared to the zymocin encoding pGKL1, the killer element pPin1-3 from *P. inositovora* displays a similar gene content, as it also contains four ORFs each with a homologous counterpart on pGKL1. However, similarity is rather weak for the γ-subunit encoding ORF4 (Klassen and Meinhardt 2003). Also indicating a more distant relation, genes are differently organized compared to pGKL1 (Fig. 5). Despite the presence of an ORF with homology to that encoding the immunity function in pGKL1, such immunity function is apparently not required for *P. inositovora*, as plasmid free derivatives do not become detectably sensitive to the toxin (Hayman and Bolen 1991). Nevertheless, it has been demonstrated that the *P. inositovora* toxin binds chitin, possesses a zymocin β-like region, functionally depends on Chs3 mediated chitin synthesis, and as a definitive sign of zymocin relatedness, on Elongator, too (Klassen and Meinhardt 2003). Thus, the *P. inositovora* toxin is functionally similar to zymocin, as is the gene load of the encoding linear plasmid.

4.1.2 Killer plasmids encoding type II toxins

The killer plasmids pPac1-2 and pWR1A from *P. acaciae* and *D. robertsiae* are closely related to each other, but their similarity to the type I toxin encoding elements pGKL1 and pPin1-3 is restricted to the ORF encoding α- and β-subunits of zymocin in pGKL1. Unlike pGKL1 and pPin1-3 they do not encode their own DNA polymerases and instead of zymocin γ, each of these elements encodes a structurally distinct intracellularly acting toxin subunit (Klassen et al. 2004).

Consistent with the absence of a zymocin γ-like function, both toxins do not require Elongator for function (Klassen et al. 2004), and thus tRNAs with Elongator-dependent mcm[5] modified wobble uridine are unlikely to be their targets, though the *trm9/kti1* mutation apparently also protects target cells from this toxin type to some degree (McCracken et al. 1994; Klassen and Meinhardt, unpublished), a finding, that may point to an additional cellular role for Trm9.

In support of a function distinct from type I (zymocin) type II toxins cause target cells to arrest in the S-phase of the cell cycle and they activate a genome surveillance mechanism, the intra-S-phase DNA damage checkpoint, which is consistent with a replication inhibitory and/or DNA damaging function (Klassen et al. 2004). The action of the *P. acaciae* toxin on the target cell was analysed in more detail, and found to induce cell death in a two-step fashion. During the first 3–4 h in toxin, target cells loose viability to approximately 30% of control levels, followed by a rather long period of constant viability until the final decline to approximately 1% occurs. The first phase is characterized by very early hyperphosphorylation of the DNA-damage checkpoint kinase Rad53 and mutation induction, suggesting that DNA-damage is involved in lethality at this stage (Klassen et al. 2004; Klassen and Meinhardt 2005). The final decline of viability in cells that survived initial DNA damage is an active process and is characterized by the appearance of apoptotic markers, such as abnormal nuclear morphology, generation of reactive oxygen species, DNA fragmentation and externalization of phosphatidylserine at the cellular surface (Klassen and Meinhardt 2005). Interestingly, the virus encoded killer toxins K1, K28 and zygocin, though significantly differing in their lethal strategies were in parallel found to activate a similar apoptotic program in target cells (Reiter et al. 2005), suggesting that apoptosis in killer toxin mediated cell death is not related to the primary mode of action of individual toxins.

4.1.3 Type I and type II toxin uptake machinery

All killer toxin encoding linear plasmids possess at least one ORF similar to the zymcoin α- and β precursor and consistently, all encoded killer toxins contain a subunit which binds to chitin in vivo, and thus requires Chs3 mediated chitin synthesis in the target cell (Jablonowski et al. 2001a; Klassen and Meinhardt 2002, 2003; Klassen et al. 2004). In addition, a hydrophobic region, similar to zymocin β is also present in each case and both, type I and type II toxins were shown to possess one effective (cell cycle arresting) subunit that acts intracellularly. Hence, there is a model emerging, in which type I and type II toxins have similar chitin binding machinery, responsible for the transport of cargo toxin subunits into target cells. Such cargo varies in type I and type II toxins functionally and structurally and, thus, the cargo transporter is assumed to be rather unspecific (Klassen et al. 2004).

The predicted proteins of pGKL1 ORF2 like genes have been compared and inspected for specific domains (Fig. 6). In addition to a cysteine rich chitin binding site, there is always a chitinase site present and consistently, both, zymocin and *P. acaciae* toxin were shown to possess chitinase activities (Butler et al. 1991b; McCracken et al. 1994). Additionally, we found one or two LysM domains being present in each case (Fig. 6).

The LysM domains are assumed to represent peptidoglycan binding sites originally identified in cell wall degrading proteins (lysins) encoded by bacteria and bacteriophages (Birkeland 1994; Pontig et al. 1999). Such sites are typically found in bacterial cell surface proteins, in which they anchor catalytic domains of cell wall degrading enzymes to their substrates by peptidoglycan binding (Bateman and Bycroft 2000).

Taking into account the similarity between the carbohydrate moiety of peptidoglycan (β-1,4-linked *N*-acetylmuramic acid and *N*-acetylglucosamine) and chitin (β-1,4 linked *N*-acetylglucosamine), it is tempting to speculate that for the chitin binding yeast killer toxins, LysM domains act in concert with the cysteine rich chitin binding domain to facilitate the observed tight (salt resistant) binding of α-β-like subunits to chitin in vitro (Jablonowski et al. 2001a; Klassen and Meinhardt 2002, 2003; Klassen et al. 2004).

The size of such chitin binding subunits in different toxins indicates that processing of the chitin binding and hydrophobic β-like regions is not essential for toxin function per se, as they are located on a joint polypeptide in toxins from *P. acaciae* and *P. inositovora* (Klassen and Meinhardt 2003; Klassen et al. 2004). Interestingly, *P. acaciae* and *D. robertsiae* plasmids pPac1-2 and pWR1A each contain two ORFs with similarity to ORF2 of pGKL1, with only one of which in each case possessing the chitin binding site (Fig. 6). In support of a functional role for the mentioned LysM domains in yeast killer toxins, such α-like non-chitin binding site carrying proteins encoded by pPac1-2 ORF3 and pWR1A ORF4 contain such LsyM domains. Hence, killer toxins from *P. acaciae* and *D. robertsiae* possibly may have an additional subunit that might support cell wall binding.

Subsequent uptake steps following cell wall binding may be similar among type I or type II toxins, as both carry a hydrophobic zymocin β-like region as well as an intracellularly acting subunit. Additionally, resistance of *S. cerevisiae kti10/pma1* mutants to both type I (*K. lactis*) and type II (*P. acaciae*) toxins (Butler et al. 1994; McCracken et al. 1994; Mehlgarten and Schaffrath 2004) suggests similar intracellular activation following uptake of the effective toxin subunit. An additional hint for similarities in early toxin action for both type I and II is based on a similar interference with mating

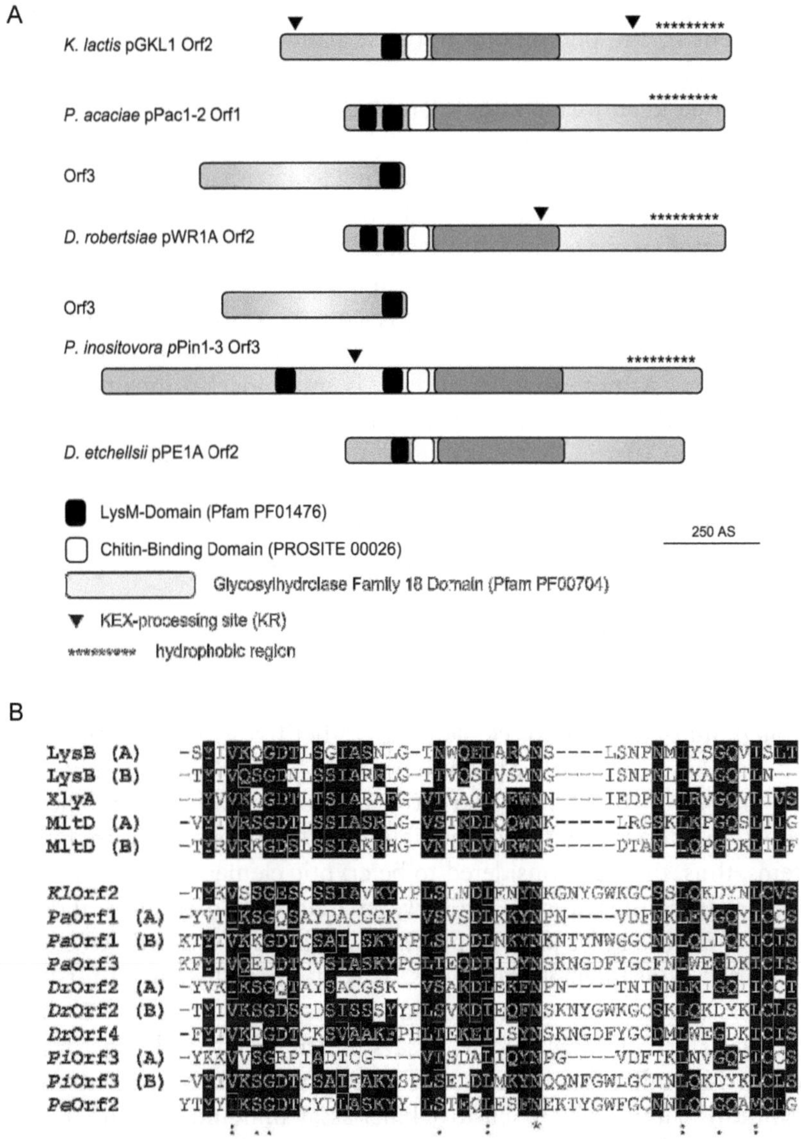

Fig. 6. Domain architecture of linear plasmid encoded proteins with similarity to the chitin-binding zymocin α-subunit. **A** Schematic representation of specific domains in α-like proteins. The position of LysM, chitin-binding and chitinase domains as well as hydrophobic C-terminal regions is indicated for each of the predicted proteins. **B** Alignment of LysM domains from zymocin-α like proteins and lysins LysB (lysin from phage phiLC3), XlyA (N-acetylmuramoyl-L-alanine amidase from *Bacillus subtilis* Prophage PBSX) and MltD (lytic murein transglycosylase D from *Escherichia coli*). Source files from NCBI: pPac1–2 Orf1 (CAE84958) and Orf3 (CAE84961), pWR1A Orf2 (CAE84954) and Orf4 (CAE84957), pPin1–3 Orf3 (CAD91890), pPE1A Orf2 (CAC34265), pGKL1 Orf2 (P09805), LysB (AAA20878), XlyA: P39800, MltD (NP_414747)

competence, and, as full toxicity was found to rely on the haploid status, involvement of the pheromone response pathway in toxicity is suggested (Klassen et al. 2006).

S. cerevisiae killer toxins with different lethal strategies encoded by dsRNA viruses are generally believed to kill target cells in a receptor mediated two step process, consisting of an initial binding to a primary receptor in the cell wall, followed by an energy dependent translocation to a secondary membrane receptor (reviewed in Schmitt and Breinig 2002). Though such a two-step process has not been investigated in the case of linear plasmid encoded killer toxins, it is likely that chitin represents the primary receptor, directly raising the question for the nature of a potential membrane receptor for these toxins. A potential candidate is mannosyl-diinositolphosphoceramide, synthesized by Kti6/Ipt1, which was found to be essential for uptake of type I toxin, but not required for cell wall binding (Zink et al. 2005). Interestingly, *kti6/ipt1* mutants resist type I toxin, but retain sensitivity to type II toxin (McCracken et al. 1994), suggesting that both toxins may interact with different (secondary) membrane receptors.

4.2 Cryptic elements

Although non-autonomous elements are rather frequent in cytoplasmic linear plasmid systems, a killer phenotype could be attributed to only a few of them (Table 1). To date, no alternative phenotype (other than killer toxin production and immunity) was attributed to the remaining mixed plasmid systems, thus, they are considered to be cryptic elements (Cong et al. 1994; Fukuhara 1995). It is noteworthy, however, that such an assignment does not a priori exclude a killer phenotype which might have escaped detection due to unsuited experimental conditions. This was recently shown for the two *Debaryomces robertsiae* linear plasmid systems (pWR1 and pWR2), initially regarded as cryptic due to plate assays which failed to detect toxin production. In both cases, killer toxin detection required concentration by ultrafiltration and a highly sensitive killer assay in liquid medium. Likewise, linear plasmids from *P. inositovora*, again initially regarded as cryptic (Ligon et al. 1990), are indeed killer elements (Hayman and Bolen 1991). Among the cryptic elements, only pPE1A is entirely sequenced (Klassen et al. 2002).

Only partial information is available for pDHL1 from *Debaryomces hansenii* (Fukuda et al. 1997), the smallest element of a triplet which is unique in its dependency on high osmolarity for stable replication (Gunge et al. 1993; Fukuda et al. 2004).

As for non-autonomous killer plasmids pWR1A and pPac1-2, cryptic pPE1A does not encode its own element specific DNA polymerase. Since the entire nucleotide sequence of the corresponding autonomous pPE1B is available (Klassen et al. 2001), only one cytoplasmically encoded DNA polymerase (encoded by pPE1B) apparently facilitates replication of both, the autonomous (pPE1B) and the non-autonomous element (pPE1A).

Interestingly, however, pPE1A carries an ORF encoding a functional zymocin α-like chitin binding protein, though a killer activity is not detectable (Cong et al. 1994; Klassen et al. 2002). Strikingly, there is evidence from available sequences of pDHL1 but also from heterologous hybridization experiments with linear plasmid pDH1A of another *D. hansenii* strain for ORFs encoding zymocin α-like proteins as well (Cong et al. 1994; Fukuda et al. 1997). Thus, expression of a functional zymocin α-like chitin binding protein may be a general feature of non-autonomous elements, and, most interestingly, such protein is not necessarily part of a killer system. The latter is especially apparent for pPE1A of *D. etchellsii*, in which genes encoding a potential intracellularly acting subunit as well as a hydrophobic region are lacking, but information for a chitin binding protein is provided (Klassen et al. 2002). It remains to be elucidated whether secreting such proteins is of any benefit for a host cell, which might contribute to understand the observed wide distribution of such genetic traits even among the cryptic elements. Provided that the principle of dependency on the autonomous elements is generally valid, one might speculate that an advantageous phenotype is likely to be associated with all of the small elements, thus justifying stable inheritance of so-called cryptic elements.

5 Outlook

Our current knowledge of the cytoplasmic replication and transcription machinery realized in yeast virus like elements is still fragmentary. Among the challenges for future research is the molecular understanding of replication-initiation, involving the processing of the TP-DNA polymerase fusion protein as well as cytoplasmic transcription initiation, termination and transcript modification. Furthermore, with ORF9p, there is at least one factor essential for cytoplasmic inheritance, which function remains enigmatic. Concerning killer toxins (type I and II), research will have to address early events in toxin action, such as binding to the cell wall receptor and the probable transport to a yet unknown membrane receptor. As also the current view of plasma-membrane passage and the intracellular journey to the final target is largely imaginary, investigation of these events will provide exciting new insights. Additionally, the target of the type II toxin and the mode of action of the respective immunity factors for both, type I and II toxins have to be elucidated.

Understanding the molecular function of virus like elements, which are closely related to harmful cytoplasmic viruses, such as vaccinia, ASFV and molluscum contagiosum virus offers the possibility to study functions conserved among the yeast virus like elements and the pox viruses, in an organism, that is, due to its safety and applicability of genetic methodologies, an attractive and promising experimental system. Moreover, killer toxins may find applications in cell biology and cell cycle research, once a precise picture of their mode of action has emerged.

Acknowledgements

We thank J.P. Paluszynski for critically reading the manuscript. Experimental work of the authors is supported by the Deutsche Forschungsgemeinschaft (DFG) Bonn, Bad Godesberg grants Me 1142/3-1 and Me 1142/5-1.

References

Bateman A, Bycroft M (2000) The structure of a LysM domain from *E. coli* membrane-bound lytic murein transglycosylase D (MltD). J Mol Biol 299:1113–1119

Banerjee H, Kopvak C, Curley D (1998) Identification of linear DNA plasmids of the yeast *Pichia pastoris*. Plasmid 40:58–60

Birkeland NK (1994) Cloning, molecular characterization, and expression of the genes encoding the lytic functions of lactococcal bacteriophage phi LC3: a dual lysis system of modular design. Can J Microbiol 40:658–665

Bisaillon M, Lemay G (1997) Viral and cellular enzymes involved in synthesis of mRNA cap structure. Virology 236:1–7

Blaisonneau J, Nosek J, Fukuhara H (1999) Linear DNA plasmid pPK2 of *Pichia kluyveri*: distinction between cytoplasmic and mitochondrial linear plasmids in yeasts. Yeast 15:781–791

Blanco L, Bernad A, Lazaro JM, Martin G, Garmendia C, Salas M (1989) Highly efficient DNA synthesis by the phage phi 29 DNA polymerase. Symmetrical mode of DNA replication. J Biol Chem 264:8935–8940

Bolen PL, Kurtzman CP, Ligon JM, Mannarelli BM, Bothast RJ (1992) Physical and genetic characterization of linear DNA plasmids from the heterothallic yeast *Saccharomycopsis crataegensis*. Antonie Van Leuwenhoek 61:195–295

Bolen PL, Eastman EM, Cihak PL, Hayman GT (1994) Isolation and sequence analysis of a gene from the linear DNA plasmid pPacl-2 of *Pichia acaciae* that shows similarity to a killer toxin gene of *Kluyveromyces lactis*. Yeast 10:403–414

Braithwaite DK, Ito J (1993) Compilation, alignment, and phylogenetic relationships of DNA polymerases. Nucleic Acids Res 21:787–802

Broyles SS, Moss B (1987) Identification of the vaccinia virus gene encoding nucleoside triphosphate phosphohydrolase I, a DNA-dependent ATPase. J Virol 61:1738–1742

Butler AR, O'Donnell RW, Martin VJ, Gooday GW, Stark MJ (1991a) *Kluyveromyces lactis* toxin has an essential chitinase activity. Eur J Biochem 199:483–488

Butler AR, Porter M, Stark MJR (1991b) Intracellular expression of *Kluyveromyces lactis* toxin γ subunit mimics treatment with exogenous toxin and distinguishes two classes of toxin-resistant mutant. Yeast 7:617–625

Butler AR, White JH, Stark MJR (1991c) Analysis of the response of *Saccharomyces cerevisiae* cells to *Kluyveromyces lactis* toxin. J Gen Microbiol 137:1749–1757

Butler AR, White JH, Folawiyo Y, Edlin A, Gardiner D, Stark MJR (1994) Two *Saccharomyces cerevisiae* genes which control sensitivity to G1 arrest induced by *Kluyveromyces lactis* toxin. Mol Cell Biol 14:6306–6316

Chen WB, Han JF, Jong SC, Chang SC (2000) Isolation, purification, and characterization of a killer protein from *Schwanniomyces occidentalis.* Appl Environ Microbiol 66:5348–5352

Chou T (2003) Ribosome recycling, diffusion, and mRNA loop formation in translational regulation. Biophysical J 85:755–773

Cong P, Shuman S (1992) Methyltransferase and subunit association domains of Vaccinia virus mRNA capping enzyme. J Biol Chem 267:16424–16429

Cong P, Shuman S (1993) Covalent catalysis in nucleotidyl transfer: a KTDG motif essential for enzyme–GMP complex formation by mRNA capping enzyme is conserved at the active sites of RNA and DNA ligases. J Biol Chem 268:7256–7260

Cong YS, Yarrow D, Li YY, Fukuhara H (1994) Linear DNA plasmids from *Pichia etchellsii, Debaryomyces hansenii* and *Wingea robertsiae.* Microbiology 140:1327–1335

De la Cruz J, Kressler D, Linder P (1999) Unwinding RNA in *Saccharomyces cerevisiae*: DEAD-box proteins and related families. Trends Biochem Sci 24:192–198

Deng L, Shuman S (1998) Vaccinia NPH-I, a DExH-box ATPase, is the energy coupling factor for mRNA transcription termination. Genes Dev 12:538–546

Dufour E, Mendez J, Lazaro JM, de Vega M, Blanco L, Salas M (2000) An aspartic acid residue in TPR-1, a specific region of protein-priming DNA polymerases, is required for the functional interaction with primer terminal protein. J Mol Biol 304:289–300

Dufour E, Rodriguez I, Lazaro JM, de Vega M, Salas M (2003) A conserved insertion in protein-primed DNA polymerases is involved in primer terminus stabilisation. J Mol Biol 331:781–794

Fichtner L, Schaffrath R (2002) *KTI11* and *KTI13, Saccharomyces cerevisiae* genes controlling sensitivity to G1 arrest induced by *Kluyveromyces lactis* zymocin. Mol Microbiol 44:865–875

Fichtner L, Frohloff F, Burkner K, Larsen M, Breunig KD, Schaffrath R (2002a) Molecular analysis of *KTI12/TOT4,* a *Saccharomyces cerevisiae* gene required for *Kluyveromyces lactis* zymocin action. Mol Microbiol 43:783–791

Fichtner L, Frohloff F, Jablonowski D, Stark MJ, Schaffrath R (2002b) Protein interactions within *Saccharomyces cerevisiae* Elongator, a complex essential for *Kluyveromyces lactis* zymocicity. Mol Microbiol 45:817–826

Fichtner L, Jablonowski D, Schierhorn A, Kitamoto HK, Stark MJ, Schaffrath R (2003). Elongator's toxin-target (TOT) function is nuclear localization sequence dependent and suppressed by post-translational modification. Mol Microbiol 49:1297–1307

Frohloff F, Fichtner L, Jablonowski D, Breuning KD, Schaffrath R (2001) *Saccharomyces cerevisiae* elongator mutations confer resistance to the *Kluyveromyces lactis* zymocin. EMBO J 20:1993–2003

Frohloff F, Jablonowski D, Fichtner L, Schaffrath R (2003) Subunit communications crucial for the functional integrity of the yeast RNA polymerase II elongator (γ-toxin target (TOT)) complex. J Biol Chem 278:956–961

Fukuda K, Maebuchi M, Takata H, Gunge N (1997) The linear plasmid pDHL1 from *Debaryomyces hansenii* encodes a protein highly homologous to the pGKL1-plasmid DNA polymerase. Yeast 13:613–620

Fukuda K, Jin-Shan C, Kawano M, Sudo K, Gunge N (2004) Stress responses of linear plasmids from *Debaryomyces hansenii.* FEMS Microbiol Lett 237:243–248

Fukuhara H (1987) The RF1 gene of the killer DNA of yeast may encode a DNA polymerase. Nucleic Acids Res 15:10046

Fukuhara H (1995) Linear DNA plasmids of yeasts. FEMS Microbiol Lett 131:1–9

Gorbalenya, AE, Koonin EV (1993) Helicases: amino acid sequence comparisons and structure-function relationships. Curr Opin Struct Biol 3:419–429

Gunge N (1995) Plasmid DNA and the killer phenomenon in *Kluyveromyces.* In: Kück, U (ed) Genetics and biotechnology (The mycota, vol 2). Springer-Verlag, Berlin, pp 189–209

Gunge N, Tokunaga M (2004) Linear DNA plasmids and killer system in *Kluyveromyces lactis.* In: Kück U (ed) Genetics and biotechnology (The mycota II, 2nd edn). Springer-Verlag, Berlin, pp 199–217

Gunge N, Tamaru A, Ozawa F, Sakaguchi K (1981) Isolation and characterization of linear deoxyribonucleic acid plasmids from *Kluyveromyces lactis* and the plasmid-associated killer character. J Bacteriol 145:382–390

Gunge N, Fukuda K, Morikawa S, Murakami K, Takeda M, Miwa A (1993) Osmophilic linear plasmids from the salt-tolerant yeast *Debaryomyces hansenii.* Curr Genet 23:443–9

Gunge N, Takata H, Matsuura A, Fukuda K (2003) Progressive Rearrangement of telomeric sequences added to both the ITR ends of the yeast linear pGKL plasmid. Biol Proceed Online 5:29–42

Hayman GT, Bolen BL (1991) Linear DNA plasmids of *Pichia inositovora* are associated with a novel killer toxin activity. Curr Genet 19:389–393

Higman HA, Bourgeois N, Niles EG (1992) The Vaccinia virus mRNA (guanine-N7-)-methyltransferase requires both subunits of the mRNA capping enzyme for activity. J Biol Chem 267:16430–16437

Higman MA, Christen LA, Niles EG (1994) The mRNA (guanine-7)methyltrans ferase domain of the Vaccinia virus mRNA capping enzyme: expression in *Escherichia coli* and structural and kinetic comparison to the intact capping enzyme. J Biol Chem 269:14974–14981

Hishinuma F, Hirai K (1991) Genome organization of the linear plasmid, pSKL, isolated from *Saccharomyces kluyveri.* Mol Gen Genet 226:97–106

Hishinuma F, Nakamura K, Hirai K, Nishizawa R, Gunge N, Maeda T (1984) Cloning and nucleotide sequence of the DNA killer plasmids from yeast. Nucleic Acids Res 12:l7581–7597

Huang B, Johansson MJ, Bystrom AS (2005) An early step in wobble uridine tRNA modification requires the Elongator complex. RNA 11:424–436

Jablonowski D, Schaffrath R (2002) *Saccharomyces cerevisiae* RNA polymerase II is affected by *Kluyveromyces lactis* zymocin. J Biol Chem 277:26276–26280

Jablonowski D, Fichtner L, Martin VJ, Klassen R, Meinhardt F, Stark MJR, Schaffrath R (2001a) *Saccharomyces cerevisiae* cell wall chitin, the potential *Kluyveromyces lactis* zymocin receptor. Yeast 18:1285–1299

Jablonowski D, Frohloff F, Fichtner L, Stark MJR, Schaffrath R (2001b) *Kluyveromyces lactis* zymocin mode of action is linked to RNA polymerase II function via elongator. Mol Microbiol 42:1095–1105

Jablonowski D, Butler AR, Fichtner L, Gardiner D, Schaffrath R, Stark MJR (2001c) Sit4p protein phosphatase is required for sensitivity of *Saccharomyces cerevisiae* to *Kluyveromyces lactis* zymocin. Genetics 159:1479–1489

Jablonowski D, Fichtner L, Stark MJ, Schaffrath R (2004) The yeast elongator histone acetylase requires Sit4-dependent dephosphorylation for toxin-target capacity. Mol Biol Cell 15(3):1459–1469

Jablonowski, D, Zink, S, Mehlgarten, C, Daum, G, Schaffrath, R (2006) $tRNA^{Glu}$ wobble uridine methylation by Trm9 identifies Elongator's key role for zymocin-induced cell death in yeast. Mol Microbiol 59:677–688
Jankowsky E, Jankowsky A (2000) The DexH/D protein family database. Nucleic Acids Res 28:333–334
Jeske S, Meinhardt F (2006) Autonomous cytoplasmic linear plasmid pPac1–1 of *Pichia acaciae*: molecular structure and expression studies. Yeast (in press)
Jeske S, Tiggemann M, Meinhardt F (2006) Yeast autonomous linear plasmid pGKL2: ORF9 is an actively transcribed essential gene with multiple transcription start points. FEMS Microbiol Lett 255:321–327
Jung GH, Leavitt MC, Ito J (1987) Yeast killer plasmid pGKL1 encodes a DNA polymerase belonging to the family B DNA polymerases. Nucleic Acids Res 15:9088
Kadare G, Haenni AL (1997) Virus encoded RNA helicases. J Virol 71:2583–2590
Kalhor HR, Clarke S (2003) Novel methyltransferase for modified uridine residues at the wobble position of tRNA. Mol Cell Biol 23:9283–92
Kämper J, Meinhardt F, Gunge N, Esser K (1989a) New recombinant linear DNA-elements derived from *Kluyveromyces lactis* killer plasmids. Nucleic Acids Res 17:1781
Kämper J, Meinhardt F, Gunge N, Esser K (1989b) In vivo construction of linear vectors based on killer plasmids from *Kluyveromyces lactis*: selection of a nuclear gene results in attachment of telomeres. Mol Cell Biol 9:3931–3937
Kämper J, Esser K, Gunge N, Meinhardt F (1991) Heterologous gene expression on the linear DNA killer plasmid from *Kluyveromyces lactis*. Curr Genet 19:109–118
Kamtekar S, Berman AJ, Wang J, Lázaro JM, de Vega M, Blanco L, Salas M, Steitz TA (2004) Insights into strand displacement and processivity from the crystal structure of the protein-primed DNA polymerase of bacteriophage phi29. Mol Cell 16:609–618
Kapp LD, Lorsch JR (2004) The molecular mechanics of eukaryotic translation. Annu Rev Biochem 73:657–704
Kempken F, Hermanns J, Osiewacz HD (1992) Evolution of linear plasmids. J Mol Evol 35:502–513
Kikuchi Y, Hirai K, Hishinuma F (1984) The yeast linear DNA killer plasmids, pGKL1 and pGKL2, possess terminally attached proteins. Nucleic Acids Res 12:5685–5692
Kikuchi Y, Hirai K, Gunge N, Hishinuma F (1985) Hairpin plasmid—a novel linear DNA of perfect hairpin structure. EMBO J 4:1881–1886
Kim EK, Jeong JH, Youn HS, Koo YB, Roe JH (2000) The terminal protein of a linear mitochondrial plasmid is encoded in the N-terminus of the DNA polymerase gene in white-rot fungus *Pleurotus ostreatus*. Curr Genet 38:283–290
Kitada K, Gunge N (1988) Palindrome-hairpin linear plasmids possessing only a part of the ORF1 gene of the yeast killer plasmid pGKL1. Mol Gen Genet 215:46–52
Kitada K, Hishinuma F (1987) A new linear DNA plasmid isolated from the yeast *Saccharomyces kluyveri*. Mol Gen Genet 2006:377–381
Kitamoto HK, Jablonowski D, Nagase J, Schaffrath R (2002) Defects in yeast RNA polymerase II transcription elicit hypersensitivity to G1 arrest induced by *Kluyveromyces lactis* zymocin. Mol Genet Genom 268:49–55
Klassen R, Meinhardt F (2002) Linear plasmids pWR1A and pWR1B of the yeast *Wingea robertsiae* are associated with a killer phenotype. Plasmid 48:142–148
Klassen R, Meinhardt F (2003) Structural and functional analysis of the killer element pPin1–3 from *Pichia inositovora*. Mol Genet Genomics 270:190–199
Klassen R, Meinhardt F (2005) Induction of DNA damage and apoptosis in *Saccharomyces cerevisiae* by a yeast killer toxin. Cell Microbiol 7:393–401
Klassen R, Tontsidou L, Larsen M, Meinhardt F (2001) Genome organization of the linear cytoplasmic element pPE1B from *Pichia etchellsii*. Yeast 18:953–961
Klassen R, Jablonowski D, Schaffrath R, Meinhardt F (2002) Genome organization of the linear *Pichia etchellsii* plasmid pPE1A: evidence for expression of an extracellular chitin

binding protein homologous to the α-subunit of the *Kluyveromyces lactis* killer toxin. Plasmid 47:224–233

Klassen R, Teichert S, Meinhardt F (2004) Novel yeast killer toxins provoke S-phase arrest and DNA damage checkpoint activation. Mol Microbiol 53: 263–273

Klassen R, Jablonowski D, Stark MJR, Schaffrath R, Meinhardt F (2006) Mating type locus control of killer toxins from *Kluyveromyces lactis* and *Pichia acaciae.* FEMS Yeast Res DOI: 10.1111/j.1567–1364.2005.00006.x

Koonin EV (1991) Similarities in RNA helicases. Nature 352:290

Larsen M, Meinhardt F (2000) *Kluyveromyces lactis* killer system: identification of a new gene encoded by pGKL2. Curr Genet 38:271–275

Larsen M, Gunge N, Meinhardt F (1998) *Kluyveromyces lactis* killer plasmid pGKL2: Evidence for a viral-like capping enzyme encoded by OFR3. Plasmid 40:243–246

Leffers H, Gropp F, Lottspeich F, Zillig W, Garrett RA (1989) Sequence, organisation, transcription and evolution of RNA polymerase subunit genes from the archaebacterial extreme halophiles *Halobacterium halobium* and *Halococcus morrhuae.* J Mol Biol 206:1–17

Ligon JM, Bolen PL, Hill DS, Bothast RJ, Kurtzman CP (1990) Physical and biological characterization of linear DNA plasmids of the yeast *Pichia inositovora.* Plasmid 2:185–194

Lu J, Huang B, Esberg A, Johansson MJ, Bystrom AS (2005) The *Kluyveromyces lactis* γ-toxin targets tRNA anticodons. RNA 11:1648–1654

Mao X, Shuman S (1994) Intrinsic RNA (guanine-7)methyltransferase activity of the Vaccinia capping enzyme D1 subunit is stimulated by the D12 subunit: identification of amino acid residues in the D1 protein required for subunit association and methyl group transfer. J Biol Chem 269:24472–24479

Martin SA, Paoletti E, Moss B (1975) Purification of mRNA guanylyltransferase and mRNA (guanine-7)methyltransferase from Vaccinia virions. J Biol Chem 250:9322–9329

McCracken DA, Martin VJ, Stark MJ, Bolen PL (1994) The linear-plasmid-encoded toxin produced by the yeast *Pichia acaciae*: characterization and comparison with the toxin of *Kluyveromyces lactis.* Microbiology 140:425–431

McNeel DG, Tamanoi F (1991) Terminal region recognition factor 1, a DNA-binding protein recognizing the inverted terminal repeats of the pGKl linear DNA plasmids. Proc Natl Acad Sci USA 88:11398–11402

Mehlgarten C, Schaffrath R (2003) Mutant casein kinase I (Hrr25p/Kti14p) abrogates the G1 cell cycle arrest induced by *Kluyveromyces lactis* zymocin in budding yeast. Mol Genet Genomics 269:188–196

Mehlgarten C, Schaffrath R (2004) After chitin docking, toxicity of *Kluyveromyces lactis* zymocin requires *Saccharomyces cerevisiae* plasma membrane H+-ATPase. Cell Microbiol 6:569–180

Meinhardt F, Rohe R (1993) Extranuclear inheritance: linear protein-primed replicating genomes in plants and microorganisms. In: Esser K, Lüttge U, Kadereit JW, Beyschlag W (eds) Progress in botany, vol 54. Springer-Verlag, Berlin, Heidelberg, New York, pp 51–70

Meinhardt F, Schaffrath R (2001) Extranuclear inheritance: cytoplasmic linear double-stranded DNA killer elements of the dairy yeast *Kluyveromyces lactis.* In: Esser K, Lüttge U, Kadereit JW, Beyschlag W (eds) Progress in botany, vol 62. Springer-Verlag, Berlin, Heidelberg, New York, pp 51–70

Meinhardt F, Kempken F, Kamper J, Esser K (1990) Linear plasmids among eukaryotes: fundamentals and application. Curr Genet 17:89–95

Niles EG, Christen L (1993) Identification of the Vaccinia virus mRNA guanyltransferase active site lysine. J Biol Chem 268:2986–24989

Niles EG, Condit R, Caro P, Davidson K, Matusick L, Seto J (1986) Nucleotide sequence and genetic map of the 16 kb Vaccinia virus *Hin*dIII D fragment. Virology 153:96–112

Niles EG, Lee-Chen S, Shuman S, Moss B, Broyles S (1989) Vaccinia virus gene D12L encodes the small subunit of the viral mRNA capping enzyme. Virology 172:513–522

Niwa O, Sakaguchi K, Gunge N (1981) Curing of the killer deoxyribonucleic acid plasmids of *Kluyveromyces lactis.* J Bacteriol 148:988–990

Otero G, Fellows J, Li Y, de Bizemont T, Dirac AM, Gustafsson CM, Erdjument-Bromage H, Tempst P Svejstrup JQ (1999) Elongator, a multisubunit component of a novel RNA polymerase II holoenzyme for transcriptional elongation. Mol Cell 3:109–118

Paoletti, E, Rosemond-Hornbeak H, Moss B (1974) Two nucleic acid-dependent nucleoside triphosphate phosphohydrolases from vaccinia virus: purification and characterization. J Biol Chem 249:3273–3280

Ponting CP, Aravind L, Schultz J, Bork P, Koonin EV (1999) Eukaryotic signalling domain homologues in archaea and bacteria. Ancient ancestry and horizontal gene transfer. J Mol Biol 289:729–745

Pring DR, Levings CS III, Hu WWL, Timothy DH (1977) Unique DNA associated with mitochondria in the "S"-type cytoplasm of male-sterile maize. Proc Natl Acad Sci USA 74:2904–2908

Rahl PB, Chen CZ, Collins RN (2005) Elp1p, the yeast homolog of the FD disease syndrome protein, negatively regulates exocytosis independently of transcriptional elongation. Mol Cell 17:841–853

Reiter J, Herker E, Madeo F, Schmitt MJ (2005) Viral killer toxins induce caspase-mediated apoptosis in yeast. J Cell Biol 168: 353–358

Rodriguez JF, Kahn JS, Esteban M (1986) Molecular cloning, encoding sequence, and expression of vaccinia nucleic acid-dependent nucleoside triphosphatase gene. Proc Natl Acad Sci USA 83:9566–9570

Rodriguez I, Lazaro JM, Blanco L, Kamtekar S, Berman AJ, Wang J, Steitz TA, Salas M, de Vega M (2005) A specific subdomain in phi29 DNA polymerase confers both processivity and strand-displacement capacity. Proc Natl Acad Sci USA 102:6407–6412

Rohe M, Schrage K, Meinhardt F (1991) The linear plasmid pMC3–2 from Morchella conica is structurally related to adenoviruses. Curr Genet 20:527–533

Romanos M, Boyd A (1988) A transcriptional barrier to expression of cloned toxin genes of the linear plasmid k1 of *Kluyveromyces lactis*: evidence that native k1 has novel promoters. Nucleic Acids Res 16:7333–7350

Schaffrath R, Meacock PA (1996) A cytoplasmic gene-shuffle system in *Kluyveromyces lactis*: use of epitope tagging to detect a killer plasmid-encoded gene product. Mol Microbiol 19:545–554

Schaffrath R, Meacock PA (2001) An SSB encoded by and operating on linear killer plasmids from *Kluyveromyces lactis.* Yeast 18:1239–1247

Schaffrath R, Meinhardt F (2004) *Kluyveromyces lactis* zymocin and other plasmid-encoded yeast killer toxins. In: Schmitt M, Schaffrath R (eds) Topics in current genetics vol 11. Microbial protein toxins, pp 133–155

Schaffrath R, Soond SM, Meacock PA (1995) The DNA and RNA polymerase genes of yeast plasmid pGKL2 are essential loci for plasmid integrity and maintenance. Microbiology 141:2591–2599

Schaffrath R, Meinhardt F, Meacock PA (1996) Yeast killer plasmid pGKL2: molecular analysis of UCS5, a cytoplasmic promoter element essential for ORF5 gene function. Mol Gen Genet 250:286–294

Schaffrath R, Meinhardt F, Meacock PA (1997) ORF7 of yeast plasmid pGKL2: analysis of gene expression in vivo. Curr Genet 31:190–192

Schaffrath R, Sasnauskas K, Meacock PA (2000) Use of gene shuffles to study the cytoplasmic transcription system operating on *Kluyveromyces lactis* linear DNA plasmids. Enzyme Microb Technol 26:664–670

Schickel J, Helmig C, Meinhardt F (1996) *Kluyveromyces lactis* killer system. Analysis of cytoplasmic promoters of linear plasmids. Nucleic Acids Res 24:1879–1886

Schmid SR, Linder P (1992) D-E-A-D protein family of putative RNA helicases. Mol Microbiol 6(3):283–291

Schmitt MJ, Breinig F (2002) The viral killer system in yeast: from molecular biology to application. FEMS Microbiol Rev 26:257–276

Schründer J, Meinhardt F (1995) An extranuclear expression system for analysis of cytoplasmic promoters of yeast linear killer plasmids. Plasmid 33:139–151

Shepherd HS (1992) Linear, non-mitochondrial plasmids of *Alternaria alternata*. Curr Genet 21:169–172

Shuman S (1989) Functional domains of Vaccinia virus mRNA capping enzyme: analysis by limited tryptic digestion. J Biol Chem 264:9690–9695

Shuman S (1995) Capping enzyme in eukaryotic mRNA synthesis. Prog Nucleic Acids Res Mol Biol 50:101–129

Shuman S, Hurwitz J (1981) Mechanism of mRNA capping by Vaccinia virus guanylyltransferase: characterization of an enzyme-guanylate intermediate. Proc Natl Acad Sci USA 78:187–191

Shuman S, Morham SG (1990) Domain structure of Vaccinia virus mRNA capping enzyme: activity of the Mr 95,000 subunit expressed in *Escherichia coli*. J Biol Chem 265:11967–11972

Shuman S, Schwer B (1995) RNA capping enzyme and DNA ligase: a superfamily of covalent nucleotidyltransferases. Mol Microbiol 17:405–410

Shuman S, Surks M, Furneaux H, Hurwitz J (1980) Purification and characterisation of a GTP-pyrophosphate exchange activity from Vaccinia virions: association of the GTP pyrophosphate exchange activity with Vaccinia mRNA guanylyltransferase-RNA (guanine-7-) methyltrans ferase complex (capping enzyme). J Biol Chem 255:11588–11598

Sor F, Fukuhara H (1985) Structure of a linear plasmid of the yeast *Kluyveromyces lactis*: compact organization of the killer genome. Curr Genet 9:147–155

Stam J C, Kwakman J, Meijer M, Stuitje AR (1986) Efficient isolation of the linear DNA killer plasmids of *Kluyveromyces lactis*: evidence for location and expression in the cytoplasm and characterization of their terminally bound proteins. Nucleic Acids Res 14, 6871–6884

Stark MJR, Boyd A (1986) The killer toxin of *Kluyveromyces lactis:* characterization of the toxin subunits and identification of the genes which encode them. EMBO J 5:1995–2002

Stark MJ, Mileham AJ, Romanos MA, Boyd A (1984) Nucleotide sequence and transcription analysis of a linear DNA plasmid associated with the killer character of the yeast *Kluyveromyces lactis*. Nucleic Acids Res 12:6011–6030

Stark MJR, Boyd A, Mileham AJ, Romanos MA (1990) The plasmid encoded killer system of *Kluyveromyces lactis:* a review. Yeast 6:1–29

Takeda M, Hiraishi H, Takesako T, Tanase S, Gunge N (1996) The terminal protein of the linear DNA plasmid pGKL2 shares an N-terminal domain with the plasmid encoded DNA polymerase. Yeast 12:241–246

Takita MA, Castilho-Valavicius B (1993) Absence of cell wall chitin in *Saccharomyces cerevisiae* leads to resistance to *Kluyveromyces lactis* killer toxin. Yeast 9:589–598

Thuriaux P, Sentenac A (1992) Yeast nuclear RNA polymerases. In: Jones EW, Pringle JR, Broach JR (eds) The molecular and cellular biology of the yeast *Saccharomyces cerevisiae*. Cold Spring Habor Press, New York, pp 1–48

Tiggemann M, Jeske S, Larsen M, Meinhardt F (2001) *Kluyveromyces lactis* cytoplasmic plasmid pGKL2: Heterologous expression of Orf3p and prove of guanylyltransferase and mRNA-triphosphatase activities. Yeast 18:815–825

Tokunaga M, Wada N, Hishinuma F (1987) Expression and identification of immunity determinants on linear DNA killer plasmids pGKL1 and pGKL2 in *Kluyveromyces lactis*. Nucleic Acids Res 15:1031–1046

Tokunaga M, Kawamura A, Hishinuma F (1989) Expression of pGKL killer 28K subunit in *Saccharomyces cerevisiae*: identification of 28K subunit as a killer protein. Nucleic Acids Res 17:3435–3446

Tokunaga M, Kawamura A, Kitada K, Hishinuma F (1990) Secretion of killer toxin encoded on the linear DNA plasmid pGKL1 from *Saccharomyces cerevisiae.* J Biol Chem 265:17274–17280

Tommasino M (1991) Killer system of *Kluyveromyces lactis*: the open reading frame 10 of the pGK12 plasmid encodes a putative DNA binding protein. Yeast 7:245–252

Tommasino M, Ricci S, Galeotti C (1988) Genome organization of the killer plasmid pGKL2 from *Kluyveromyces lactis.* Nucleic Acids Res 16:5863–5978

Turmel M, Bellemare G, Lee RW, Lemieux C (1986) A linear DNA molecule of 5z9 kilobase-pairs is highly homologous to the chloroplast DNA in the green algae *Chlamydomonas moewusii.* Plant Mol Biol 6:313–319

Walker JE, Saraste M, Runswick MJ, Gay NJ (1982) Distantly related sequences in the α- and β-subunits of ATP synthetase, myosin, kinases and other ATP-requiring enzymes and a common nucleotide binding fold. EMBO J 1:945–951

Wilson DW, Meacock PA (1988) Extranuclear gene expression in yeast: evidence for a plasmid encoded RNA-polymerase of unique structure. Nucleic Acids Res 16:8097–8112

Winkler GS, Petrakis TG, Ethelberg S, Tokunaga M, Erdjument-Bromage H, Tempst P, Svejstrup JQ (2001) RNA polymerase II elongator holoenzyme is composed of two discrete subcomplexes. J Biol Chem 276:32743–32749

Wittschieben BO, Otero G, de Bizemont T, Fellows J, Erdjument-Bromage H, Ohba R, Li Y, Allis CD, Tempst P, Svejstrup JQ (1999) A novel histone acetyltransferase is an integral subunit of elongating RNA polymerase II holoenzyme. Mol Cell 4:123–128

Worsham PL, Bolen PL (1990) Killer toxin production in *Pichia acaciae* is associated with linear DNA plasmids. Curr Genet 18:77–80

Zink S, Mehlgarten C, Kitamoto HK, Nagase J, Jablonowski D, Dickson RC, Stark MJR Schaffrath R (2005) Mannosyl-diinositolphospho-ceramide, the major yeast plasma membrane sphingolipid, governs toxicity of *Kluyveromyces lactis* zymocin. Eukaryot Cell 4:879–889

Stefanie Jeske,
Friedhelm Meinhardt,
Roland Klassen
Institut für Molekulare Mikrobiologie
und Biotechnologie
Westfälische Wilhelms-Universität Münster
Correnssstr. 3
D-48149 Münster
Germany
Tel.: 049 2 51 / 83-3 35 59
Fax: 049 2 51 / 83-3 83 88

Population Genetics: Evolutionary Features of Asexual Species

Jürgen Tomiuk and Wolfgang Köhler

"*A species consists of a population or group of populations that shares a common evolutionary fate through time. This definition has the advantage of being applicable to both living and extinct groups and to sexual and asexual organisms. Moreover, a species unit can be held together not only through gene flow but also through developmental, genetic, and ecological constraints*" (The Evolutionary Species Concept; Templeton 1989, p. 4).

1 Introduction

The correct definition of a species is fundamental for our understanding and interpretation of results obtained from population biological studies (Mayr 1988). However, the term *species* is a human figment whose definition also depends on the genetic transmission of the particular case (Hoffmann and Esser 1978, Mayr 1988, Templeton 1989). In this chapter we adopt the evolutionary species concept (Templeton 1989), in which species with diverse reproduction modes are considered, e.g. sexual species and their derived asexual lineages.

Evolutionary processes that can affect the genetic structure of populations and may determine the fate of asexual lineages are intensively discussed. For example, assuming that the reproductive success of competing dioecious sexuals and asexuals is identical, theory predicts selection against sexual reproduction because of a 2-fold reproductive advantage of the asexuals (a frequently used argument in animals; Maynard Smith 1971). In nature, however, just the opposite is observed, and sexual reproduction is favoured. Which forces counterbalance the reproductive advantage of asexuals? In this respect, populations of asexual lineages are sometimes considered dead ends of evolution because of freezing their genotypes. Clones can only response genetically to new environmental conditions through rarely occurring mutational events that provide the basis for adaptation processes. However, selection might operate in favour for optimal asexual genotypes that can be transmitted stably from one to the next generation, whereas in sexuals gene

Progress in Botany, Vol. 68

combinations assigning high fitness to their carrier are broken up in meiosis and cannot be conserved over generations. The often used arguments in favour for sexual reproduction are higher adaptability in a changing environment through permanent reorganization of gene combinations (Jaenike 1978, Hamilton 1982), an effective spread of new favourable mutations (Halkett et al. 2005a), and DNA repair mechanisms during meiosis (DNA repair hypothesis; see, e.g. Michod et al. 1988). Indeed, sexual organisms might have an advantage through their higher adaptability in a world of ever-changing biotic interactions (Red Queen Hypothesis; van Valen 1973). Finally, we can also consider a fitness concept that is based on the number of genes transferred by an individual to its offspring. Under this aspect, most asexuals have a success rate of 100%, but sexuals are limited to an upper threshold of 50% (Williams 1975).

Here, we first introduce some main topics and peculiarities of asexual reproduction and subsequently we focus on models analysing the dynamics of genetic diversity. Finally, population genetic analyses of asexuals and problems arising with the study of asexual polyploids are discussed. Some of the theoretical population genetic approaches can be adopted from diploid sexuals, but most of them have limits when used for polyploids or complexes comprising sexuals and asexual lineages with different ploidy levels.

2 Modes and origin of asexuality

Asexuality without the formation of gametes, i.e. cell duplication, is certainly the evolutionarily oldest reproduction mode. Sexual reproduction, meiosis and fertilization, first evolved in eukaryotes (e.g. see Stearns 1987). Later on, gametic asexuality as a new form of asexual reproduction has occurred in sexually reproducing populations (Suomalainen et al. 1987). Gametic asexuality considers individual development from a single ovum (Owen 1849; see Suomalainen et al. 1987, p. 1). In plants, seed formation without fertilization is called agamospermy, and in animals parthenogenesis describes the development of a zygote without a fusion of egg and spermatozoon (Suomalainen et al. 1987). The distinction between agametic and gametic asexuality eliminates false comparisons between "gametic" and vegetative progeny in which types of reproductions, ecologically distinct roles have evolved between progeny that grow close to the mother plant (ramets or shoots) and progeny produced for wider spatial or temporal dispersal, the gametic asexually produced seeds. In most asexually reproducing plants, seeds, the dispersal phase of the life cycle, are produced by clonal transmission of the mother's genotype. Thus, here we can use a direct contrast between ecologically equivalent individuals, differing in their transmission system, but not in their ecological requirements (for discussion, see Williams 1975).

Many eukaryotes can use two reproduction modes, being either sexual or asexual. Especially, in plants, sexual seed production is often accompanied with some forms of clonal propagation, e.g. agamospermy, stolons, gemmae, corms, bulbils, runners, rhizomes (Ceplitis and Bengtsson 2004). In the latter cases, the offspring are genetically identical to the parental genotype, except for mutations. Nevertheless, the means of passing genetic information from the parental generation to their offspring has some pivotal consequences for the genetic variability in the progeny, genetic population structures of populations and adaptability of populations. In most asexually reproducing species, mother and daughter are genetically identical, but in some asexual lineages the mechanism of chromosome transmission is leading to a reorganization of genomes. First, we must distinguish between agametic and gametic asexuality. Agametic asexuality, e.g. fission and budding, always results in genetically identical individuals. However, in gametic asexuality we have to discriminate again between two different modes of reproduction, apomixis and automixis. In apomictic plants three subgroups can be observed (Asker and Jerling 1992, Archetti 2004): (i) plants such as *Taraxacum* spp. can pass a regular meiotic process but suppress the reductional division (meiotic diplospory). In these cases, recombination can take place; (ii) in mitotic diplospory, the lack of meiosis-I creates a mitotic-like process during which recombination is absent; and (iii) in two *Hieracium* species, *H. piloselloides* and *H. aurantiacum*, for example, somatic cells function like spores (apospory) which are formed without recombination (Asker and Jerling 1992, Bicknell et al. 2000).

Figure 1 llustrates the outcome of diverse apomictic reproduction modes and considers the consequences of recombination. In the case of mitotic diplospory without recombination (Fig. 1, case a) genotypes are identical across generation and mutations only can change genomic structures. Recombinational rearrangements of chromosome segments generate an increasing number of homozygous loci (Fig. 1, case b). The consequence of recombination can be the loss of complementation in individuals (Archetti 2004) and, without selection, the loss of genetic variability in populations. The reduction of individual heterozygosity may results in lower viability through inbreeding-like effects and reduced adaptability. In polyploids, however, the loss of complementation through recombination is weakened through compensatory effects from more homologous and homeologous chromosomes, respectively.

Automixis is considered a rare reproduction mode in nature because a loss of heterozygosity can occur (Maynard Smith 1978; for examples, see

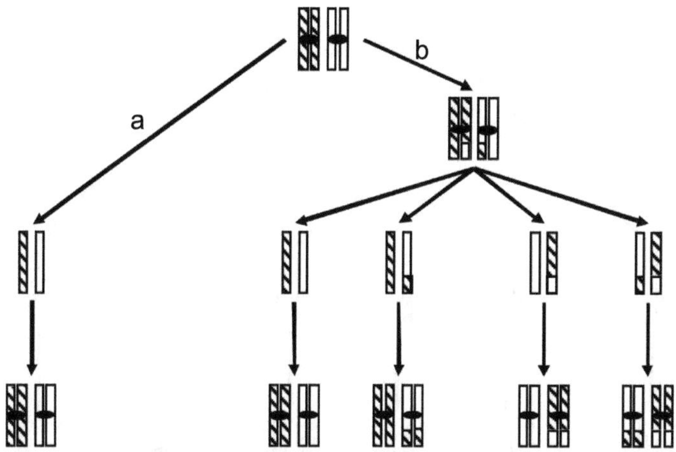

Fig. 1. Suppression of the reductional division (apomixis). Case a: the loss of meiosis I is associated with a lack of recombination. Offspring are genetically identical to their mother. Case b: Homologous chromosomes pair during meiosis I and recombination can take place but cell division is suppressed

Suomalainen et al. 1987). Automictic organisms produce gametes through meiotic divisions. Parental chromosome numbers can then be restored by (i) fusion of a gamete with its genetic complement (Fig. 2, case a) and (ii) random fusion of two gametes (Fig. 2, case b), but (iii) autodiploidization of haploid gametes can also restore diploidy, resulting in complete homozygosity (Fig. 2, case c), e.g. in the fruitfly *Drosophila mercatorum* (Templeton 1979). The first process conserves parental genotypes over generations, i.e. the offspring genotypes are identical to the maternal ones. Random fusion of gametes, however, has drastic consequences for the genetic structure of diploid populations and results can be genetically analogous to selfing: homologous chromosomes can be completely identical and, therefore, loss of individual fitness through homozygosity of harmful alleles and loss of heterozygosity in populations may be inevitable (Asher 1970).

Many plants have the potential to reproduce either sexually or asexually, i.e. asexuality can be obligate or facultative (Suomalainen 1950). Apomictic plants can have some sexual offspring, and they often produce pollen hybridizing with their related sexuals (Verduijn et al. 2004a). In contrast, some sexuals are sporadically able to produce viable asexual lineages (e.g. *D. mercatorum*; Templeton 1978). Furthermore, many asexuals are hybrids of two or several ancestral sexual species (Parker 2002). In these hybrid asexuals, chiasma formation and chromosomal rearrangements are often suppressed and heterozygosity associated with adaptability can be maintained (e.g. Stalker 1956; Asher 1970). New asexual lineages can occur independently

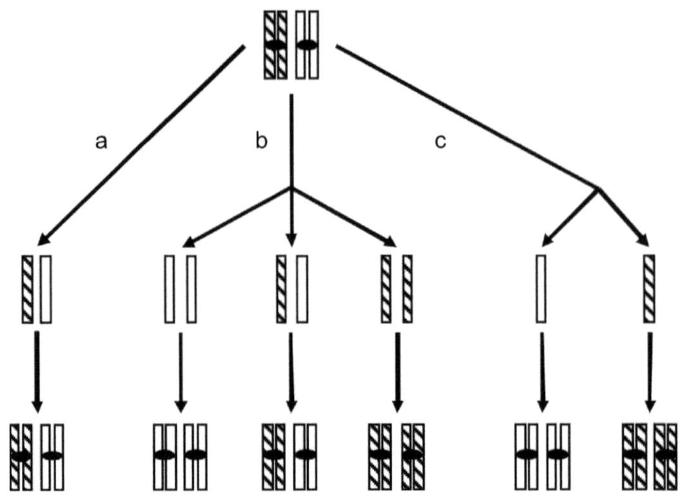

Fig. 2. Meiosis I and meiosis II are passed (automixis; in this scheme recombinational events are not considered). Case a: an egg cell fuses with its genetically complementary polar body. Offspring are genetically identical to their mother. Case b: an egg cell fuses with any of its polar bodies. Case c: autodiploidization of egg cells

several times (polyphyletic), but a single event (monophyletic) can also be the starting point of asexuality. The genetic variability of asexual populations, however, differs considerably if they originate from multiple or single asexual founders. Nevertheless, apomixis in plants is considered (i) adaptive because of its occurrence in different clades and (ii) a reproductive assurance against instable conditions in future. However, molecular processes initiating asexual reproduction in sexual populations are far from being fully understood, and in most cases the occurrence of asexuality can only be explained descriptively (White 1973). Nevertheless, in a few cases, the genetic basis of asexuality could be analysed, e.g. monogene inheritance of apomixis in two apomictic *Hieracium* species, the triploid *H. piloselloides* and the aneuploid *H. aurantiacum*, was observed. The dominant character of the gene for apomixis could be disclosed in progeny from crosses between both species and the related sexual *H. pilosella* (Bicknell et al. 2000).

3 Origin of polyploidy

The generation of polyploid genomes is more apparent than the origin of apomixis. Basically three mechanisms are discussed: (i) polyploidization within species (autoploidization), (ii) hybridization between related species with different ploidy levels and (iii) polyploidization of hybrid genomes (alloploidization). Again polyploidy can occur mono- or polyphyletically within populations. In the North-American Brassicaceae *Arabis holboellii*, for example, variability of chloroplast haplotypes suggests that polyploidy has repeatedly and independently arisen (Sharbel and Mitchel-Olds 2001). Autoploidization most likely explains tetra- and hexaploidy in *Vaccinium species* (Lyrene et al. 2003). Sexual ploidization and gene flow among the different lineages can result in the observed cytotype diversity. Similarly, in *Taraxacum* species, tetraploid lineages can emerge in populations of triploids (van Baarlen et al. 2000), and multiple hybridization events may explain the high genetic variability observed in these apomicts. Grimalelli et al. (1998) studied the *Tripsacum* agamic complex in which polyploid lineages reproduce apomictically and diploid ones are sexuals. They postulated that genes controlling apomixis are linked with a segregation distorter-type system promoting the elimination of the apomixis allele when transmitted through haploid gametes. Their model may explain a relationship between apomixis and polyploidy. Considering geographic patterns of distribution and the association between polyploidy and asexuality in plants, hybridization between related species is indeed considered a common concept to increase fitness in nature (Kearney 2005), i.e. hybrids and polyploids are more tolerant against environmental harsh conditions. In this context, D'Souza et al. (2005) found in the planarian flatworm *Schmidtea polychroa* contrary to the assumption of a positive correlation between fitness and ploidy level, reduced fitness in tetraploids compared with their related triploids (triploids have 58% more offspring than tetraploids).

4 Population genetics

4.1 Clonal diversity

The genotyping of individuals provides information about the reproduction mode in populations. In sexuals, we expect genotype frequencies to be close to Hardy–Weinberg proportions for selectively neutral loci whereas in asexuals deviations from Hardy–Weinberg proportion are expected. Furthermore, in asexuals genotypes of loci located even on different chromosomes do not segregate randomly, except for automixis with random gamete fusion (see Fig. 2, case b), i.e. genotype combinations across loci can be in linkage disequilibrium. Some loci might be fixed and others can be heterozygous in all individuals of an asexually reproducing population, i.e. asexuals can maximize heterozygosity of individual loci but loose allelic variation across all loci. Indeed, low allelic variation and the pattern of heterozygosity were suggested to indicate clonality in *Acacia aneura* (Andrew et al. 2003). Almost all apomictic individuals of the aphid *Rhopalosiphum padi* were heterozygous at a combination of nuclear loci (Delmotte et al. 2003). These findings forced the educated guess of a multiple hybrid origin of asexual lineages that can be subdivided into facultative and obligate asexuals specialized to different host plants (Halkett et al. 2005b). Chapmann et al. (2000) described high genetic variation in *Pilosella officinarum* that was introduced to New Zealand and until recently considered apomictic. The conclusion was that the high amount of genetic diversity can be explained most likely by some sexual reproduction. Hörandl et al. (2000) studied enzyme variability of 13 polymorphic loci in two species of the *Ranunculus auricomus* complex (diploid *R. notabilis*; tetraploid *R. variabilis*). *R. notabilis* was found to be similar to other sexual relatives, and therefore considered an ancestral sexual species of the *R. auricomus* complex. Differently, the excess of heterozygotes in *R. variabilis* confirmed the assumption of apomixis with a hybrid origin and a lack of meiosis. However, Barraclough et al. (2003) studied theoretical diversification patterns in sexual and asexual organisms. They concluded that asexuals can display discrete clusters similar to those found in sexual organisms, and the only way that sexuals have stronger diversification patterns than asexuals is when they are forced by faster rates of adaptive change. In this respect, ecological differences are often seen as important factors enabling stable coexistence of plant species with different ploidy level. Meirsmans et al. (2003), however, argued that gene flow between neighbouring plants can explain congruence of plants with different cytotypes.

4.2 Geographic parthenogenesis/apomixis

The causes for the distinctive features in the geographic distribution of sexually and asexually reproducing sister species in the northern hemisphere has been of permanent interest in ecology for about 80 years. Vandel (1928) first mentioned the geographically distribution patterns of asexual arthropod species and their closest sexual relatives. He coined the term "*geographic parthenogenesis*" for the phenomenon that in European complexes of closely related species with asexual and sexual reproduction; the asexual lineages have a more northern distribution whereas sexual reproduction can predominantly be found in southern populations. The ecological differentiation of sister species with different reproduction modes has been considered to determine the distribution patterns. In animals ecological and geographical parameters are described that could characterize the distribution area of sexuals and asexuals (Glesener and Tilman 1978; Bell 1982; Parker 2002). It seems, for example, that asexuality is favoured in high latitudes and elevation, respectively, in habitats disturbed by humans, and lastly in marginal or island populations. Summarizing environments of asexuals seem to be more stressful or disturbed than those of their sexual relatives (Glesener and Tilman 1978; Bell 1982; Suomalainen et al. 1987; Parker 2002). The association of clonal dispersal with human activities, i.e. human-assisted spread of asexuals, also points into the same direction (Roth 1974; Hughes 1996; see for review, Parker and Niklasson 2000).

Bierzychudek (1987) reviewed hypotheses that can explain the geographic distribution of asexually reproducing animals and their sexual relatives for their applicability to and validity for plants. In general, four concepts are discussed:

(i) It was first stated that asexuals and selfers have an enhanced ability to colonize virgin environments (Baker 1965). An idea that is also closely related to the conception of "reproductive assurance", i.e. there is no need for asexuals to find mates. Especially in marginalized populations with a low density, for example the more northern, sexual reproduction might bear the risk that females cannot reproduce.

(ii) Permanent gene flow from a sexually reproducing main population into marginal habitats can prevent adaptation to suboptimal conditions. Consequently, asexuals that are not affected by such negative genetic input can displace the maladapted sexual relatives (Antonovics 1968). Negative selection against sexuals in habitats with harsh climatic conditions may create the south–north gradient of sexual abundance (Peck et al. 1998).

(iii) However, another selection scenario assumes that biotic interactions affect the success of either sexuals or asexuals (Levin 1975, Glesener and Tilman 1978, Jaenike 1978). The evolutionary arms race among interacting species, e.g. host-parasite, predator–prey or competitors favour sexuals in central and south European habitats in which the biotic nexuses become more and more complex.

(iv) The *General Purpose Genotype* hypothesis was first postulated by Baker (1965) to explain the adaptations of weedy species of plants, in general, both selfers and apomictics. Parker et al. (1977) adopted this idea to explain the colonizing success of parthenogenetic cockroaches *(Pycnoscelus surinamensis)*. They argued that in a sexual population generating permanently asexual individuals, selection will favour those clones with the broadest ecological tolerance over several generations. In this respect, clones with high tolerance against wide range of abiotic conditions can be evolutionary more successful than their sexual relatives which cannot inherit the entity of an optimal genome to the next generation. Accordingly, experimental studies provided strong evidence for the selection of general purpose genotypes in composite plants.

Apomictic *Antennaria parvifolia* (Asteraceae) were less sensitive to environmental conditions, and their survival and biomass significantly exceeded that of their sexual relatives (Bierzychudek 1989). Vrijenhoek (1979) proposed a different evolutionary model in order to explain the successful clonal invasion of ecological niches. In his *Frozen Niche Model*, a parthenogenetic population consists of a set of microspecies, each specialized to the predetermined niche conditions. Environment and adaptation to niches determine clonal diversity and the success of either sexual or related asexual competing species (Vrijenhoek and Pfeiler 1997). The general purpose genotype should be robust against environmental effects combined with a long-lasting success whereas the frozen in niche genotype might be susceptible against environmental disturbances resulting in temporal instability.

Furthermore, another hypothesis has to be added to the above four: metapopulation structures can favour asexual reproduction (Ladle et al. 1993). Thus, in marginal populations, where metapopulation structures are more likely to determine the fate of populations, this hypothesis might explain the abundance of asexual groups in general.

In plants, only significant associations between latitude and elevation, respectively, and the geographic distribution of sexuals and related asexuals could be detected (Bierzychudek 1987). Most important, besides the reproductive mode, the ploidy level must also be considered, since an increased ploidy level can be observed in most asexually reproducing plants (Stebbins 1980; Parker 2002). Polyploidy may be associated closely with an increased tolerance against extreme environmental stress (e.g. Suomalainen 1969; Templeton 1982; but for an exception, see above D'Souza et al. 2005). Lynch (1984) argued that geographic distribution patterns of sexuals and asexuals can be explained by rather the existence of general-purpose genotypes than an increased colonization ability of clones. Nevertheless, in plants a conglomerate of both might be true: general purpose genotypes can be the better colonizers, combined with an increased fitness of polyploids. The two hypotheses, therefore, are difficult to keep apart from each other (Bierzychudek 1987).

Geographic distribution patterns can be observed in plants that are similar to those in animals (Bierzychudek 1987; Meirsmans et al. 2003). A well-known example are dandalions (*Taraxacum* spp.), where sexually reproducing individuals become rarer in the northern European populations (e.g., Menken

1995; Verduijn et al. 2004b). As polyploidy in plants is closely associated with apomictic reproduction, distribution patterns of asexuals are very congruent with those of higher ploidy levels (for examples, see Bierzychudek 1987). However, there is no general concept that related species with different reproduction mode and polyploidy levels are strictly geographically separated. In central and northern Europe, individuals of *Taraxacum* species coexist that are either diploid sexuals or triploid and tetraploid asexuals (van Baarlen et al. 2000; Meirmans et al. 2003; Verduijn et al. 2004b). In animals, populations of the weevil *Othiorhynchus scaber* that are a mixture of polyploid parthenogens and diploid sexuals are also found in the Alps (Austria) and Slovenia but only polyploid parthenogenetic forms are found in northern areas (Poland, Finland, and Sweden; Suomalainen et al. 1987; Stenberg et al. 1997, 2003a,b).

4.3 Models

Recent experimental and theoretical studies described ecological conditions and genetic mechanisms that can explain co-occurrence of different cytotypes and, therefore, disclose the limited validity of the "*minority cytotype exclusion principle*" postulated by Levin (1975). Husband (2000), for example, argued that in *Chamerion angustifolium* (Onagraceae; syn. *Chamaenerion a.* and *Epilobium a.*, respectively) the exclusion principle works through infertility of triploids originating from crosses between diploids and tetraploids, but positive assortative mating can maintain the rare tetraploid cytotype. In an experimental population, he found frequency-dependent selection on diploid and tetraploid individuals induced by bees visiting the inflorescences. The proportion of flights between tetraploids increased as their frequency decreased. Similarly, despite a significant correlation between ploidy level and elevation, coexistence of diploid sexuals and triploid asexuals was described for dandalions (Chapman et al. 2003). Gene flow between neighbouring plants of different ploidy levels was considered important to prevent ecological differentiation to elevation of diploid and triploids. Indeed, many apomictic plants have still maintained male function and can cross with their related sexuals (Britton and Mogie 2001). Furthermore, the crossing between asexual and sexual lineages with different ploidy levels (van der Hulst 2003, Verduijn et al. 2004a), migration and adaptation (Peck et al. 1998) can favour the coexistence of cytotypes, but also could have created the observed patterns of geographic distribution of sexual and asexual subspecies. Nevertheless, distinct occurrence and coexistence of species, respectively, are the consequence of multifactorial evolutionary processes. The exclusion of universal valid principles indicates the importance of carefully considering the life cycle parameters of species and the ecology of habitats under study. Some cyclic parthenogens, e.g. aphid species, show exactly an opposite trend in Europe, i.e. they reproduce sexually in northern areas but can permanently reproduce asexually in southern areas with mild climatic conditions (Hales et al. 1997). The high dependency of aphid species on their host plants dominates their life cycle. Furthermore, aphids introduced from northern to southern climates do not have the photoperiodic cues to produce sexuals. Thus the reverse pattern of geographic parthenogenesis in aphids may also reflect an ecological induced suppression of the sexual phase of their life cycle.

Models concentrate on three major issues that are closely linked to each other: (i) the dominance of sexual reproduction in nature, (ii) the competition between sexuals and their derived asexual lineages, and (iii) the origin and continuance of clonal diversity. Theoretical approaches and computer simulations have been performed to get a better insight into these problems, and our understanding of processes, which determine the success or the decline of either reproduction mode, has improved with the increasing number of studies (e.g. Maynard Smith 1978; Stearns 1989). Therefore, we here only present shortly some recent studies.

In asexuals that inherit their genomes as indivisible packages, mutations are accumulated all the time and detrimental mutations cannot be eliminated from clonal lineages. At the beginning of the mutational meltdown (e.g. Lynch et al. 1993), polyploidy serves to mask the effects of deleterious mutations but in consequence of the perpetual negative genetic input, extinction of clonal lineages appears to be unavoidable (according to Lynch et al. 1993). The so-called Muller's Ratchet (Muller 1932) turns faster in smaller populations and theoretically limits the long-term evolutionary continuity of asexual lineages (for review, see Lynch et al. 1993). This could explain the dominance of sexual reproduction. Differently to the former scenario of stable genome structures within clonal lineages, Archetti (2004) considered effects of crossing over in asexually reproducing populations. The loss of allelic complementation through recombination causes obviously an adverse knock-on effect in asexuals which can replace sexual reproduction under a wide range of conditions only if asexuality is associated with higher ploidy levels (Archetti 2004). Accordingly, demographic parameters of the respective organisms under study must also be considered if we analyse the evolution of asexual lineages (Kondrashow and Crow 1991; Bengtsson 1992).

The total population size and the number of reproducing individuals are the main objectives in analyses of the adaptability, the extinction risk, and the conservation of species. In sexual organisms, the effective population size specifies the number of reproducing individuals in a population (Hartl and Clark 1997), and it can be annexed to different genetic aspects: (i) the population size needed to maintain genetic diversity through mutation in a finite population (eigenvalue effective size), (ii) the loss of genetic variation from one to the next generation in a inbreeding population (inbreeding effective size), and (iii) the stability of allele frequencies in evolutionary time (variance effective size). Modelling diploid populations that consist of either sexually or asexually reproducing individuals with transition matrix of inbreeding and coancestry coefficients, the effects of asexuality on effective population size was considered (Yonezawa et al. 2004). The effective size of populations with individuals having the potential of changing their reproduction strategy was not identical but asymptotic to the eigenvalue effective size in sexuals. However, in reality environmental conditions can strongly affect the reproductive parameters of individuals, especially in sedentary organisms as plants, and the asymptotic state is only occasionally realized. Yonezawa et al. (2004), therefore, propose using the instantaneous rather than the effective size in evolutionary studies of populations with a high proportion of asexual individuals.

The evolutionary consequences of variable rates of clonal reproduction on population structures have been analysed with selectively neutral markers in an island model (Balloux et al. 2003). Using analytical and stochastic simulation approaches, principles of *F*-statistics were applied to diploid populations. Interestingly, asexual population can maintain a higher average degree of heterozygosity across loci through fixed heterozygosity, but a lower number of different genotypes in populations. The structures of populations consisting of a mixture of sexual and asexual individuals, however, are almost not distinguishable from those of solely sexually reproducing ones (Balloux et al. 2003). This may also point to statistical problems arising with identifying clonal reproduction in populations, i.e. estimating clonal diversity (see section 4.4).

Some experimental studies focusing on effects of parasites on their host, point to the importance of the reproduction mode in sparse populations where mates are difficult to find (Lively et al. 1990; Moritz et al. 1991; Lively 1992; Lively and Howard 1994; Brown et al. 1995). Indeed, using computer simulations Ladle et al. (1993) found that in metapopulations, asexual organisms can compensate for the advantage of sex and recombination. In temporary small populations, sexual organisms are exposed to an extended risk of inbreeding, loss of genetic variation by genetic drift, and a lowered chance of finding mates. Accordingly, Peck et al. (1998) explained the distinct geographic distribution patterns of sexual and asexual races by simulating migration, local adaptation and fertility. Furthermore, Lynch (1984) proposed the hypothesis of destabilizing hybridization, i.e. the fitness is lowered in offspring that result from hybridization between sexuals and their derived asexuals. As a consequence of this scenario, sexuals and asexuals can only exist allopatrically. Beyond the evolutionary exchange of blows between sexuals and asexuals, factors that determine clonal diversity within populations are considered. Modelling the reproductivity of clones in a spatial and/or temporal parameter range, the significance of life cycle parameters, e.g. fitness, dormancy and diapause, respectively, but also random processes are rated for their effects on clonal coexistence. Stable clonal coexistence under alternating environmental conditions is predicted by deterministic models where clones can escape strong competition through specialization (Tuljapurkar and Istock 1993; Hedrick 1995, Tomiuk et al. 2004a). These models explain oscillating changes of clonal frequencies in experimental populations (Weeks and Hofmann 1998) and seasonal changes of clones of the spearwinged fly in Swedish populations (Niklasson et al. 2004). However, these models consider only interactions between two clones but most natural populations consist of a mixture of diverse clonal lineages. Consequently, the question remains whether (i) most genetic variation is selectively neutral or (ii) the diversity of ecological niches to which clones are adapted is until yet not recorded? Therefore, we should also direct our attention to the attributes of clones that enable them to invade an existing clonal community and enlarge its clonal diversity.

4.4 Statistics

One major concern in studying the life cycle of plants is to determine the magnitude of asexual reproduction in natural populations in order to determine the evolutionary potential of populations. In the absence of morphological differences and in avoidance of rearing experiments, only genetic analyses can provide reliable information about clonal reproduction (Stenberg et al. 2003). For this purpose, our population genetical tools, e.g. testing for Hardy–Weinberg proportions or linkage disequilibrium between loci, can be adopted in diploid organisms. The deviations from Hardy–Weinberg proportions at individual loci or multilocus genotypes might indicate for the presence of asexual lineages in a population. Assuming sexual reproduction and based on a binomial approach (Parks and Werth 1993), Stenberg et al. (2003) propose statistics and software (*MLGSIM*), which estimate the likelihood that a multilocus genotype can be observed several times in a population. In European aspen (*Populus tremula*) this statistics disclosed higher clonal diversity than formerly described morphotypes (Suvanto et al. 2005). Similarly, tests for linkage disequilibrum between loci (null hypothesis: alleles of different loci are combined by chance) can be used to detect asexual reproduction in populations. However, none of the proposed estimators for linkage disequilibrium was able to accurately measure the proportion of clonal reproduction in populations (de Meeûs and Balloux 2004).

Now we consider statistical approaches to compare and analyse the genetic variation in sexually reproducing populations with that of asexuals. Certainly, multilocus fingerprints can easily disclose clonal and genetic variability even in species with various ploidy level, independently from the reproduction mode. However, information about parameters of dynamic evolutionary processes such as temporal and spatial changes in populations through mutation, selection, genetic drift and migration are limited through dominant–recessive patterns of multilocus phenotypes. Most population genetic statistics is based on allele and heterozygote frequencies and, therefore, codominant markers should be studied. Henceforth, we focus on statistical tools applied to genetic variation that results from studies of marker systems which are often used for the genotyping of individuals at single loci, e.g. allozymes, microsatellites and single nucleotide substitutions (SNPs). Using estimates of allele frequencies from such loci, many estimation procedures of parameters, e.g. the expected degree of heterozygosity and the genetic similarity of individuals, groups or species, that are used in population genetic studies of sexuals (Hartl and Clark 1997), can also be applied without any modification to genetic data of diploid asexuals (Halkett et al. 2005). Recently, Halkett et al. (2005) reviewed and discussed available statistical methods for the analyses of population structures and recovering clonal reproduction in populations with diploid asexuals and sexuals.

Basic data analysis of population genetic studies may first consider allele and genotype variation. In diploid organisms, gene counting and the estimating of genotype frequencies are simple and basic statistics, regardless of the reproductive mode. Differently, difficulties can arise in the analysis of genotypic data of polyploids where the number of identical genes at a locus cannot unequivocally identified (Tomiuk and Loeschcke 1991). For example, allozymes of allodiploid oilseed rape (*Brassica napus*) coded by genes that are located on homeologous chromosomes AACC (A-chromosomes originate from *B. campestris* and C-chromosomes from *B. oleracea*), reveal electrophoretic pattern that are difficult to interpret genetically (Chen et al. 1989; Mündges et al. 1989). Furthermore, our conclusions from genetic analyses can be biased in the presence of null or defect alleles.

4.5 Measures of genetic diversity

4.5.1 Clonal diversity

Correspondingly to the commonly used measure of allelic diversity (the effective number of alleles), a measure of diversity in asexually reproducing populations is given by the probability that two identical genotypes can be found, i.e. the effective number of clones

$$C_e = \frac{1}{\Sigma\, g_i^2}, \qquad (1)$$

where g_i is the frequency of the *i*-th genotype with $i = 1,....,n$ and $\Sigma gi = 1$. Evidently, an estimate of C_e can be done with all kind of data recording reliably differences among clones, i.e. by morphology, fingerprinting, genotyping of individual loci or sequencing. The measure of effective clone number can be applied to populations and species with diverse ploidy levels and allows comparison of populations with different distributions of clone frequency (see Hartl and Clark 1997, p. 176).

4.5.2 Allelic and genotypic diversity

In polyploid species, basic descriptive statistics inform, for example, about the observed frequency of heterozygotes but corresponding expectations are difficult to postulate. Evidently, information about the transition from sexual to asexual reproduction is necessary to frame an expected genetic structure. Our expectations can strongly differ if asexuals have either a mono- or polyphyletic origin, and also the evolution of polyploidy through autoploidy, alloploidy and/or hybridization must be taken into account when a theoretical evolutionary scenario will be designed.

4.5.3 Population structures

F-Statistics provides information about genetic differences among populations (F_{ST}), mating behaviour within populations (F_{IS}), and migration among subpopulations. Ceplitis (2001) applied classic *F*-statistics to populations of *Allium vineale* that consist of both, sexually and asexually reproducing individuals. He argues that the weedy habit has favoured bulbil reproduction and local adaptation was facilitated to sexual genotypes. In populations consisting of asexual, sexual and selfing individuals, the most relevant evolutionary parameter is the co-ancestry between individuals (de Meeûs and Balloux 2005). Accordingly, they modified the classical F-statistics and proved that structures of populations with individuals having diverse reproduction modes can completely be explained by either F_{ST} or F_{IS}, where $F_{ST} = -F_{IS}/(1-F_{IS})$. Furthermore, modified F-statistics also provides an measure for the number of migrants among populations $Nm = -(1 + F_{IS}) / 4F_{IS}$.

In asexual organisms, the total genetic variation of populations can also be subdivided into its diverse temporal and spatial components by a modified analysis of variance (AMOVA, Excoffier et al. 1992). The analysis of molecular variance was originally developed for mtDNA haplotypes and estimates the amount of genetic variation by analysis of variance. Nevertheless, the procedure can easily be applied to genetic data of diploid asexual populations and some free software is available that can handle genetic data (see for review, Halkett et al. 2005).

4.5.4 Differences between species

The evolutionary comparison of species complexes consisting of sexuals and diploid asexuals can easily be done by classical population genetic tools, e.g. the *genetic distance* (Nei 1972). However, this endeavour poses a challenge if diverse polyploid asexual races are involved (see for review, Tomiuk et al. 2004b). The origin of polyploidy, i.e. mono-via polyphyletic origin, hybridization and auto- via alloploidization, must be considered in our phylogenetic models if we want to estimate genetic distances between species/races and to get reliable information about the evolutionary time. Birky (1996) analysed the information value of phylogenetic trees in asexual eukaryotes. Indeed, the evolutionary history of asexual lineages can enhance the complications in the analyses and the interpretations of phylogenetic trees.

5 Conclusions

Statistical analyses based on gene frequencies have limits with polyploid species. Most of the above described procedures focus on the analysis of diploid species from which reliable estimates of gene frequencies are made available. However, the number of individual alleles contributing to phenotype patterns cannot be determined precisely for some polyploid heterozygotes. Consequently, an approach to statistically handle genetic data of polyploids is the use of phenotypes that can be associated with one or several genotpyes (Tomiuk and Loeschcke 1991). Beyond the problems arising with polyploid asexual species in many statistical procedures, the assumption of independent loci is not fulfilled in asexuals that have frozen genotypes over generations. Pure mutation models, excluding random drift, seem to best describe processes along an evolutionary time scale.

Our understanding of evolutionary processes in asexual species has increased considerably during the last decades, but there is still a large gap in our facilities to treat statistically genetic data of polyploid species, e.g. estimates of rates of gene flow among polyploids and sexuals, estimates of clonal variation in dependence on the genesis of asexual lineages. For example, no well-defined null hypothesis for the clonal diversity of polyploid populations can be given without quantitative information on the processes causal for the origin of asexual polyploid lineages, i.e. we cannot assess whether an observed clonal diversity is higher or lower than an expected one. Summarizing, the available statistics are useful to analyse diploid species but there is an urgent need for the development of population genetic statistics that can be applied to polyploid asexual species.

Acknowledgements

We thank E.D. Parker Jr for many valuable comments on an earlier draft of this chapter.

References

Andrew RL, Miller JT, Peakall, R, Crips PM, Bayer RJ (2003) Genetic, cytogenetic and morphological patterns in a mixed mulga population: evidence for apomixis. Aust Syst Biol 16:69–80

Archetti M (2004) Recombination and loss of complementation: a more than two–fold cost for parthenogenesis. J Evol Biol 17:1084–1097

Asher JH Jr (1970) Parthenogenesis and genetic variability. II. One–locus models for various diploid populations. Genetics 66:369–391

Asker SE, Jerling L (1992) Apomixis in plants. CRC Press, Boca Raton, Florida, USA

Baker HG (1965) Characteristics and modes of origin of weeds. In: Baker HG, Stebbins GL (eds) Genetics of colonizing species. Academic Press, New York, pp 137–172

Balloux F, Lehmann L, de Meeûs T (2003) The population genetics of clonal and partially clonal diploids. Genetics 164:1635–1644

Barraclough TG, Birky CW Jr, Burt A (2003) Diversification in sexual and asexual organisms. Evolution 57:2166–2172

Bell G (1982) The Masterpiece of Nature. The evolution and genetics of sexuality. Croom Helm, London

Bengtsson BO (1992) Deleterious mutations and the origin of meiotic ploidy cycle. Genetics 131:741–744

Bicknell RA, Borst NK, Koltunow AM (2000) Monogenic inheritance of apomixis in two *Hieracium* species with distinct developmental mechanisms. Heredity 84:228–237

Bierzychudek P (1987) Patterns in plant parthenogenesis. In: Stearns SC (ed) The evolution of sex and its consequences. Birkhäuser Verlag, Basel, Switzerland. pp 197–217
Bierzychudek P (1989) Environmental sensitivity of sexual and apomictic *Antennaria*: do apomicts have general-purpose genotypes. Evolution 43:1456–1466
Birky CW (1996) Heterozygosity, heteromorphy, and phylogenetic trees in asexual eukaryotes. Genetics 144:427–437
Britton NF, Mogie M (2001) Poor male function favours the coexistence of sexual and asexual relatives. Ecol Lett 4:116–121
Brown SG, Kwan S, Shero S (1995) The parasitic theory of sexual reproduction: Parasitism in unisexual and bisexual geckos. Proc R Soc Lond B Biol Sci 260:317–320
Ceplitis A (2001) The importance of sexual and asexual reproduction in the recent evolution of *Allium vineale*. Evolution 55:1581–1591
Ceplitis A, Bengtsson BO (2004) Genetic variation, disequilibrium and natural selection on reproductive traits in *Allium vineale*. J Evol Biol 17:302–311
Chapman HM, Parh D, Oraguzie N (2000) Genetic structure and colonizing success of a clonal, weedy species, *Pilosella officinarum* (Asteraceae). Heredity 84:401–409
Chen BY, Heneen WK, Simonsen V (1989) Comparative and genetic studies of isozymes in resynthesized and cultivated *Brassica napus* L., *Brassica campestris* L., and *B. alboglabra* Baitey. Theor Appl Genet 77:673–679
de Meeûs T, Balloux F (2004) Clonal reproduction and linkage disequilibrium in diploids: a simulation study. Infect Genet Evol 4:345–351
de Meeûs T, Balloux F (2005) *F*–statistics of clonal diploids structured in numerous demes. Mol Ecol 14:2695–2702
Delmotte F, Sabater–Muñoz B, Prunier–Leterme N, Latorre A, Sunnucks P, Rispe C, Simon J–C (2003) Phylogenetic evidence for hybrid origins of sexual lineages in an aphid species. Evolution 57:1291–1303
D'Souza. TG, Strohas M, Michiels NK (2005) The effect of ploidy level on fitness in parthenogenetic flatworms. Biol J Linn Soc Lond 85:191–198
Excoffier L, Smouse PE, Quattro JM (1992) Analysis of molecular variance inferred from metric distances among DNA haplotypes: application to human mitochondrial DNA restriction data. Genetics 131:479–491
Glesener RR, Tilman D (1978) Sexuality and the components of environmental uncertainty: clues from geographic parthenogenesis in terrestrial animals. Am Nat 112:459–673
Grimanelli D, Leblanc O, Espinosa E, Perotti E, González de León D, Savidan Y (1998) Non-Mendelian transmission of apomixis in maize–*Tripsacum* hybrids caused by a transmission ratio distortion. Heredity 80:40–47
Hales DF, Tomiuk J, Wöhrmann K, Sunnucks P (1997) Evolutionary and genetic aspects of aphid biology: a review. Eur J Entomol 94:1–55
Halkett F, Simon JC, Balloux F (2005a) Tackling the population genetics of clonal and partially clonal organisms. Trends Ecol Evol 20:194–201
Halkett F, Plantegenest M, Prunier-Leterme N, Mieuzet L, Delmotte F, Simon J-C (2005b) Admixed sexual and facultatively asexual aphid lineages at mating sites. Mol Ecol 14:325–336
Hamilton WD (1982) Sex versus nonsex versus parasite. Oikos 35:282–290
Hartl DL, Clark AG (1997) Principles of population genetics. Sinauer Associates, Sunderland, Mass., USA
Hedrick PW (1995) Genetic polymorphism in a temporally varying environment: Effects of delayed germination or diapause. Heredity 75:164–170
Hoffmann P, Esser K (1978) Genetics of speciation in the basidiomyceotous genus *Polyporus*. Theor Appl Genet 53;273–282
Hörandl E, Greilhuber J, Dobeš C (2000) Isozyme variation and ploidy level within the apomictic *Ranunculus auricomus* complex: evidence for a sexual progenitor species in southeastern Austria. Plant Biol 2:53–62

Hughes RN (1996) Evolutionary ecology of parthenogenetic strains of the prosobranch snail, *Potamopyrgus antipodarum* (Gray) = *P. jenkinsi* (Smith). Malac Rev Suppl 6:101–113
Husband BC (2000) Constraints on polyploid evolution: a test of the minority cytotype exclusion principle. Proc R Soc Lond B Biol Sci 267:217–223
Jaenike J (1978) An hypothesis to account for the maintenance of sex within populations. Evol Theor 3:191–194
Kearney M (2005) Hybridization, glaciation and geographical parthenogenesis. Trends Ecol Evol 20:495–502
Kondrashow AS, Crow JF (1991) Haploidy or diploidy: Which is better? Nature 351:314–315
Ladle RJ, Johnstone RA, Judson OP (1993) Coevolutionary dynamics of sex in metapopulation: escaping the Red Queen. Proc R Soc Lond B Biol Sci 253:155–160
Levin DA (1975) Minority cytotype exclusion in local plant populations. Taxon 24:35–43
Lively CM (1992) Parthenogenesis in freshwater snail: reproductive assurance versus mixed mating. Evolution 46:907–913
Lively CM, Howard, RS (1994) Selection by parasites for clonal diversity and mixed mating. Phil Trans R Soc Lond B Biol Sci 346:271–281
Lively CM, Craddock C, Vrijenhoek RC (1990) Red queen hypothesis supported by parasitism in sexual and clonal fish. Nature 344:864–866
Lynch M (1984) Destabilizing hybridization, general–purpose genotypes and geographic parthenogenesis. Q Rev Biol 59:257–290
Lynch M, Bürger R, Butcher D, Gabriel W (1993) The mutational meltdown in asexual populations. J Hered 84:339–344
Lyrene PM, Vorsa N, Ballington JR (2003) Polyploidy and sexual polyploidization in the geneus *Vaccinium*. Euphytica 133:27–36
Maynard Smith J (1971) The origin and maintenance of sex. In: Williams GC (ed) Group selection. Aldine Atherton, Chicago, pp 163–175
Maynard Smith J (1978) The evolution of sex. Cambridge University Press, Cambridge, England
Mayr E (1988) Towards a new philosophy of biology—observations of an evolutionist. The Belknap Press of Harvard University Press, Cambridge, Mass., USA
Meirsmans PG, Vlot EC, den Nijs JCM, Menken SBJ (2003) Spatial ecological and genetic structure of a mixed population of sexual diploid and apomictic triploid dandelions. J Evol Biol 16:343–352
Menken SBJ, Smit E, den Nijs HCM (1995) Genetical population structure in plants: Gene flow between diploid sexual and triploid asexual dandelions (*Taraxacum* section *Ruderalia*). Evolution 49:1108–1118
Michod RE, Wojciechowski MF, Hoelzer MA (1988) DNA repair and the evolution of transformation in the bacterium *Bacillus subtilis*. Genetics 118:31–39
Mündges H, Diederichsen E, Köhler W (1989) Comparisons of isozyme patterns in resynthesized amphihaploid rapeseed (*Brassica napus*) and their parental species*Brassica campestris* and *Brassica oleracea*. Plant Breed 103:258–261
Muller HJ (1932) Some genetic aspects of sex. Am Nat 66:118–138
Nei M (1972) Genetic distance between populations. Am Nat 106:283–292
Moritz C, McCallum H, Donnellan S, Roberts JD (1991) Parasite load in parthenogenetic and sexual lizards (*Heteronotia binoei*): support for the Red Queen hypothesis. Proc R Soc Lond B Biol Sci 244:145–149
Niklasson M, Tomiuk J, Parker ED Jr (2004) Maintenance of clonal diversity in *Dipsa bifurcata* (Fallén, 1810) (Diptera: Lonchopteridae). I. Fluctuating seasonal selection moulds long-term coexistence. Heredity 93:62–71
Owen R (1849) On parthenogenesis or successive production of procreating individuals from a single ovum. John van Voorts, London
Parker ED Jr (2002) Geographic parthenogenesis in terrestrial invertebrates: Generalist or specialist. In: Hughes RN (ed) Reproductive biology of invertebrates. Vol. XI. Progress in asexual reproduction. Wiley, New York, pp 93–114

Parker ED Jr, Niklasson M (2000) Genetic structure and evolution in parthenogenetic animals. In: Singh RS, Krimbas CB (eds) Evolutionary genetics: from molecules to morphology, vol. 1. Cambridge University Press, Cambridge, USA, pp 456–474
Parker ED Jr, Selander RK, Hudson RO, Lester LJ (1977) Genetic diversity in colonizing parthenogenetic cockroaches. Evolution 31:836–842
Parks JC, Werth CR (1993) A study of spatial features of clones in a population of bracken fern, *Pteridium aquilinum* (Dennstaedtiaceae). Am J Bot 80:537–544
Peck JR, Yearsley JM, Waxman D (1998) Explaining the geographic distributions of sexual and asexual populations. Nature 391:889–892
Roth LM (1974) Reproductive potential of bisexual *Pycnoscelus indicus* and clones of its parthenogenetic relative, *Pycnoscelus surinamensis.* Ann Entomol Soc Am 67:215–223
Sharbel TF, Mitchel–Olds T (2001) Recurrent polyploid origins and chloroplast phylogeography in the *Arabis holboellii* complex (Brassicaceae). Heredity 87:59–68
Stalker HD (1956) On the evolution of parthenogenesis in *Lonchoptera* (Diptera). Evolution 10:345–359
Stearns SC (1987) The evolution of sex and its consequences. Birkhäuser Verlag, Basel, Switzerland
Stebbins GL (1950) Variation and evolution in plants. Columbia University Press, New York, USA
Stenberg P, Terhivuo J, Lokki J, Saura A (1997) Clone diversity of tetraploid *Otiorhynchus scaber* in northern Europe. Hereditas 126:169–172
Stenberg P, Lundmark M, Saura A (2003a) MLGSim: a program for detecting clones using a simulation approach. Mol Ecol Notes 3:329–331
Stenberg P, Lundmark M, Knutelski S, Saura A (2003b) Evolution of clonality and polyploidy in a weevil system. Mol Biol Evol 20:1626–1632
Suomalainen E (1950) Parthenogenesis in animals. Adv Genet 3:193–253
Suomalainen E, Saura A, Lokki J (1987) Cytology and evolution in parthenogenesis. CRC Press, Boca Raton, Fla., USA
Suvanto L, Latva–Karjanmaa TB (2005) Clone identification and clonal structure of the European aspen (*Populus tremula*). Mol Ecol 14:2851–2860
Templeton AR (1979) The parthenogenetic capacities and genetic structures of sympatric populations of *Drosophila mercatorum* and *Drosophila hydei.* Genetics 92:1283–1293
Templeton AR (1982) The prophecies of parthenogenesis. In: Dingle H, Hegmann JP (eds) Evolution and genetics of life histories. Springer Verlag, New York, USA. pp 75–101
Templeton A (1989) The meaning of species and speciation: a genetic perspective. In: Otte D, Endler JA (eds) Speciation and its consequences. Sinauer Associates, Sunderland Massachusetts, USA, pp 3–27
Tomiuk J, Loeschcke V (1991) A new measure of genetic identity between populations of sexual and asexual species. Evolution 45:1685–1694
Tomiuk J, Niklasson M, Parker ED Jr (2004a) Maintenance of clonal diversity in *Dipsa bifurcata* (Fallén, 1810) (Diptera: Lonchopteridae). II. Diapause stabilizes clonal coexistence. Heredity 93:72–77
Tomiuk J, Roedenbeck I, Köhler W (2004b) Biodiversity in anthropogenic landscapes – population genetics and ecological modelling. In: Esser K, Lüttge U, Beyschlag W, Murata J (eds) Progress in botany 66. Springer Verlag, New York–Heidelberg, pp 112–143.
Tuljapurkar S, Istock C (1993) Environmental uncertainty and variable diapause. Theor Popul Biol 43:251–280
van Barlen P, van Dijk PJ, Hoekstra RF, de Jong JH (2000) Meiotic recombination in sexual diploid and apomictic triploid dandelions (*Taraxacum oficinale* L.). Genome 43:827–835
Vandel A (1928) La Parthénogenèse géographique. IV. Polyploïdie et distribution géographique. Bull Biol France–Belgique 74:94–100
van der Hulst RGM, Mes THM, den Nijs JCM, Bachmann K (2000) Amplified fragment length polymorphism (AFLP) markers reveal that population structure of triploid dandelions (*Taraxacum officinale*) exhibits both clonality and recombination. Mol Ecol 9:1–8

van Valen L (1973) A new evolutionary law. Evol Theor 1:1–30
Verduijn MH, van Dijk PJ, van Damme JMM (2004a) The role of tetraploids in the sexual–asexual cycle in dandelions (*Taraxacum*). Heredity 93:390–398
Verduijn MH, van Dijk PJ, van Damme JMM (2004b) Distribution, phenology and demography of sympatric sexual and asexual dandelions (*Taraxacum officinale* L.): Geographic parthenogenesis on a small scale. Biol J Linn Soc Lond 82:205–218
Vrijenhoek RC (1979) Factors effecting clonal diversity and coexistence. Am Zool 19:787–797
Vrijenhoek RC, Pfeiler E (1997) Differential survival of sexual and asexual *Poeciliopsis* during environmental stress. Evolution 51:1593–1600
Weeks AR, Hoffman AA (1998) Intense selection of mite clones in a heterogeneous environment. Evolution 52:1325–1333
White MJD (1973) Animal cytology and evolution. Cambridge University Press, Cambridge, England
Williams GC (1975) Sex and evolution. Princeton University Press, New York, USA
Yonezawa K, Ishii T, Nagamine T (2004) The effective size of mixed sexually and asexually reproducing populations. Genetics 166:1529–1539

Professor Dr. Jürgen Tomiuk
University of Tübingen
Institute of Human Genetics
Division of General Human Genetics
Wilhelmstrasse 27
D-72074 Tübingen
Germany
Phone: +49-(0)7071-29-76883
Fax: +49-(0)7071-29-5233
e-mail: juergen.tomiuk@uni-tuebingen.de

Wolfgang Köhler
Institute of Biometry and Population Genetics
University of Giessen
Germany

Plant Breeding: Assessment of Genetic Diversity in Crop Plants and its Exploitation in Breeding

Wolfgang Friedt, Rod Snowdon, Frank Ordon and Jutta Ahlemeyer

1 Introduction

Today, about 500,000 living plant species are known, of which approximately 240,000 species belong to the angiosperms. The number of cultivated plants comprises only a very small portion of all these species. An analysis of FAO food data for 146 countries has come to the conclusion that a total of 103 species contribute 90% of national per capita supplies of food plants (Prescott-Allen and Prescott-Allen 1990). Within each of the cultivated species, domestication has limited the genetic diversity to useful genotypes, adapted to the needs of humans as well as local growing conditions. Along with the industrialization in the nineteenth century, the diversity of domestic crops decreased dramatically at the species and intraspecific levels, leading to an irreversible loss of cultivated and wild species, subspecies, landraces, former varieties and single genes. One of the primary losses is the diversity of landraces once grown on-farm. At the same time, the need for genetic resources in order to develop new and improved varieties started to increase. However, based on still existing diversity, novel genetic variation can be created by sexual hybridization within and across species.

2 Assessment of genetic diversity

For the assessment of diversity in plant populations different marker systems have been established on morphological, physiological and DNA levels. Each of the various marker systems utilized for the characterization of similarities and differences between individuals has its specific strengths and constraints, e.g. regarding the number of available markers, the polymorphism per marker, the mode of inheritance or the genomic location of the markers. Hence, to obtain unbiased estimates of the genetic diversity within a population, attention has to be paid to the choice of the marker system utilized as well as of the statistical methods applied once an appropriate data set is obtained.

Progress in Botany, Vol. 68

2.1 Genetic markers

During recent years markers that detect diversity directly on the DNA level have become more and more important and have replaced morphological markers and the formerly widely utilized isozyme marker systems almost completely. Unlike morphological markers, genetic markers are unaffected by the conditions in which the plants are grown and usually show a larger number of loci and a higher level of polymorphisms than isozyme markers. Polymorphisms in the nucleotide sequence can be revealed by different techniques. The most commonly used marker systems to probe the architecture of genomes are currently RFLPs (restriction fragment length polymorphisms), RAPDs (random amplified polymorphic DNAs), AFLPs (amplified fragment length polymorphisms), SSRs (simple sequence repeats; microsatellites) and SNPs (single nucleotide polymorphisms; Table 1).

RFLP markers are applications of the Southern blotting procedure and were first described by Botstein et al. (1980). Genomic DNA is digested with appropriate restriction endonucleases and the generated fragments are electrophoretically separated, transferred to a membrane and hybridized to a specific cloned DNA sequence either derived from genomic DNA or cDNA. The obtained length polymorphisms of the DNA fragments are either due to deletions and insertions or to point mutations in the recognition sites of the restriction enzymes. The codominantly inherited RFLP markers have successfully been used for the assessment of genetic diversity in various crop plants such as, e.g. maize (Melchinger et al. 1990; Messmer et al. 1991), oilseed rape (Diers and Osborn 1994; cf. Snowdon and Friedt 2004), turnip rape (McGrath and Quiros 1992), tomato (Miller and Tanksley 1990) and cucumber (Horejsi and Staub 1999).

RAPDs are DNA fragments amplified by PCR using short random primers (Welsh and McCelland 1990). RAPDs are very easy and quick to assay, but show

Table 1. Overview of the most commonly utilized genetic marker systems for the assessment of diversity in crop plants

	RFLP	RAPD	AFLP	SSR	SNP
Codominance	Yes	No	No	Yes	Yes
Level of polymorphisms	Medium	High	High	High	Low
Coding	Known	Unknown	Unknown	No	Known
Reproducibility	High	Low	High	High	High
Technical demands	Medium	Low	Medium	Medium	High

relatively low reproducibility between different laboratories and are dominantly inherited. RAPD markers have been used to determine genetic diversity for example in wheat (Vierling and Nguyen 1992), barley (Dawson et al. 1993; Baum et al. 1997), common bean (Beebe et al. 2000), and Lima bean (Fofana et al. 1997).

The AFLP technique is a combination of restriction digestion and PCR amplification (Vos et al. 1995). After digestion with two different restriction enzymes adaptors are ligated to the ends of the restriction fragments. In a subsequent PCR reaction with primers complementary to the adaptors, but with a 3′ overhang, a subset of these restriction fragments is amplified and visualized after separation on a sequencing gel. AFLPs show a high genomic abundance and generate a large number of informative bands per reaction, but are technically demanding and relatively expensive. AFLPs have been utilized to access diversity in a wide range of crop species, e.g. wheat (Barrett et al. 1998), soybean (Maughan et al. 1996), and sunflower (Hongtrakul et al. 1997; Rönicke et al. 2004, 2005).

SSRs or microsatellites are tandem repeats of a very short nucleotide motif (1–5 basepairs) that can be amplified with PCR primers specific to the flanking regions of these repeats. SSRs are easy and fast to assay, and are robust and highly reproducible. SSRs have become an important tool in crop germplasm management, and have been used, for example, to evaluate cultivar variation in rice (Olufowote et al. 1997), soybean (Rongwen et al. 1995), and barley (Russell et al. 1997).

One of the most recent developments for the assessment of genetic diversity is the utilization of single DNA base differences between homologous DNA fragments. These so-called SNPs provide an abundant source of DNA polymorphisms in plants (Henikoff and Comai 2003, Rafalski 2002, Gupta et al. 2001). Various different high-throughput techniques have been established that allow rapid and relatively inexpensive detection of SNPs. The most widely used methods for the identification of SNPs rely on hybridization, primer extension or sequencing followed by a visualisation step based on, e.g. electrophoresis, chip technology or mass spectrometry. SNPs have been used for genotyping purposes in different crop plants, e.g. barley (Wenzl et al. 2004), soybean (Zhu et al. 2003) and maize (Batley et al. 2003).

2.2 Measures of genetic association

For the assessment of genetic diversity within a plant population the choice of an appropriate measure to determine the association between individuals and the utilization of a suitable statistical procedure to assign accessions to different sub-populations is as important as the decision for a marker system. Broadly, two different methods to classify and order accessions according to the obtained polymorphisms in the nucleotide sequence are distinguished. One option is to calculate a pairwise distance matrix by applying one of various measures of genetic association to the data set and use the obtained association matrix as an input to display the relatedness of the individuals graphically either by clustering or by ordination in reduced space. A second option is to use a model-based clustering method to infer population structure and to assign individuals to populations.

Plant breeders usually want to determine the genetic diversity between different varieties or individual lines within a species. Some commonly used measures of genetic association between two individuals are listed in Table 2. Most widely employed are the similarity measures (S) that directly use the binary matrix obtained in marker analysis as, for example, Jaccard's coefficient S_J (Jaccard 1908), Dice's coefficient S_D (Dice 1945) or the simple matching coefficient S_{SM} (Sokal and Michener 1958). These coefficients calculate the ratio of the number of bands that are present in both individuals to the total number of bands and range from 0 for no similarity at all to 1 for complete identity. While S_{SM} takes into account all possible combinations of presence or absence of a band in both or either of the two individuals, S_J and S_D do not involve pairs in which a band is absent in both individuals. The later two measures are therefore appropriate for marker systems in which the absence of a band in both individuals is not meaningful.

For marker systems in which not only the presence or absence of a band can be scored, but the number of alleles at a locus can be determined, frequency-based genetic measures can be applied. The most commonly known distance measure to

Table 2. Measures of association between two individuals. *a* is the number of bands present in both individuals, *b* and *c* are the number of bands present either of the two individuals and *d* is the number of bands absent in both individuals. p_{ij} and q_{ij} refer to the allele-frequencies of the *j*th allele at the *i*th locus were *n* is the number of alleles at the *i*th locus and *m* the number of loci

Measure	Coefficient	Reference
Simple matching S_{SM}	$\frac{a+d}{a+b+c+d}$	Sokal and Michener 1958
Jaccard S_J	$\frac{a}{a+b+c}$	Jaccard 1908
Dice S_D	$\frac{2a}{2a+b+c}$	Nei and Li 1979
Euclidean distance D_E	$\sqrt{\sum_{i=1}^{m}\sum_{j=1}^{u_i}(p_{ij}-q_{ij})^2}$	
Modified Rogers' D_{MR}	$\frac{1}{\sqrt{2m}}\sqrt{\sum_{i=1}^{m}\sum_{j=1}^{u_i}(p_{ij}-q_{ij})^2}$	Wright 1978; Goodman and Stuber 1983

be utilized for frequency-based data is the Euclidean distance D_E, which is simply an application of Pythagoras' formula in m-dimensional space, where m refers to the number of loci under investigation. A disadvantage of this simple measure is that its range varies with the number of loci of which it is calculated and thus makes it difficult to compare results from different studies. This shortcoming can be overcome by dividing the Euclidean distance by the square root of $2m$. The derived distance measure is called Modified Rogers' distance D_{MR} (Wright 1978; Goodman and Stuber 1983) and ranges between 0 and 1. D_{MR} is especially suited for investigations on plant material with a high degree of heterozygosity. It should be mentioned that for any similarity measure ranging from 0 to 1 the corresponding distance measure can be calculated as its one-complement.

2.3 Methods to order and display genetic variation

For a clearly arranged graphical display of the genetic associations between individuals, either a cluster analysis can be conducted on the association matrix with results usually displayed in form of a dendrogram, or an ordination in reduced space can be carried out that represents the relationships between individuals in a scatter plot with two or three axes. Whereas cluster analysis displays all the fine relations between individuals, an ordination projects the multidimensional associations between individuals on only two or three axes that represent a large fraction of the variability between the individuals.

The objective of a cluster analysis is to gather individuals with similar properties into the same group. To decide whether individuals are similar enough to be allocated to the same cluster several multivariate techniques have been described. The most basic and intuitive clustering methods are simple linkage (Sneath 1957) and complete linkage (Sørensen 1948). The simple linkage clustering algorithm links two individuals at a specific level of partition in a way that an individual linked to a group displays a similarity with one of the members of the group that at least equals the level of partition. By contrast, in complete linkage the individual has to display the given level of similarity not only with one member of the cluster but with all of its members. While the single linkage method produces elongated clusters with loose chaining and is unsuitable for the isolation of poorly separated clusters, complete linkage leads to rather globular clusters with small diameters. Another clustering algorithm commonly applied is the UPGMA (unweighted pair-group method using arithmetic averages) clustering (Sneath and Sokal 1973). As in linkage clustering, the most similar pair of individuals is grouped together first. Then in a subsequent step, the similarity values of these two individuals with each of the remaining individuals are averaged leading to a new, smaller association matrix. In UPGMA clustering, only the relationships between groups are considered while information about the association between pairs of individuals is lost. Another clustering method that can either be applied directly on allele-frequency data or on a distance matrix is Ward's minimum variance method (Ward 1963). Here, the distance measure between two clusters is the ANOVA sum of squares between the two clusters added up over all individuals. At each step the within-cluster sum of squares is minimized over all partitions obtainable by merging two clusters from the previous step. Ward's method is strongly biased towards producing clusters with roughly the same number of observations and is also very sensitive to outliers (Sokal and Michener 1958).

During recent years, it has become more and more popular to display the relationships between individuals by means of an ordination method like principle component analysis (PCA), principle coordinate analysis (PCoA) or multidimensional scaling (MDS). These techniques can be used to reduce the information of a multidimensional data set in a way that it can be displayed in a scatter plot with only two or three axes. The PCA (Hotelling 1933; Rao 1964) is a linear transformation that rotates the axes in multidimensional space in a way that the major part of variance of the data set comes to lie on the first axis. The second axis is orthogonal to the first and accounts for the second greatest variance and so on. This method is very powerful, but defined for data with multinormal and thus unskewed distributions. Data derived from genetic marker analysis usually do not fulfil this assumption. In these cases, PCoA (Torgerson 1958; Rao 1964; Gower 1966) or MDS (Shepard 1962, 1966; Kruskal 1964a,b) can be applied. Both methods are useful for data of any mathematical kind. Unlike PCA in PCoA and MDS the ordination is not performed directly on the data matrix, but on a distance matrix derived from the original data by the application of any suitable association measure. Whereas in PCoA the goal is, like in PCA, to find new axes in a way that these are associated with a maximum of variability and at the same time preserve the distances between the individuals, MDS represents the individuals in a small and specific number of dimensions, where similar individuals are plotted close to one another and dissimilar individuals are placed far apart.

3 Diversity of crop plants as determined by evolution, domestication and breeding

For many of today's crop plants, domestication started 9000–13,000 years ago by selection of wild plants with certain traits meeting the needs of man (Diamond 2002). The repeated cultivation and maintenance of these selected plants under site-specific conditions led to landraces highly adapted to respective growing conditions and production methods. In some areas with extremely harsh conditions these landraces are still of certain relevance, while starting in Europe in the first half of the nineteenth century landraces were in many parts of the world gradually replaced by higher yielding cultivars generated by classical breeding methods (Smartt and Simmonds 1995).

These modern plant breeding strategies have led to a multiplication of yield and considerable quality improvements in many important crop plants. At the same time, the continuous selection for specific traits and higher yields has caused a loss in diversity in modern breeding material compared to landraces or the wild ancestors of cultivated plants. The domestication and the subsequent breeding processes represent two genetic bottlenecks restricting genetic variability. On the other hand, genetic diversity is a precondition for breeding. Without a broad base of heterogeneous plant material, it is impossible for plant breeders to produce cultivars that meet the changing needs regarding adaptation to growing conditions, resistance to biotic and abiotic stresses, product yield, or specific quality requirements. Therefore, the most efficient way to further improve the performance of crop varieties is still to have access to a large and diverse pool of genetic diversity.

In the following paragraphs, the development and breeding history of two distinct crop plants, i) barley and ii) rapeseed, will be described in detail. These two species are typical representatives of i) a self-pollinating cereal with a long-lasting history and ii) an outcrossing oil crop of more recent origin.

3.1 Barley (*Hordeum vulgare*)

Barley (*Hordeum vulgare* ssp. *vulgare*) is one of the oldest and most important crop plants. Due to its adaptability to a broad range of growing conditions barley is one of the world's most widespread cereals. In 2004 up to 57.0 million ha were planted with barley gaining an average yield of about 2.72 tons per ha (FAO 2005). In a few regions such as Ethiopia, Peru and Tibet, barley is still mainly used for food. However, by far the greatest share of the world's barley production is used for animal feed (about 85%), followed by malting barley for beer and whisky production.

3.1.1 Evolution and history of barley

The progenitor of today's cultivated barley is its wild relative *Hordeum vulgare* spp. *spontaneum* that nowadays can still be found in southern Turkey, the Aegean region, Israel, Jordan, Iran and central Asia, including Afghanistan and the Himalayan region. Like for many of today's crop plants, the domestication process started about 10,000–12,000 year ago in the region of the Fertile Crescent. This region, from Israel and Jordan to southern Turkey, Iraqii Kurdistan and south-western Iran, was first suggested to be the centre of origin by Vavilov and later by Harlan because of the broad diversity of wild as well as cultivated barleys and archaeological remains of barley grains found in this region (Harlan and Zohary 1966; Knüpffer et al. 2003). This idea was more recently confirmed by a molecular marker study on 317 wild and 57 cultivated barley lines (Badr et al. 2000). On the basis of 400 AFLP polymorphic loci, the genetic distances between these accessions were calculated and led to the conclusion that wild populations from Israel–Jordan are molecularly more similar to the cultivated gene pool than any other wild barleys. These molecular results strongly suggest a monophyletic origin of today's cultured barley in the Israel–Jordan area.

Starting from this region, the cultivation of barley expanded during prehistoric time. The neighbouring regions Anatolia and Iraq where reached first. There is archaeological evidence that barely was cultivated during the 6th millennium BC in Greece and 1 millennium later in Spain and the Lower Rhine region. During the same period of time, barley cultivation expanded to the Mediterranean and North African costal regions and Ethiopia. In prehistoric Egypt, barley was the most important cereal plant (Smartt and Simmonds 1995). The first domesticated barleys distributed during this process were two-rowed and covered forms. Naked as well as six-rowed barleys first appeared by 6500 BC in the regions of today's Israel and Iraq, respectively (Zohary and Hopf 1994). One of the most important traits for cultivation was the non-brittleness of rachis to ensure efficient harvest of grains. In addition to high productivity of seed material, different quality characteristics and tolerance to various types of biotic and abiotic stresses might have been important for selection during the migration process and changed several of the plants characteristics irreversibly as barley became a cultivated crop (von Bothmer et al. 2003).

Barley was first brought to America by Columbus during his second voyage in 1493, but remained of minor importance in the New World for a long time. Nowadays the USA and Canada contribute to the world's acreages with about 5.7 million ha, while with about 0,9 million ha barley production in South America is from a global point of view of minor importance (FAO 2005).

During the first half of the nineteenth century in Europe, the well-established landraces were further improved by means of single plant and mass selection processes. Further improvements in quality and yield were made by the utilization of recombination through cross-breeding. These processes led to considerable improvements in the performance of the breeding material in the UK, France, Germany and the Danube monarchy. In these agriculturally active regions, landraces disappeared during the early decades of the twentieth century and were replaced by high-yielding cultivars (Fischbeck et al. 2003). The magnitude of the genetic gain achieved by these intensive breeding activities was determined in several studies (Riggs et al. 1981; Wych and Rasmusson 1983; Bulman et al. 1993; Jedel and Helm 1994; Muñoz et al. 1998) and varies on average between 16 kg ha^{-1} y^{-1} for the period between

1920 and 1984 in the USA (Boukerrou and Rasmusson 1990) and 74 kg ha^{-1} y^{-1} between 1960 and 1980 in Italy (Martiniello et al. 1987). Results of field tests carried out on 64 six-rowed and 49 two-rowed winter barley cultivars that were registered in Germany during the last 40 years estimate the genetic gain in yield at 54.6 kg ha^{-1} y^{-1} (r^2=0.567) for the six-rowed cultivars and at 37.5 kg ha^{-1} y^{-1} (r^2=0.621) for the two-rowed cultivars (Ordon et al. 2005). For the actual state of barley variety development in Europe, see also Friedt and Rasmussen (2003).

3.1.2 Genetic diversity in wild and cultivated barley

The influence of the domestication process, followed by the distribution of barley from its center of origin to all parts of the world and the recent intensive breeding process on the diversity of barley can be assessed by means of molecular markers. A number of studies have been carried out to estimate the diversity within and between wild barley populations, to compare the genetic diversity in wild and cultivated barely and assess the diversity within today's modern, high-yielding barley cultivars.

Several studies found a high genetic diversity in today's wild barley populations originating from the Fertile Crescent and its neighboring regions by means of different DNA markers. The level of genetic diversity was higher within populations than between regions or between populations within a region (Brown et al. 1978; Nevo 1998), and in many cases genetic diversity was associated with eco-geographical (Song and Henry 1995) and climatic conditions in such a way that wild barley populations were more diverse in regions with climatically more stressful conditions (Pakniyat et al. 1997).

In most of the studies comparing the genetic diversity of cultivated barley with the currently found genetic diversity in its wild progenitor, the diversity within the genepool of cultivated barley was lower. Analysing SSR polymorphisms in 207 accessions of wild and cultivated barley, Saghai Maroof et al. (1994) found a significantly higher diversity in wild barley than in cultivated barley at three of the four loci under investigation. An analysis with 16 RFLP probes covering nearly all barley chromosomes showed a higher number of polymorphisms in the wild barleys than in the cultivated barleys. These analyses supported earlier results obtained by a study of ribosomal DNA spacer-length polymorphisms at two loci (Saghai Maroof et al. 1990).

To estimate the genetic diversity present in the gene pool of today's modern high-yielding barley cultivars, several studies have been carried out on cultivars originating mainly from North America and Europe. The results of these studies are difficult to compare, because of the utilization of different marker systems and different measures of genetic diversity. In an analysis carried out on 104 accessions of cultivated barley from all major barley growing areas of the world tested with four SSR primer combinations, the allele diversity was strongly dependent on the specific locus. For the four individual loci, the diversity was calculated as 0.11, 032, 0.88 and 0.94, respectively with an average diversity index (DI; Nei 1973) of 0.56 (Saghai Maroof et al. 1994). In 28 North American spring barley cultivars that were analysed by means of 100 genomic as well as cDNA derived RFLP probes, 57% of the cloned sequences detected polymorphisms. The average DI based on the probes that detected polymorphisms was assessed as 0.419 for the genomic probes and 0.762 for the cDNA derived clones (Dahleen et al. 1997). In an analysis of 25 European spring barley cultivars with 681 AFLP markers derived form eight primer combinations 62.1% of the markers were polymorphic. The genetic similarity among all genotypes, calculated as S_D, was extremely high, and ranged between 0.901 and 0.978 with an average of 0.932 (Schut et al. 1998). Casas et al. (1998) investigated 37 European spring and winter cultivars with 32 RFLP probes combined with three restriction enzymes for polymorphisms, and found an average S_D of 0.70 between the cultivars with a range between 0.533 and 0.957. In a set of 48 spring and winter cultivars that were registered between 1925 and 1988 in Germany, an analysis with 23 RFLP probes in combination with three restriction enzymes found 43% of polymorphisms in the whole set and 34% and 30% in subsets of spring and winter barleys, respectively (Graner et al. 1990).

To get an even more detailed insight into the influences modern plant breeding had on the genetic diversity during the last 4 decades, 64 six-rowed and 49 two-rowed cultivars that were registered during this time in Germany and gained some importance were analysed using a set of 30 SSR markers distributed evenly over the whole genome (Ordon et al. 2005). The 30 SSRs corresponded to 169 different alleles and the genetic similarity S_D was estimated on average at 0.51 on the whole set and at 0.56 and 0.58 for the six-rowed and two-rowed cultivars, respectively. In accordance to these results both subsets of cultivars showed on average a similar level of genetic diversity with a DI of 0.434 for the six-rowed and 0.418 for the two-rowed cultivars (Table 3). By grouping the cultivars according to their year of release, differences in the changes of genetic diversity between the two subsets were found. While the genetic diversity stayed constant over time in the six-rowed material, the diversity of the two-rowed cultivars was increased over recent decades. These results show that modern plant breeding not only caused a considerable yield gain in German two-rowed winter barley cultivars, but at the same time increased the genetic diversity in this crop.

Table 3. Genetic diversity (DI; Nei 1973) of six and two rowed German winter barley cultivars in relation to the year of release

	1959–2003	<1985	1985–1995	>1995
Six-rowed cultivars	0.434	0.432	0.419	0.392
Two-rowed cultivars	0.418	0.320	0.407	0.422

3.2 Oilseed rape (*Brassica napus*)

3.2.1 Evolution and history of brassicas

Rapeseed (*Brassica napus* L.), an allotetraploid (genome AACC, $2n=38$) member of the Brassicaceae (Cruciferae) family originated through spontaneous interspecific hybridization between turnip rape (*B. rapa* L., syn. *campestris*, AA, $2n=20$) and cabbage (*B. oleracea* L., CC, $2n=18$), resulting in the new amphidiploid species. Since no wild *B. napus* forms are known, it is assumed that the species arose relatively recently, in the Mediterranean region where both of the parental species concurred. The related amphidiploids Indian mustard (*B. juncea*, AABB, $2n=36$) and Abyssinian or Ethiopian mustard (*B. carinata*, genome BBCC, $2n=34$) arose in the same manner after crosses of black mustard (*B. nigra*, genome BB, $2n=16$) with *B. rapa* and *B. oleracea*, respectively. Rapeseed (*B. napus*) is today the most widely cultivated crop species of the crucifer family. The species is divided into two subspecies, i) the swedes (*B. napus* ssp. *napobrassica*) and ii) *B. napus* ssp. *napus*, including winter and spring oilseed, fodder and vegetable forms. The latter include the distinct leaf rapes (*B. napus* ssp. *napus* var. *pabularia*), which used to be common as a winter-annual vegetable (Snowdon et al. 2006).

Brassica vegetables and oilseeds were among the earliest systematically cropped plants. There are indications that a vegetable crucifer was cultivated already 10,000 years ago. In India records have been identified suggesting that oilseed brassicas (probably *B. rapa*) were used as early as 4000 BC, and later spread into China and Japan. Swedes were already known in Europe at Roman times, and utilization (probably of *B. rapa*) for oil purposes in northern Europe has probably begun around the thirteenth century. By the sixteenth century, rapeseed was the major source of lamp oil in Europe, but significant cultivation of the crop was only recorded from the eighteenth century on (Snowdon et al. 2006). During the course of the past 3 decades, oilseed rape has become a major international crop (Lühs and Friedt 1994). Today, it is the most important oilseed crop in Europe and only soybean has a greater importance worldwide (rapeseed approximately 27 million ha in 2005). Oilseed rape production is dominated by China (7.2 million ha), Canada (5.2 million ha) and Western Europe (Germany: 1.4 million ha), but the crop is also significant in Australia (1,080,000 ha), the Indian subcontinent and Eastern Europe (FAO 2005).

3.2.2 Rapeseed cultivation, oil quality and applications

The use of rapeseed oil for lamp fuel was largely superseded by petroleum since the end of the nineteenth century, and only the high quality of rapeseed fat as a lubricant for industrial machinery guaranteed continued production of the crop throughout the twentieth century. Oil from early rapeseed varieties was characterized by a high content of bitter tasting erucic acid (*cis* 13-docosenoic acid, 22:1n-9), which can lead to cardiac damage and related health problems. Therefore, rapeseed production in Europe only peaked significantly during the wars in the twentieth century when rapeseed oil was used especially for the production of margarine. The poor reputation of rapeseed oil as a foodstuff was overcome only by the development of "0" and "00" rapeseed varieties in the 1970s. The first major breakthrough came with the initial "0-quality" cultivars with erucic acid levels <2% in the seed lipids (Snowdon et al. 2006), whereas earlier rapeseed cultivars contained up to 50% erucic acid. Due to major improvements in seed analysis techniques fatty acid mutants had been identified. A spontaneous mutant of the German spring rapeseed cv. 'Liho' led to the release of a first erucic acid-free variety in Canada in the early 1970s. However, the value of the crop was still limited by the presence of high quantities of glucosinolates in the seed, which made the rapeseed extraction meal unsuitable as a feed for monogastric animals (the digestion of glucosinolates results in the release of toxic by-products). In 1969 the Polish low-glucosinolate spring rape variety 'Bronowski', was discovered and this provided the basis for international breeding activities to introduce this complex trait (at least three recessive genes for low glucosinolates) into high-yielding 0-material. The first 00-quality spring rapeseed variety, 'Tower', was release in 1974 and the advance of oilseed rape (canola) began. Modern rapeseed oil is used for various purposes. Besides its use as a highly nutritional food oil, rapeseed oil also provides a raw material for many other products ranging from rapeseed methyl ester (biodiesel) to industrial lubricants and hydraulic oils, tensids for detergent and soap production and biodegradable plastics.

Oilseed rape is cultivated in Europe and Asia predominantly as a winter form, whereas in Canada, northern Europe and Australia only spring forms are suitable. The differentiation into winter versus spring forms is due to the presence or absence of vernalization requirement for the onset of flowering. While spring oilseed rape is not winter-hardy and therefore is sown in spring, winter oilseed rape is sown in autumn and survives the winter in a leaf rosette. Shooting and flowering occurs in late spring, with pod development and ripening taking place over a period of about 6–8 weeks until mid-summer. As a member of the Brassicaceae, *B. napus* has a typical radial flower with four normally yellow petals. Flowers have one pair of lateral stamens with short filaments and four median stamens with longer filaments. In contrast to the majority of *B. rapa* and *B. oleracea*, its diploid progenitors, *B. napus* is a (facultative) outcrossing species with a variable degree of self-pollination. When insect pollinators such as bees and bumble bees, are abundant, a greater proportion of cross-pollination can occur, and by directed fertilization it is possible to obtain up to 100% outcrossing (e.g. using male-sterility systems for hybrid varieties). The world average rapeseed yield of about 1.68 metric tons per hectare (2005) covers a wide range, from 3.35 t/ha in the European Union (25), 1.04 in Australia, 1.57 in China, and 1.64 in Canada to 0.9 t/ha in India (FAO 2005). These differences are due to variety type and cultivar, climate and soil as well as agricultural inputs (seed quality, fertilizers, and agrochemicals).

3.2.3 Brassica species, wide hybridization and cytological status

After Morinaga (1934) and U (1935) had discovered that amphidiploid (allotetraploid) *Brassica* species originated from diploid progenitors, cytological studies became a leading role in genome analysis in the Brassicaceae. During the last few decades, classical cytogenetics have been replaced by DNA-based techniques and genomics. In earlier time, interspecific or intergeneric crosses, giving insight into genome relationships among *Brassica* species could only be studied by chromosome counts in somatic cells or meiotic studies of chromosome pairing and segregation. A first major achievement was the identification of the chromosome number for *B. rapa* by Takamine (1916), followed by the synthesis and analysis of *Raphanobrassica* by Karpechenko (1927). Then, the work of Morinaga (1934) and U (1935) initiated a better understanding of genome homology in the Brassicacaeae. The development of ovary culture and embryo rescue techniques in the 1950s along with chromosome pairing analyses enabled enormous progress in the study of genome homologies. The first detailed cytological description of *Brassica* somatic chromosome structure was published by Röbbelen (1960). Studies of chromosome pairing among a huge number of species by Harberd (1972), led to the classification of cytodemes describing homologous genomes. Since there is extensive genome homology or homeology throughout the entire *Brassica* coenospecies, it is possible to broaden rapeseed gene pools far beyond the species boundary by the introgression of

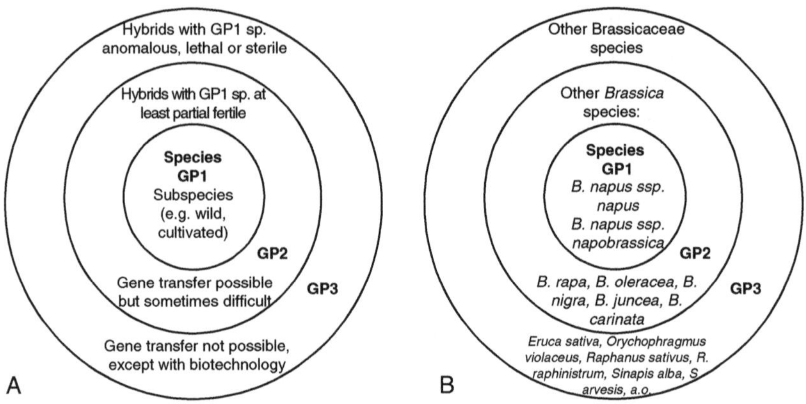

Fig. 1. Gene pool concept representing biological species, including subspecies e.g. wild and cultivated plants (GP1), their closest relatives, usually belonging to the same genus (GP2), and more distant species (GP3). **A** The general concept. **B** Gene pool concept for Brassicaceae (Hammer et al. 2003, adapted)

novel genes or alleles (Fig. 1). Consequently, related *Brassica* species and their relatives among the Brassicaceae represent a huge pool of potential gene donors for improving agronomic and quality traits of oilseed rape by breeding.

In more recent times, advance in the molecular cytogenetic technique of fluorescence *in situ* hybridization (FISH), allow the direct chromosomal localization of labelled DNA probes. FISH techniques have the potential for more reliable chromosome identification in *Brassica*, but also for the integration of genetic and physical maps, for ordering molecular markers and measuring physical genome distances, and for structural and functional chromosome analysis. FISH methods for the accurate localization of repetitive DNA sequences at chromosomal sub-arm level, particularly ribosomal DNA sequences, have provided karyotypes for *B. napus* and its progenitor species and the identification of A and C genome chromosomes in the amphidiploid species (Armstrong et al. 1998; Fukui et al. 1998; Snowdon et al. 2002). FISH hybridization of BAC clones to *B. oleracea* and *B. rapa* chromosomes represents a first step towards integration of physical and genetic maps with the karyogrammes of the diploid species and their amphidiploid hybrid *B. napus* (Howell et al. 2002, Jackson et al. 2000). Such comparative fibre-FISH mapping results support evidence that chromosomal duplications, rather than regional expansion due to accumulation of repetitive sequences in the intergenic regions, played the major role in the evolution of the diploid *Brassica* genomes. The use of total genomic DNA as a FISH probe (genomic in-situ hybridization, or GISH) is especially useful for diagnostic studies of the amount and integration of foreign chromatin in interspecific and intergeneric plant hybrids (Heslop-Harrison and Schwarzacher 1996). Hybrids between high-yielding rapeseed cultivars and related species are relatively easily produced and have often been used to develop new lines containing introduced traits such as novel pest or disease resistances. Great advances in interspecific hybridization have resulted from the application of in vitro techniques for the generation of viable offspring from interspecific and intergeneric hybrids (Lühs et al. 2002). Identification of alien DNA in wide crosses has been achieved by quantification of chromosome content by flow cytometry and by tracing chromosome and DNA transfer using molecular markers (Sabharwal and Dolezel 1993). Visualization of alien chromatin in interspecific hybrids using in situ hybridization techniques, on the other hand, potentially enables pinpointing of introgressions to specific chromosomes (Heslop-Harrison and Schwarzacher 1996; Snowdon et al. 1997).

Intergeneric sexual hybrids between *B. napus* and *Sinapis arvensis* containing novel genes for resistance against blackleg disease on chromosome additions and introgressions were analysed via GISH by Snowdon et al. (2000). Selfed BC_3 progenies included fertile plants exhibiting high seedling and adult plant resistance associated with the presence of an acrocentric addition chromosome from *S. arvensis.* Furthermore, some individuals with adult plant resistance but cotyledon susceptibility were observed to have a normal *B. napus* karyotype with no visible GISH signals, indicating introgression lines carrying at least a subset of the *S. arvensis* resistance genes. Schelfhout et al. (2006) used a B-genome specific centromeric repeat sequence as a PCR and FISH marker to characterize B-genome introgressions in sexual progeny from *B. napus*×*B. juncea* crosses exhibiting various traits of agronomic interest, including resistance against blackleg disease and pod shattering. Genotypes with normal *B. napus* karyotype were identified in which the minisatellite sequence could be detected by PCR although no FISH signals were observed, indicating small chromosomal introgressions that carried the gene of interest. Voss et al. (2000) generated intergeneric crosses between spring oilseed rape and nematode-resistant oil radish (*R. sativus*) genotypes, using embryo rescue to overcome incompatibility barriers. In three backcross (BC) generations, highly resistant progeny with a minimal number of *R. sativus* chromosomes were selected by resistance testing accompanied by GISH analysis. This strategy led to the identification of a resistant BC_3 plant with a monosomic, acrocentric addition chromosome. This individual was backcrossed once again to produce a stable disomic addition line; however, efforts to introgress the resistance on a stable introgression failed. Similarly, Peterka et al. (2004) also generated oilseed rape interspecific hybrid lines containing nematode resistance on a monosomic *R. sativus* addition chromosome. Here, *R. sativus* chromatin was identified by PCR and FISH with a

Raphanus-specific centromeric repeat sequence. In this case also, however, no intergenomic transfer of the resistance was reported. Fahleson et al. (1994) analysed somatic hybrids between *Eruca sativa* and *B. napus* using *in situ* hybridization with two *E. sativa*-specific repetitive DNA sequences accompanied by GISH. One of the repetitive sequences showed 100% similarity with a part of the *E. sativa* rDNA intergenic spacer, and localized to the three pairs of *E. sativa* rDNA loci, whereas the other clone was a tandemly repeated element located close to the telomeres on at least ten *E. sativa* chromosomes. Analysis of progenies derived from the somatic hybrids revealed the presence of *E. sativa* DNA; however, no intergenomic translocations could be detected by GISH, although the somatic hybrid progeny contained one to two complete *E. sativa* chromosomes. Together, these results emphasize the fact that chromosome translocations among non-homologous genomes are more likely in the presence of homeologous chromosome pairing allowing intergenomic recombination. Genome homeology at the chromosomal level is expected to be more extensive between oilseed rape and its closer relatives, and this is confirmed by the relative ease of transferring agronomic traits to *B. napus* from *B. nigra*, *B. juncea*, *B. carinata* and *Sinapis* species, in comparison with the difficulties observed in more distant crosses. On the other hand, successful transfer of genes of interest in intertribal asymmetric hybrids has also been demonstrated on a number of occasions and may indicate unknown or partial genome homologies. Interesting results in this respect were obtained by Wang et al. (2004), who produced sexual progenies of asymmetric somatic hybrids between *B. napus* and *Crambe abyssinica* in an effort to improve the fatty-acid composition of oilseed rape seed. Through meiotic GISH, these authors were able to identify intergenomic chromatin bridges and detect asynchrony between the *B. napus* and *C. abyssinca* meiotic cycles. Lagging, bridging and late disjunction of univalents derived from *C. abyssinica* were observed, whereas analysis of cleaved amplified polymorphic sequence (CAPS) markers derived from the *FAE1* gene showed novel patterns different from the *B. napus* recipient in some hybrid offspring. This indicated the existence of novel allelic variation in the interspecific hybrids that presumably arose from introgression from crambe chromatin to one or more *B. napus* chromosomes. Some of the recombinant offspring contained significantly greater amounts of seed erucic acid than the *B. napus* parent, demonstrating that it is possible to transfer agronomic traits from distantly-related crucifers into elite oilseed rape material. In another example, Winter et al. (2003) used mitotic GISH to characterize recombination lines containing genes for blackleg resistance from *Moricandia arvensis.* Resistant lines were identified which exhibited a normal *B. napus* karyotype but carried the *Moricandia* resistance genes on putative chromosome introgressions. Although such crosses can exhibit significant linkage drag and hence must be viewed as very basic material from a breeding perspective, such pre-breeding is of enormous interest in terms of broadening the genetic variability for particular traits where little variation is available within *B. napus* itself.

Due to its partial allogamy, oilseed rape can be treated both as a self-pollinating or an outcrossing species. Backcrossing has been used to transfer genetically simple traits such as low erucic acid and glucosinolate content into adapted breeding material. *Brassica napus* is also one of the crop species most amenable to the application of biotechnology. For instance, it is possible to reproducibly obtain haploid and subsequently doubled-haploid (DH) plants through anther and/or microspore culture (cf. Weber et al. 2005). Major advantages of haploidy technique are simpler genetic segregations and the rapid fixation of rare recessive genotypes. Thus, utilization of microspore culture allows a substantial acceleration of the breeding cycle. DH production has become common practice in commercial breeding rapeseed programs and has already resulted in numerous licensed cultivars. Besides haploid techniques, wide hybridization using embryo rescue or protoplast fusion is also used to create novel genetic variation. However, once a useful property has been identified in a basic breeding stock, e.g. a mutant line or germplasm from a wild relative, it still takes many years to develop cultivars possessing this novel desired trait. Here, marker-assisted selection has become a significant impact on the efficiency of plant breeding practice. In cases where conventional approaches have not been sufficient, further improvements are achievable by genetic engineering.

3.2.4 Exploitation of novel genetic variation for rapeseed breeding

In modern plant breeding, it becomes more and more important to have suitable and efficient tools for an effective discrimination of genotypes and breeding lines. Many studies have demonstrated the use of molecular marker techniques for the analysis of genetic variation in oilseed rape: Becker et al. (1995) compared cultivars and resynthesized lines (RS; generated from interspecific hybridization between suitable forms of *Brassica rapa* L. and *B. oleracea*) by isozyme and RFLP markers and concluded that RS rapeseed are a suitable resource for broadening the genetic base of the species; Song et al. (1995) described the rapid genome changes in synthetic *Brassica* polyploids and discussed the evolutionary implications arising from the ability of polyploid species to generate extensive genetic diversity in a short period of time; Thormann et al. (1994) used RFLP and RAPD markers to determine genetic distances in and between cruciferous species; Halldén et al. (1994) characterized *B. napus* breeding lines by RFLP and RAPD techniques, while Diers and Osborn (1994) compared RFLP patterns in 61 winter and spring rapeseed genotypes and concluded that the two forms constitute two genetically distinct groups. The relationship between genetic distance and heterosis in oilseed rape was investigated by Diers et al. (1996) using RFLP markers, and by Riaz et al. (2001) with sequence-related amplified polymorphisms

(SRAP). Plieske and Struss (2001) were able to clearly differentiate winter and spring rapeseed in a cluster analysis using SSR markers. Furthermore, RAPDs were used by Mailer et al. (1994) for discrimination of rapeseed cultivars, and by Demeke et al. (1992) for taxonomic analyses in *Brassica* species. Lombard et al. (2000) used AFLPs to genotype winter rapeseed and to estimate genetic similarities and demonstrated the effectiveness of AFLP markers for genetic distinction of cultivated rapeseed types.

Many cultivars still represent a kind of (pure) lines (also named "open pollinated" varieties, OP varieties) derived by breeding schemes designed for self-fertilizing crops, i.e. pedigree selection or modifications thereof. However, an increasing number of cultivars nowadays are single cross (F_1) hybrids derived from a male-sterile female and a corresponding fertile male restorer line. For example, the current German list of winter rapeseed cultivars comprises 56 varieties, 14 of which are restored hybrids (BSA 2005). Leading cultivars such as Oase and Trabant represent the line and the hybrid type, respectively, and are both characterized by a very high oil yield. Both OP and hybrid cultivars show an enormous genetic variability, since they can give rise to highly diverse progeny due to genetic segregation, particularly when extremely different parents were used for hybrid production. However, breeding for agronomic and economic value tends to biased exploitation of genetic variation. Therefore, it is necessary to continuously integrate diverse germplasm into the breeding process to maintain a high level of genetic diversity. Here, resynthesis of novel *B. napus* genotypes through artificial crosses between the diploid parents, assisted by embryo rescue, has repeatedly been shown to be useful for broadening the genetic basis of rapeseed. Both progenitor species exhibit an extremely broad genetic and phenotypic diversity that is the potential for a huge variety of different RS rapeseed forms (Chen and Heneen 1989, Song et al. 1993, Lühs et al. 2000). The relatively high extent of intergenomic recombination between A and C genome chromosomes in early generations of RS rape further increases the potential for creation of novel genotypes through resynthesis (Song et al. 1993; Lydiate et al. 1995). In a detailed study, Seyis et al. (2003) have compared RS rapeseed lines originating from crosses between Yellow Sarson (*B. rapa* ssp. *trilocularis*, high erucic acid and high glucosinolate contents, and different cauliflower (*B. oleracea L.* convar. *botrytis*, BK) varieties using AFLP markers (Fig. 2). Genetic diversity was compared with a collection of diverse spring oilseed and fodder rape types from different countries (Australia, Canada, Denmark, France, Germany, Sweden, New Zealand) by PCoA analysis. A total of 165 RS lines were tested in field trials together with 40 European spring oilseed and fodder rape cultivars. All RS lines were self-fertile, possessed traditional high-erucic and high-glucosinolate seed quality, had moderate or no vernalization requirement and week winter hardiness (cf. Seyis et al. 2003). For genetic analysis, a set of three AFLP primer combinations was used, scoring a total of 467 polymorphic bands in the complete set of *B. napus* RS lines and spring rapeseed varieties. The 40 cultivars studied formed a distinct cluster in the PCoA analysis (Fig. 2). As expected, the RS lines proved to be rather closely related due to their

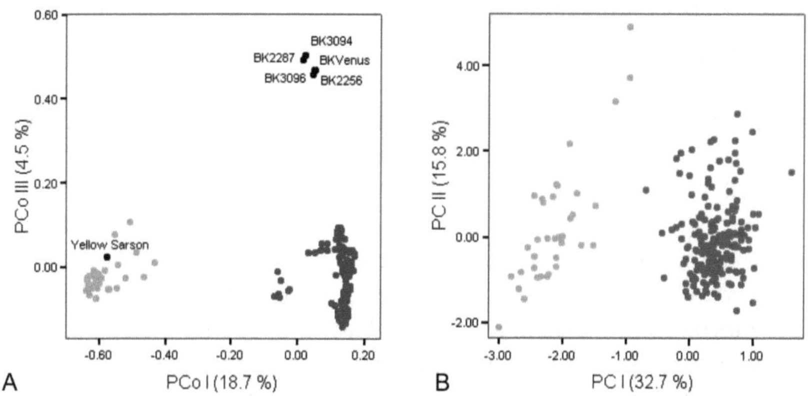

Fig. 2. Genotypic and phenotypic relatedness among resynthesized rapeseed lines (dark grey), their parents (black) and a set of spring rapeseed cultivars (light grey) **A** PCoA analysis based on 467 polymorphic AFLP fragments showing the genetic relatedness. **B** PCA analysis based on 19 different morphologic and agronomic traits. Both analyses show a clear cut separation on the first axis between resynthesized rapeseed lines and the cultivars (Seyis et al. 2003; modified)

common *B. rapa* parent Yellow Sarson. However, in addition to the molecular data the phenotypic data from field evaluations of morphologic and agronomic traits including plant height, leaf morphology, days to flowering, flowering period, time of maturity, vegetation period and seed yield components showed a clear differentiation between the different RS families (Fig. 2), which in turn could be clearly differentiated from natural spring rapeseed cultivars, fodder types and Canadian oil types.

Altogether, the results of different studies show that a vast, exploitable genetic variation exists both in spring and winter rapeseed materials. In addition, both groups represent clearly differentiated genetic pools which form the basis for the identification of superior combinations for breeding of, both, line and hybrid varieties. However, the limited geographic range of rapeseed and intensive breeding particularly focussing on canola (00) quality has led to a comparatively narrow genetic basis of current breeding materials. In this context, RS material can significantly increase the available gene pool and provide novel basic germplasm for further improvements of agronomic and quality traits, i.e. grain yield, oil and meal composition. Furthermore, RS rapeseed is potentially of great value for hybrid breeding, since heterotic effects tend to be higher in crosses of genetically distant versus more related materials. On the other hand, RS genotypes are generally not suitable to be used directly for oilseed rape hybrids, because they usually display inferior seed quality traits such as low seed oil content, high erucic acid and glucosinolate contents as well as other undesirable quality traits. The large genetic distance between RS lines and rapeseed cultivars is of particular interest in terms of increasing potential heterosis via hybrid development. In order to investigate the heterotic potential for yield components, selected RS lines from the material described by Seyis et al. (2003) were used to develop experimental hybrids with male-sterile breeding lines. In field trials at three locations, several hybrids based on RS lines gave a higher yield potential than check cultivars (cf. Seyis et al. 2002; Spiller et al. 2004).

The enormous potential of resynthesic basic material for rapeseed breeding will make efforts dealing with germplasm conservation and well-directed use of the diploid *Brassica* parents more important in the future. On the other hand, the initial yield potential of RS rapeseed (usually spring type) is low. Therefore, the use of such forms and the novel genetic variability thus created must be directed, particularly with regard to seed quality and yield, to facilitate its integration into high-yielding winter rapeseed breeding material. Specifically, this can be achieved by producing semi-synthetic rapeseed forms via backcrosses to adapted cultivars or by developing high-yielding hybrids. Obviously, the establishment of new gene pools based on novel *B. napus* is limited by its inferior agronomic performance and seed quality (high erucic, high glucosinolates), hence this approach must be considered under more long-term perspectives. However, the discovery of low-erucic acid mutants among *B. oleracea* accessions and the development of RS rapeseed forms via interspecific crosses with interesting 0- or 00-quality *B. rapa* genotypes will open the possibility to use such new rapeseed material as a genetic source for quality and yield improvement in oilseed rape (Lühs et al. 2000, 2002; Seyis et al. 2002). As demonstrated, molecular markers such as AFLPs and SSRs assist in the evaluation of RS lines in terms of describing their genetic distance in relation to existing breeding material, in order to enrich the available gene pool for breeding of the crop species. Based on the broad genetic diversity in the basic *Brassica* materials, including wild species, undomesticated accessions, breeding lines and released varieties, novel rapeseed cultivars representing a great genetic diversity will continue to be valuable sources for rapeseed breeding.

4 Conclusions

Modern plant breeding methods have enabled rapid breeding progress leading to a multiplication of product yield of practically all crop plants. Due to the bottleneck function of domestication and continuous selection for higher yield along with specific resistance and quality traits genetic diversity in modern cultivars has decreased as compared with old varieties, landraces or wild ancestors of cultivated plants. However, genetic diversity is a prerequisite for breeding progress regarding major characteristics such as adaptation to growing conditions, resistance to biotic and abiotic stresses or quality requirements. Therefore, the creation of novel genetic variation is a major, continuous task of plant breeders to build the basis for further improvements in the performance of crop varieties. As shown in this chapter, plant breeders have many options to widen genetic diversity in respective plant species, like barley and rapeseed. In particular novel genetic variation can still be created by sexual hybridisations within and across species, i.e. (i) intraspecific crosses between distinct genotypes within a species i.e. subspecies, landraces, varieties, cultivars or inbred lines, (ii) interspecific crosses between related species within a genus leading to fertile hybrids, e.g. *Brassica* sp., *Hordeum* sp. or *Triticum* sp., (iii) intergeneric crosses between species belonging to different but related genera, e.g. *Brassica* x *Orychophragmus*, *Brassica*×*Raphanus*, *Festuca*×*Lolium*, *Triticum*×*Secale* (Triticale).

In addition to that, cell fusion allows the creation of hybrids which would be extremely difficult to create or not achievable at all by sexual hybridization, e.g. products of intergeneric somatic hybridisation between distantly related cereals such as rice (*Oryza sativa*) and barley (*Hordeum vulgare*). However, if the distance is too large, the fusion product may still be sterile, as in the case of rice–barley hybrids (Kisaka et al. 1998). Again, the results will be more promising in the case of closer related genera or even hybridisations between species belonging to the same genus, like in the case of asymmetric fusions between cells of *Helianthus annuus* and *H. maximiliani*, leading to partial hybrids and selected sunflower lines with enhanced resistance against the fungal pathogen *Sclerotinia sclerotiorum* (Rönicke et al. 2004).

In addition, genetic engineering allows the creation of new genetic variation since useful genes for plant breeding are already abundant (cf. Koornneef and Stam 2001). Numerous respective examples in the literature include improvements of important agronomic and quality traits such as fatty acid composition (e.g. Stoll et al. 2005, 2006) and tocopherol content (e.g. Kumar et al. 2005; Raclaru et al. 2006). Along with increasing knowledge of plant biochemistry and physiology the options for broadening genetic variation and plant improvement by breeding are expected to grow substantially.

References

Armstrong SJ, Fransz P, Marshall DF, Jones GH (1998) Physical mapping of DNA repetitive sequences to mitotic and meiotic chromosomes of *Brassica oleracea* var. *alboglabra* by fluorescence in situ hybridization. Heredity 81:666–673

Badr A, Muller K, Schafer-Pregl R, El Rabey H, Effgen S, Ibrahim HH, Pozzi C, Rohde W, Salamini F (2000) On the origin and domestication history of barley (*Hordeum vulage*). Mol Biol Evol 17:499–510

Barrett BA, Kidwell KK, Fox PN (1998) Comparison of AFLP and pedigree-based genetic diversity assessment methods using wheat cultivars from the Pacific Northwest. Crop Sci 38:1271–1278

Batley J, Mogg R, Edwards D, O'Sullivan H, Edwards KJ (2003) A high-throughput SNuPE assay for genotyping SNPs in the flanking regions of *Zea mays* sequence tagged simple sequence repeats. Mol Breeding 11:111–120

Baum BR, Nevo E, Johnson, DA, Beiles A (1997) Genetic diversity in wild barley (*Hordeum spontaneum* C. Koch) in the Near East: a molecular analysis using Random Amplified Polymorphic DNA (RAPD) markers. Genet Res Crop Evol 44:147–157

Becker HC, Engqvist GM, Karlsson B (1995) Comparison of rapeseed cultivars and resynthesized lines based on allozyme and RFLP markers. Theor Appl Genet 91:62–67

Beebe S, Skroch PW, Tohme J, Duque MC, Pedraza F, Nienhuis J (2000) Structure of genetic diversity among common bean landraces of Middle American origin based on correspondence analysis of RAPD. Crop Sci 40:264–273

Botstein D, White RL, Skolnick M, Davis RV (1980) Construction of a genetic map in man using restriction fragment length polymorphisms. Am J Hum Genet 32:314–331

Boukerrou L, Rasmusson DD (1990) Breeding for high biomass yield in spring barley. Crop Sci 30, 31–35

Brown AHD, Nevo E, Zohary D, Dagan O (1978) Genetic variation in natural populations of wild barley (*Hordeum spontaneum*). Genetica 49:97–108

BSA (2005) Beschreibende Sortenliste. Deutscher Landwirtschaftsverlag. Hannover

Bulman P, Mather DE, Smith DL (1993) Genetic improvement of spring barley cultivars grown in eastern Canada from 1910 to 1988. Euphytica 71:35–48

Casas MA, Igartua E, Valles MP, Molina-Cano JL (1998) Genetic diversity of barley cultivars grown in Spain by RFLP, similarity and co ancestry coefficients. Plant Breeding 117:429–436

Chen Y, Heneen WK (1989) Resynthesized *Brassica napus* L.: A review of its potential in breeding and genetic analysis. Hereditas 111:255–263

Dahleen LS, Hoffman DL, Dohrmann J, Gruber R, Franckowiak J (1997) Use of a subset of doubled-haploid lines for RAPD interval mapping in barley. Genome 40:626–632

Dawson IK, Chalmers KJ, Waugh R, Powell W (1993) Detection and analysis of genetic variation in *Hordeum spontaneum* populations from Israel using RAPD markers. Mol Ecol 2:151–159

Demeke T, Adams RP, Chibbar R (1992) Potential taxonomic use of random amplified polymorphic DNA (RAPD): a case study in Brassica. Theor Appl Genet 84:990–994

Diamond J (2002) Evolution, consequences and future of plant and animal domestication. Nature 418:700–707

Dice LR (1945) Measures of the amount of ecologic association between species. Ecology 26:297–302

Diers BW, Osborn TC (1994) Genetic diversity of oilseed *Brassica napus* germplasm based on restriction fragment length polymorphisms. Theor Appl Genet 88:662–668

Diers BW, McVetty PBE, Osborn TC (1996) Relationship between heterosis and genetic distance based on RFLP markers in oilseed rape (*Brassica napus* L.). Crop Sci 36:79–83

Fahleson J, Eriksson I, Landgren M, Stymne S, Glimelius K (1994) Intertribal somatic hybrids between *Brassica napus* and *Thlaspi perfoliatum* with high conten of the *T. perfoliatum*-specific nervonic acid. Theor Appl Genet 87:795–804

FAO (2005) FAOSTAT: Agricultural Data. *http://apps.fao.org*, last accessed: October 2005

Fischbeck G (2003) Diversification through breeding. In: von Bothmer R, van Hintum T, Knüpffer H, Sato K (eds) Diversity in barley (*Hordeum vulgare*). Elsevier Science BV, Amsterdam, The Netherlands. pp. 29–52

Fofana B, Vekemans X, du Jardin P, Baudoin JP (1997) Genetic diversity in Lima bean (*Phaseolus lunatus* L.) as revealed by RAPD markers. Euphytica 95: 157–165

Friedt W, Rasmussen M (2003) Modern European barley cultivars: genetic progress in resistance, quality and yield. In: Maèr, P. Faccioli, A.M. Stanca (eds) From biodiversity to genomics: breeding strategies for small grain cereals in the third millennium. Exp Inst Cereal Res, Fiorenzuola d'Arda, Italy, pp 73–78.

Fukui K, Nakayama S, Ohmido N, Yoshiaki H, Yambe M (1998) Quantitative karyotyping of three diploid *Brassica* species by imaging methods and localizaion of 45SrDNA loci on the identified chromosomes. Theor Appl Genet 96, 325–330

Goodman MM, Stuber CW (1983) Races of maize: VI. Isozyme variation among races of maize in Bolivia. Maydica 28:169–187

Gower JC (1966) Some distance properties of latent root and vector methods used in multivariate analysis. Biometrika 53:325–338.

Graner A, Siedler H, Jahoor A, Herrmann RG, Wenzel G (1990) Assessment of the degree and the type of restriction fragment length polymorphism in barley (*Hordeum vulgare*). Theor Appl Genet 80:826–832

Gupta PK, Roy JK, Prasad M (2001) Single nucleotide polymorphisms: a new paradigm for molecular marker technology and DNA polymorphism detection with emphasis on their use in plants. Curr Sci 80:524–535

Halldén C, Nilsson NO, Rading I, Sall T (1994) Evaluation of RFLP and RAPD markers in a comparison of *Brassica napus* breeding lines. Theor Appl Genet 88:123–128

Hammer K, Arrowsmith N, Gladis T (2003) Agrobiodiversity with emphasis on plant genetic resources. Naturwissenschaften 90:241–250

Harberd DJ (1972) A contribution to the cyto-taxonomy of *Brassica* (*Cruciferae*) and its allies. Bot J Linn Soc 65:1–23

Harlan JR, Zohary D (1966) Distribution of wild wheats and barley. Science 153: 1074–1080

Henikoff S, Comai L (2003) Single-nucleotide mutations for plant functional genomics. Annu Rev Plant Biol 54:375–401

Heslop–Harrison JS, Schwarzacher T (1996) Genomic southern and *in situ* hybridization for plant genome analysis. In: Jauhar PP (ed) Methods of genome analysis in plants. CRC Press, Boca Raton, Florida, pp 163–179

Hongtrakul V, Huestis GM, Knapp SJ (1997) Amplified fragment length polymorphisms as a tool for DNA fingerprinting sunflower germplasm: genetic diversity among oilseed inbred lines. Theor Appl Genet 95:400–407

Horejsi T, Staub JE (1999) Genetic variation in cucumber (*Cucumis sativus* L.) as assessed by random amplified polymorphic DNA. Genet Res Crop Evol 46:337–350

Hotelling H (1933) Analysis of a complex of statistical variables into principle components. J Educ Psychol 24:417–520.

Howell EC, Barker GC, Jones GH, Kearsey MJ, King GJ, Kop EP, Ryder CD, Teakle GR, Vicente JG, Armstrong SJ (2002) Integration of the cytogenetic and genetic linkage maps of *Brassica oleracea.* Genetics 161:1225–1234

Jaccard P (1908) Nouvelles recherches sur la distribution florale. Bull Doc Vaudoise Sci Nat 44:223–270

Jackson SA, Cheng ZK, Wang ML, Goodman HM, Jiang JM (2000) Comparative fluorescence *in situ* hybridization mapping of a 431–kb *Arabidopsis thaliana* bacterial artificial chromosome contig reveals the role of chromosomal duplications in the expansion of the *Brassica rapa* genome. Genetics 156:833–838

Jedel P, Helm JH (1994) Assessment of western Canadian barleys of historical interest: I. Yield and agronomic traits. Crop Sci 34:922–927

Karpechenko GD (1927) Polyploid hybrids of *Raphanus sativus* X *Brassica oleracea.* Appl Bot Genet Plant Breed 17:305–410

Kisaka H, Kisaka M, Kanno A, Kameya T (1998) Intergeneric somatic hybridization of rice (*Oryza sativa* L.) and barley (*Hordeum vulgare* L.) by protoplast fusion. Plant Cell Reports 17:362–367

Knüpffer H, Terentyeva I, Hammer K, Kovaleva O, Sato K (2003) Ecogeographical diversity—a Vavilovian approach. In: von Bothmer R, van Hintum T, Knüpffer H,Sato K (eds) Diversity in barley (*Hordeum vulgare*). Elsevier Science BV, Amsterdam, The Netherlands, pp 53–76

Koornneef M, Siam P (2001) Changing paradigms in plant breeding. Plant Physiol 125:156–159

Kruskal JB (1964a) Multidimensional scaling by optimising goodness of fit to a nonmetric hypothesis. Psychometrica 29:1–27

Kruskal JB (1964b) Nonmetric multidimensional scaling: a numerical method. Psychometrica 29:115–129

Kumar R, Raclaru M, Schüaeler T, Gruber J, Sadre R, Lühs W, Zarhloul KM, Friedt W, Enders D, Frentzen M, Weier D (2005) Characterisation of plant tocopherol cyclases and their overexpression in transgenic *Brassica napus* seeds. FEBS Lett 579:1357–1364

Lombard V, Baril CP, Dubreuil P, Blouet F, Zhang D (2000) Genetic relationships and fingerprinting of rapeseed cultivars by AFLP: consequences for varietal registration. Crop Sci 40:1417–1425

Lühs W, Friedt W (1994) Major oil crops. In: Murphy DJ (ed) Designer oil crops. VCH Verlagsgesellschaft, Weinheim, Germany. pp 5–71

Lühs W, Seyis F, Voss A, Friedt W (2000) Genetics of erucic acid content in *Brassica oleracea* seed oil. Czech J Genet Plant Breed 36:116–120

Lühs W, Seyis F, Snowdon R, Baetzel R, Friedt W (2002) Genetic improvement of *Brassica napus* by wide hybridisation. GCIRC Bull 18:227–234

Lydiate D, Sharpe AG, Parkin I (1995) Colinearity and homoelogous recombination between the A and C genomes of *Brassica napus*. Proc 9th Int Rapeseed Congress (GCIRC) Vol 4. Cambridge, UK, pp 1122–1124

Mailer RJ, Scarth R, Fristensky B (1994) Discrimination among cultivars of rapeseed (*Brassica napus* L.) DNA polymorphisms amplified from arbitrary primers. Theor Appl Genet 87:697–704

Martiniello P, Delogu G, Oboardi M, Boggini G, Stanca AM (1987) Breeding progress in grain yield and selected agronomic characters of winter barley (*Hordeum vulgare* L.) over the last quarter of a century. Plant Breed 99:289–294

Maughan PJ, Saghai Maroof MA, Buss GR, Huestis GM (1996) Amplified fragment length polymorphism (AFLP) in soybean: species diversity, inheritance, and near-isogenic line analysis. Theor Appl Genet 93:392–401

McGrath JM, Quiros CF (1992) Genetic diversity at isozyme and RFLP loci in *Brassica campestris* as related to crop type and geographical origin. Theor Appl Genet 83:783–790

Melchinger AE, Lee M, Lamkey KR, Hallauer AR, Woodman WL (1990) Genetic diversity for restriction fragment length polymorphisms and heterosis for two diallel sets of maize inbreds. Theor Appl Genet 80:488–496

Messmer MM, Melchinger AE, Lee M, Woodman WL, Lee EA, Lamkey KR (1991) Genetic diversity among progenitors and elite lines from the Iowa Stiff Stalk Synthetic (BSSS) maize population: comparison of allozyme and RFLP data. Theor Appl Genet 83:97–107

Miller JC, Tanksley SD (1990) RFLP analysis of phylogenetic relationships and genetic variation in the genus *Lycopersicon*. Theor Appl Genet 80:437–448

Morinaga T (1934) Interspecific hybridization in *Brassica*. VI. The cytology of F_1 hybrids of *Brassica juncea* and *B. nigra*. Cytologia 6:62–67

Muñoz P, Voltas J, Araus JL, Igartua E, Romagosa I (1998) Changes over time in the adaptation of barley releases in northeastern Spain. Plant Breed 117:531–535

Nei M (1973) Analysis of gene diversity in subdivided populations. Proc Natl Acad Sci USA 70:3321–3323

Nevo E (1998) Genetic diversity in wild cereals: regional and local studies and their bearing on conservation ex situ and in situ. Genet Res Crop Evol 45:355–370

Olufowote JO, Xu Y, Chen X, Park WD, Beachell HM, Dilday RH, Goto M, McCouch SR (1997) Comparative evaluation of within cultivar variation of rice (*Oryza sativa* L.) using microsatellite and RFLP markers. Genome 40:370–378

Ordon F, Ahlemeyer J, Werner K, Köhler W, Friedt W (2005) Molecular assessment of genetic diversity in winter barley and its use in breeding. Euphytica 146:21–28

Pakniyat H, Powell W, Baird E, Handley LL, Robinson D, Scrimgeour CM, Nevo E, Hackett CA, Caligari PDS and Forster BP (1997) AFLP variation in wild barley (*Hordeum spontaneum* C. Koch) with reference to salt tolerance and associated ecogeography. Genome 40:332–341

Peterka H, Budahn H, Schrader O, Ahne R, Schütze W (2004) Transfer of resistance against the beet cyst nematode from radish (*Raphanus sativus*) to rape (*Brassica napus*) by monosomic chromosome addition. Theor Appl Genet 109:30–41

Plieske J, Struss D (2001) Microsatellite markers for genome analysis in *Brassica*. I. Development in *Brassica napus* and abundance in Brassicaceae species. Theor Appl Genet 102:689–694

Prescott–Allen R, Prescott–Allen C (1990) How many plants feed the world? Conserv Biol 4:365–374

Raclaru M, Gruber J, Kumar R, Sadre R, Lühs W, Zarhloul KM, Friedt W, Frentzen M, Weier D (2006) Increase of the tocochromanol content in transgenic *Brassica napus* seeds by overexpression of key enzymes involved in prenylquinone biosynthesis. Mol Breeding (in press)

Rafalski A (2002) Applications of single nucleotide polymorphisms in crop genetics. Curr Opin Plant Biol 5:94–100

Rao CR (1964) The use and interpretation of principal component analysis in applied research. Sankhyaá Ser A 26:329–358

Riaz A, Li G, Quresh Z, Swati MS, Quiros CF (2001) Genetic diversity of oilseed *Brassica napus* inbred lines based on sequence-related amplified polymorphism and its relation to hybrid performance. Plant Breeding 120:411–415

Riggs TJ, Hanson PR, Start ND, Miles DM, Morgan CL, Ford MA (1981) Comparison of spring barley varieties grown in England and Wales between 1880 and 1980. J Agric Sci Camb 97:599–610

Röbbelen G (1960) Beiträge zur Analyse des *Brassica*-genomes. Chromosoma 11:205–228

Rongwen J, Akkaya MS, Bhagwat AA, Lavi U, Cregan PB (1995) The use of microsatellite DNA markers for soybean genotype identification. Theor Appl Genet 90:1432–2242

Rönicke S, Hahn V, Friedt W (2005) Molecular characterisation of sunflower inbred lines and identification of *Sclerotinia* resistance genotypes. Plant Breeding 124:376–381

Rönicke S, Hahn V, Horn R, Gröne I, Brahm L, Schnabl H, Friedt W (2004) Interspecific hybrids of sunflower as a source of *Sclerotinia* resistance. Plant Breeding 123:152–157

Russell J, Fuller J, Young G, Thomas B, Taramino G, Macaulay M, Waugh R, Powell W (1997) Discriminating between barley genotypes using microsatellite markers. Genome 40:442–450

Sabharwal PS, Dolezel J (1993) Interspecific hybridization in *Brassica*: application of flow cytometry for analysis of ploidy and genome composition in hybrid plants. Biol Planta 35:169–177

Saghai Maroof MA, Allard RW, Zhang Q (1990) Genetic diversity and ecogeographical differentiation among ribosomal DNA alleles in wild and cultivated barley. Proc Natl Acad Sci 87:8486–8490

Saghai Maroof MA, Biyashev RM, Yang GP, Zhang Q, Allard RW (1994) Extraordinarily polymorphic microsatellite DNA in barley: Species diversity, chromosomal locations, and population dynamics. Proc Natl Acad Sci USA 91:5466–5470

Schelfhout CJ, Snowdon RJ, Cowling WA, Wroth JM (2006) Tracing B-genome introgressions in *B. napus* x *B. juncea* interspecific hybrids. Theor Appl Genet (in press)

Schut JW, Qi X, Siam P (1997) Association between relationship measures based on AFLP markers, pedigree data and morphological traits in barley. Theor Appl Genet 95:1161–1168

Seyis F, Friedt W, Snowdon RJ, Lühs W (2002) Erweiterung der genetischen Basis von Raps (*Brassica napus* L.) durch Resynthese. Vortr Pflanzenzüchtg 54:493–496

Seyis F, Snowdon RS, Lühs W, Friedt W (2003) Molecular characterization of novel resynthesized rapeseed (*Brassica napus*) lines and analysis of their genetic diversity in comparison with spring rapeseed cultivars. Plant Breeding 122:473–478

Shepard RN (1962) The analysis of proximities: multidimensional scaling with an unknown distance function. Psychometrica 27:125–139

Shepard RN (1966) Metric structure in ordinal data. J Math Psychol 3:287–315

Smartt J, Simmonds NW (1995) Evolution of crop plants. Longman Scientific and Technical, UK

Sneath PHA (1957) The application of computers to taxonomy. J Gen Microbiol 17:201–226

Sneath PHA, Sokal RR (1973) Numerical taxonomy—the principles and practice of numerical classification. WH Freeman, San Francisco

Snowdon RJ, Friedt W (2004) Molecular markers in *Brassica* oilseed breeding: current status and future possibilities. Plant Breeding 123:1–8

Snowdon RJ, Köhler W, Friedt W, Köhler A (1997) Genomic in situ hybridization in *Brassica* amphidiploids and interspecific hybrids. Theor Appl Genet 95:1320–1324

Snowdon RJ, Winter H, Diestel A, Sacristan MD (2000) Development and characterisation of *Brassica napus–Sinapis arvensis* addition lines exhibiting resistance to *Leptosphaeria maculans*. Theor Appl Genet 101:1008–1014

Snowdon RJ, Friedrich T, Friedt W, Köhler W (2002) Identifying the chromosomes of the A- and C-genome diploid *Brassica* species *B. rapa* (syn. *campestris*) and *B. oleracea* in their amphidiploid *B. napus*. Theor Appl Genet 104:533–538

Snowdon RJ, Lühs W, Friedt W (2006) *Brassica* oilseeds. In: Singh R (ed) Genetic resources, chromosome engineering, and crop improvement, vol 3, chapter 7: Oilseed crops. CRC Press, Boca Raton, Fla. (in press)

Sokal RR, Michener CD (1958) A statistical method for evaluating systematic relationships. Univ Kans Sci Bull 38:1409–1438

Song W, Henry R (1995) Molecular analysis of the DNA polymorphism of wild barley (*H. spontaneum*) germplasm using the polymerase chain reaction. Gen Res Crop Evol 42:273–281

Song GK, Lu P, Tang K, Osborn TC (1995) Rapid genome change in synthetic polyploids of *Brassica* and its implication for polyploid evolution. Proc Natl Acad Sci USA 92: 7719–7723

Song K, Lu P, Tang K, Osborn TC (1993) Development of synthetic *Brassica* amphidiploids by reciprocal hybridization and comparison to natural amhidiploids. Theor Appl Genet 86:811–821

Sørensen T (1948) A method of establishing groups of equal amplitude in plant sociology based on similarity of species content and its application to analysis of the vegetation on Danish commons. Biol Skr 5:1–34

Spiller T, Lühs W, Baetzel R, Snowdon RJ, Friedt W (2004) Entwicklung von verbesserten Populationen für die Qualitätszüchtung und die QTL–Kartierung von Leistungseigenschaften bei Winterraps. Vortr Pflanzenzüchtg 64:79–81

Stoll C, Lühs W, Zarhloul KM, Friedt W (2005) Genetic modification of saturated fatty acids in oilseed rape (*Brassica napus*). Eur J Lipid Sci Technol 107:244–248

Stoll C, Zarhloul MK, Brummel M, Spener F, Lühs W, Friedt W (2006) Knockout of KASIII regulation changes fatty acid composition in canola (*Brassica napus* L.). Eur J Lipid Sci Technol 108 (in press)

Takamine N (1916) Über die ruhenden und die präsynaptischen Phasen der Reduktionsteilung. Bot Mag 30:293–303

Thormann CE, Ferreira ME, Camargo LEA, Tivang JG, Osborn TC (1994) Comparison of RFLP and RAPD markers to estimating genetic relationships within and among cruciferous species. Theor Appl Genet 88:973–980
Torgerson WS (1958) Theory and methods of scaling. Wiley, New York
U N (1935) Genomic analysis of *Brassica* with special reference to the experimental formation of *B. napus* and peculiar mode of fertilization. Jpn J Bot 7:389–452
Vierling RA, Nguyen HT (1992) Use of RAPD markers to determine the genetic diversity of diploid, wheat genotypes. Theor Appl Genet 84:835–838
von Bothmer R, Sato K, Komatsuda T, Yasuda S, Fischbeck G (2003) The domestication of cultivated barely. In: von Bothmer R, van Hintum T, Knüpffer H and Sato K (eds) Diversity in barley (*Hordeum vulgare*).Elsevier Science BV, Amsterdam, The Netherlands, pp 53–76
Vos P, Hogers R, Bleeker M, Reijans M, van de Lee T, Hornes M, Frijters A, Pot J, Peleman J, Kuiper M, Zabeau M (1995) AFLP: a new technique for DNA fingerprinting. Nucleic Acids Res 23:4407–4414
Voss A, Snowdon RJ, Lühs W, Friedt W (2000) Intergeneric transfer of nematode resistance from *Raphanus sativus* into the *Brassica napus* genome. Acta Hort 539:129–134
Wang YP, Snowdon RJ, Rudloff E, Wehling P, Friedt W, Sonntag K (2004) Cytogenetic and molecular characterization of progeny from asymmetric somatic hybrids between *Brassica napus* and *Crambe abyssinica*. Genome 47:666–673
Ward JH (1963) Hierarchical grouping to optimise an objective function. J Amer Statist Assoc 58:236–244
Weber S, Ünker F, Friedt W (2005) Improved doubled haploid production protocol for *Brassica napus* using microspore colchicine treatment *in vitro* and ploidy determination by flow cytometry. Plant Breeding 124:511–513
Welsh J, McClelland M (1990) Fingerprinting genomes using PCR with arbitrary primers. Nucleic Acids Res 18:7213–7218
Wenzl P, Carling J, Kudrna D, Jaccoud D, Huttner E, Kleinhofs A, Kilian A (2004) Diversity arrays technology (DArT) for whole-genome profiling in barley. Proc Natl Acad Sci 101:9915–9920
Winter H, Diestel A, Gärtig S, Krone N, Sterenberg K, Sacristan MD (2003) Transfer of new blackleg resistances into oilseed rape. Proc 11th Intl Rapeseed Congr Vol 1, Copenhagen, Denmark. pp. 19–21
Wright S (1978) Evolution and genetics of populations, Vol IV. University of Chicago Press
Wych RD, Rasmusson DC (1983) Genetic improvement in malting barley cultivars since 1920. Crop Sci 23:1037–1040
Zhu YL, Song QJ, Hyten DL, Van Tassell CP, Matukumalli LK, Grimm DR, Hyatt SM, Fickus EW, Young ND, Cregan PB (2003) Single-nucleotide polymorphisms in soybean. Genetics 163:1123–1134
Zohary D, Hopf M (1994) Domestication of plants in the old world. Clarendon Press, Oxford

Wolfgang Friedt,
Rod Snowdon,
Jutta Ahlemeyer,
IFZ, Research Centre for Bio Systems
Land Use and Nutrition
Department of Crop Science
and Plant Breeding
Justus-Liebig-University of Giessen
Heinrich-Buff-Ring 26-32
35392 Giessen, Germany

Frank Ordon
Institute of Epidemiology and
Resistance Resources
Federal Centre for Breeding Research
on Cultivated Plants
Theodor-Roemer-Weg 4
06449 Aschersleben
Germany

Plant Breeding: Antisense ODN Inhibition in in vitro spike cultures as a powerful Diagnostic Tool in Studies on Cereal Grain Development

Christer Jansson, Chuanxin Sun, Seedhabadee Ganeshan and Ravindra N. Chibbar

1 Introduction

The study of grain development is perhaps one of the most complex and confounding processes in cereals. In addition to carbon partitioning into the grain, other plant metabolic processes, tissues, organs and environmental factors influence the development of the grain. Due to the importance of cereals as a food/energy source, there has been tremendous effort over the past several decades to understand the grain development process in order to enhance productivity and cater to the food requirements of the ever-growing world population. The significant influence of the environment on grain development has hindered the proper understanding of the factors that play key roles in this process. Therefore, development of a system, wherein the confounding effects of the environment and tissues/organs could be neutralized would be valuable in understanding grain development/filling. Furthermore, with the environment being kept constant, the effects of monitored alterations of physico-chemical factors on grain development can be studied. Field and growth chamber experiments, although valuable, have the disadvantage of being governed by uncontrollable variables. Thus, development of an in vitro approach that would allow for a coordinated, concerted and targeted approach to study grain development would be useful. Even though results from such studies cannot be directly extrapolated to actual field situations, a basic understanding of the grain developmental process can be obtained under various treatments and used as a model on which to build further understanding of grain development under actual growth conditions. However, even with an in vitro approach, every effort should be made to tailor the system to be as close as possible to an in vivo situation.

This chapter has the 2-fold purpose of describing the cereal in vitro spike culture as a model system for cereals, and the antisense ODN technology as an emerging means for studying cereal grain development.

Progress in Botany, Vol. 68

2 Antisense ODN inhibition

Antisense oligodeoxynucleotides (ODNs) are naturally occurring in both prokaryotes and eukaryotes where they partake in gene regulation and defence against viral infections (Vanhee-Brossollet and Vaquero 1998; Lehner et al. 2002). Antisense ODNs are short (12–25 nt-long) stretches of single-stranded ODNs that hybridize to the cognate mRNA in a sequence-specific manner, thereby inhibiting gene expression. The mechanisms for antisense ODN inhibition are not fully understood but it is generally considered that the ODN either sterically interferes with translation or promotes transcript degradation by RNase H activation (Gewirtz et al. 1998; Shi and Hoekstra 2004; Fig. 1).

The most effective mechanism of antisense ODN inhibition of gene expression is attributed to RNase H-mediated target-mRNA degradation (Crooke 2000). The enzyme RNase H (EC 3.1.26.4) is a non-specific ribonuclease that cleaves the 3′-O-P-bond of RNA in a DNA:RNA duplex to produce 3′-hydroxyl and 5′-phosphate-terminated products. In contrast to

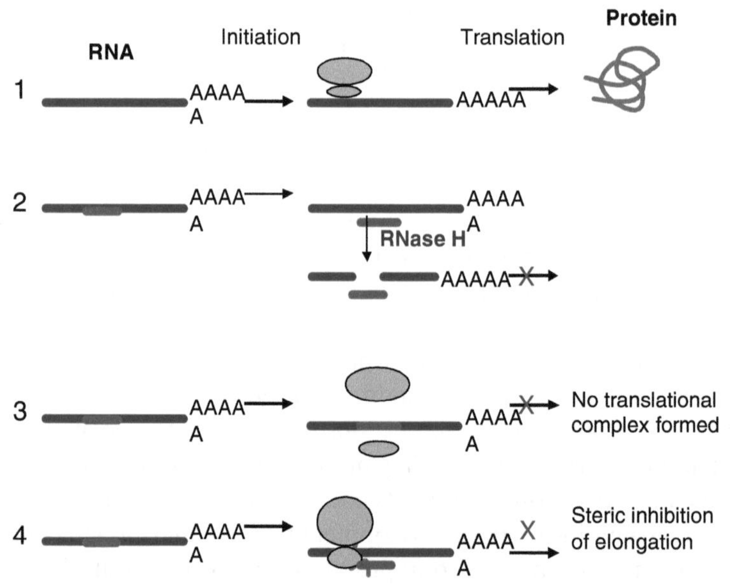

Fig. 1. Different modes by which antisense ODN inhibition can suppress gene expression. The antisense effect of gene expression (1) can be obtained by RNase activation (2) or by translational arrest by sterically interfering with either with formation of the initiation complex (3) or the elongation step (4)

other ribonucleases, such as RNase A or RNase T1, RNase H leaves a 3′-phosphorylated product. The two fragments generated by the RNase endonucleolysis are a 5′ product possessing a cap structure but lacking a poly(A) tail and a 3′ product possessing a poly(A) tail but lacking the cap structure. It is generally assumed that both resulting mRNA cleavage products are rapidly degraded in the cell, thus resulting in an untranslatable message; the lack of the poly(A) tail of the 5′ product leads to subsequent 3′-5′ exonucleolytic attack whereas the 3′ product may be a substrate for 5′-3′ exonucleolysis. Members of the RNase H family are found in most organisms, from (eu)bacteria and archae to eukaryotes. RNase H is present also in retroviruses, where it is part of the viral reverse transcriptase enzyme. During DNA replication in eukaryotes, RNase H is responsible for digestion of the RNA primer, allowing completion of the newly synthesized DNA.

In laboratory work, RNase H is used to specifically degrade RNA in DNA:RNA hybrids, e.g. to destroy the RNA template after first-strand complementary DNA (cDNA) synthesis by reverse transcription.

3 Antisense ODN inhibition as a technology for gene silencing

One of the hallmarks in the initiation of the development of antisense ODN technology for the purposes of molecular biology and medical therapy was the demonstration in 1978 that synthetic ODNs complementary to Raos sarcoma virus could inhibit virus replication in tissue cultures of chick embryo fibroblasts (Zamecnik and Stephenson 1978). Since then, the antisense ODN technology has been widely used in animal sciences and as an important emerging therapeutic approach in clinical medicine (Gewirtz et al. 1998; Dagle and Weeks 2001; Hu et al. 2002; Shi and Hoekstra 2004; Yang et al. 2004).

3.1 Construction of antisense ODNs

To obtain the desirable effects, the ODNs need to meet certain physical requirements. First, ODNs need to be able to traverse the plasma membrane and then hybridize with the cognate target transcript. Second, the ODNs should exert as little non-sequence-related toxicity as possible. Third, the ODNs should remain stable in the extracelleular and intracellular environment in which they are located. While natural DNA has been used in many studies (see Gewirtz et al. 1998 and references therein), it is becoming increasingly common to use DNA that has been modified to better withstand attacks from endonucleases and exonucleases. Two such modifications are accomplished by replacing one of the non-bonding oxygen atoms in the phosphate group with either a methyl or a sulfur group, resulting in methylphosphonate or phosphorothioate ODNs, respectively. Another ODN modification is offered by the peptide nucleic acid (PNA).

As opposed to natural ODNs, which are negatively charged, methylphosphonates are neutral, and therefore lipophilic. Thus, it has been postulated that these modified ODNs might be more efficiently imported into cells as compared to non-modified ODNs. However, utilization of methylphosphonates in antisense ODN studies has been hampered by several shortcomings (Gewirtz et al. 1998). First, and most importantly, methylphosphonate ODNs do not allow RNase H-mediated digestion of the RNA in the ODN:RNA duplex, and their antisense effect is most likely instead attributed to translational blocking. Second, since they are lipophilic the methylphosphonates are difficult to get into solution. Third, these ODNs are chiral with respect to the methylphosphonate bond and hence exist as racemic mixtures for any given conformation. This probably lowers their affinity to the target RNA.

In contrast to the methylphosphonates, phosphorothioate ODNs are successfully employed in both basic biological research and clinical medicine (Gewirtz et al. 1998; Dagle and Weeks 2001; Shi and Hoekstra 2004). Phosphorothioates are compatible with RNase H attacks of the ODN:RNA hybrid and they are relatively nuclease resistant. Furthermore, their negative charge renders them water soluble. However, since the outer surface of the plasma membrane is usually negatively charged, the polyanionic nature of the phopshorothioates also impairs cellular uptake. In addition, phosphorothioate ODNs have been found to electrostatically interact with proteins, e.g. various growth factors and DNA polymerase (Brown et al. 1994; Gewirtz et al. 1998; Stein 1999; Shi and Hoekstra 2004). Finally, high concentrations of phosphorothioates inhibit RNase H activity (Gewirtz et al. 1998). A great number of strategies have been set up to further improve the efficacy of the phosphrothioate ODNs. One such attempt is to use end-capped phosphorothioates, where the 5′ and 3′ linkages are sulfated (Gewirtz et al. 1998).

In PNAs, the phosphodiester linkage is replaced by a peptide backbone composed of *N*-(2-aminoethyl)-glycine units (Gewirtz et al. 1998; Shi and Hoekstra 2004). PNA-based ODNs are completely nuclease resistant and retain the RNA-binding capacity. They are also achiral and show minimum binding to proteins. However, like methylphosphonates, PNAs do not activate or recruit RNase H (Lomakin and Frank-Kamenetskii 1998; Pitts and

Corey 1998). Another limitation with PNAs is that they cannot freely permeate plasma membranes, but have to be delivered with artificial vectors.

In addition to methylphosphonates, phosphorothioates and PNAs, there are a great number of other modifications possible to improve the functional and structural properties of antisense ODNs (Gewirtz et al. 1998; Shi and Hoekstra 2004).

3.2 Delivery and intracellular trafficking of antisense ODNs

A successful antisense ODN approach requires a convenient carrier system that transfers ODNs efficiently into the cells and, for RNase H-mediated antisense effects, a subsequent efficient uptake into the nucleus. In animal experiments, naked natural and phosphorothioate ODNs are usually not taken up by the cells, since both the ODNs and the outside of the plasma membrane carry a net negative charge. In some studies (Shi and Hoekstra 2004), an antisense effect has been obtained with free naked ODNs provided they were applied at a relatively high concentration (10–20 μM). Methylphosphonate derivatives, on the other hand, are uncharged, and it has been reported that they enter cells via diffusion (Gewirtz et al. 1998).

Problems with membrane permeability can be circumvented by microinjection, electroporation or membrane permeabilization with chemical agents (Shi and Hoekstra 2004). However, since these methods are either harmful to cells or laborious, they are rarely used for routine purposes.

Alternative methods for delivery of ODNs to the cell rely on the use of liposomes, polymers and peptides as vectors (Shi et al. 2001; Hu et al. 2002; Shi and Hoekstra 2004; Yang et al. 2004). For example, cationic liposomes readily accommodated DNA via electrostatic interactions in a complex structure, referred to as lipoplexes (Oberle et al. 2000; Maurer et al. 2001). Liposome and polymer carriers form complexes with the ODNs and enhance their absorption to the plasma membrane. The complexed ODNs are then internalized by endosyctosis. Following endocytosis, the ODNs are released from the endosomes and can reach the nucleus. The mechanisms of endosomal release of complexed ODNs are not clear but it seems as if it is promoted by destabilization of the endosomal membrane.

4 Antisense ODN inhibition in in vitro spike cultures as a powerful strategy in studies on cereal grain development

With a notion of being able to control supply and composition of assimilates into developing wheat grains, Donovan and Lee (1977) developed a detached wheat spike culture approach in liquid medium containing sucrose, amino acids and salts. Spikes of about 10 cm in length at 8 or 20 days post-anthesis were cultured for up to 12 days and grain weight, total nitrogen, starch, DNA content and metabolic and storage nitrogen contents were determined. Their results indicated that grain weight, nitrogen and starch were comparable to field-grown plants. In a subsequent study, the effect of nitrogen source on such detached wheat spikes were studied, with the result that weight and starch content of the grains were unaffected, but there was an increase in nitrogen content with increased nitrogen supplement (Donovan and Lee 1978). Further modifications and variations of this technique have subsequently been reported mostly with regards to carbohydrate and nitrogen source supplements to the culture medium (Barlow et al. 1983; Singh and Jenner 1983; Ma et al. 1996). The spike culture approach has also been used with barley. Giese et al. (1983) used different levels of nitrogen for culture of 8 days post-anthesis spikes and after 20 days observed higher protein contents at higher nitrogen levels compared to controls. Similarly, protein and carbohydrate accumulation was studied in spike-cultured grains of normal and high-lysine mutants of barley on media containing different levels of nitrogen and sucrose levels (Mather and Giese 1984). It was found that the spikes of normal barley in culture accumulated higher levels of N in response to increasing N levels, but had lower endosperm dry weight and starch than plants grown under normal conditions. Using a similar approach spike culture of barley has been studied in relation to endosperm protein synthesis and accumulation in response different levels of nitrogen supplied as ammonium nitrate to 8 days post-anthesis spikes (Corke and Atsmon 1988, 1990). In one of these studies, Corke and Atsmon (1988) showed that cultivated barley (*H. vulgare* cv Ruth) endosperm protein could increase to above 30% with nitrogen supplement compared with endosperm of plants grown under normal conditions, in which endosperm protein content was 14%. In this situation, it was concluded that it was most likely that nitrogen uptake by the root was the limiting factor in in vivo grown plants accounting for the low protein content. The culture of detached wheat spikelets has also been attempted. Trione et al. (1989) cultured 7 days pre-anthesis wheat spikelets to maturity of the embryos. Although their study suggested a close correlation between in vivo and in vitro floret and grain physiology, recourse to embryo rescue for most of the developing spikelets was necessary for further development and germination of the embryo. Millet and Jenner (1991) cultured 15–20 days post-anthesis wheat endosperms on liquid medium supplemented with (^{14}C)-sucrose and (^{3}H)-glutamine and observed starch and protein accumulation for up to a week, albeit at a lower rate compared to normally developing grains.

Although these fundamental studies since the early 1980s have added to our understanding of carbohydrate and protein accumulation in the grain to some extent, we are still far from fully elucidating the complex process of grain development. This has mainly been due to the spike culture having been started post-anthesis and being stopped before the grain attained full maturity. There was therefore a need to further refine this culture from a pre-anthesis stage to full maturity of the grain, so that grain development could be followed over that duration. Furthermore, with technological advances over the past 10 years, the spike culture can be viewed with renewed interest as an ideal in vitro system for studying grain development employing transcriptomics, metabolomics and proteomics tools. From using the spike culture only for monitoring carbohydrate and protein accumulation in the developing grain in response to different levels of N or sucrose, the spike culture can additionally be used as a system to study the effects of abiotic factors on grain development as well as to dissect the developmental pathways during grain development in conjunction with the above mentioned tools. The immature spike culture has successfully been refined for both wheat and barley for such studies (Chibbar et al. 2005). The spikes were excised prior to anthesis and deprived of all leaves, including the flag leaf. Anthesis occurred about 7 days after initiation of culture in a medium containing sucrose and L-glutamine as defined by Dalal et al. (1999), and attained maturity in about 30 days, with mature and fertile seed set. The immature wheat spike cultures have been optimized to study the effects of salinity, pH and cadmium accumulation on grain development. Study of these factors is often confounded by the presence of soil solutes in the rhizosphere. With the influence of the leaves and roots being ruled out, the extent of the influence of these abiotic factors on grain development could be assessed. Salinity, for example, decreased seed weight, but had no effect on the number of seeds per spike and on the ratio of starch to protein. At low pH there was an increase in seed weight, starch and protein contents. There was, however, a reduction in the number of seeds. High pH did not affect grain development. If these studies were conducted in vivo in growth chambers or field conditions, with the number of uncontrollable variables, data interpretation would have been more difficult. However, conclusions drawn from the in vitro experiments should be used cautiously until parallel experiments with the same variables are tested on intact plants, wherein leaves from hydroponically grown plants at the same stage of spike culture initiation can be removed and grain development studied.

The in vitro spike cultures of wheat, barley and other cereals potentially offer an excellent system for the utilization of the powerful antisense ODN. However, antisense ODN inhibition has been an under-exploited strategy for plant tissues, although the prospects for plant cells in suspension cultures to take up single-stranded ODNs was reported over a decade ago (Tsutsumi et al. 1992). In 2001, two reports from Malhó and coworkers (Moutinho et al. 2001a,b) demonstrated the use of cationic-complexed antisense ODNs to suppress expression of genes encoding pollen-signaling proteins in pollen tubes from the lilly *Agapanthus umbellatus.* For the uptake of DNA pollen tubes represent a unique system, since the growing tip is surrounded by a loose matrix of hemicellulose and pectins, exposing the plasma membrane (Moutinho 2001a) and the first uptake of ODNs by pollen tubes was reported as early as 1994 (Estruch et al. 1994).

Complexed ODNs are unlikely to be efficiently taken up by spike cultures. Therefore, a breakthrough in the employment of antisense ODN

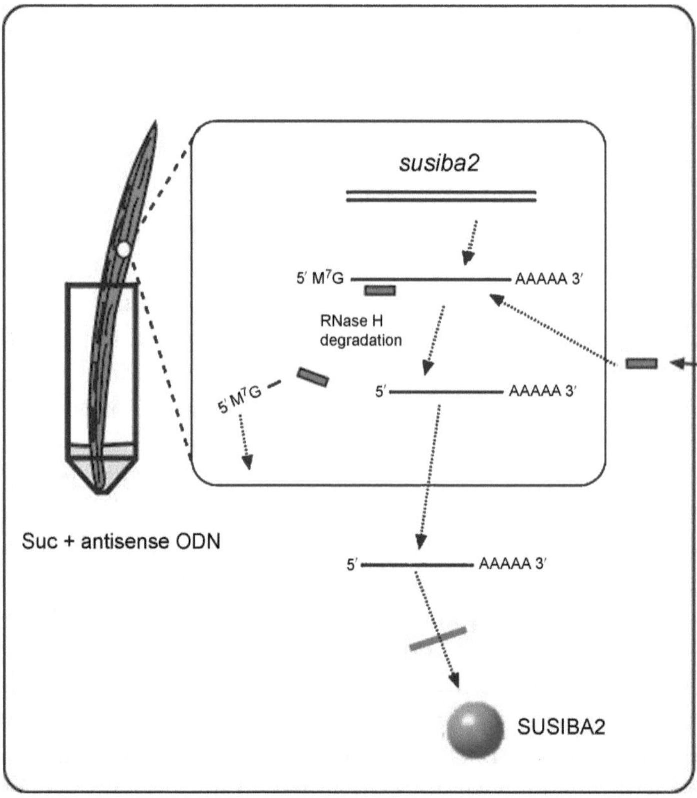

Fig. 2. A model representing the action of the *suisba2* antisense ODN in barley leaves. The antisense ODN enters the nucleus and binds to the complementary mRNA. The RNA strand of the duplex region is degraded by RNase H into two fragments. The capped (m^7G) 5′ fragment lacking a poly(A) tail is rapidly degraded, whereas the uncapped 3′ fragment is stable due to its poly(A) tail. The 3′ fragment enters the cytosol but fails to be translated since it lacks the 5′ cap. *Suc* sucrose. Modified from Sun et al. (2005)

inhibition as a powerful approach in plant biology was recently presented by Sun et al. (2005) in their work on intact barley leaves. As was illustrated by confocal microscopy and fluorescently labelled ODNs, naked ODNs were taken up through the leaf petiole and efficiently imported into the plant cell and the nucleus. The work portrayed in that study demonstrate the applicability of antisense ODN inhibition in plant biology, e.g. as a rapid antecedent to time-consuming transgenic studies, and that it operates through RNase H degradation (Fig. 2). Sun et al. used the antisense ODN strategy to demonstrate the importance of the SUSIBA2 transcription factor (Sun et al. 2003) in regulation of starch synthesis, and to depict a possible mechanism for sugar signalling in plants and how it might confer endosperm-specific gene expression during seed development. The antisense ODN experiments by Sun et al. (2005) also illustrated that transcription factor inhibition offers the potential to simultaneously target several genes for a given pathway, and thus suggests a novel strategy for metabolic engineering of starch synthesis.

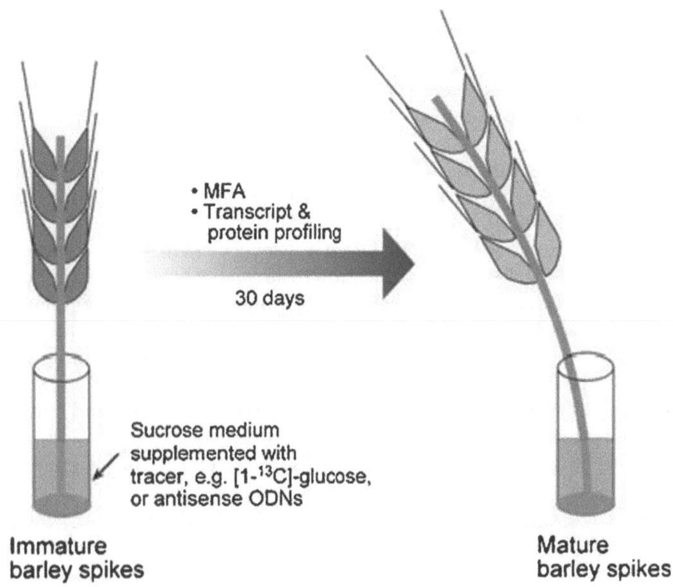

Fig. 3. Experimental setup with barley spike cultures. Excised immature barley spikes are placed in tubes with culture medium for 30 days and post-anthesis metabolic changes in the endosperm of dissected seeds are monitored at regular intervals until maturity. Due to the simple and defined nature of this system, it offers the opportunity to variously supplement the medium with factors of interest, such as [1-^{13}C]-glucose for metabolic flux analyses (MFA), or antisense ODNs

5 Conclusions and perspectives

Combining the antisense ODN technology as reported by Jansson and coworkers with the in vitro cereal spike cultures developed by Chibbar's group should prove an important experimental system in the studies on cereal grain development (Fig. 3). In preliminary experiments, the uptake of ODNs into endosperm cells of in vitro spike cultures of barley was illustrated by using *susiba2* (Sun et al. 2003) antisense ODNs to inhibit starch branching activities in the grain (Sun et al. unpublished). The mechanisms by which naked ODNs are able to permeate across plant plasma membranes remain to be elucidated.

Acknowledgements

We greatly acknowledge the financial support from The Swedish Foundation for International Cooperation in Research and Higher Education (STINT) and from Genome Canada.

References

Barlow EWR, Donovan GR, Lee JW (1983) Water relations and composition of wheat ears grown in liquid culture: effect of carbon and nitrogen. Aust J Plant Physiol 10:99–108

Brown DA, Kang SH, Gryaznov SM, DeDionisio L, Heidenreich O, Sullivan S, Xu X, Nerenberg MI (1994) Effect of phosphorothioate modification of oligodeoxynucleotides on specific protein binding. J Biol Chem 269:26801–26805

Chibbar RN, Båga M, Ganeshan S, Hucl PJ, Limin A, Perron C, Ratnayaka I, Kapoor A, Verma V, Fowler DB (2005) Carbohydrate modification to add value to wheat grain. In: Chung OK, Lockhart GL (eds) International Wheat Quality Conference Proceedings, Grain Industry Alliance, Manhattan, Kansas, USA, pp 75–84

Corke H, Atsmon D (1988) Effect of nitrogen nutrition on endosperm protein systhesis in wild and cultivated barley grown in pikes culture. Plant Physiol 87:523–528

Corke H, Atsmon D (1990) Endosperm protein accumulation in wild and cultivated barley and their cross grown in spike culture. Euphytica 48:22–31

Crooke ST (2000) Progress in antisense technology: the end of the beginning. Methods Enzymol 313:3–45

Dagle JM, Weeks DL (2001) Oligonucleotide-based strategies to reduce gene expression. Differentiation 69:75–82

Dalal M, Vajaya Lakshmi KVS, Khanna-Chopra R, Bharti S (1999) Ear culture as a technique to overcome hybrid necrosis in wheat. Plant Cell Tiss Org Cult 59:151–154

Donovan GR, Lee JW (1977) The growth of detached wheat heads in liquid culture. Plant Sci Lett 9:107–113

Donovan GR, Lee JW (1978) Effect of the nitrogen source on grain development in detached wheat heads in liquid culture. Aust J Plant Physiol 5:81–87

Estruch JJ, Kadwell S, Merlin E, Crossland L (1994) Cloning and characterization of a maize pollen-specific calcium-dependent calmodulin-independent protein kinase. Proc Natl Acad Sci USA 91:883–841

Gewirtz AM, Sokol DL, Ratajczak MZ (1998) Nucleic acid therapeutics: state of the art and future prospects. Blood 92:712–736

Giese H, Anderson B, Doll H (1983) Synthesis of the major storage protein, hordein, in barley. Planta 159:60–65

Hu Q, Bally MB, Madden TD (2002) Subcellular trafficking of antisense oligonucleotides and down-regulation of bcl-2 gene expression in human melanoma cells using a fusogenic liposome delivery system. Nucl Acids Res 30:3632–3641

Lehner B, Williams G, Campbell RD, Sanderson CM (2002) Antisense transcripts in the human genome. Trends Genet 18:63–65

Lomakin A, Frank-Kamenetskii MD (1998) A theoretical analysis of specificity of nucleic acid interactions with oligonucleotides and peptide nucleic acids (PNAs). J Mol Biol 276: 50–70

Ma Y-Z, MacKown CT, van Sanford DA (1996) Kernel growth of in vitro cultured wheat spikes of cultivars with divergent source-sink limitations. Plant Sci 119:135–148

Mather DE, Giese H (1984) Protein and carbohydrate accumulation in normal and high-lysine barley in spike culture. Physiol Plant 60:75–80

Maurer N, Wong KF, Stark H, Louie L, McIntosh D, Wong T, Scherrer P, Semple SC, Cullis PR (2001) Spontaneous entrapment of polynucleotides upon electrostatic interaction with ethanol-destabilized cationic liposomes. Biophys J 80:2310–2326

Millet E, Jenner CF (1991) Accumulation of label from radioactive precursors into dry matter by wheat endosperm cultured in vitro. Physiol Plant 83:591–596

Moutinho A, Camacho L, Haley A, Pais MS, Trewavas A, Malhó R (2001a) Antissense perturbation of protein function in living pollen tubes. Sex Plant Reprod 14:101–104

Moutinho A, Hussey PJ, Trewavas AJ, Malhó R (2001b) cAMP acts as a second messenger in pollent tube growth and reorientation. Proc Natl Acad Sci USA 98:10481–10486

Oberle V, Bakowsky UU, Zuhorn IS, Hoekstra D (2000) Lipoplex formation under equilibrium conditions reveals a three-step mechanism. Biophys J 79:1447–1454

Pitts AE, Corey DR (1998) inhibition of human telomerase by 2′-*O*-methyl-RNA. Proc Natl Acad Sci USA 95:11549–11554

Shi F, Hoekstra D (2004) Effective intracellular delivery of oligonucleotides in order to make sense of antisense. J Con Rel 97:189–209

Shi F, Nomden A, Oberle V, Engberts BFN, Hoekstra D (2001) Efficient cationic lipid-mediated delivery of antisense oligonucleotides into eukaryotic cells: down-regulation of the corticotropin-releasing factor receptor. Nucl Acids Res 29:2079–2087

Singh BK, Jenner CF (1983) Culture of detached ears of wheat in liquid culture: modification and extension of the method. Aust J Plant Physiol 10:227–236

Stein CA (1999) Two problems in antisense biotechnology: in vitro delivery and the design of antisense experiments. Biochim Biophys Acta 1489:45–82

Sun C, Palmqvist S, Olsson H, Borén M, Ahlandsberg S, Jansson C (2003) A novel WRKY transcription factor, SUSIBA2, participates in sugar signaling in barley by binding to the sugar-responsive elements of the *iso1* promoter. Plant Cell 15:2076–2092

Sun C, Höglund A-S, Olsson H, Mangelsen E, Jansson C (2005) Antisense oligodeoxynucleotide inhibition as a potent strategy in plant biology: identification of SUSIBA2 as a transcriptional activator in plant sugar signaling. Plant J 44:128–138

Trione ED, Stockwell VO (1989) Development of detached wheat spikelets in culture. Plant Cell Tiss Org Cult 17:161–170

Tsutsumi N, Kanayama K, Tano S (1992) Suppression of alpha-amylase gene expression by antisense oligodeoxynucleotide in barley cultured aleurone layers. Jpn J Genet 67:147–154

Vanhee-Brossolet C, Vaquero C (1998) Do natural antisense transcripts make sense in eukaryotes? Gene 211:1–9

Yang L, Li J, Zhou W, Yuan X, Li S (2004) Targeted delivery of antisense oligodeoxynucleotides to folate receptor-overexpressing tumor cells. J Con Rel 95:321–331

Zamecnik PC, Stephenson ML (1978) Inhibition of Rous sarcoma virus replication and cell transformation by a specific oligodeoxynucleotide. Proc Natl Acad Sci USA 75:280–284

Christer Jansson,
Professor in Molecular Cell Biology - Head of Department
Dept of Plant Biology & Forest Genetics
Uppsala Genetic Center
The Swedish University of Agricultural Sciences (SLU)
P. O. Box 7080
SE-750 07 Uppsala
Sweden
Email: christer.jansson@vbsg.slu.se
Website: http://www-cjlab.slu.se
Phone (work): +46-(0)18-673313
Phone (mobile): +46-(0)709-624469
Fax: +46-(0)18-673279

Chuanxin Sun
Department of Plant Biology & Forest Genetics
The Swedish University of Agricultural Sciences
Uppsala
Sweden

Seedhabadee Ganeshan,
Ravindra N Chibbar
Department of Plant Sciences
University of Saskatchewan
Saskatoon
Saskatchewan
Canada S7N 5A8

Physiology

Characean Algae: Still a Valid Model System to Examine Fundamental Principles in Plants

Markus Braun, Ilse Foissner, Hinrich Lühring, Hendrik Schubert
and Gerhard Thiel

We dedicate this article to the late Geoff Findlay, a pioneer in modern plant electrophysiology

1 Introduction

The large cells of the *Characeae* have served for many decades as the model system *par excellence* to study basic physiological and cell biological phenomena in plants. One reason for this model character was given by the many physiological and structural similarities between these highly evolved green algae and higher plants. More important for the model character, however, was the unique geometry of the internodal cells. Single *Chara* cells can be as long as 10 cm with a diameter of 1 mm. These dimensions favoured the first use of intracellular microelectrodes in plants. Later, they basically imposed no limits to experimenters using even multiple microelectrodes simultaneously in one cell. The geometry of the cells also provided an invitation to perform some of the most incredible kind of microsurgery. Cells were deprived of their tonoplast in order to study the effect of this organelle on other cellular functions. Furthermore, chloroplasts could be removed and re-supplied to individual cells in order to examine the correlation between ion fluxes across the chloroplast membrane and the regulation of plasma membrane transport.

Because of the advantageous geometric features, many of the key techniques in modern plant physiology were first developed, tested and calibrated in characean cells. They were the first to be studied with modern voltage clamp techniques and were used for recording the first quantitative measurements on cytosolic free Ca^{2+} with various methods. In addition, the turgor pressure probe was initially established in *Chara*. Verification of the recordings was disburdened by the fact that *Chara* already represented a well-studied model system for the understanding of water relations in plants.

Progress in Botany, Vol. 68

Thanks to this *avant gard* development of techniques, many of the fundamental concepts in plant physiology, including membrane excitation, the presence of an H^+-ATPase in the plasma membrane, the mechanism of cyclosis, the importance of statolith-cytoskeleton interactions for gravitropism and many more concepts, were discovered first in *characeaen cells*.

Technical developments in the last 2 decades seem at first glance to have made the *Chara* model system redundant. The discovery of the patch clamp technique, in which *Chara* again played a significant role, now allows us to study the transport features of nearly all higher plant cells of interest directly. Fluorescent techniques and expression of Ca^{2+}-sensor proteins in higher plants no longer require large cells. In addition, many of the techniques developed for *Chara* are now miniaturized to such a level that they can be applied with no problem to any plant cell in question. Most of all, however, the progress in molecular biology and the complete sequencing of the *Arabidopsis* and rice genomes have provided the ideal tools to examine the function and expression of proteins directly in higher plants. These tools are not available (yet) for *Chara*.

The critical question, therefore, is whether the *Characeae* still offer in contemporary research any benefits as model system. The present review summarizes some recent work concerning fundamental principles in plant biology. The contributions provide new insights in the molecular and cellular mechanisms of gravity sensing and gravity-oriented growth, wound response, signal transduction and sex differentiation and will show that the *Characeae* are more than ever an indispensable source of information. The interest in these algae derives not at least from the phylogenetic position of the Charales as the precursors of land plants (Graham 1993; Kranz et al. 1995; Lewis and McCourt 2004). And, as in the past, the variety of specific features, like the geometry, the single cell character of the internodes, rhizoids and other characean cell types, the excellent access to plant membranes and organelles etc., still bears great potential for future research.

2 *Chara* as a model for pattern formation

The formation of complex structures in multicellular organisms is a key interest in modern biology. The genetic identity of the cells participating in the formation of cell patterns alone does not provide an explanation for the generation of the spatial structures. The question, therefore, is how patterns can emerge from an initially more or less structureless system, and which factors influence this pattern formation. Many lessons on this can be learned from single cell systems, because breaking of the cell symmetry occurs within a single cell. A classical example is the formation of alkaline and acidic bands in the *Characeae* (Lucas and Smith 1973). This formation is tightly coupled to the photosynthetic system of the cells. However, also for stimulus induced pattern formation, *Chara* offers an interesting model system. It is well known that wounding induces a polarization of cellular repair activity in the area of injury. Also differential growth involved in gravitropic responses can be seen in the context of pattern formation since a locally perceived stimulus induces differential growth activity within the same cell.

3 Rhizoids and protonemata of the green alga *Chara* as unicellular model systems for gravity sensing and polarized growth

Gravity is one of the most important environmental stimuli that plants use to adapt to their environment. They cannot escape this stimulus, and use it in a most beneficial way to optimize exploitation of resources. Gravity sensing mechanisms evolved early in the history of plants; they are already used by water-living algae and became a *conditio sine qua non* for the evolution of higher plants which left the water and conquered the space above and below the surface of the earth. Knight (1806) was the first to discover that plants use gravity as a guide for the orientation of their organs and later in the nineteenth century, gravity sensing of higher plants roots was found to take place in a specialized region, the root tip (Ciesielski 1872). In 1900, Nemec and Haberlandt independently identified specialized cells, so-called statocytes, in the shoot endodermis and the root cap as the primary gravity-sensing sites, which contain sedimentable starch-filled amyloplasts that function as statoliths (starch-statolith hypothesis). Gravity-induced sedimentation of statoliths was proposed to represent the early physical step of gravitropism (susception) that precedes graviperception and the gravitropic response.

Whereas our knowledge on hormone-dependent gravitropic response mechanisms is rapidly increasing (Blancaflor 2002; Ottenschläger et al. 2003; Aloni et al. 2004; Blilou et al. 2005), the cellular and molecular basis of the decisive early phases of gravity susception and gravity perception is not understood. So far, only opposite hypotheses exist which to explain how the vectorial information of a physical displacement of statoliths is perceived by cellular components which create a physiological signal and elicit the gravitropic signalling pathway. Elements of the cytoskeleton, the highly dynamic filamentous networks of microtubules and actin microfilaments, have been postulated to be involved in gravity sensing, but their role in the gravitropic signalling pathways is still enigmatic, since experimental findings are contradictory (Sievers et al. 1991a, 2002; Kiss 2000; Blancaflor 2002; Hou et al. 2004).

Characean algae provide two well established model cell types, rhizoids and protonemata, which are increasingly used to study cellular mechanisms and the molecular basis of gravitropism (Sievers et al. 1996; Braun 1997; Braun and Wasteneys 2000; Braun and Limbach 2006). The tube-like cells of the rhizoids with diameters of up to 30 μm are produced by nodal cells of the green thallus and rapidly expand into the surrounding medium by tip growth. They are more easily accessible for experimental approaches than the gravity-sensing cells in higher plants. The gravitropic signal-transduction and response pathways are short and limited to the apical region of a single cell. Positively gravitropic (downward growing) rhizoids have a root-like function and anchor the algal thallus in the sediment (Fig. 1). From the morphological point of view, protonemata are very

Fig. 1. Rhizoids and protonemata of characean algae are tip-growing cells which originate from nodal cells of the green thallus. Rhizoids grow in the direction of gravity (positive gravitropism, on the right) to anchor the thallus in the sediment. Protonemata are produced in the absence of blue light, e.g. when the thallus was buried in the sediment, and grow upward against the direction of gravity (negative gravitropism, on the left) back into light where they terminate tip growth, divide and regenerate the green thallus. The arrow denotes the direction of gravity. (Modified after Braun and Limbach 2006)

similar cells; however, they respond negatively gravitropic (upward growing). Protonemata are produced only in the absence of light, e.g. when the thallus is buried in the sediment (Fig. 1). As soon as a protonema cell has penetrated the substrate and has reached the light, tip growth is arrested and a complex series of cell divisions are initiated that lead to the regeneration of the green thallus (Hodick 1993; Braun and Wasteneys 1998a).

The arrangement, function and dynamics of the actin cytoskeleton and microtubules have been thoroughly investigated in both characean cell types. Microtubules maintain the prominent polar cytoplasmic zonation and the subapical organelle distribution, but they are not present in the apex and are not involved in the primary steps of gravitropic sensing and are not essential for polarized growth (Braun and Sievers 1994; Braun and Wasteneys 1998b). The actin cytoskeleton, however, has been recognized as a key player in the mechanisms of gravity sensing and gravity-oriented polarized growth (Fig. 2). The multiple functions of the actin microfilament system are controlled by numerous actin-binding proteins (Braun et al. 2004). By interacting with myosins, actin microfilaments regulate the positioning of the $BaSO_4$-crystal-filled vesicles which serve as statoliths

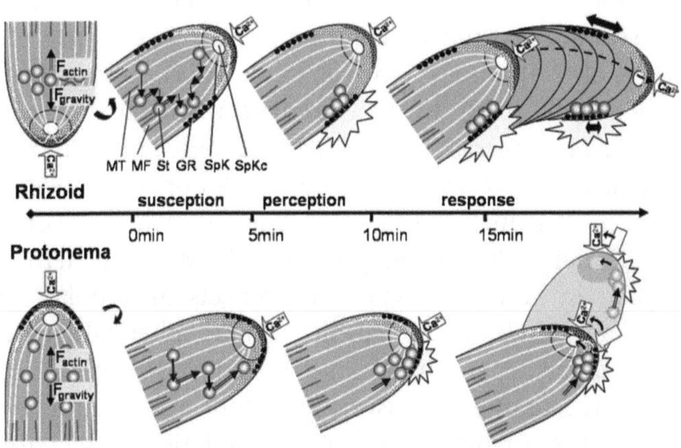

Fig. 2. Illustration of gravity sensing mechanisms in characean rhizoids and protonemata. In tip-downward growing rhizoids (upper row), the statolith (St) position results from net-basipetally acting actomyosin forces (F_{actin}) compensating gravity ($F_{gravity}$). Upon reorientation, statoliths sediment onto the lower cell flank. Net-acropetally acting actomyosin forces in the basal part of the statolith regions and in the subapical region prevent statoliths from leaving the apical region and transport the sedimenting statoliths onto membrane-bound gravireceptors (GR) which are restricted to a narrow, beltlike area of the plasma membrane 10–35µm from the tip. The Spitzenkörper (Spk) remains arrested at the tip and the calcium gradient (indicated by darker and lighter grey dotted area) is always highest at the tip. Statolith sedimentation causes a local reduction of cytosolic Ca^{2+} that results in differential extension of the opposite cell flanks (double-headed arrows)

(Hejnowicz and Sievers 1981; Sievers et al. 1991b; Braun and Sievers 1993; Buchen et al. 1993; Braun 1996a, 2002; Cai et al. 1997). Upon gravistimulation, the actomyosin system directs sedimenting statoliths to the gravisensitive region of the plasma membrane where the gravitropic signalling cascade is elicited resulting in the reorientation of the growth direction (Fig. 2, see also Sievers et al. 1996; Braun 2002).

The following sections address recent breakthroughs characean rhizoids and protonemata have provided as unicellular model cell systems in our understanding of the cellular and molecular mechanisms of gravitropic sensing in plants.

3.1 Cytoskeleton dynamics and polarized growth

So-called Spitzenkörper, apical body or "clear zones" describe a vesicle-rich region in the apex of tip-growing cell types. This is the case in pollen tubes, root hairs, fungal hyphae, moss chloronemata and caulonemata, protonemata of ferns and algae and rhizoids of mosses, ferns and algae. The tip region is characterized by an accumulation of secretory vesicles and the exclusion of other organelles such as dictyosomes, mitochondria and in most cases also endoplasmic reticulum cisternae (Geitmann and Emons 2000; Hepler et al. 2001; Lovy-Wheeler et al. 2005). F-actin has not been found or shown to be scarce in the extreme tip (for review, see Geitmann and Emons 2000). In contrast to these actin patterns, in characean rhizoids and protonemata extensive arrays of distinct actin microfilaments penetrate all cytoplasmic regions including the extreme apex which accomodates the tip growth machinery (Braun and Wasteneys 1998b). The actin cytoarchitecture is similar in both cells types, and reflects their polar cytoplasmic organization. Fine actin bundles focus in a unique spherical actin area in the centre of the Spitzenkörper (Braun and Wasteneys 1998b). This area contains a dense aggregate of endoplasmic reticulum membranes (Bartnik and Sievers 1988), and is surrounded by an accumulation of secretory vesicles which deliver cell wall material into the apex (Fig. 2). The position of the Spitzenkörper defines the centre of growth, the plasma membrane area where incorporation of vesicles is maximal (Hejnowicz et al. 1977; Sievers et al. 1979; Braun 1996b). Myosins mediate the transport of secretory vesicles along actin microfilaments towards the tip, where they accumulate, perform shuttle-like movements and eventually incorporate into the plasma membrane releasing new cell wall material (Braun 1996a).

Recently, the distribution of vesicles in the apical region was reinvestigated in high-pressure frozen and freeze-substituted rhizoids (to be published elsewhere). High-pressure freeze fixation yields much better results than chemical fixation with respect to ultra-structural preservation. By using the novel innovative technique of dual axis electron tomography (Mastronarde 1997, Ladinsky et al. 1999) for high resolution ultrastructural analysis, two different types of secretory vesicles as well as microvesicles were found evenly distributed in the apical region of rhizoids. In contrast, clathrin-coated vesicles were exclusively located in close vicinity of the apical plasma membrane. These vesicles are supposedly involved in endocytotic processes which mediate recycling of excessive membrane material and turn-over of membrane-bound proteins like ion channels. When the concerted action of exocytotic and endocytotic processes was disturbed by inhibitor-induced disruption of the actin cytoskeleton, the tip-focussed distribution pattern of calcium channels and the steep, tip-high gradient of cytoplasmic free calcium dissipated and tip growth stopped (Braun and Richter 1999). The calcium gradient dictates the incorporation pattern of secretory vesicles and also spatiotemporally controls the activity of actin-binding proteins. These signalling mechanisms, which include actin and calcium, contribute to a complex feedback regulation pathway that controls tip growth, a special type of polarized growth.

The multiple functions and the dynamic nature of the actin cytoskeleton in rhizoids and protonemata are coordinated by the concerted action of numerous actin-binding proteins. The actin-binding proteins identified in rhizoids and protonemata so far are mostly identical to those that have been shown to regulate cytoarchitecture and function of the actin microfilament system in other tip-growing cell types (Geitmann and Emons 2000; Hepler et al. 2001). However, the cytoskeletal arrangement in characean rhizoids and protonemata is unique and seems to be closely linked to the mechanisms of gravity sensing and gravity-oriented growth. Spectrin-like epitopes, actin-depolymerizing factor (ADF) and profilin specifically accumulate in the centre of the Spitzenkörper (Braun et al. 2004). Spectrin-like proteins most likely participate in the structural integrity of the ER aggregate by forming crosslinks between ER membranes and actin microfilaments (Braun 2001). Furthermore, spectrins are known to provide a mechanism for recruiting specific subsets of membrane proteins and to form functional microdomains in animal cells and, thus, they might help to create the particular physiological conditions for the mechanisms of gravity-sensing and polarized growth (Braun 2001, and references therein). Molecular studies failed to identify spectrins in *Chara* and spectrin-like proteins have not been identified in the *Arabidopsis* genome (for review, see Drobak et al. 2004). However, immunolocalization of spectrin-like epitopes in *Chara* and immunocytochemical analyses implicate the existence of an actin-binding protein that at least shares similar functional domains and an almost identical molecular mass with spectrins (Braun 2001).

The accumulation of the actin-binding proteins ADF and profilin in the centre of the Spitzenkörper indicates high actin-turnover rates and an actin polymerizing function of this central area (Braun et al. 2004). Strong evidence comes from cytochalasin D-induced disruption of the actin cytoskeleton, which causes a complete dissociation of the centre of the Spitzenkörper. Immunolocalization of actin, ADF, profilin and the ER aggregate dissipate and tip growth terminates. Removal of the inhibitor is followed by the reorganization of the actin cytoskeleton that starts with the reappearance of a dense actin array in the outermost tip (Braun et al. 2004). As soon as actin microfilaments radiate out, the actin array rounds up and is repositioned in the centre of the apical dome. This process is accompanied by the re-accumulation of ER membranes, the reappearance of ADF and profilin and followed by the resumption of tip growth (Braun et al. 2004). The results suggest that the centre of the Spitzenkörper functions as an amazingly localized apical actin polymerization site that has not been found in any other tip-growing cell type. It is tempting to speculate that the complexly coordinated, highly dynamic actin architecture in the rapidly extending tip is functionally related to the fundamental role of the actomyosin system in the different phases of the gravitropic signalling pathways and the gravity-oriented polarized growth.

The actin-crosslinking protein fimbrin has been localized in the apical and subapical region, suggesting that this actin-binding protein is involved in the formation of the mainly axially oriented dense actin meshwork (Braun et al. 2004). Immunolocalization of the actin-bundling protein villin is restricted to the basal zone, where actin microfilaments are separated according to their polarities and two populations of thick actin cables are formed which generate the rotational cytoplasmic streaming around the large central vacuole (Braun et al. 2004).

3.2 Cytoskeletal basis of the gravity-sensing apparatus

In tip-downward growing rhizoids the statoliths are actively kept at a distance of 10–35 fm basal to the tip. By exerting net basipetal forces, the actomyosin system prevents statoliths from settling into the tip. In tip-upward growing protonemata, actomyosin prevents statoliths from sedimenting towards the cell base by acting net acropetally (Hodick et al. 1998; Braun et al. 2002). Inhibitor studies have shown that disrupting the actin cytoskeleton in rhizoids and in protonemata did not only stop tip growth, but also caused statoliths to fall into the tip or towards the nucleus, respectively, following the direction of gravity (Hejnowicz and Sievers 1981; Bartnik and Sievers 1988; Sievers et al. 1996). After removal of the drug, statoliths were readily repositioned and tip growth continued.

Microgravity and simulated weightlessness represent suitable conditions for unravelling the role of gravity and the role of actomyosin-based forces in the complex regulation of statolith positioning (Buchen et al. 1993, 1997; Cai et al. 1997; Hoson et al. 1997; Braun et al. 2002). When the influence of gravity was abolished during the microgravity phases of parabolic flights of TEXUS (Technologische Experimente unter Schwerelosigkeit) in sounding rockets (Buchen et al. 1993) and during rotation on the three-dimensional and the fast-rotating clinostat (Hoson et al. 1997; Braun et al. 2002), actomyosin forces generated a displacement of statoliths against the former direction of gravity. This observation justified the conclusion that in normal, vertically-oriented rhizoids and protonemata, the statoliths are kept in a dynamic equilibrium position by actomyosin forces which exactly compensate the effect of gravity on the statoliths (Fig. 2). Interestingly, during long-term microgravity conditions of Space-Shuttle missions IML-2 (Second International Microgravity Laboratory) and S/MM05 (Fifth Shuttle-to-Mir Mission), the absence of gravity did not result in a random distribution of statoliths in rhizoids. Instead, after an initial basipetal transport at the beginning of microgravity, the statoliths spread over the entire statolith region, never leaving this cell area (Braun et al. 2002).

Detailed analysis of the movements of statoliths in microgravity and of statoliths which were displaced in the different cell regions by optical laser tweezers (Braun 2002) or by centrifugation revealed the surprising complexity of the transport system by which actomyosin forces control statoliths positioning (Braun et al. 2002). Individual acropetal and basipetal movements of statoliths were observed in both cell types, indicating that statoliths interact with the mainly axially oriented actin microfilaments with opposite polarities. When statoliths were centrifuged into the subapical region, a statolith transport back to the original position was observed that is not notably influenced by gravity (Sievers et al. 1991b; Braun and Sievers 1993). Active transport occurs along actin microfilaments and statoliths do not sediment onto the lower cell flank until they have reached the statolith region near the tip where statolith sedimentation is not constrained by microtubules (Braun and Sievers 1994).

Taken together, the data reveal that in the cell regions basal and apical from the statolith region the actomyosin component is always the strongest that points towards the statolith region. This ensures that statoliths are always kept in or are retransported to their original position. In the statolith region, however, gravity plays a critical role as an additional passive transport component that contributes to the positioning of statoliths. During vertical growth both forces that act on statoliths, i.e. gravity and active actomyosin transport components, are precisely balanced so that the statoliths are kept in a dynamically stable position without any net-transport (Braun et al. 2002). Changes in the orientation of the cell with respect to the direction of gravity or changing the amount of the acceleration must inevitably result in a displacement of statoliths.

There are indications from experiments in microgravity that statoliths also interact with actin in higher plant statocytes. Statoliths move in the direction against the originally acting gravity force (Volkmann et al. 1991) until they reach a new non-random steady state position (Driss-Ecole et al. 2000). Their sedimentation has also been reported to be modulated by actomyosin forces (Perbal et al. 2004). However, consistent evidence for the role of actin in graviperception is still missing.

3.3 Critical role of actomyosin in gravity susception

The actomyosin forces described above, acting oppositely on statoliths in rhizoids and protonemata, have important implications on how fast and where statoliths sediment. This highlights the critical role of actin in the process of gravisensing (Fig. 2). Upon a change in the orientation of the cell with respect to the gravity vector sedimenting statoliths are directed to distinct graviperception sites, which are the only regions of the plasma membrane where gravitropic signalling can be triggered. The graviperception site is confined to a narrow belt-like region 10-35 fm behind the tip in rhizoids and to the plasma membrane of the apical dome (5-10 μm behind the tip) in protonemata. Forcing statoliths to sediment outside these areas by optical laser tweezers or centrifugation did not result in a gravitropic response (Braun 2002).

Microgravity experiments (Buchen et al. 1997) and optical laser tweezers experiments (Leitz et al. 1995) have shown that, in the lateral direction, the statolith position is only weakly controlled by the actomyosin system in both cell types; the force needed to move statoliths towards the apex is greater than the force to move the statoliths towards the flank. Recently, the forces acting on statoliths in lateral direction were characterized in detail by microgravity experiments during two MAXUS (enlarged version of TEXUS) sounding rocket flights. It was demonstrated that, in vertically downward growing rhizoids, lateral acceleration forces in a range of 0.1 *g* were sufficient to displace statoliths towards the membrane-bound gravireceptors. In conclusion, the molecular forces acting on a single statolith in lateral direction were determined to be in a range of 2×10^{-14} N (Limbach et al. 2005). When rhizoids are reoriented by 90°, the sedimenting statoliths mainly follow the gravity vector and settle onto the lower cell flank of the statolith region where graviperception takes place and the graviresponse is initiated. However, when cells were rotated in angles different from 90°, statoliths did not simply follow the gravity vector (Hodick et al. 1998). Instead, even in inverted cells, statoliths were actively redirected against gravity and were guided to the confined graviperception site in the statolith region.

Gravistimulation of protonemata causes an actin-mediated acropetal displacement of sedimenting statoliths into the apical dome where they sediment onto the gravisensitive plasma membrane area close to the tip (Fig. 2, see also Hodick et al. 1998). During the upward bending of protonemata, the statoliths periodically sediment along the gravity vector and leave the graviperception site, which deactivates the gravireceptor and is reflected by phases of straight growth. Actomyosin-mediated transport of statoliths back to the gravisensitive membrane area reinitiates gravitropic bending until the vertical orientation is resumed (Fig. 2).

3.4 How statoliths activate gravireceptors

It was shown that statoliths have to be fully settled on membrane-bound gravireceptors in order to trigger graviperception and to induce the gravitropic signalling cascade (Braun 2002). Lateral movements of statoliths which do not lead to a contact with the plasma membrane do not induce a curvature response. Most recently, experiments have been performed during parabolic flights on board of the A300 Zero-G aircraft to elucidate the mode of gravireceptor activation in characean rhizoids (Limbach et al. 2005). Statoliths, which were weightless but still in contact with the plasma membrane, were able to activate the membrane-bound gravireceptor. In conclusion, it could be ruled out that the pressure exerted by the weight of statoliths is required for gravireceptor activation. This finding was supported by control experiments on ground which demonstrated that increasing the weight of sedimented statoliths by lateral centrifugation did not enhance the gravitropic response. However, graviperception was terminated within seconds when the contact of statoliths with the plasma membrane was interrupted by inverting gravistimulated cells. These results provide evidence that graviperception in characean rhizoids relies on direct contact, allowing yet unknown components on the surface of the statoliths to interact with membrane-bound receptors rather than on pressure or tension exerted by the weight of statoliths (Limbach et al. 2005). A mechanoreceptor was postulated basically because the gravitropic responses of many plant organs seem to obey the sine law of gravitropism (Galland 2002). The pressure statoliths exert on receptors at different gravistimulation angles would explain the sinusoidal dependency, but the observation that the number of statoliths which settle on the receptor area of the plasma membrane and activate receptors in characean rhizoids increases with the steepness of the angle can equally well account for this dependency. At the moment, however, this picture does only apply for the rhizoids of characean algae. Experimental results giving insight in the mode of gravireceptor activation are still missing to date.

3.5 Calcium, cytoskeleton and gravitropic responses

The nature of the receptor and the immediate downstream physiological steps of graviperception in rhizoids and protonemata remain to be clarified, but there is an increasing amount of data illuminating the physiological mechanisms by which the direction of the growth is reoriented.

The smooth downward curvature response of a rhizoid is best described as "bending by bowing", whereas the response of a protonema was described as "bending by bulging" (Braun 1996b), referring to the bulge that appears on the upper cell flank indicating the drastic upward shift of cell growth. The Spitzenkörper and, in consequence, also the centre of maximal growth is displaced upon gravistimulation of protonemata by intruding statoliths (Fig. 2). Rhizoids can be forced to respond to some extent like protonemata, but only by pushing statoliths aymmetrically into the apical dome with optical tweezers or by centrifugal forces >50 *g* (Braun 2002). There is evidence from centrifugation experiments (Braun 1996b; Hodick and Sievers 1998) and from attaching particles to the surface of gravitropically responding rhizoids (Sievers et al. 1979) that the position of the growth center at the cell tip is relatively stable and that the Spitzenkörper is more tightly anchored by cytoskeletal forces in rhizoids than in protonemata.

The idea that the specific properties of the actin cytoskeleton which are responsible for Spitzenkörper anchorage are depending on calcium is strongly supported by calcium imaging demonstrating a drastic shift of the steep tip-high calcium gradient towards the upper flank during initiation of the graviresponse in protonemata, but not in rhizoids (Braun and Richter 1999). In accordance with this observation, dihydropyridine fluorescence indicating the tip-focussed distribution of putative calcium channels was also found to be displaced towards the upper flank in graviresponding protonemata (Braun and Richter 1999) which was also not found in rhizoids. The results suggest that the early asymmetric distribution of the calcium gradient in protonemata results either from statolith-induced repositioning of calcium channels or, more likely, might be caused by differential activation and/or inhibition of apical calcium channels. This leads to an asymmetric influx of calcium and, thus, alters the exocytosis pattern and causes an asymmetric incorporation of calcium channels which then establishes the new polarity and the new growth direction. The asymmetric influx of calcium could also mediate the repositioning of the Spitzenkörper and the growth centre by differentially regulating the actin-anchorage or the activity of actin-associated proteins along the shifting calcium gradient (Braun and Richter 1999).

Support for the proposed gravitropic response mechanisms in protonemata comes from immunofluorescence labelling of spectrin-like proteins in the actin-rich area which contains the ER aggregate in the center of the Spitzenkörper. The labelling, which localizes to the median cell axis during vertical growth, is drastically displaced towards the upper flank, the site of future outgrowth, during initiation of the graviresponse in protonemata, clearly before curvature is recognizable (Braun 2001). In contrast, the same labelling in rhizoids remains symmetrically positioned in the apical dome throughout the graviresponse. The findings confirm that a repositioning of the Spitzenkörper is involved in the negative graviresponse of protonemata but not in the positive graviresponse of rhizoids (Fig. 2; Braun 2001). The tendency of protonemata to reorient towards the former growth axis after only short gravistimulation indicates that the new growth axis induced by the upward shift of the Ca^{2+} gradient is rather labile and may require actin cytoskeletal anchorage to stabilize the new growth direction (Braun and Richter 1999; Braun 2001).

Recently, calcium imaging indicated that the impact of statolith sedimentation in rhizoids seems to be limited to a local decrease in the concentration of cytosolic Ca^{2+} at the lower subapical cell flank (personal communication, S. Gilroy, Penn State University, USA), which most likely results from the local inhibition of calcium channels in the area of statolith sedimentation. The subsequent reduction of the rate of exocytosis of secretory vesicles causes differential growth of the opposite cell flanks resulting in the positively gravitropic curvature (Fig. 2; Sievers et al. 1979).

Although calcium is likely to play a role also in gravity sensing of higher plants, several studies failed to show gravity-induced changes in cytosolic Ca^{2+} in higher plant statocytes (Legué et al. 1997). This result may be due to the limited accessibility of the cells and/or the techniques that are unable to show very small or highly localized changes (Boonsirichai et al. 2002).

4 Photosynthesis-dependent interactions of cortical organelles

Photosynthesis requires the interaction of chloroplasts, mitochondria and peroxisomes for regeneration of metabolites and detoxification (Padmasree et al. 2002). The optical conditions in photosynthetically active tissues and organs often do not allow detailed investigation of organelle interaction in higher plant cells. In the characean internode chloroplasts are firmly anchored in the stationary cortex. Cortical mitochondria and peroxisomes are motile but sandwiched between the plasma membrane and the chloroplast layer. This cytoplasmic organization and the presence of plasma membrane microdomains provide excellent conditions for the study of organelle dynamics, distribution, interaction and signalling.

The characean internodes are characterized by alternating bands of acid and alkaline pH, which differ in photosynthetic activity in spite of a homogenous distribution of chloroplasts (Lucas and Smith 1973; Bulychev and Vredenberg 2003). This pattern has long been explained as a pure membrane phenomenon. Recent data, however, highlight the importance of the three dimensional organisation of the cells for pattern formation. It occurs that the pH-banding pattern is reflected by spatio-temporal variations in the subcellular distribution of cortical mitochondria (and probably also peroxisomes). Acid regions, which have a high photosynthetic activity, contain significantly more cortical mitochondria than alkaline regions where the rate of photosynthesis is low. The accumulation of mitochondria at the acid bands requires interaction with both cortical actin filaments and microtubules indicating signalling between chloroplasts and the cortical cytoskeleton (Foissner 2004).

Recent findings also uncovered an intimate interplay between transport phenomena in the plasma membrane and photosynthetic activity (Bulychev et al. 2004). Combined microscopic recordings of pH and photosynthesis revealed that a propagated action potential along the axis of a *Chara* internode modulated the periodic pH pattern along the cell axis. This suggests that the mechanism underlying pattern formation is interacting with the processes involved in excitation. A model has been proposed according to which excitation in the alkaline cell regions may initiate a pathway which modulates membrane processes at the thylakoid membrane (Bulychev et al. 2004). The excursions in the concentration of free Ca^{2+} in the cytoplasm during excitation and their interplay with cation fluxes at chloroplast membrane could be relevant for this communication.

5 Wound-induced cell polarization

The characean internodes are diffusely elongating cells where exocytosis of wall forming vesicles occurs over the whole surface. Local wounding causes reorganization of the cytoplasm and, eventually, local deposition of a wound wall (Foissner and Wasteneys 2000). Such experiments help to identify fundamental processes involved in the establishment of transient polarity, membrane differentiation, exocytosis and endocytosis, organization and dynamics of the cytoskeleton.

In characean internodes, wounds can be induced by local UV illumination, by puncturing and chemical treatment. Chemically induced deposition of wound walls requires high external pH and therefore occurs at the alkaline regions of the internodal cells (Foissner 1989).

UV illumination for up to 10 min causes the detachment of cortical chloroplasts without deposition of a wound wall. The resulting "window" allows better observation of the subcortical actin bundles and the streaming endoplasm (Kamitsubo 1972).

Puncture wounds are sealed by solid vacuolar inclusions which prevent further loss of cytoplasm (Foissner 1988a). Onto this wound plug, a wound wall is secreted by exocytosis of Golgi-derived vesicles. The large size of these vesicles with diameters up to 500 nm allows the visualization of single exocytotic events by high resolution video microscopy (Foissner et al. 1996). Surplus membrane is recycled via endocytotic coated vesicles. Formation, fine structure and chemistry of these "fibrillar wound walls" are similar to the normal secondary cell wall where cellulose microfibrils are embedded in an amorphous, pectin-containing matrix (Foissner 1992). In *Nitella flexilis*, formation of fibrillar wound walls can also be induced by treatment of internodal cells with high concentrations of $CaCl_2$ (Foissner 1990).

When internodal cells are treated with substances which act as calcium ionophores (chlorotetracycline or A 23187), secretory vesicles fuse not only with the plasma membrane but also with each other and with cisternae of the endoplasmic reticulum (Foissner 1988b, 1990). Endocytosis via coated vesicles is inhibited, and the resulting "amorphous wound walls" therefore contain not only the contents of the fusing organelles, but also their membranes. Cellulose microfibrils are not formed but callose, a polysaccharide typical for wounds of higher plants, is present. The amorphous wound walls are further characterized by a high concentration of calcium delivered by the endoplasmic reticulum (Foissner 1998). The secretion of calcium sequestering cisternae of the endoplasmic reticulum could be an effective mechanism to lower the concentration of free Ca^{2+} in the cytoplasm.

The actin cytoskeleton plays an important role during wound healing (Foissner and Wasteneys 2000). After injury, subcortical actin bundles which are attached to the inner side of the stationary cortical chloroplasts and which provide the tracks for myosin-generated endoplasmic streaming (review by Grolig and Pierson 2000) are replaced by a meshwork of randomly organized actin filaments that guarantees the delivery of secretory vesicles towards the wound. After wound wall secretion is complete, continuous actin bundles and active cytoplasmic streaming are regenerated (Foissner et al. 1996; see also Kamitsubo 1972 and Williamson and Hurley 1986 for reorganization of the actin cytoskeleton at windows). Since chloroplasts are absent at wounds these actin bundles are able to interact with the regenerated cortical microtubules (Foissner and Wasteneys 1994, 1999).

6 *Chara* as model for Ca^{2+}-mediated signal-response coupling

Transient excursions in the concentration of free Ca^{2+} in the cytoplasm $[Ca^{2+}]_{cyt}$ are without doubt a major component of many signal response coupling mechanisms in plants (Reddy 2001). Physiologically relevant signals trigger an elevation of $[Ca^{2+}]_{cyt}$ from the low resting level in order to control cellular responses ranging from the control of ion channel activity to the regulation of gene expression. Two prominent examples for such $[Ca^{2+}]_{cyt}$ mediated signal response coupling in *Chara*, i.e. gravitropism and wound response, have been mentioned above.

In many cases, however, it is still difficult to understand the full role of $[Ca^{2+}]_{cyt}$ in most of these signal cascades, because this second messenger is stimulated by a multitude of physiological signals. However, the output of each individual signalling cascade is still specific. A proposed solution to this puzzling situation is that each signal may lead to a specific "signature" of the $[Ca^{2+}]_{cyt}$ response (McAinsh and Hetherington 1998). Such a "signature" may be determined by the amplitude of the $[Ca^{2+}]_{cyt}$ response and/or its kinetics. Indeed monitoring of $[Ca^{2+}]_{cyt}$ in different plant cells in response to relevant physiological signals has shown that the concentration of the second messenger can exhibit sustained elevation. Alternatively, it may oscillate with defined periods (Allen et al. 2001) where both amplitude and oscillation frequency may encode the information for specifics physiological responses (McAinsh and Hetherington 1998; Allen et al. 2001).

The mechanisms underlying these complex $[Ca^{2+}]_{cyt}$ responses in plants are not known. This ignorance is mostly due to the fact that the molecular reactions underlying Ca^{2+} mediated signal response coupling in plants are not well understood. There is evidence that Ca^{2+} influx via plasma membrane channels is involved and that also Ca^{2+} release from internal stores is relevant (Reddy 2001). The coupling of the two processes may cause, via the so-called calcium-induced calcium release, the complex $[Ca^{2+}]_{cyt}$ oscillations observed in plants (Reddy 2001).

A source of quantitative understanding of Ca^{2+} signalling comes form investigations on voltage stimulated $[Ca^{2+}]_{cyt}$ signalling in *Chara* as a model system for Ca^{2+} mediated stimulus response coupling in plants (Thiel et al. 2002).

Chara cells are electrically excitable. This electrical excitation in *Chara* is intimately associated with a rise in $[Ca^{2+}]_{cyt}$ (Thiel et al. 2002). The latter step is essential for the activation of the Cl^- channels, e.g the conductance, which depolarize the membrane. It has been a matter of debate whether the transient rise in $[Ca^{2+}]_{cyt}$ during excitation originates from influx via channels in the plasma membrane or whether this rise in $[Ca^{2+}]_{cyt}$ is the result of release from internal stores (Thiel et al. 2002; Tazawa and Kikuyama 2003). There are still good reasons to believe that a channel mediated minute influx of Ca^{2+} is involved in the very early phase of excitation (Thiel et al. 1993, 2002; Tazawa andKikuyama, 2003), but experimental results clearly show that the bulk rise in $[Ca^{2+}]_{cyt}$ during excitation is due to a release of Ca^{2+} from internal stores (Wacke and Thiel 2001). The result of these measurements reveal that the electrically stimulated elevation of $[Ca^{2+}]_{cyt}$ in *Chara* has all the hallmarks of an excitable system: stimuli below a defined threshold result in no measurable increase in $[Ca^{2+}]_{cyt}$; only after passing this narrow threshold will any stimulation evoke the full extend of the $[Ca^{2+}]_{cyt}$ response (Wacke and Thiel 2001). Such an excitable system of a $[Ca^{2+}]_{cyt}$ response can no longer be explained by simple voltage dependent Ca^{2+} influx across channels in the plasma membrane.

Recent experiments using the well-defined *Chara* system foster the hypothesis that the process of Ca^{2+}-excitation is based on a voltage dependent production of a long lived second messenger (Wacke and Thiel 2001; Wacke et al. 2003). This yet unknown second messenger, which might be inositol (1,4,5) tris phosphate (IP_3) (Biskup et al. 1999), triggers the release of Ca^{2+} from the internal stores; in this way, it initiates the excitation process. The key dynamic features of the second messenger and its relation to voltage stimulation can be obtained from experiments in which Ca^{2+} mobilization in *Chara* is examined under the influence of defined graded electrical stimuli. Based on these experimental data a kinetic model was developed, which was able to simulate the experimental observations with respect to triggering excitation with electrical stimuli (Wacke and Thiel 2001).

A further step towards understanding the dynamics of the whole process was obtained by combining the aforementioned model on voltage dependent IP_3 production with a well-established model from animal cells. This animal cell model describes the release of Ca^{2+} from internal stores in relation to the gating of the IP_3 receptor channel, i.e. the channel which allows Ca^{2+} release from the internal stores into the cytoplasm. A kinetic four-state model worked out by Tang and coworkers (1996) is sufficient to explain the bulk of the experimental observations on IP_3 generated Ca^{2+} mobilization in animal cells including complex dynamic processes and $[Ca^{2+}]_{cyt}$ oscillations. The minimal model (Fig. 3) developed by Tang et al. (1996) reflects the fact that the receptor has three distinct binding sites. These sites can either be empty (R), occupied by one IP_3 molecule (RI), by one IP_3 molecule plus one Ca^{2+}-ion (RIC_+) and finally by an additional Ca^{2+} ion (RIC_+C_-). The binding order of the ions and molecules to the receptor is not free but proceeds sequentially from R over RI, RIC_+ to RIC_+C_- and back. The channel is only active in the state RIC_+ which accounts for the observation that low concentrations of Ca^{2+} stimulate and high concentrations inhibit an IP_3 generated Ca^{2+} release from internal stores.

The validity of such a model for the explanation of Ca^{2+}-excitation in *Chara* can be tested on the basis of the predictions derived from this particular model. It predicts a complex behaviour of $[Ca^{2+}]_{cyt}$ mobilization under periodic stimulation. A testable prediction is that $[Ca^{2+}]_{cyt}$ response includes higher-order phase locking and irregular responses upon increased stimulation frequency. When the electrically stimulated transient increase in $[Ca^{2+}]_{cyt}$ was monitored the $[Ca^{2+}]_{cyt}$ excursions behaved as predicted from the model (Wacke et al. 2003a,b). Furthermore, since the elevation of

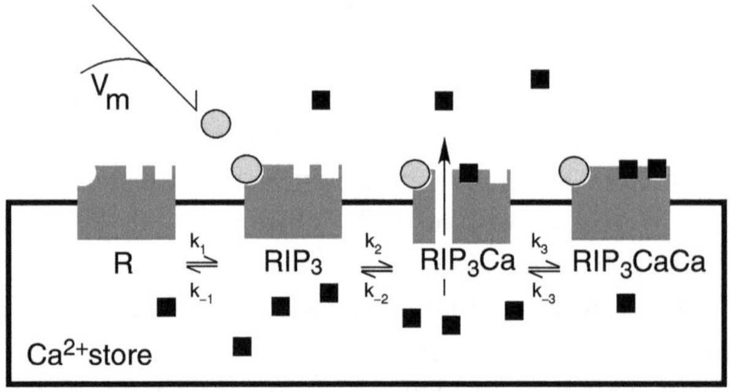

Fig. 3. A schematic view of the voltage stimulated IP_3 production and activation of the IP_3 receptor channel. The receptor has three distinct binding sites which can be empty (R), occupied in a sequential order by one IP_3 molecule (RIP^3), plus one Ca^{2+} ion (RIP_3Ca) or two Ca^{2+} ions (RIP_3CaCa). Ca^{2+} release from internal stores occurs only in the state PIP_3Ca. IP_3 and Ca^{2+} are represented by *grey circles* and *black squares* respectively

$[Ca^{2+}]_{cyt}$ underlies membrane excitation, the model predicts that also action potentials are triggered under periodic stimulation in a frequency dependent manner. Again the experimental tests were in accordance with the model predictions. Figure 4 shows the results of recordings of the membrane potential in a *Chara* cell under periodic stimulation. As predicted from the model a rise in stimulation frequency resulted in an increasing order of phase locking. At very high frequencies the response became irregular.

The examination of the robustness of the model also successfully passed another test. The simulations predict that the kinetics of $[Ca^{2+}]_{cyt}$ elevation is depending on the magnitude of the resting $[Ca^{2+}]_{cyt}$ concentration. The lower the resting concentration prior to stimulation, the more immediate the $[Ca^{2+}]_{cyt}$ response to stimulation. This assumption can be tested by recording the kinetics of electrical membrane excitation in light versus dark adapted *Chara* cells. It is known that a transfer of *Chara* from light into dark results in a slowly progressing elevation of the $[Ca^{2+}]_{cyt}$ resting concentration (Plieth 1995; Miller and Sanders 1987). On the basis of this rise in $[Ca^{2+}]_{cyt}$, the model on membrane excitation predicts that the depolarization of the

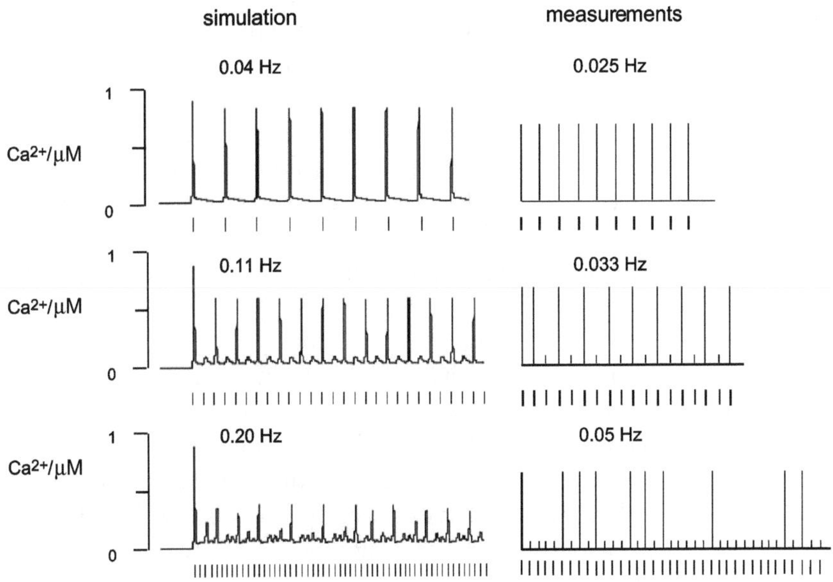

Fig. 4. Simulated $[Ca^{2+}]_{cyt}$ transients (left panel) and measured APs in *Chara* (right panel) in response to periodic stimulation. Electrical stimulation is indicated by small *bar* below graphs. Successful or non-successful stimulations of an AP are represented by *long* or *short lines* respectively. Experimental details and simulation parameters as given in Wacke et al. (2003)

action potential occurs faster in dark than in light adapted cells (Baudenbacher et al. 2005). Measurements of the kinetics of membrane excitation by means of electrical or magnetic recordings show that the cells behave exactly as predicted (Baudenbacher et al. 2005). A transfer of cells from the light into the dark resulted in a progressive shortening of a delay time prior to the steep depolarization. The same shortening of a delay time is predicted from the model simulations.

Collectively, these data support the good quality of the model for the explanation of Ca^{2+} mediated excitation in *Chara*. It can be anticipated that the same quantitative model also presents an excellent basis for the understanding of even more complex Ca^{2+} signatures in *Chara* and higher plants. Spontaneous or triggered oscillatory changes in $[Ca^{2+}]_{cyt}$ can be induced in this model by tuning some of the parameters. In the most simple scenario, the model can be altered such that the steady state solution is unstable and a stable periodic solution occurs (Othmer 1997).

7 The genus *Chara* as a model for sex differentiation

Most of the angiosperms develop hermaphroditic ("perfect") flowers, only a few of them separate the sexes of their reproductive organs (Juarez and Banks 1998). Ainsworth et al. (1998) estimated that 4% of higher plants are dioecious, representing male and female individuals, carrying male or female flowers only. Another estimated 7% are monoecious plants, exhibiting a spatial separation of the male and female flowers carried by the same individual (Ainsworth et al. 1998). This large variation in sex expression with respect to where, when and how the decision about the sex differentiation occurs indicates that there are several different sex determining mechanisms in plants (Juarez and Banks, 1998).

Until recently, only a few examples of dioecious plants with sex-chromosome systems of both types, active Y-chromosome (e.g. *Silene latifolia*, Lardon et al. 1999) or X-autosome dosage (e.g. *Rumex acetosa*, Ainsworth et al. 1999) have been described. In the majority of dioecious plants regulation of gene activity (e.g. by hormonal regulation as in *Mercurialis annua*, Durand and Durand 1991a,b; for review, see Chailakhyan, 1979) rather than the presence of sex chromosomes determines the development of reproductive structures (Ainsworth et al. 1998).

The genus *Chara* consists of both dioecious as well as monoecious species. Irrespective of the fact that some of the *Chara* species are able to propagate vegetatively by means of specialized cells rich in reserve compounds like bulbils, or by thallus fragments containing omnipotent nodal cells, all recent species of the order Charales lack, in contrast to most of the other algae, asexual reproduction systems by means of zoo- or autospores (Krause, 1997).

The sexual reproduction system of the Charales is highly developed and unique for plants. It consists of the female archegonia and the male antheridia, the latter ones representing the highest level of organization of male gametangia in autotrophs (Schussnig 1954).

All Charales lack any kind of generation cycle. In general, the plants are haploid; the zygote is the only diploid phase. Before propagation, meiosis occurs resulting in an oospore containing four nuclei. Three of the nuclei are left in a large cell which stays in the oospore, these nuclei degenerate and do not contribute to the further development of the plant. The remaining uninuclear cell becomes the initial cell and leaves the oospore at the apical end. The next division of this initial cell is already accompanied by a differentiation, one of these cells representing the first cell of the rhizoid and the other one the first cell of the thallus. Both cells are haploid. These first divisions and all following ones are mitotic. In addition, amitotic development of "polyenergide? internodal cells, containing up to 2000 nuclei, has been described as a unique phenomenon in the order Charales (Shen 1967). The development of the sexual reproduction organs, however, starts from uninuclear cells remained in an omnipotent status which are located in the nodi.

Compared with higher plants or animals, very little is known about the mechanism of sex determination in the order Charales. Being haploid, the presence of sex chromosomes can be excluded for monoecious species. But also in the dioecious *Chara tomentosa* no karyotypic differences between male and female individuals could be found (Kunachowicz et al. 2001). These authors explicitly point to the fact that this does not exclude the presence of sex chromosomes per se. Sex chromosomes can be homomorphic as in *Asparagus officinalis* (Marziani et al. 1999), or the genes involved in sex determination might be dispersed on several chromosomes. On the other hand, since Ernst (1901) first reported cases of pseudo-hermaphrodity of *Nitella syncarpa*, in which spermatogenesis occurred in the archegonia of female individuals of this dioecious species, it can be assumed that the genetic information leading to antheridia formation is present in female individuals of this species as well.

In addition, a striking difference in the DNA methylation level between male (9.1%) and female (5.9%) *Chara tomentosa* reported by Kunachowicz et al. (2001) as well as the results of Olszewska et al. (1997), showing pronounced changes in the DNA methylation level during spermatogenesis in the monoecious *Chara vulgaris*, suggest that an epigenetic control mechanism is involved in sex determination of *Chara*.

An analysis of the effects of hypomethylation, as done for, e.g. *Melandrium album* (Vyskot 1999), might help to gain further insight in the regulation mechanisms leading to the formation of antheridia and archegonia of monoecious *Chara* species.

A special case with respect to sex determination is *Chara canescens* which, according to Wood and Imahori (1964), consists of a monoecious and a dioecious sub-species. The latter one can be further split into an apomictic (parthenogenetic) and a bisexual reproducing race (Krause 1997). The first detailed investigations about the mechanism of parthenogenesis of *Chara canescens* by Ernst (1916) already revealed that the apomictic race consists of diploid individuals which do not need fertilization by spermatids. Moreover, in several experiments Ernst (1916) could show that these apomictic individuals are unable to get fertilized and consist of female individuals only. Two hypotheses about the origin of the apomictic form of *Chara canescens* were formulated by Ernst (1918): (i) apomictic *Chara canescens* are the result of interspecific hybridisation or (ii) osmotic shock suppresses meiosis during propagation of the zygospore and results in diploid apomictic *Chara canescens* individuals. Neither the interspecific cross-fertilization experiments of Ernst (1918) nor AFLP-based genetic analysis showing strong similarity of parthenogenetic and bisexual populations (own unpublished results) support the first hypothesis. With respect to the second hypothesis Ernst (1918) stated that he was able to produce parthenospores by hyper- and hypoosmotic shock in preliminary experiments. However, no detailed description of the experimental procedure as well as their results could be found and thus the experiment should be repeated.

Unfortunately, no recent occurrence of the monoecious subspecies can be traced. The site where the herbarium specimens originate from is destroyed and it might be that this subspecies is extinct. However, a comparison of the apomictic and the bisexual race might allow further insight in the processes leading to the development of parthenogenesis as well as the sex differentiation mechanisms in the order Charales since the sex of the apomictic individuals is locked in the female state.

8 Outlook

Complex systems such as pattern formation and development in higher plants can only be understood if the function and integration of individual components of these processes are fully known. This requires well-defined and simple model systems to start with. The present review has presented recent examples on how some basic signal transduction cascades, mechanisms of pattern formation and sex determination can be studied in great depth in the unicellular algae of the *Characeae*. Together with molecular tools, which are currently developed in some laboratories, these algae will also in the future present an excellent model system to uncover some of the basic principles in plant biology.

Acknowledgements

This work was financially supported by the DFG, DLR on behalf of the Bundesministerium für Bildung und Forschung. Special thanks to the Space-Shuttle crews of STS 65 and STS 81, the teams of DLR (German Aerospace Center), ESA (European Space Agency), NASA (National Aeronautics and Space Administration), EADS ST (European Aeronautics Defense and Space Company, Space Transportation), Kayser-Threde and Novespace. We thank Andreas Sievers and Mark Hütt for discussions and helpful comments.

References

Ainsworth CC, Parker J, Buchanan-Wollaston V (1998) Sex determination in plants. Curr Top Dev Biol 38:167–223

Ainsworth CC, Lu J, Winfield M, Parker J (1999) Sex determination by X:autosome dosage: Rumex acetosa (sorrel). In: Ainsworth CC (ed) Sex determination in plants, BIOS Sci Publ, Oxford, pp 121–136

Allen GJ, Chu SP, Harrington CL, Schumacher K, Hoffmann T, Tang YY, Grill E, Schroeder JI (2001) A defined range of guard cell calcium oscillation parameters encodes stomatal movements. Nature 411:1053–1057

Aloni R, Langhans M, Aloni E, Ullrich CI (2004) Role of cytokinin in the regulation of root gravitropism. Planta 220:177–182

Bartnik E, Sievers A (1988) In-vivo observation of a spherical aggregate of endoplasmic reticulum and of Golgi vesicles in the tip of fast-growing *Chara* rhizoids. Planta 176:1–9

Baudenbacher F, Fong LE, Thiel G, Wacke M, Jazbinsek V, Holzer JR, Stampfl A, Trontelj Z (2005) Intracellular axial current in *Chara corallina* reflects the altered kinetics of ions in cytoplasm under the influence of light. Biophys J 88:690–697

Blilou I, Xu J, Wildwater M, Willemsen V, Paponov I, Friml J, Heidstra R, Aida M, Palme K, Scheres B (2005) The PIN auxin efflux facilitator network controls growth and patterning in *Arabidopsis* roots. Nature 433:39–44

Biskup B, Gradmann D, Thiel G (1999) Calcium release from IP_3-sensitve stores initiates action potential in *Chara*. FEBS Lett 453:72–76

Blancaflor EB (2002) The cytoskeleton and gravitropism in higher plants. J Plant Growth Regul 21:120–136

Boonsirichai K, Guan C, Chen, R, Masson PH (2002) Root gravitropism: an experimental tool to investigate basic cellular and molecular processes underlying mechanosensing and signal transmission in plants. Annu Rev Plant Physiol Plant Mol Biol 53: 421–447

Braun M (1996a) Immunolocalization of myosin in rhizoids of *Chara globularis* Thuill Protoplasma 191:1–8

Braun M (1996b) Anomalous gravitropic response of *Chara* rhizoids during enhanced accelerations. Planta 199:443–450

Braun M (1997) Gravitropism in tip-growing cells. Planta 203: S11-S19
Braun M (2001) Association of spectrin-like proteins with the actin-organized aggregate of endoplasmic reticulum in the Spitzenkörper of gravitropically tip-growing plant cells. Plant Physiol 125:1611–1620
Braun M (2002) Gravity perception requires statoliths settled on specific plasma-membrane areas in characean rhizoids and protonemata. Protoplasma 219: 150–159
Braun M, Limbach C (2006) Rhizoids and protonemata of characean algae – model cells for research on polarized growth and plant gravity sensing. Protoplasma (in press)
Braun M, Richter P (1999) Relocalization of the calcium gradient and a dihydropyridine receptor is involved in upward bending by bulging of *Chara* protonemata, but not in downward bending by bowing of *Chara* rhizoids. Planta 209:414–423
Braun M, Sievers A (1993) Centrifugation causes adaptation of microfilaments; studies on the transport of statoliths in gravity sensing *Chara* rhizoids. Protoplasma 174:50–61
Braun M, Sievers A (1994) Role of the microtubule cytoskeleton in gravisensing *Chara* rhizoids. Eur J Cell Biol 63: 289–298
Braun M, Wasteneys GO (1998a) Reorganization of the actin and microtubule cytoskeleton throughout blue-light-induced differentiation of characean protonemata into multicellular thalli. Protoplasma 202:38–53
Braun M, Wasteneys GO (1998b) Distribution and dynamics of the cytoskeleton in graviresponding protonemata and rhizoids of characean algae: exclusion of microtubules and a convergence of actin filaments in the apex suggest an actin-mediated gravitropism. Planta 205:39–50
Braun M, Wasteneys GO (2000) Actin in characean rhizoids and protonemata. Tip growth, gravity sensing and photomorphogenesis. In: Staiger CJ, Baluska F, Volkmann D, Barlow PW (eds) Actin: a dynamic framework for multiple plant cell functions. Kluwer, Dordrecht, pp 237–258
Braun M, Buchen B, Sievers A (2002) Actomyosin-mediated statolith positioning in gravisensing plant cells studied in microgravity. J Plant Growth Regul 21:137–145
Braun M, Hauslage J, Czogalla A, Limbach C (2004) Tip-localized actin polymerization and remodeling, reflected by the localization of ADF, profilin and villin, are fundamental for gravity-sensing and polarized growth of characean rhizoids. Planta 219:379–388
Buchen B, Braun M, Hejnowicz Z, Sievers A (1993) Statoliths pull on microfilaments. Experiments under microgravity. Protoplasma 172:38–42
Buchen B, Braun M, Sievers A (1997) Statoliths, cytoskeletal elements and cytoplasmic streaming of *Chara* rhizoids under reduced gravity during TEXUS flights. In: Life sciences experiments performed on sounding rockets (1985–1994). Nordwijk, ESA Publications Division, ESA-SP 1206, pp 71–75
Bulychev A, Vredenberg W (2003) Spatio-temporal patterns of photosystem II activity and plasma-membrane proton flows in *Chara corallina* cells exposed to overall and local illumination. Planta 218:143–151
Bulychev AA, Kamzolkina NA, Luengviriya J, Rubin AB, Muller SC. (2004) Effect of a single excitation stimulus on photosynthetic activity and light-dependent pH banding in Chara cells. J Membr Biol 202:11–19.
Cai W, Braun M, Sievers A (1997) Displacement of statoliths in *Chara* rhizoids during horizontal rotation on clinostats. Acta Biol Exp Sinica 30:147–155
Chailakhyan MK (1979) Genetic and hormonal regulation of growth, flowering and sex expression in plants. Am J Bot 66:717–736
Ciesielski T (1872) Untersuchungen über die Abwärtskrümmung der Wurzel. Beitr Biol Pflanz 1:1–30
Driss-Ecole D, Jeune B, Prouteau M, Julianus P, Perbal G (2000) Lentil root statoliths reach a stable state in microgravity. Planta 211:396–405
Drobak BK, Franklin-Tong VE, Staiger CJ (2004) The role of actin the cytoskeleton in plant cell signaling. New Phytol 163:13–30
Durand B, Durand R (1991a) Sex determination and reproductive organ differentiation in Mercurialis. Plant Sci 80:49–65

Durand B, Durand R (1991b) Male sterility and restored fertility in annual mrcuries, relations with sex differentiation. Plant Sci 80:107–118

Ernst A (1901) Über Pseudohermaphroditismus und andere Miβbildungen der Oogonien bei *Nitella syncarpa* (Thuill.) Kützing. Flora od allg bot Zeitg 88:1–36

Ernst A (1916) Experimentelle Erzeugung erblicher Parthenogenesis. Z f ind Abstammungs- und Vererbungslehre 17:203–250

Ernst A (1918) Bastardisierung als Ursache der Apogamie im Pflanzenreich. G Fischer-Verlag, Jena, 673 pp

Foissner I (1988a) The relationship of echinate inclusions and coated vesicles on wound healing in *Nitella flexilis* (Characeae). Protoplasma 142:164–175

Foissner I (1988b) Chlortetracycline-induced formation of wall appositions (callose plugs) in internodal cells of *Nitella flexilis* (Characeae). J Phycol 24:458–467

Foissner I (1989) PH-dependence of chlortetracycline(CTC)-induced plug formation in *Nitella flexilis* (Characeae). J Phycol 25:313–318

Foissner I (1990) Wall appositions induced by ionophore A 23187, $CaCl_2$, $LaCl_3$, and nifedipine in characean cells. Protoplasma 154:80–90

Foissner I (1992) Effects of dichlorobenzonitrile on the formation of cell wall appositions (plugs) in internodal cells of *Chara corallina* Klein ex. Willd, em. R.D.W. and *Nitella flexilis* (L.) Ag. New Phytol 121:447–455

Foissner I (1998) Localization of calcium ions in wounded characean internodal cells. New Phytol 139:449–458

Foissner I (2004) Microfilaments and microtubules control the shape, motility, and subcellular distribution of cortical mitochondria in characean internodal cells. Protoplasma 224:145–157

Foissner I, Wasteneys GO (1994) Injury to *Nitella* internodal cells alters microtubule organization but microtubules are not involved in the wound response. Protoplasma 182:102–114

Foissner I, Wasteneys GO (1999) Microtubules at wound sites of *Nitella* internodal cells passively co-align with actin bundles when exposed to hydrodynamic forces generated by cytoplasmic streaming. Planta 208:480–490

Foissner I, Wasteneys GO (2000) Actin in characean internodal cells. In: Staiger C, Baluska DF, Volkmann D, Barlow P (eds) Actin: a dynamic framework for multiple plant cell functions. Kluwer Academic Publisher, Dordrecht, Boston, London, pp 259–274

Foissner I, Lichtscheidl IK, Wasteneys GO (1996) Actin-based vesicle dynamics and exocytosis during wound wall formation in characean internodal cells. Cell Motil Cytoskel 35:35–48

Galland P (2002) Tropisms of *Avena* coleoptiles: sine law for gravitropism, exponential law for photogravitropic equilibrium. Planta 215:779–784

Geitmann A, Emons AM (2000) The cytoskeleton in plant and fungal cell tip growth. J Microsc 198:218–245

Graham LE (1993) Origin of land plants. Wiley, New York, 287 pp

Grolig F, Pierson ES (2000) Cytoplasmic streaming: From flow to track. In: Staiger C, Baluska F, Volkmann D, Barlow PW (eds) Actin: a dynamic framework for multiple plant cell functions. Kluwer Academic Publishers, Dordrecht, Boston, London, pp 165–181

Haberlandt G (1900) Über die Perzeption des geotropischen Reizes. Ber Dtsch Bot Ges 18: 261–272

Hejnowicz Z, Sievers A (1981) Regulation of the position of statoliths in *Chara* rhizoids. Protoplasma 108:117–137

Hejnowicz Z, Heinemann B, Sievers A (1977) Tip growth: pattern of growth rate and stress in the *Chara* rhizoid. Zeitschr Pflanzenphysiol 81:409–424

Hepler PK, Vidali L, Cheung AY (2001) Polarized cell growth in higher plants. Annu Rev Cell Dev Biol 17:159–187

Hodick D (1993) The protonema of *Chara fragilis* Desv.: Regenerativeformation, photomorphogenesis, and gravitropism. Bot Acta 106:388–393

Hodick D, Sievers A (1998) Hypergravity can reduce but not enhance the gravitropic response of *Chara globularis* protonemata. Protoplasma 204:145–154
Hodick D, Buchen B, Sievers A (1998) Statolith positioning by microfilaments in *Chara* rhizoids and protonemata. Adv Space Res 21:1183–1189
Hoson T, Kamisaka S, Masuda Y, Yamashita M, Buchen B (1997) Evaluation of the three-dimensional clinostat as a simulator of weightlessness. Planta 203:S187–S197
Hou G, Kramer VL, Wang Y-S, Chen R, Perbal G, Gilroy S, Blancaflor EB (2004) The promotion of gravitropism in *Arabidopsis* roots upon actin disruption is coupled with the extended alkalinization of the columella cytoplasm and a persistent lateral auxin gradient. Plant J 39:113–125
Juarez C, Banks JA (1998) Sex determination in plants. Curr Opin Plant Biol 1:68–72
Kamitsubo E (1972) A "window technique" for detailed observation of characean cytoplasmic streaming. Exp Cell Res 74:613–616
Kiss JZ (2000) Mechanisms of the early phases of plant gravitropism. Crit Rev Plant Sci 19:551–573
Knight TA (1806) On the direction of the radicle and germen during the vegetation of seeds. Phil Trans R Soc 99:108–120
Kranz HD, Miks D, Siegler ML, Capesius I, Sense CW, Huss VA (1995) The origin of land plants: phylogenetic relationships among charophytes, bryophytes, and vascular plants inferred from complete small-subunit ribosomal RNA gene sequences. J Mol Evol 4:74–84
Krause W (1997) Charales (Charophyceae). (Süsswasserflora von Mitteleuropa, Bd. 18), G. Fischer-Verlag, Jena, 202 pp
Kunachowicz A, Luchniak P, Olszewska MJ, Sakowicz T (2001) Comparative karyology, DNA methylation and restriction pattern analysis of male and female plants of the dioecious alga *Chara tomentosa* (Charophyceae). Eur J Phycol 36:29–34
Ladinsky MS, Mastronarde DN, McIntosh JR, Howell KE, Staehelin LA (1999) Golgi structure in three dimensions: functional insights from the normal rat kidney cell. J Cell Biol 144:1135–1149
Lardon A, Aghmir A, Georgiev S, Monéger F, Negrutiu I (1993) The Y chromosome of white campion: sexual dimorphism and beyond. In: Ainsworth CC (ed) Sex determination in plants. BIOS Sci Publ, Oxford, pp 89–100
Legué V, Blancaflor E, Wymer C, Perbal G, Fantin D, Gilroy S (1997) Cytoplasmic free Ca^{2+} in *Arabidopsis* roots changes in response to touch but not gravity. Plant Physiol 114:789–800
Leitz G, Schnepf E, Greulich KO (1995) Micromanipulation of statoliths in gravity-sensing *Chara* rhizoids by optical tweezers. Planta 197:278–288
Lewis LA, McCourt RM (2004) Green algae and the origin of land plants. Am J Bot 91:1535–1556
Limbach C, Hauslage J, Schaefer C, Braun M (2005) How to activate a plant gravireceptor. Early mechanisms of gravity sensing studied in characean rhizoids during parabolic flights. Plant Physiol 139:1030–1040
Lovy-Wheeler A, Wilsen KL, Baskin TI, Hepler PK (2005) Enhanced fixation reveals the apical cortical fringe of actin filaments as a consistent feature of the pollen tube. Planta 221:95–104
Lucas WJ, Smith FA (1973) The formation of alkaline and acid regions at the surface of *Chara corallina* cells. J Exp Bot 24:1–14
McAinsh MR, Hetherington AM (1998) Encoding specificity in Ca^{2+} signalling systems. Trends Plant Sci 3:32–36
Mastronarde DN (1997) Dual-axis tomography: an approach with alignment methods that preserve resolution. J Struct Biol 120:343–352
Marziani G, Caporali E, Spada A (1999) Search for genes involved in asparagus sex determination. In: Ainsworth CC (ed) Sex determination in plants. BIOS Sci Publ, Oxford, pp 149–162
Miller AJ, Sanders D (1987) Depletion of cytosolic free calcium induced by photosynthesis. Nature 326:397–400

Nemec B (1900) Über die Art der Wahrnehmung des Schwerkraftreizes bei den Pflanzen. Ber Dtsch Bot Ges 18:241–245
Olszewska MJ, Gernand D, Godlewski M, Kunachowicz A (1997) DNA methylation during antheridial filament development and spermiogenesis in *Chara vulgaris* (Charophyceae) analysed by in situ nick-translation driven by methylation-sensitive restriction enzymes. Eur J Phycol 32:287–291
Othmer HG (1997) Signal transduction and second messenger Systems. In: Othmer HG, Adler FR, Lewis MA, Dallon J (eds) Case studies in mathematical modelling-ecology, physiology and cell biology. Prentice Hall, Englewood Cliffs, N.J.
Ottenschläger I, Wolff P, Wolverton C, Bhalerao RP, Sandberg G, Ishikawa H, Evans M, Palme K (2003) Gravity-regulated differential auxin transport from columella to lateral root cap cells. Proc Natl Acad Sci USA 100:2987–2991
Padmasree K, Padmavathi L, Raghavendra AS (2002) Essentiality of mitochondrial oxidative metabolism for photosynthesis: optimization of carbon assimilation and protection against photoinhibition. Crit Rev Biochem Mol Biol 37:71–119
Perbal G, Lefrance A, Jeune B, Driss-Ecole D (2004) Mechanotransduction in root gravity sensing cells. Physiol Plant 120:303–311
Plieth C (1995) Estimation of ion concentrations and their variations in cells and tissues of green plants using image analysis as ratiometric fluorescence microscopy and laser Doppler anemometry. PhD thesis University of Kiel, Germany
Reddy ASN (2001) Calcium: silver bullet in signalling. Plant Sci 160:381–404
Schussnig B (1954) Grundriß der Protophytologie, G Fischer, Jena, 310 pp
Shen EYF (1967) Amitosis in Chara. Cytology 32:481–488
Sievers A, Heinemann B, Rodriguez-Garcia MI (1979) Nachweis des subapikalen differentiellen Flankenwachstums im *Chara*-Rhizoid während der Graviresponse. Zeitschr Pflanzenphysiol 91:435–442
Sievers A, Buchen B, Volkmann D, Hejnowicz Z (1991a) Role of the cytoskeleton in gravity perception. In: Lloyd CW (ed) The cytoskeletal basis for plant growth and form. Academic Press, London, pp 169–182
Sievers A, Kramer-Fischer M, Braun M, Buchen B (1991b) The polar organization of the growing *Chara* rhizoid and the transport of statoliths are actin-dependent. Bot Acta 104:103–109
Sievers A, Buchen B, Hodick D (1996) Gravity sensing in tip-growing cells. Trends Plant Sci 1:273–279
Sievers A, Braun B, Monshausen GB (2002) The root cap: structure and function. In: Waisel Y, Eshel A, Kafkafi U (eds) Plant roots—the hidden half, ed 3. Marcel Dekker, New York, pp 33–47
Tang Y, Stephenson J, Othmer HG (1996) Simplification and analysis of models of calcium dynamics based on $InsP_3$-sensitive calcium channel dynamics. Biophys J 70:246–263
Tazawa M, Kikuyama M. (2003) Is Ca^{2+} release from internal stores involved in membrane excitation in characean cells? Plant Cell Physiol 44:518–526
Thiel G, Homann U, Gradmann D (1993). Microscopic elements of electrical excitation in *Chara*: transient activity of Cl^- channels in the plasma membrane. J Membrane Biol 134:53–66
Thiel, G, Wacke M, Foissner I (2002): Ca^{2+} mobilization from internal stores in electrical membrane excitation in *Chara*. Prog Bot 64:217–233
Volkmann D, Buchen B, Hejnowicz Z, Tewinkel M, Sievers A (1991) Oriented movement of statoliths studied in a reduced gravitational field during parabolic flights of rockets. Planta 185:153–161
Vyskot B (1999) The role of DNA methylation in plant reproductive development. In: Ainsworth CC (ed) Sex determination in plants. BIOS Sci Publ, Oxford, pp 101–120
Wacke M, Thiel G (2001) Electrically triggered all-or-none Ca^{2+} liberation during action potential in the giant alga *Chara*. J Gen Physiol 118:11–21
Wacke M, Hütt M-T, Thiel G (2003) Patterns of membrane excitation in *Chara* cells in response to periodic stimulation. Nova Acta 332:225–238

Wacke M, Thiel G, Hütt M-T (2003) Ca^{2+} dynamics during membrane excitation of green alga *Chara:* model simulations and experimental data. J Membrane Biol 191:179–192
Williamson RE, Hurley UA (1986) Growth and regrowth of actin bundles in *Chara*: bundle assembly by mechanisms differing in sensitivity to cytochalasin. J Cell Sci 85:21–32
Wood RD, Imahori K (1965) A revision of the Characeae. J Cramer, Weinheim, 904 pp

Markus Braun
Gravitational Biology Group
Institute of Plant Molecular Physiology
and Biotechnology (IMBIO)
University of Bonn
Kirschallee 1
53115 Bonn
Germany

Hinrich Lühring
Research Center Jülich
Phytosphere Institute
Wilhelm-Johnen-Straße
52425 Jülich
Germany

Hendrik Schubert
University of Rostock
Inst. for Biol. Sciences
Einsteinstr. 3
18051 Rostock
Germany

Ilse Foissner
University of Salzburg
Dep. Cell Biology
Inst. of Plant Physiology
Hellbrunnerstr. 34
A-5020 Salzburg
Austria

Gerhard Thiel
Darmstadt University of Technology
Institute of Botany
Schnittspahnstr. 3
64287 Darmstadt
Germany

Receptors for the Five Classical Plant Hormones

Hartwig Lüthen, Michael Böttger and Daniel Schenck

1 35 years of hunting receptors: we never got what we expected

Textbooks suggest that plant growth and development is controlled by a set of five major groups of plant hormones. These classical hormones (Kende and Zevaart 1997), auxins, gibberellins, cytokinins, ethylene and abscisic acid (ABA), have been discovered in within half a century (auxins: 1928, gibberellins: 1935, cytokinins 1955, ethylene: 1959, ABA: 1970). In the last 3 decades, the mode of action of these substances became a thrilling target for biochemical, physiological and molecular research. Hormone signalling was much better understood in animals than in plants, so plant physiologists tried their hand at using the strategies proven in the success stories of animal signalling research. Since any signalling chain starts with a receptor, identifying these proteins appeared to be a good starting point. This track was, however, unexpectedly stony. Looking back, most of the problems in the early work were due to the fact that plants were not animals at all!

Serious hormone receptor research began when Rainer Hertel characterized the first membrane-associated auxin binding activities (Hertel et al. 1970; Dohrmann et al. 1978). In the 1980s and early 1990s, the search was inspired by the discovery of signalling chains in animals involving G-protein-coupled 7-transmembrane receptors (e.g. Millner et al. 1996). It was generally assumed that such receptors were also involved in plant signalling. Hormones would bind to the plant analogues of the receptor, G protein dissociation and activation would occur, followed by the formation of 2nd and 3rd messengers. Inositol triphosphate (IP_3) and calcium ions, sometimes even cAMP, were suggested. Today, we know that while human genomes contain thousands of 7-transmembrane receptors, only one gene coding for such a protein has been found in the *Arabidopsis* genome. Although this receptor may be important for ABA signalling, the sheer numbers indicate that signalling through heterotrimeric G proteins, although being widespread in animals is quite exotic in plants.

Progress in Botany, Vol. 68

The 1990s saw a first result: the identification of two-component receptors for both ethylene and cytokinin. But what a surprise! Genes for these receptors are not found in the human and animal genomes, but are commonplace in bacteria. Suddenly "plant lifestyle appeared to be much more prokaryotic than expected" (Lohrmann and Harter 2002).

In the 1980s and 1990s, most workers in the field postulated plant hormone receptors to be integral plasma membrane proteins. Then the stunning news broke that the two-component ethylene receptors were located at the endoplasmic reticulum membrane. Very recently, in 2005, a number of breakthrough papers were published identifying intracellular receptors for auxin and gibberellic acid. These nuclear receptors are directly involved in regulating expression of hormone-induced genes at the promoter level.

We will restrict the present review to those receptors involved in perception of the five classical plant hormones. A vast number of other receptor proteins discovered in the plant genome will not be covered here.

2 Intracellular hormone receptors acting at the level of gene expression

2.1 Intracellular auxin perception by TIR1 and other F-box proteins

In the 1980s and 1990s, auxin perception was generally regarded as a process localized at the cell surface. A putative auxin receptor, auxin binding protein 1 (ABP1), had been identified. Details on ABP1 have been covered in depth in our earlier reviews (Lüthen et al. 1999; Christian et al. 2006). Briefly, ABP1 is a 22 kDa auxin binding protein with a KDEL-ER retention signal at the C-terminus (Löbler and Klämbt 1985; Hesse et al. 1989; for an overview on the history of ABP1 research, see Napier and Perrot-Rechenmann 2002). There is evidence that some ABP1 is nevertheless excreted to the cellular surface (Diekmann et al. 1995). After binding auxin, ABP1 is supposed to interact with an unidentified transmembrane docking protein which transduces the signal to the cell (Klämbt 1990). Structural data have meanwhile been obtained, and the physiological significance of extracellular ABP1 as an auxin receptor is clearly indicated by several recent publications (e.g. Chen et al. 2001, 2003; Steffens et al. 2001; Christian et al. 2003; Yamagami et al. 2004).

Already back in the 1990s, there was some physiological evidence pointing to an additional intracellular auxin receptor. It was shown that coleoptiles grew in the absence of extracellular auxin if enough auxin was accumulated inside the cell and auxin efflux was blocked (Vesper and Kuss 1990; Claussen et al. 1996). These data suggested that extracellular auxin perception was insufficient for triggering auxin-induced elongation growth.

In 2005, two groups independently identified Transport Inhibitor Response1 (TIR1) as an intracellular auxin receptor (Dharmasiri et al. 2005a; Kepinski and Leyser 2005). The discovery came from the analysis of auxin-induced gene expression. Promotors of auxin-induced genes harbour auxin response elements (AuxREs) with the recognition sequence TGTCTC (Ballas et al. 1993). A class of transcription factors, the auxin response factors (ARFs), binds to this sequence. ARFs can either form dimers or interact with so-called Aux/IAA proteins. The transcription of an auxin-induced gene is blocked if a complex of an ARF and an Aux/IAA is bound to an AuxRE in the promoter region. Expression of auxin-induced genes is triggered by a rapid degradation of Aux/IAA proteins.

The analysis of the mechanism of Aux/IAA degradation resulted in the discovery of the Seventeen Kilodalton Protein (SKP1) Cullin F-Box^{tir} (SCF^{TIR}) protein complex, an E3 ubiquitin ligase containing a cullin and the F box protein TIR1. Very recently, it was independently shown by two groups that TIR1 binds auxin directly (Dharmasiri et al. 2005a; Kepinski and Leyser 2005). Obviously, auxin binding to TIR1 directly affects the activity of the SCF^{TIR} complex. The substrate Aux/IAA is ubiquitinylated and labelled for degradation by the proteasome. This rapidly eliminates Aux/IAA and leads to a removal of the transcriptional block of auxin-induced genes (Fig. 1). TIR1 is a member of a family of ***Arabidopsis*** auxin binding F-box proteins (AFB1 to AFB3). Quadruple knockouts have been recently generated and analysed. They do not show auxin-induced gene expression and displayed abnormal phenotypes in embryogenesis (Dharmasiri et al. 2005b). This proves that the intracellular auxin receptors control expression of auxin-induced genes and have a major function in controlling plant development.

2.2 ABP1 or TIR1?

It appears that TIR1 and the AFBs are identical to the intracellular auxin receptor auxin that physiologists were seeking. On the other hand, ABP1 seems directly to modulate ATPase (Rück et al. 1993) and perhaps other membrane transporters important for growth control. Stimulation of elongation growth clearly requires such rapid effects *and* expression of genes. It has been recently shown in maize coleoptiles that K^+ channel activity controls growth (Claussen et al. 1997) and that expression of ZMK1, a potassium channel gene in the maize coleoptile, is induced by auxin (Philippar et al. 1999). ABP1 may also be relevant for the redistribution and cycling of auxin efflux carrier molecules. At that level, there is obvious cross-talk between the TIR/AFB pathway and ABP1 signalling. It is also possible that

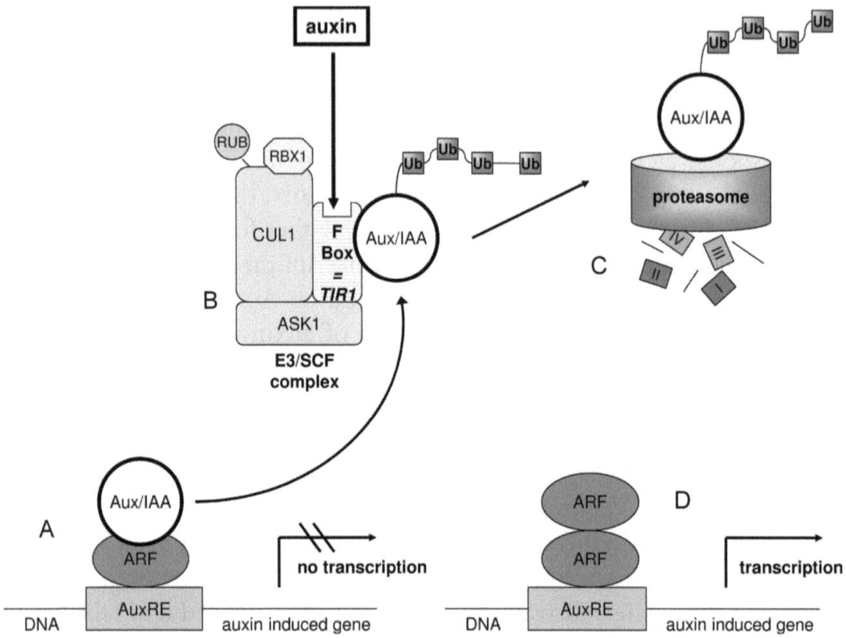

Fig. 1. The F-box protein TIR1 as an intracellular auxin receptor controlling the expression of auxin response genes. In the absence of auxin Aux/IAA and ARFs bind to auxin response elements in the promotors of auxin-induced genes, blocking their expression. **A** TIR1 is a part of the nuclear SCF^{TIR} complex and binds auxin **B**. The SCF^{TIR} complex catalyses and controls the ubiquitinylation of Aux/IAA proteins, labelling them for degradation by the proteasome **C**. In the absence of free Aux/IAA the equilibrium will be shifted to a situation where ARFs dimerize at the DNA **D**. This will remove the block of gene expression

ABP1 also influences TIR1-based expression of auxin-induced genes in some way, but this mode of interaction has not yet been clarified. In the appreciation of the breakthrough discovery, TIR1 was sometimes called *the* (one and only) auxin receptor. We do not feel that this is justified now. With both extracellular and intracellular auxin receptor being identified, it is time to address their function in control of the numerous auxin responses.

2.3 GID1, an intracellular receptor for the gibberellins

Gibberellins (GA) constitute another class of plant hormones involved in stem elongation. Their second major role is the mobilization of storage substances like carbohydrates during the germination of seeds. When a caryopse germinates, a gibberellin signal diffuses from the embryo across the endosperm to the aleuron layer. The aleuron cells respond to that signal by the induction of gibberellin responsive α-amylase genes. α-Amylase protein is formed in vast amounts and excreted to the endosperm, where it catalyses the breakdown of starch to smaller saccharides.

In recent years, it has become clear that gibberellin responsive genes are induced by a similar mechanism to auxin-induced genes. This insight came from the discovery of a class of regulatory proteins used in GA signalling that have been identified in several plant species. It was termed the DELLA subfamily, after a consensus motif called DELLA. Typical members are *SLR*1 in rice (Ikeda et al. 2001) and *GAI*, *RGA*, *RGL1*, *RGL2*, *RGL3* in *Arabidopsis* (Peng et al. 1997; Fleck and Harberd 2002), and *D8* in maize (Harberd and Freeling 1989; Peng et al. 1999; Peng and Harberd 2002). Mutants of these genes are either GA-insensitive, or display a constitutive response phenotype. It has been shown that at least some DELLA proteins are rapidly degraded in response to GA, suggesting that they are blockers of transcription of GA regulated genes (Itoh et al. 2002). As in the case of SCF^{TIR}, there exists a protein complex involved in ubiquitinylation, including an F-box protein called GID2. A mutation in the *Gid2* gene yields GA-insensitive phenotypes.

Very recently, a GA-insensitive mutant has been isolated from rice, the *gid1* mutant, displaying a severe dwarf phenotype and lacking GA-dependent induction of α-amylase (Ueguchi-Tanaka et al. 2005). In contrast, overexpression of *GID1* yields gibberellic acid hypersensitive plants. GID1 shows sequence homologies to animal hormone sensitive lipases (HSL), but is lacking amino acids critical for enzyme activity. GST-GID interacts with biologically active gibberellins, but not with inactive GAs, with a dissociation constant in a micromolar range. GST-GID with mutated amino acid sequence corresponding to the three *gid1*-alleles did not bind gibberellins. In a yeast-two-hybrid assay, it could be shown that GID1 interacts with the rice DELLA protein SLR1, and that this interaction is GA-dependent. These data are in line with the idea that GID1 is an intracellular gibberellin receptor. After binding a gibberellin molecule GID1 interacts with SLR1, which then becomes degradable by the SCF^{GID2} proteasome pathway (Fig. 2). SLR1 functions (much like the Aux/IAA proteins in the case of auxin signalling) as a transcriptional blocker for GA-dependent genes. The precise mechanism of this process is still under investigation.

3 Membrane-bound receptors

3.1 Two-component receptors in ethylene signalling

During the 1990s, a new strategy was applied for hunting genes relevant for ethylene signalling. For these screens, a well-known ethylene response was used, the ethylene triple response. When seedlings are grown in the dark in

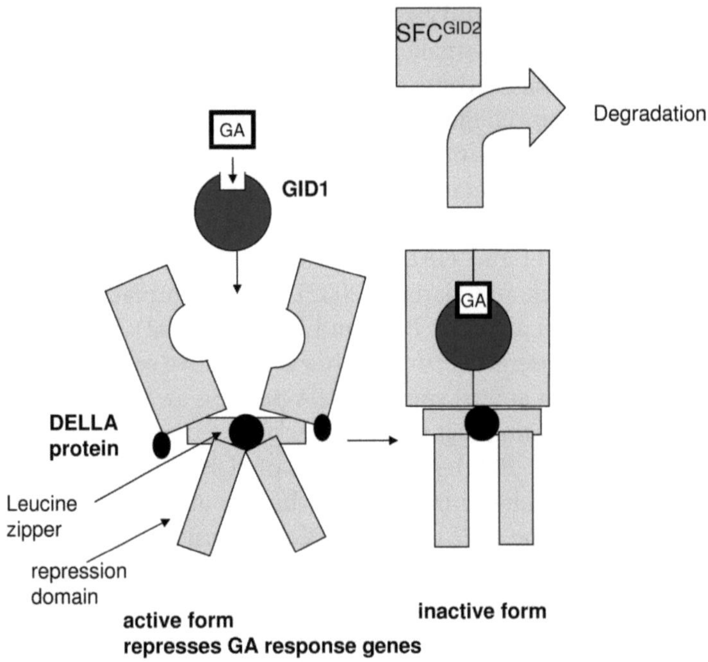

Fig. 2. Model of gibberellin perception. GA binds to GID1. The GA-GID1 attaches to a DELLA protein, a transcription factor bound to the promotor region of a GA-response gene. This will inactivate the DELLA-protein and promote its ubiquitinylation by the SCF^{GID2} complex, finally resulting in its degradation

an atmosphere containing ethylene, it is observed that the elongation growth is reduced, the epicotyl becomes thicker than in untreated controls, and the hypocotyl hook remains closed. This striking effect made it easy to screen for plants responding abnormally to ethylene. A number of ethylene-insensitive mutants (lacking the ethylene phenotype when treated with the hormone) were isolated, as well as mutants showing a constitutive triple response in the absence of ethylene. Some of these mutants were linked to ethylene biosynthesis, while others were ethylene signal transduction mutants. From the analysis of the latter, a number of genes coding for ethylene receptors and for downstream elements of the signalling chain were identified.

3.1.1 ETR1 as a prototype ethylene receptor

The *ethylene resistant* 1 (*etr1)* mutants showed dominant ethylene insensitivity not only in the triple response, but also in a number of other ethylene effects as well as in the expression of ethylene response genes (Bleecker et al. 1988). The *ETR1* gene was soon speculated to encode for a receptor, since leaves of *etr1* plants showed a strongly reduced ethylene binding compared to wild type leaves.

Positional cloning was used to isolate the gene (Chang et al. 1993; Chang and Meyerowitz 1995). It was soon found that ETR1 occurred as a dimer (Schaller et al. 1995), linked by disulfide bridges. Expressing wild type *ETR1* generated ethylene binding in yeast cells, this was not observed when mutant *ETR1* was expressed (Schaller and Bleecker 1995), pointing at ETR1 as an ethylene receptor. There was a strong relationship between ethylene binding and ethylene insensitivity by mutant forms of the ETR1 receptor when expressed in plants (Hall et al. 1999). Analyses of the sequence soon revealed homologies to proteins involved in the prokaryotic two-component signalling systems.

3.1.2 ETR1 is reminiscent to bacterial two component signalling systems

Two-component signalling is a widespread mechanism controlling various responses in bacteria, among them chemotaxis to attractants and repellants and the induction of porin genes triggered by changes in medium osmolarity (for review, see Stock et al. 1990). Briefly, transmembrane sensor proteins, normally integrated in the outer membrane, perceive the signal (Fig. 3A). They are often arranged as dimers. Their cytosolic part includes a histidine kinase domain. After autophosphorylation, they are able to transfer a phosphoryl group to a second protein containing an aspartic acid side chain, the so-called response regulator. In the simplest cases, the phosphorylated response regulator protein can bind to the promotors of response genes and control their expression (Fig. 3B). More complex modes of two-component signalling have been found. Some bacterial response regulators are directly fused to the sensor protein (Fig. 3B). In these cases, phospho-relay systems as well as competitive regulation modes have been suggested (Stock et al. 1990).

ETR1 is a typical two component sensor protein (Fig. 3C). Ethylene binding is located in the transmembrane domains and involves the participation of a copper ion (Rodríguez et al. 1999). The copper ion can be easily replaced by a silver ion, which eliminates ethylene binding (Zhao et al. 2002). This is consistent with the well established function of Ag^+ as inhibitor of ethylene action. The *etr1-1* mutant gene product binds neither copper nor ethylene. Taken together with the recently established localization of ETR1 at the endoplasmic reticulum, and not at the plasma

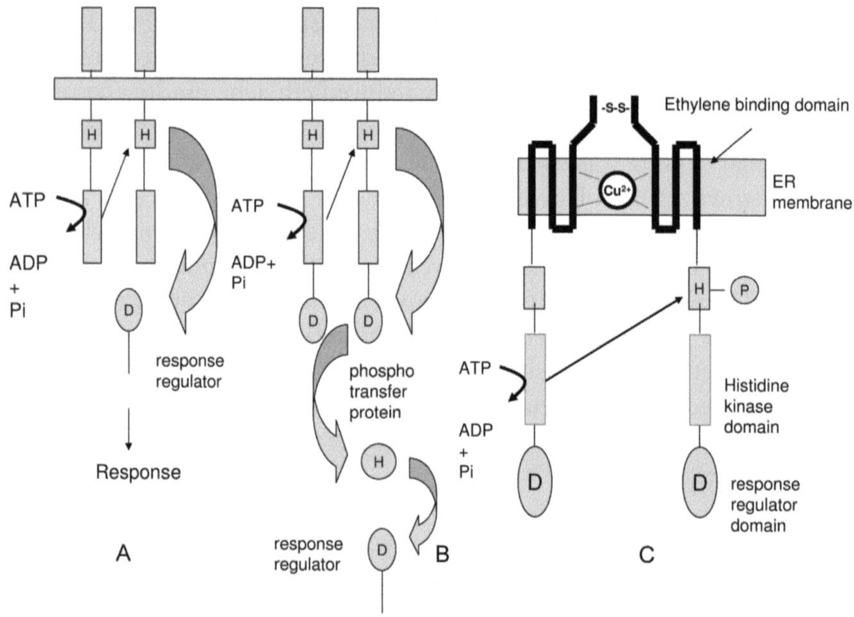

Fig. 3. **A** Simple two-component signal transduction as it is found in bacterial systems. After autophosphorylation, an aspartate-containing response regulator protein is phosphorylated triggering downstream effects (e.g. by serving as a transcription factor). **B** A phospho-relay system can complicate this scheme. A response regulator directly fused to the sensor kinase (as it is the case in the ethylene receptor ETR1) transmits the phosphate to a histidine-containing phospho-transfer protein, which relays it to an aspartate-containing response regulator protein. Variations of these schemes are operating in bacterial signal transduction, osmotic sensing in yeasts and in ethylene and cytokinin signalling in plants. **C** The plant ethylene receptor ETR 1 is a two-component sensor histidine kinase. It forms dimers linked by disulfide bridging. Ethylene binding involves a copper ion. Autophosphorylation occurs at a histidine (*H*) residue, but is probably not essential for signalling. The protein also contains a fused response regulator harbouring an aspartate residue (labelled as *D*)

membrane (Chen et al. 2002), the requirement of copper for ethylene binding to ETR1 easily explains that mutants of the putative copper transporter RAN1 (Responsive to ANtagonist1) display altered ethylene sensitivity (Woeste and Klieber 2000).

ETR1 contains a putative histidine kinase domain with a high degree of homology to bacterial two component sensor kinases. The histidine 353 is the putative phosphorylation site. Gamble et al. (1998) could demonstrate that ETR1, when expressed in yeast, was autophosphorylated. Versions of ETR1 not containing the putative kinase domain or lacking the histidine 353 lacked autophosphorylation activity.

3.1.3 A further two component ethylene receptors have been identified

In *Arabidopsis* four other two-component ethylene receptors have been identified: ETR2, ERS1, ERS2 and EIN4. They all bind ethylene with similar affinity, and mutation of any of them confers dominant ethylene insensitivity. ETR1, ETR2 and EIN4 possess a response regulator directly fused to the protein, ERS1 and ERS2 do not. Generally, two subfamilies have been defined, with ETR1 and ERS1 belonging to subfamily 1, and EIN4, ETR2 and ERS2 to subfamily 2 (Hua et al. 1998). There is a high degree of homology in the region of ethylene binding in the N-terminal transmembrane domains. In family 1 proteins, the histidine kinase domains display a strong homology to those of bacterial two-component sensors. This is not the case in family 2 receptors. The histidines are not at the predicted position. In ERS2, the misplaced histidine is even replaced by an arginine residue. The family 2 kinase appears to be degenerated and possibly not functional (Hua et al. 1998; Sakai et al. 1998).

Data from Wang et al. (2003) confirm that histidine kinase is not required for ethylene signalling even in family 1 receptors. They generated an *etr1;ers1* loss-of-function double mutant (containing only functional family 2 receptors), which exhibits a strong constitutive ethylene response phenotype. They could be rescued by expressing wild type ETR1. However, transforming the double mutant with a kinase-inactivated genomic clone of ETR1 resulted in the same effect. It can be speculated if kinase activity other than histidine kinase has taken over (as is the case in the phytochrome, where bacterial histidine kinase has been evolutionary replaced by serine-threonine kinase activity) or that ethylene signalling does not require receptor kinase activity at all.

In a number of plants, especially in tomato, several ethylene receptors (LeETR1 to 6) have been identified (for reviews, see Klee and Tieman 2002 and Klee 2004). They are very similar to the *Arabidopsis* family of ethylene receptors. LeETR3 is also referred to as NR (Never Ripe), clearly demonstrating the importance of ethylene receptors for the control of fruit ripening.

3.1.4 How do two-component receptors transduce the ethylene signal?

A central downstream element of the ethylene signalling chain is the protein kinase CTR1 (Constitutive Triple Response). The mutant *ctr1* has a constitutive ethylene phenotype. CTR1 is a mitogen-activated protein kinase kinase kinase (MAPKKK), suggesting the involvement of a MAP-kinase cascade in ethylene signalling. In 2003, Ouaked et al. showed that the MAPKK associated to ethylene signalling is SIMKK, and the MAPKs are MPK6 and MPK13.

Recently, it was demonstrated that CTR1 is localized at the endoplasmic reticulum (Gao et al. 2003), as are the ethylene receptors. Since MAKKK are generally cytoplasmic enzymes, it appears that CTR1 and the ethylene receptors form complexes and that ETR1 transmits the signal to CTR1 by direct interaction. Direct evidence that CTR1 is part of an ethylene receptor signalling complex was obtained by co-purification of the ethylene receptor ETR1 with a tagged version of CTR1 from an *Arabidopsis* membrane extract. The histidine kinase activity of ETR1 is not required for its association (Gao et al. 2003).

An astonishing fact in ethylene signalling is that single ethylene binding mutations in one receptor are dominant insensitive (even though the other family members are functional), while loss-of-function mutations cause a constitutive ethylene response. This can be explained by the following model: In the absence of ethylene, the ethylene receptors activate CTR1. The downstream signalling chain represses the induction of ethylene responses (Fig. 4A). Ethylene operates an OFF-switch in this system (Fig. 4B). The activity of CTR1 will be abolished, and the block will be removed, resulting in an induction of ethylene response genes. This model also explains that the *etr1* mutant is ethylene insensitive (Fig. 4C), and that this cannot be overcome by the simultaneous presence of other wild type receptors. A defective CTR1 kinase in the ctr1 mutant will cause a constitutive ethylene response phenotype, since the ethylene responses can no longer be repressed, even in the absence of the hormone (Fig. 4D).

The CTR1-MAP-kinase cascade may not be the only pathway relevant for ethylene signalling. There are several response regulator genes in plants (to be discussed further in the cytokinin section of this review). Recently the Harter group linked one of these regulators, ARR2 to ethylene signalling (Hass et al. 2004). Since the same response regulator is also involved in cytokinin signalling, it may be suggested that it integrates signalling pathways originating from various hormones.

3.2 Cytokinins are also perceived by two-component type receptors

Cytokinins regulate cell proliferation and tissue differentiation in concert with auxin. In 2005, the 50 year anniversary of the discovery of the first cytokinin was celebrated (Miller et al. 1955, Amasino 2005), as plant scientists were finishing up their search for cytokinin receptors. The first line of

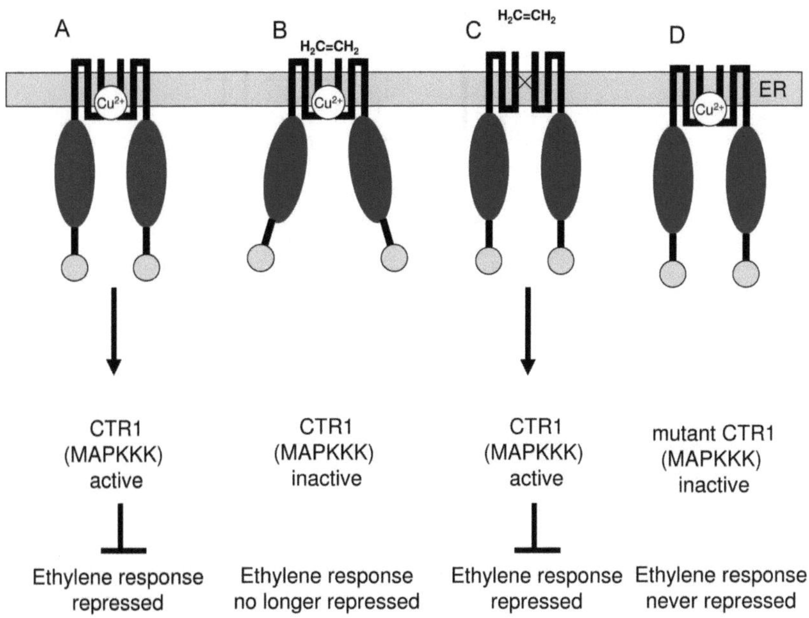

Fig. 4. Mode of action of ETR1, a two-component receptor for the ethylene signal. In the absence of ethylene, **A** ETR1 activates the MAP-kinase kinase kinase CTR1. The resulting signalling chain blocks downstream ethylene responses. Binding of ethylene involves a copper ion in ETR1 and inactivates CTR1, which will unleash ethylene responses **B**. In the *etr1-1* mutant **C** copper binding is impaired which results in an inability of binding ethylene and in dominant ethylene insensitivity. The *ctr1* mutant **D** displays a constitutive ethylene response phenotype, since ethylene responses cannot be repressed even in the absence of ethylene

evidence linking two-component signalling to cytokinins was the discovery of the sensor histidine kinase CKI1 in a screen of activation-tagged *Arabidopsis* mutants. When this histidine kinase gene was overexpressed in *Arabidopsis*, cytokinin independent callus growth was observed (Kakimoto 1996). Activation of cytokinin primary response gene promoters was observed when CKI1 was transiently expressed in *Arabidopsis* protoplasts, but the response was independent of cytokinins (Hwang and Sheen 2001). Today, the general view seems to be that CKI1 is somehow linked to cytokinin signalling, but no primary cytokinin receptor.

CRE1, another two-component sensor kinase in *Arabidopsis*, has been shown to be a genuine cytokinin receptor (Fig. 5A). The *CRE1* gene was identified in a screen of mutagenized plants for mutants with reduced cytokinin sensitivity in tissue culture (Inoue et al. 2001; Suzuki et al. 2001). CRE1 is a classical two-component receptor. Cytokinin binding occurs in the N-terminal CHASE domain, which consists of two predicted

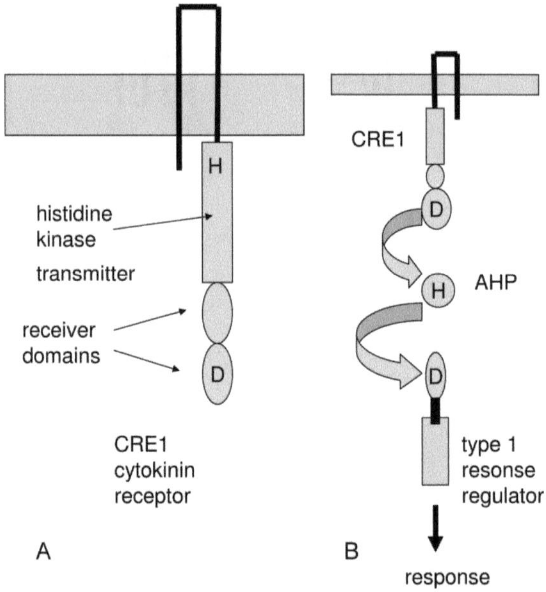

Fig. 5. The two-component receptor CRE1 as a prototype cytokinin receptor. **A** CRE1 is an integral (plasma?) membrane protein. Histidine kinase and two receiver domains are indicated. **B** Mode of action of the CRE1 receptor. The phospho-transfer protein AHP is involved in a phospho-relay and shuttles a phosphate ion to a type 1 response regulator, triggering downstream cytokinin responses

transmembrane helices linked by a predicted extracellular domain. There is a canonical histine kinase domain and a histidine phosphorylation site. As in ETR1 a response regulator complete with the telltale aspartate residue is fused directly to the kinase. Between the kinase domain and the terminal response regulator there is a second (probably non-functional) response regulator-like domain.

The cytokinin receptor nature of CRE1 could be elegantly proven by heterologous expression of CRE1 in yeasts (Fig. 6, Inoue et al. 2001). Yeasts possess a two-component receptor SLN1 (Synthetic Lethal OF N-end Rule 1) which is required for osmoregulation. It transfers its signal by phospho-relay through a protein called YPD1 (Tyrosine Phosphatase Dependent 1) and a response regulator SSK1 (SSK stands for Suppressor of Sensor Kinase). SSK1 then suppresses SSK2, a MAPKKK. Deletions in SLN1 are lethal. If CRE1 is expressed in the background of this SLN1 deletion mutant, viability can be restored, but only in the presence of active cytokinins. In short, the plant cytokinin receptor can feed a signal into the signal transduction chain for osmoregulation in yeasts—a stunning achievement! Similar experiments were done using *E. coli* sensor kinases by

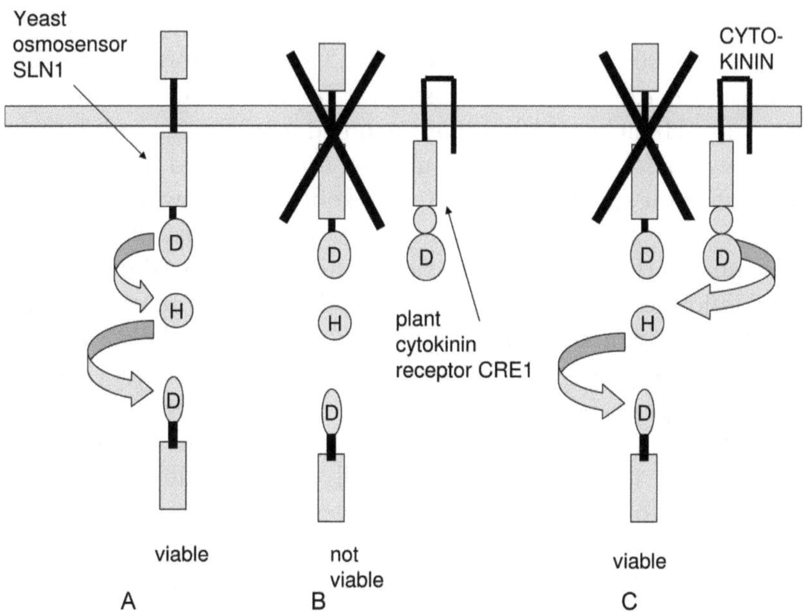

Fig. 6. Experimental demonstration of the cytokinin receptor role of CRE1 by heterolog expression in a yeast system. **A** Yeast two-component signalling system for osmotic responses, the receptor kinsase SLN1 serving as an osmo-sensor. **B** The *sln1* mutant is not viable, even if the plant cytokinin receptor is expressed in the yeast. **C** Treatment with cytokinin rescues the mutant, demonstrating that the yeast signalling system can accept signals from the plant cytokinin receptor and that the receptor function of CRE1 is specific for cytokinins. Similar experiments have also been successfully performed using bacterial two component signalling systems

Suzuki et al. (2001) and in *S. pombe* (Yamada et al. 2001). Membranes from *S. pombe* expressing CRE1 displayed cytokinin binding.

A mutant of CRE1 in *Arabidopsis* is *wooden leg* (*wol*). This mutant displays a reduction in the number of vascular initials due to the lack of asymmetric cell divisions that normally occur in the late torpedo stage of embryogenesis (Mähönen et al. 2000). CRE1 expression occurs primarily in roots. There is no shoot phenotype in intact plants, although shoot regeneration in vitro in response to cytokinin treatment is strongly reduced.

3.2.1 More two-component cytokinin receptors for *Arabidopsis* and the rest of the plant world

AHK2 and AHK3 (AHK stands for *Arabidopsis* Histidine Kinase, CRE1 is sometimes referred to as AHK4) are additional two-component receptors which appear to be active cytokinin receptors (Nishimura et al. 2004). Suzuki et al. (2001) could show that AHK3 responds to cytokinin in the heterologous *E. coli* system. Addressing the *in planta* role, Nishimura et al. (2004) and Higuchi et al. (2004) studied expression and knockout phenotypes. AHK2 and AHK3 are ubiquitously expressed in various plant tissues, also in the shoot. While *CRE1* knockout plants show no marked phenotype on shoot and leave development, *cre1 ahk3* and *ahk2 ahk3* plants display a semi-dwarf phenotype as to shoots, such as a reduced leaf size and a reduced inflorescence stem length. A dramatic reduction in growth, probably due to a loss in meristematic activity, was shown for *cre1 ahk2 ahk3* triple mutants. This suggests that all three histidine kinase sensors are active cytokinin receptors *in planta.* Higuchi et al. (2004) investigated several single, double and triple mutants in various cytokinin assays (root growth inhibition, root formation inhibition, cell division, callus greening and induction of cytokinin primary response genes). They were small in size, infertile, and meristem size and activity was greatly reduced. Analysis of double and single mutants reveals significant redundancy of receptor function. A similar very thorough study of cytokinin receptor mutants confirming and extending this view was recently published by the Schmülling group (Riefler et al. 2006).

Similar cytokinin receptors have been isolated from maize (ZmHK1, ZmHK2, ZmHK3a and ZmHK3b). Their overall structure is similar to the *Arabidopsis* proteins. Cytokinin responsiveness has been verified in the *E.coli* system (Yonekura-Sakakibara et al. 2004). Interestingly, in contrast to the *Arabidopsis* receptors, the maize counterparts responded to cis-zeatin, which is generally not considered an active cytokinin.

3.2.2 How do two-component receptors transduce the cytokinin signal?

The downstream cytokinin signalling chain is meanwhile quite well understood (Fig. 4B). The fact that the cytokinin receptors can feed a signal into bacterial and yeast signal transduction chains was a clue that the architecture must be similar. In both systems, the signal is transferred by a histidine phosphotransfer protein to the aspartate residue of a response regulator in a phospho-relay mechanism. Similar phosphotransfer proteins could be identified in *Arabidopsis*, coded by the *AHP* (*Arabidopsis histidine containing phosphotransmitter*) genes. In the heterologous *E. coli* system, it could be shown that coexpression of the AHPs inhibited reporter gene expression, suggesting that the plant AHPs can compete with the bacterial histidine phosphotransfer protein. Two of these AHPs have been shown to be translocated to the nucleus in response to cytokinin (Hwang and Sheen 2001). Interaction of AHPs and cytokinin receptors has also been detected in yeast two hybrid studies. All data support the view that the AHPs take up the signal from the receptors and transmit it to a nuclear response regulator.

Two types of response regulators (RRs) have been found: type B RRs contain an output domain whereas type A RRs do not. Type B RRs have been shown to be activated by AHP interaction and can directly control expression of cytokinin response genes, among them type A RRs (for review see Hutchinson and Kieber 2002, Lohrmann and Harter 2002; Kakimoto 2003; Schmülling 2004). It appears that type A RRs can downregulate the primary cytokinin response in a feedback loop or otherwise modulate downstream cytokinin effects.

3.3 Candidates for the ABA receptor

At present (January 2006), the perception of ABA is the least understood of all classical plant hormones, in spite of the fact that one of the classical ABA responses, the closure of guard cells, offers an ideal target for all methods of electrophysiology and cellular biology. Downstream ABA signalling has been successfully explored with these techniques. However, the receptor question is just beginning to be addressed. Up to now, two receptor proteins have been found that appear to be linked to ABA signalling, but their nature as an ABA receptor has not been demonstrated up to now.

3.3.1 A receptor-like kinase as a possible ABA receptor

Receptor tyrosine kinases (RTKs) play a pivotal role in animal and human signal transduction, especially as receptors for growth factors (epidermal growth factor, platelet derived growth factor, for review, see Frantl et al. 1993). The insulin receptor is an RTK. Upon perceiving a signal, they autophosphorylate, and specific proteins bind to the resulting phosphotyrosine residues (e.g. through SH2 domains). Complex adaptors are assembled linking the receptors to downstream signalling elements, for instance Rous Abdominal Sarcoma (RAS)-like small G proteins (Sprang 1997) or to protein kinase B (Nicholson and Anderson 2002).

Based on sequence homology the so-called receptor-like-kinases (RLKs) have been discovered. In plant genomes they are widespread (600 or 1200 members in the *Arabidopsis* and rice genomes, respectively, but only 1–6 members in animals). It has been shown that RLK do possess serine-threonine kinase rather than tyrosine kinase activity (Zhang 1998, Shui and Bleecker 2001, for a recent review see Shui et al. 2004). They appear to regulate cell division, differentiation, organ polarity and are involved in a variety of responses to pathogen attack.

One such Receptor Protein Kinase, RPK1 in *Arabidopsis*, has been recently related to ABA signalling (Osakabe et al. 2005). RPK1 knockout and antisense plants display various phenotypes linked to ABA, especially strongly reduced ABA sensitivity during germination, growth and stomatal closure. The expression of several ABA response genes is much reduced in the mutants. However, ABA binding has not yet been reported. Therefore, we do not yet know of RPK1 is an ABA receptor or rather linked to ABA signalling in a more indirect way.

3.3.2 At last: A plant G protein coupled receptor may be involved in ABA signalling

G protein coupled receptors have been linked to ABA signalling for more than a decade. It is known that potassium and chloride channels modulated by ABA in guard cells are also subject to known effectors of G proteins like GTP-γ-S, pertussis toxin and cholera toxin. Knowing the fact that at least the latter two only function with heterotrimeric G proteins and that heterotrimeric G proteins only function with 7-transmembrane receptors (G protein coupled receptors, GPCR) the involvement of GPCR in ABA signalling appears to be a realistic possibility. However, circumstantial evidence was also brought up for the involvement of GPCR in the signalling chains of nearly every other hormone throughout the 1980s.

The complete elucidation of the *Arabidopsis* genome was a serious blow to all these speculations. It was shown that both GPCRs and heterotrimeric G proteins are extremely rare in plants. Up to now, there is only one candidate GPCR in the complete *Arabidopsis* genome, the gene *GCR1* (Plakidou-Dymock et al. 1998). There is only one single G protein alpha subunit (GPA1), one single G protein beta subunit (AGB1) and two G protein γ-subunits in the *Arabidopsis* genome (AGG1 and AGG2). This illustrates that G protein coupled signalling, besides the fact that it is a commonplace in animals, is a quite exotic signalling mode in plants.

It was recently shown in a ubiquitin split study by Pandey and Assmann (2004) that the one and only *Arabidopsis* 7-transmembrane receptor GCR1 and the one and only alpha subunit GPA are interaction partners. In the same study, it was demonstrated that *gcr1* mutants displayed a slight ABA hypersensitivity in a number of ABA responses (root growth inhibition, stomatal closure, inhibition of stomatal opening) and an increased ABA-induced expression of some ABA related genes. Water loss and water stress symptoms were less severe in the mutants compared to the wild type. All this suggests that GCR1 is somehow linked to ABA signalling. However, the ABA hypersensitivity and the quite mild phenotypes may indicate that this is not a primary ABA receptor, although it appears to be somehow linked to ABA signalling. The real substrate for the GCR receptor remains unknown.

4 Perspective

By 2006, receptors for all but one of the classical plant hormones have been identified. In many cases, there are several receptor genes, indicating redundancy and the involvement of several signal transduction pathways. Thus, the identification of downstream signalling elements will not be sufficient. Like in animal systems there will be a intricate network and a lot of crosstalk that remains to be elucidated. In the case of auxin perception there are two very different receptors (ABP1 and the TIR1 family), and it remains to be clarified which responses are triggered by which receptor.

References

Amasino R (2005) 1955: Kinetin arrives. The 50th anniversary of a new plant hormone. Plant Physiol 138:1177–1184

Ballas N, Wong LM, Theologis A (1993) Identification of the auxin resonponsive element, AuxRE, in the primary indoleaecetic acid-inducible gene, PS-IAA4–5, of pea (*Pisum sativum*). J Mol Biol 233:580–596

Bleecker AB, Estelle MA, Sommerville C (1988) Insensitivity to ethylene conferred by a dominant mutation in *Arabidopsis thaliana.* Science 241:1086–1089

Chang C, Meyerowitz EM (1995) The ethylene hormone receptor in *Arabidopsis*: a eukaryotic two-component system. Proc Natl Acad Sci USA 92:4129–4133

Chang C, Kwok SF, Bleecker AB, Meyerowitz EM (1993) *Arabidopsis* ethylene response gene ETR1: Similarity of product to two-component regulators. Science 262:539–544

Chen JG, Ullah H, Young JC, Sussman MR, Jones AM (2001) ABP1 is required for organized cell elongation and division in *Arabidopsis* embryogenesis. Genes Dev 15:902–911

Chen JG, Shimomura S, Sitbon F, Sandberg G, Jones AM (2003) The role of auxin-binding protein 1 in the expansion of tobacco leave cells. Plant J 28:607–617

Chen Y-F, Randlett MD, Findell JL, Schaller GE (2002) Localisation of the ethylene receptor ETR1 to the endoplasmic reticulum of *Arabidopsis.* J Biol Chem 277:19861–19866

Christian M, Steffens B, Schenck D, Lüthen H (2003) The *diageotropica* mutation of tomato disrupts a signalling chain using extracellular auxin binding protein 1 as a receptor. Planta 218:309–314

Christian M, Schenck D, Böttger M, Steffens B, Lüthen H (2006) New insight into auxin perception, signal transduction and transport. Prog Bot 67:217–244

Claussen, M, Lüthen H, Böttger M (1996) Inside or outside? Localization of the receptor relevant to auxin-induced growth. Physiol Plant 98:861–867

Claussen M, Lüthen H, Blatt M, Böttger M (1997) Auxin-induced growth and its linkage to potassium channels. Planta 201:227–231

Dharmasiri N, Dharmasiri S, Estelle M (2005a) The F-box protein TIR1 is an auxin receptor. Nature 435:441–445

Dharmasiri N, Dharmasiri S, Weijers D, Lechner E, Yamada M, Hobbie L, Ehrismann JS, Jürgens G, Estelle M (2005b) Plant development is regulated by a family of auxin receptor F box proteins. Dev Cell 9:109–119

Diekmann W, Venis MA, Robinson DG (1995) Auxins induce clustering of the auxin-binding protein at the surface of maize coleoptile protoplasts. Proc Natl Acad Sci USA 92:3425–3429

Dohrmann U, Hertel R, Kowalik H (1978) Properties of auxin binding sites in different subcellular fractions from maize coleoptiles. Planta 140:97–106

Fleck B, Harberd NP (2002) Evidence that the *Arabidopsis* nuclear gibberellin signalling protein GAI is not destabilised by gibberellin. Plant J 32:935–947

Frantl WJ, Johnson DE, Williams LT (1993) Signalling by receptors of tyrosine kinase. Annu Rev Biochem 62:453–481

Gamble RL, Coonfield ML, Schaller GE (1998) Histidine kinase activity of the ETR1 ethylene receptor from *Arabidopsis.* Proc Natl Acad Sci USA 95:7825–7829

Gao Z, Chen YF, Randlett MB, Zhao XC, Findell JL, Kieber JJ, Schaller GE (2003) Localization of the Raf-like kinase CTR1 to the endoplasmic reticulum of *Arabidopsis* through participation in ethylene receptor signaling complexes. J Biol Chem 278: 34725–34732

Hall AE, Chen QG, Findell JL, Schaller GE, Bleecker AB (1999) The relationship between ethylene binding and dominant insensitivity conferred by mutant forms of the ETR1 ethylene receptor. Plant Physiol 121:291–299

Harberd NP, Freeling M (1989) Genetics of dominant gibberellin-insensitive dwarfism in maize. Genetics 121:827–838

Hass C, Lohrmann J, Albrecht V, Sweere U, Hummel F, Yoo SD, Hwang I, Zhu T, Schäfer E, Kudla J, Harter K (2004) The response regulator 2 mediates ethylene signalling and hormone signal integration in *Arabidopsis.* EMBO J 23:3290–3302

Hertel R, Thompson KS, Russo VFA (1970) In vitro auxin binding to particulate fractions from corn coleoptiles. Planta 107:325–340

Hesse T, Feldwisch J, Balshüsemann D, Bauw G, Puype M, Vanderckove J, Löbler M, Klämbt D, Schell J, Palme K (1989) Molecular cloning and structural analysis of a gene from *Zea mays* L. coding for a putative receptor for the plant hormone auxin. EMBO J 8:2453–2461

Higuchi M, Pischke M, Mähönen AP, Miyawaki K, Hashimoto Y, Seki M, Kobayashi M, Shinozaki_K, Kato T, Tabata S, Helariutta Y, Sussman MR, Kakimoto T (2004) *In planta* functions of the *Arabidopsis* cytokinin receptor family. Proc Natl Acad Sci USA101:8821–8826

Hua J, Sakai H, Nourizadeh S, Chen QG, Bleecker AB, Ecker JR, Meyerowitz EM (1998) EIN4 and ERS2 are members of the putative ethylene receptor gene family in *Arabidopsis.* Plant Cell 10:1321–1332

Hutchison CE, Kieber JJ (2002) Cytokinin signalling in *Arabidopsis.* Plant Cell 14 supplement: S47–S59

Hwang I, Sheen J (2001) Two component circuitry in *Arabidopsis* cytokinin signal transduction. Nature 413:383–389

Ikeda A, Ueguchi-Tanaka M, Sonoda Y, Kitano H, Koshioka M, Futsuhara Y, Matsuoka M, Yamaguchi, J. (2001) *Slender rice*, a constitutive response mutant, is caused by a null mutation of the *SLR1* gene, an ortholog of the height-regulating gene *GAI/RGA/RHT/D8.* Plant Cell 13, 999–1010

Inoue T, Higuchi M, Hashimoto Y, Seki M, Kobayashi M, Kato T, Tabata S, Shinozaki K, Kakimoto T (2001) Identification of CRE1 as a cytokinin receptor from *Arabidopsis.* Nature 409:1060–1063

Itoh H, Ueguchi-Tanaka M, Sato Y, Ashikari M, Matsuoka M (2002) The gibberellin signaling pathway is regulated by the appearance and disappearance of SLENDER RICE1 in Nuclei. Plant Cell 14:57–70

Kakimoto T (1996) CKI1, a histidine kinase implicated in cytokinin signal transduction. Science 274:982–985

Kakimoto T (2003) Perception and signal transduction of cytokinins. Annu Rev Plant Biol 54:605–627

Kende H, Zeevaart JAD (1997) The five "classical" plant hormones. Plant Cell 9:1197–1210

Kepinski S, Leyser O (2005) The *Arabidopsis* F-box protein TIR1 is an auxin receptor. Nature 435:446–451

Klämbt D (1990) A view about the function of auxin binding proteins at the plasma membrane. Plant Mol Bio 14:1045–1050

Klee H (2004) Ethylene signal transduction. Moving beyond *Arabidopsis.* Plant Physiol 135:660–667

Klee H, Tieman D (2002) The tomato ethylene receptor gene family: form and function. Physiol Plant 115:366–341

Löbler M, Klämbt D (1985) Auxin binding protein from coleoptile membranes of corn (*Zea mays* L.). I. Purification by immunological methods and characterization. J Biol Chem 260:9848–9853

Lohrmann J, Harter K (2002) Plant two-component signalling systems and the role of response regulators. Plant Physiol 128:363–369

Lüthen H, Claussen M, Böttger M (1999) Growth: Progress in auxin research. Prog Bot 60:315–340

Mähönen AP, Bonke M, Kauppinen L, Riikonen M, Benfey PN, Helariutta Y (2000) A novel two-component hybrid molecule regulates vascular morphogenesis of the *Arabidopsis* root. Genes Dev 14:2938–2943

Miller CO, Skoog F, Von Saltza MH, Strong F (1955) Kinetin, a cell division factor from deoxyribonucleic acid. J Am Chem Soc 77:1392

Millner PA, Clausier BE (1996) G-protein coupled receptors in plant cells. J Exp Bot 47: 983–992

Napier RM, David KM, Perrot-Rechenmann C (2002) A short history of auxin-binding proteins. Plant Mol Biol 49:339–348

Nicholson KM, Anderson NG (2002) The protein kinase B/Akt signalling pathway in human malignancy. Cell Signal 14:381–395

Nishimura C, Ohashi Y, Sato S, Kato T, Tabata S, Ueguchi C (2004) Histidine kinase homologs that act as cytokinin receptors possess overlapping functions in the regulation of shoot and root growth in *Arabidopsis.* Plant Cell 16:1365–1377

Osakabe Y, Marauyama K, Seki M, Satou M, Shinozaki K, Yamaguchi-Shinozaki (2005) Leucine-rich-repeat receptor-like kinase 1 is a key membrane-bound regulator of abscisic acid early signaling in *Arabidopsis.* Plant Cell 17:1105–1119

Ouaked F, Rozhon W, Lecourieux D, Hirt H (2003) A MAPK pathway mediates ethylene signalling in plants. EMBO J 22:1282–1288

Pandey S, Assmann SM (2004) The Arabidopsis putative G protein-coupled receptor GCR1 interacts with the G protein alpha subunit GPA1 and regulates abscisic acid signaling. Plant Cell 16(6):1616–1632

Peng J, Harberd NP (2002) The role of GA-mediated signalling in control of seed germination. Curr Opin Plant Biol 5:376–381

Peng J, Carol P, Richards DE, King KE, Cowling RJ, Murphy GP, Harberd NP (1997) The *Arabidopsis GAI* gene defines a signaling pathway that negatively regulates gibberellin responses. Genes Dev 11:3194–3205

Peng JR, Richards DE, Hartley NM, Murphy GP, Devos KM, Flintham JE, Beales J, Fish LJ, Worland AJ, Pelica F, Sudhakar D, Christou P, Snape JW, Gale MD, Harberd NP (1999) Green revolution' genes encode mutant gibberellin response modulators. Nature 400: 256–261

Philippar K, Fuchs I, Lüthen H, Hoth S, Bauer CS, Haga K, Thiel G, Ljung K, Sandberg G, Böttger M, Becker D, Hedrich R (1999) Auxin-induced K^+ channel expression represents an essential step in coleoptile growth and gravitropism. Proc Natl Acad Sci USA 96:12186–12191

Plakidou-Dymock S, Dymock D, Hooly R (1998) A higher plant seven-transmembrane receptor that influences sensitivity to cytokinin. Curr Biol 8:315–324

Riefler M, Novak O, Strnad M, Schmülling T (2006) *Arabidopsis* cytokinin receptor mutants reveal functions in shoot growth, senescence, seed size, germination, root development and cytokinin metabolism. Plant Cell 18:40–54

Rodrígues FI, Esch JJ, Hall AE, Binder BM, Schaller GE, Bleecker AB (1999) A copper cofactor for the ethylene receptor ETR1 from *Arabidopsis.* Science 283:996–998

Rück A, Palme K, Venis MA, Napier RM, Felle HH (1993) Patch clamp analysis extablishes a role for an auxin binding protein in the auxin stimulation of plasma membrane current in *Zea mays* protoplasts. Plant J 4:41–46

Sakai H, Hua J, Chen QG, Chang C, Bleecker AB, Medrano LJ, Meyerowitz EM (1998) ETR2 is an ETR1-like gene involved in ethylene signaling in *Arabidopsis.* Proc Natl Acad Sci USA: 95:5812–5817

Schaller GE, Bleecker AB (1995) Ethylene binding sites generated in yeast expressing the *Arabidopsis* ETR1 gene. Science 270:1809–1811

Schaller GE, Ladd AN, Lanahan MB, Spanbauer JM, Bleecker AB (1995) The ethylene response mediator ETR1 from *Arabidopsis* forms a disulfide linked dimer. J Biol Chem 270:12526–12530

Schmülling T (2004) Cytokinin. In: Lennarz W, Lane MD (eds) Encyclopedia of biological chemistry. Academic Press/Elsevier Science, Amsterdam

Shui SH, Bleecker AB (2001) Receptor-like kinase from *Arabidopsis* form a monophyletic gene family related to animal receptor kinases. Proc Natl Acad Sci USA 98:10763–10768

Shui SH, Karlowski WM, Pan R, Tzeng YH, Mayer KFX, Li WH (2004) Comparative analysis of the receptor-like kinase family in *Arabidopsis* and rice. Plant Cell 16:1220–1234.

Sprang SR (1997) G protein mechanism: Insight from structural analysis. Annu Rev Biochem 66:639–678

Steffens B, Feckler C, Palme K, Christian M, Böttger M, Lüthen H (2001) The auxin signal for protoplast swelling is perceived by extracellular ABP1. Plant J 27:591–599

Stock JB, Stock AM, Mottonen JM (1990) Signal transduction in bacteria. Nature 344:395–400

Suzuki T, Miwa K, Ishikawa K, Yamada H, Aiba H, Mizuno T (2001) The *Arabidopsis* sensor kinase, AKH4, can respond to cytokinin. Plant Cell Physiol 42:107–113

Ueguchi-Tanaka M, Ashikari M, Nakajiima M, Itoh H, Katoh E, Kobayashi M, Chow T-Y, Hsing YC, Kitano H, Yamaguchi I, Matsuoka M (2005) *GIBBERELLIN-INSENSITIVE DWARFT*1 encodes a soluble receptor. Nature 437:693–698

Vesper MJ, Kuss CL (1990) Physiological evidence that the primary site of auxin action is an intracellular site. Planta 182:486–491

Wang W, Hall AE, O'Malley R, Bleecker A (2003) Canonical histidine kinase activity of transmitter domain of the ETR1 ethylene receptor from *Arabidopsis* is not required for signal transmission. Proc Natl Acad Sci USA 100:352–357

Woeste KE, Klieber JJ (2000) A strong loss-of-function mutation in RAN1 results in constitutive activation of the ethylene response pathway as well as in a rosette-lethal phenotype. Plant Cell 12:443–455

Yamada H, Suzuki T, Terada K, Takei K, Ishikawa K, Miwa K, Yamashino T, Mizuno T (2001) Plant Cell Physiol 42:1017–1023

Yamagami M, Haga K, Napier RM, Iino M (2004) Two distinct signaling pathways participate in auxin-induced swelling of pea epidermal protoplasts. Plant Physiol 134:735–747

Yonekura-Sakakibara K, Kojima M, Yamaya T, Sakakibara H (2004) Molecular characterization of cytokinin-responsive histidine kinases in maize. Differential ligand preferences and response to cis-zeatin. Plant Physiol 134:1654–1661

Zhang X (1998) Leucine-rich-repeat receptor-like-kinases in plants. Plant Mol Biol 16:301–311

Zhao XC, Qu X, Mathews DE, Schaller GE (2002) Effect of ethylene pathway mutations upon expression of the ethylene receptor ETR1 from *Arabidopsis*. Plant Physiol 130:1983–1999

PD Dr. Hartwig Lüthen
Prof. Dr. Michael Böttger
Dipl. Biol. Daniel Schenck
Biozentrum Flottbek und Botanischer Garten
Universität Hamburg
Ohnhorststr. 18
22609 Hamburg
e-mail: h.luthen@botanik.uni-hamburg.de

Spatiotemporal Patterns and Distributed Computation—A Formal Link between CO_2 Signalling, Diffusion and Stomatal Regulation

Ulrich Lüttge and Marc-Thorsten Hütt

1 Inorganic carbon themes and perspectives for signalling

The global change debate on the background of continuously increasing atmospheric CO_2 concentrations (Ca)[*)] has placed a wealth of botanical inorganic carbon (C)[*)] themes in the limelight of international research interest. What is the response of photosynthesis? Is it enhanced or down regulated? What is less frequently and broadly considered is the role of pedospheric Ca (e.g. Cramer et al. 1999). Recently, Warren (2004) has listed a number of points that must be born in mind when C-signalling is considered. This is first of all the old diffusion story of Ca–Ci–Cc[*)], where various resistances in series determine the vertical flow of carbon from the atmosphere (Ca) into the internal leaf air spaces (Ci) and to the carboxylation sites of photosynthesis (Cc) (Parkhurst 1994). The limitations posed by conductance at the various levels to CO_2 movement involve diffusion in the gas phase as well as in the liquid phase, and facilitated or active transport across membranes of both CO_2 and bicarbonate, HCO_3^- (e.g. Mimura et al. 1993; Price et al. 2002). An HCO_3^- -transporter might even be related to the well known red blood cell anion transporter (Drechsler et al. 1993, 1994; Sharkia et al. 1994). Aquaporins facilitate transmembrane transport of CO_2 (Uehlein et al. 2003; Hanba et al. 2004; *Progress in Botany*, review: Kaldenhoff 2005). Such transport processes are heavily involved in internal CO_2-concentrating mechanisms in cyanobacteria (Badger et al. 2002) and algae. Where such CO_2-concentrating mechanisms are absent, e.g. in chloroplasts of higher plants (Price et al. 1994) and other algae, a particularly high activity of the carboxylating enzyme ribulose-bis-phosphate carboxylase/oxygenase (RubisCO) may be an alternative (Palmquist et al. 1994). The biochemical

[*)]Throughout this review we use C to refer to inorganic carbon and Ca and Ci, respectively, for external and internal inorganic carbon.

Progress in Botany, Vol. 68

pathways of C_4-photosynthesis (Leegood 2002) and crassulacean acid metabolism (CAM) (Lüttge 2002) provide different ways of maintaining the internal CO_2 concentrating.

In this review, it cannot be our aim to cover these various issues in detail and the vast relevant literature systematically. Mentioning these points above with some admittedly arbitrarily chosen references may serve as kind of a background for our attempt to cover the much more restricted perception we have of Ci signalling. It is clear that stomata are not the only target for Ci signalling, but even for stomatal Ci responses our knowledge is astonishingly limited. We will show the structural and physiological basis for Ci signalling, which becomes overt in the spatiotemporal performance of leaves of higher plants and is a currently emerging new field. Along the lines of Peak et al. (2004), we will interpret spatiotemporal patterns (e.g. in stomatal dynamics) as the result of a distributed computation of an optimal metabolic state. We will briefly summarize how data analysis tools motivated by non-linear dynamics and complexity theory may help understand such processes, where spatiotemporal degrees of freedom are exploited for a system-wide optimization. Furthermore, we will comment on how properties of spatiotemporal patterns (in stomatal conductance, photosynthetic activity or internal CO_2 concentrations) can be related to cellular properties and, consequently, may provide indirect support, e.g., for certain regulatory pathways.

2 Ci signalling to stomatal guard cells

According to a long established consensus stomatal guard cells sense Ci and not Ca (Mott 1990), where high Ci concentrations elicit stomatal closure. This has also been reviewed recently (Schroeder et al. 2001; Vavasseur and Raghavendra 2005). However, we do not even know the exact relevant critical concentrations. Generally, it is taken that $Ci > Ca$ causes stomatal closure and $Ci < Ca$ stomatal opening (Assmann 1999).The guard cells of C_4 plants appear to be more sensitive to Ci than those of C_3 plants (Huxman and Monson 2003). Ci responses are very variable between species and there are even differences due to acclimation within given species (Mansfield et al. 1990; Frechilla et al. 2002). Apparently stomata of C_3 and CAM plants have a similar CO_2 sensitivity particularly in the range of 0–36 Pa, and a half saturation constant of 19.6 Pa was observed in both C_3 and CAM plants (Jewer et al. 1985). The phytohormone ABA, which is an elicitor of stomatal closure, cytoplastic Ca^{2+}, and blue and red light are major factors modulating Ci sensitivity (Vavasseur and Raghavendra 2005). ABA increases the sensitivity of guard cells to Ci in C_3 plants (Leymarie et al. 1998; Huxman and

Monson 2003) and is involved in the signalling, where the photosynthetic modulation of the availability of NADPH which is necessary for ABA synthesis using the xanthophyll violoxanthin as a precursor is also involved (Tallman 2004).

The guard cell chloroplasts are considered to be at least one possible location for Ci sensing mediated by the xanthophyll zeaxanthin (Zeiger et al. 2002). The levels of zeaxanthin are reciprocal to CO_2 concentration and may modulate CO_2 dependent changes of stomatal apertures in the light (Zhu et al. 1998). Carboxylating enzymes could be Ci sensors in stomatal guard cells. There is a long lasting argument whether (Gotov et al. 1988; Rother et al. 1988; Vaughn and Vaughn 1988; Reckmann et al. 1990; Cardon and Berry 1992; Lawson et al. 2003) or not (Schnabl 1981; Shimazaki and Zeiger 1985; Birkenhead and Willmer 1986; Outlaw 1989) RubisCO is present in guard cells. RubisCO could be a Ci sensor, not only via CO_2 as its substrate, but also via activation by carboxylation of the RubisCO protein (RubisCO-activase) (Portis 1992). Phosphoenolpyruvate carboxylase (PEPC) is present in guard cells and known to contribute to stomatal osmoregulation via production of malate (Schulz et al. 1992; Klockenbring et al. 1998).

Perhaps ion channel activation is also involved in Ci sensing (Brearley et al. 1997; Hanstein 2002; Hanstein and Felle 2002). In *Vicia* guard cells, the chloroplasts transduce lowered Ci to an activation of stomatal opening via an ion uptake mechanism that depends on chloroplastic photosynthetic electron transport and shares downstream components of the blue light signal transduction cascade (Olsen et al. 2002). Hedrich and Marten (1993) and Hedrich et al. (1994) have proposed a malate hypothesis for stomatal Ci sensing, where CO_2 is fixed by dark reactions via PEPC producing malate via NAD(P)H dependent reduction of the oxaloacetate formed and malate then acts on anion channels in the plasma membrane, mediating stomatal movements. However, this hypothesis has several shortcomings and is not generally accepted (Assmann 1999; Schroeder et al. 2001; Vavasseur and Raghavendra 2005). Hedrich and coworkers used 1% CO_2 in their experiments, a much too high non-physiological concentration. Internal CO_2 concentrations in the percent range are only observed in CAM plants during the day when nocturnally stored vacuolar malate is remobilized and decarboxylated. This causes stomatal closure (phase III of CAM; Osmond 1978) but as noted above, we do not know the actual Ci levels required for stomatal closure and CAM plants even appear to have the same Ci sensitivity as C_3 plants.

Hanstein (2002) underlines how little we know about the signalling pathway from Ci to stomatal guard cell movements. Indeed, having read the latest review on the topic (Vavasseur and Raghavendra 2005), one is left with the strong disappointment that we know practically nothing about the molecular identity of the primary Ci sensor.

3 Lateral CO_2-diffusion in leaves

The traditional Ca–Ci–Cc diffusion story (see section 1 above) focuses on vertical transport (Parkhurst 1994) and relates to leaf anatomy and thickness (Hanba et al. 1999; Terashima et al. 2001). Lateral diffusion is also considered in the context, but only over very limited short distances, i.e. between neighbouring substomatal cavities (Terashima 1992; Parkhurst 1994). However, lateral CO_2-diffusion in leaves over larger distances is a newly developing field of interest for various reasons, and it is the basis of intra-leaf Ci signalling.

3.1 The heterobaric/homobaric leaf concept

The interest in lateral diffusion currently revives the heterobaric/homobaric leaf concept (Neger 1912, 1918), i.e. whether gas partial pressures in a leaf are heterogeneous or homogeneous. This depends on the relative size of air spaces of leaf aerenchyma but especially on leaf vein anatomy. Heterobaric leaves are characterized by bundle sheath extensions which span the distance between the upper and the lower epidermis. This creates physical barriers for lateral gas diffusion because bundle sheaths mostly lack intercellular air spaces (Neger 1912, 1918). Thus, the mesophyll air space is compartmentalized allowing a spatiotemporal heterogeneity of gas partial pressures within a leaf. Conversely, in the homobaric leaves such restrictions are absent.

Situations can be more complex, however. In a generally homobaric leaf like that of tobacco (*Nicotiana tabacum*), it was found recently that the major lateral veins separate the mesophyll into larger patches between which there are diffusion barriers, while the mesophyll between the large lateral veins is homobaric (Jahnke and Krewitt 2002). The leaves of the C_3/CAM-intermediate species *Clusia minor* have a large intracellular air space of 9.3% of total leaf volume and look like homobaric leaves. However, the large photosynthetically active chlorenchyma at the adaxial side of the central leaf vein has very densely packed cells with very little intracellular air space in between, and is isolated from the rest of the leaf to the extent that even the interveinal leaf parenchyma and the central vein chlorenchyma, respectively, can simultaneously perform different modes of photosynthesis, i.e. C_3-photosynthesis and CAM, respectively (Duarte and Lüttge 2006a). Based on venation anatomy the leaves of *Kalanchoë* are homobaric. However, their mesophyll cells are all almost spherical and isodiametric and so densely packed that the intercellular air space is only 3% of total leaf volume (Duarte et al. 2005) and diffusion resistance is so large that the leaves must be considered as functionally heterobaric (Maxwell et al. 1997).

3.2 Stomatal patchiness

Stomatal patchiness is due to heterogeneous distribution of CO_2 partial pressure in heterobaric leaves (Terashima 1992) and is based on the Ci signalling on stomatal guard cell movements (section 2). It has been reviewed in this series by Beyschlag and Eckstein (1997). One of the consequences of stomatal patchiness is heterogeneity of photosynthetic activity over the leaf.

3.3 Lateral gas diffusion over large distances within leaves

It appears that initially lateral diffusion of CO_2 over larger distances within leaves has been studied out of concerns about artefacts in gas exchange measurements where the leaves are clipped on a gas exchange cuvette (Jahnke and Krewitt 2002; Pieruschka et al. 2005; Jahnke and Pieruschka 2006). Lateral diffusion can occur over distances of at least 8 mm in leaves and when it is substantial, this leads to errors in such measurements due to partial darkening of the leaves by the ring of the leaf clip causing artificial heterogeneity and lateral flow of gas into and out of the clamped part of the leaves.

However, from such technical considerations also highly interesting physiological studies evolved. Pieruschka et al. (2006) darkened parts of homobaric and heterobaric leaves by attaching adhesive non-transparent and gas-impermeable tape which at the same time also blocked stomatal gas exchange. Using chlorophyll fluorescence imaging, they could show that in homobaric leaves respiratory CO_2 can diffuse from the darkened parts into the illuminated parts and stimulate photosynthesis there, which was most active at the border between the darkened and the illuminated parts. The ecophysiological implications of this effect in larger leaves, especially in tropical forests with an important light fleck regime, are obvious. Morison et al. (2005) used silicon grease to prevent stomatal gas exchange on patches of leaves. By chlorophyll fluorescence imaging, they found that CO_2 could not diffuse laterally into the greased areas by more than ca. 0.3 mm. However, they had illuminated the entire leaves and the silicon grease is transparent, so that CO_2 diffusing towards the areas where atmospheric CO_2 supply was blocked was fixed photosynthetically en route. Duarte et al. (2005) have performed a similar experiment, but they used the obligate CAM species *Kalanchoë daigremontiana,* where in the daytime phase III of CAM stomata are closed and the decarboxylation of nocturnally accumulated malate leads to very high internal CO_2 partial pressures, e.g. up to 0.5% (see Lüttge 2002). In this case, also using chlorophyll fluorescence imaging they could clearly demonstrate lateral CO_2 diffusion from non-greased into greased parts of the leaves. Evidently, of course, not only conductance but also concentration gradients are relevant parameters in the resulting event of diffusion. *K. daigremontiana* has functionally heterobaric leaves (section 3.1).

4 Lateral diffusion and Ci signalling for synchronization/desynchronization of photosynthetic activity within leaves

As noted above (section 3.3), chlorophyll fluorescence imaging (CFI) has been used to study heterogeneity of photosynthetic activity over entire leaves produced experimentally using tape or silicon grease. CFI is also used effectively in studying natural patchiness of photosynthetic activity over leaves and the spatiotemporal heterogeneities and dynamics (Raschke et al. 1990; West et al. 2005). An approach based on a nearest neighbour matrix algorithm allows to obtain a quantitative measure of heterogeneity (Hütt and Neff 2001). This provides a wealth of new insights into the spatiotemporal dynamics of photosynthesis in leaves and the role of lateral gas diffusion in signalling.

4.1 Heterogeneity and the role of photorespiration in leaves performing C_3 photosynthesis

Clusia minor is a C_3/CAM intermediate plant where the modes of photosynthesis actually performed depend on environmental cues and can also be induced experimentally (Lüttge 2006). The relative quantum use efficiency of photosystem II (PS II), Φ_{PSII}, as revealed by CFI, is a measure of the use of irradiance and excitation energy of photosynthesis. It was observed that in the C_3 state of *C. minor* it was rather homogenous over the entire leaf throughout the light period (Duarte 2006, Duarte and Lüttge 2006a). In these experiments, photorespiration was also assessed by supplying 20 min pulses of air with only 1% O_2 at intervals while photosynthesis parameters were recorded.

This caused non-photorespiratory conditions so that the difference between CO_2 uptake under 21% O_2 and 1% O_2 indicated the rate of photorespiration. During the 1% O_2 pulses internal CO_2 partial pressure, Ci, as well as Φ_{PSII} decreased, which is explained by the increased use of CO_2 due to increased carboxylation activity of RubisCO at low O_2 and the fact that photorespiration is known to have a particularly high energy demand (Osmond and Grace 1995; Heber et al. 2001; Heber 2002). During the 1% O_2 pulses heterogeneity of Φ_{PSII} over the leaves showed a drastic increase. It has been noted above (section 3.1) that *C. minor* has large intercellular air spaces and basically is a homobaric leaf. The results of Duarte (2006; Duarte and Lüttge 2006a) show, however, that patchiness develops at lowered Ci and in the absence of photorespiration. Hence, quite obviously photorespiration stabilizes photosynthetic energy use over the leaves. The high heterogeneity observed under these conditions demonstrates that both CO_2 and O_2, i.e. most likely the CO_2/O_2 ratio, are involved in lateral signalling and synchronization/desynchronization of photosynthetic energy use in different parts (patches) of the leaf.

4.2 Heterogeneity in leaves performing CAM

Heterogeneity of photosynthetic energy use was also studied in leaves performing CAM using the obligate CAM-species *K. daigremontiana* and the C_3/CAM intermediate *C. minor* in the CAM state. In both species heterogeneity of Φ_{PSII} was strongly related to the CAM phases (Rascher et al. 2001; Rascher and Lüttge 2002; Duarte and Lüttge 2006a). It was high in the transition phase between the dark period and the light period (phase II) and strongly increased at the end of phase III and in phase IV in the afternoon when closed stomata opened again and Ci changed from being determined by internal mobilization of organic acids and became dependent on stomatal CO_2 diffusion again. In phase II Ci is kept low by both PEPC and RubisCO being active in CO_2 fixation. In phase IV, Ci is low due to carboxylation activity of RubisCO. In phase III, Ci is very high due to internal CO_2 supply from organic acids (Lüttge 2002; Duarte et al. 2005). Thus, both in the functionally heterobaric *K. daigremontiana* and in the anatomically more homobaric *C. minor* Φ_{PSII} heterogeneity strongly reflects Ci and is determined by the changing metabolic states with competitive energy demand of CO_2 fixation for organic acid synthesis and vacuolar compartmentation and of RubisCO. This is underlined by the absence of heterogeneity in the C_3-state of *C. minor* under photorespiratory conditions at 21% O_2 (section 3.1). In both the C_3 and the CAM state heterogeneity is low at high Ci and high Φ_{PSII} (Duarte and Lüttge 2006a).

These observations provide clear evidence for lateral Ci signalling involved in synchronization/desynchronization of photosynthetic activity over the leaves and its dependence on both structural traits determining resistance (or conductance) and concentration gradients, i.e. the essential elements of diffusion equations. Indeed, as soon as one formulates a (general) mathematical model of the essential metabolic processes, one can in principle apply the general framework of reaction-diffusion equations (Turing 1952) with different diffusion terms in order to test hypotheses about the signalling substance and to estimate diffusion constants. The involvement of Ci and C diffusion in these phenomena is underlined by the observation of wave fronts running within minutes to hours over the leaves of *K. daigremontiana* during the change from phase III to phase IV of CAM, which can proceed in opposite directions, and when they meet extinguish rather than superimpose upon each other (Rascher and Lüttge 2002). Chlorophyll fluorescence imaging has also shown waves running over the leaves in short ultradian oscillations in C_3-leaves (West et al. 2005).

4.3 Heterogeneity during endogenous circadian rhythmicity of photosynthesis

In the obligate CAM species *K. daigremontiana* there are long lasting endogenous circadian oscillations of gas exchange under constant external conditions (e.g. Lüttge and Beck 1992). In the C_3/CAM-intermediate species *C. minor* both in the C_3 and in the CAM state endogenous rhythmicity is also observed, which is strongly dampened out, however, after a small number of periods in both modes of photosynthesis (Duarte 2006; Duarte and Lüttge 2006b). In *K. daigremontiana* it was found that rhythmicity was lost above an upper temperature threshold and that this effect was reversible when temperature was lowered again into the rhythmic domain. However, this reversibility was only observed when the reduction of temperature was rapid. When temperature was reduced by the same degree but rather slowly in time reversibility was not obtained (Rascher et al. 1998). This indicated that many individual oscillators were involved in different parts of the leaves possibly even in each cell of *K. daigremontiana* and that a strong signal was needed to synchronize them. Temperature is thought here to affect tonoplast permeability and malate compartmentation, and thus indirectly Ci, and drastic temperature changes will elicit strong Ci signals.

In *K. daigremontiana*, heterogeneity of Φ_{PSII} was low in the first three periods of endogeous oscillations and then strongly increased and began to oscillate. Individual patches were desynchronized as Φ_{PSII} declined, i.e. during putative decline of Ci and resynchronized as Φ_{PSII} rose again (Rascher et al. 2001). In *C. minor* in both modes of photosynthesis heterogeneity under constant conditions and irradiance was low at 21% O_2 and did not oscillate (Duarte 2006). With 1% O_2 endogenous oscillations were revealed where heterogeneity was quite high in their peaks. In *K. daigremontiana* the maxima of heterogeneity were reached at maximum CO_2 uptake, J_{CO2}, from the atmosphere, and in *C. minor* with 1% O_2 maximum heterogeneity was seen during the rise of J_{CO2}, i.e. when Ci was most likely low due to limitation by stomatal diffusion. These dynamics were not primarily due to endogenous stomatal rhythmicity but to metabolism dependent shifts in Ci. In *K. daigremontiana*, detailed correlation analysis showed that oscillations of leaf conductance for water vapour in the oscillations lagged behind oscillations of Ci by about 15 min, i.e. they were regulated downstream from Ci (Bohn et al. 2001).

4.4 Interpreting spatiotemporal patterns beyond explicit models

In spite of its intrinsic lack of quantitative comparison with a precise mathematical model from our point of view the study by Peak et al. (2004) constitutes a huge progress in conceptually understanding the spatiotemporal behaviour of stomatal guard cells. By observing that stomata, basically, have to compute an opening state optimal on the global (leaf-wide) level using only local (cell-cell) communication the authors compare stomatal dynamics with a model system from complexity theory, namely cellular automata (CA).

Let us look at this type of modelling in more detail. A cellular automaton is a framework for simulating spatiotemporal patterns arising from local interactions. Figure 1 summarizes this concept. In the case of one spatial dimension (1D CA) one has a chain of elements, each of which can be in a particular state out of a state space Σ. Update rules translate neighbourhood constellations at time t into the state of the central element at the next time step $t+1$ (see Fig. 1A). By consecutively applying the update rules one can simulate the time evolution of such a chain of elements corresponding to this particular selection of update rules. The strength of cellular automata lies in the simplicity of both the concept and the state space of a specific model, which usually contains only few distinct states. Consequently, the idea of cellular automata is not to describe a biological phenomenon as accurately as possible, but to grasp the essence of a system in terms of few degrees of freedom. In this sense, cellular automata represent an extreme case of minimal model for biological situations.

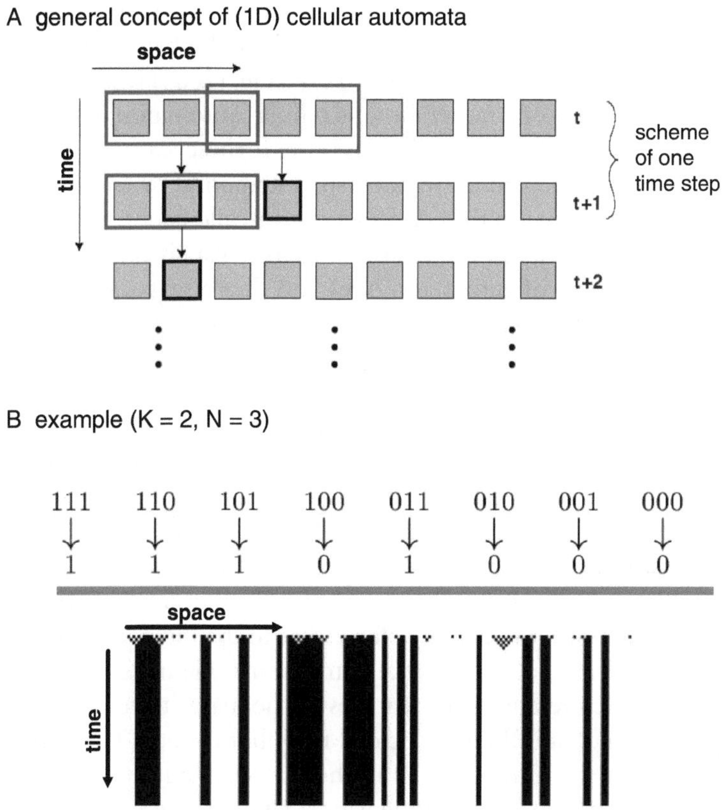

Fig. 1. Concept of cellular automata (CA) illustrated for one spatial dimension **A** and an example of a CA simulation for a binary state space and a three-element neighbourhood **B**. For this selection of update rules, one obtains fixed-point dynamics after a short transient

As this brief summary of cellular automata already shows, this model operates on a discrete space, a discrete time and on a finite and discrete state space. Figure 1B gives an example of a simple cellular automaton. This example considers three-element neighbourhoods (N=3) and a binary (two-element) state space (K=2), e.g. containing only 0 and 1. In this case 2^3=8 neighbourhood constellations have to be specified. The example given in Fig. 1B after a short transient leads to a steady-state pattern.

This general framework of cellular automata has received a lot of scientific attention as a laboratory for studying complex systems (see, e.g. Wolfram 2002). At the same time, specific cellular automata have proven useful as minimal models for aspects of biological pattern formation ranging from simplest Turing patterns to the characteristic spiral wave patterns observed in excitable media. But even against this background, it is not readily understood, how this CA-type of models could mimic the behaviour of stomatal dynamics. The interesting approach by Peak et al. (2004) has been to analyse the statistical features of a rather famous cellular automaton designed to perform a specific computational task and compare these features with the computational task they record in stomatal spatiotemporal dynamics.

The cellular automaton they consider is a so-called *density classifier*. In its simplest form the computational task of density classification starts from a random distribution of zeroes and ones and then uses local rules to determine the state with the highest global density. Note that with the help of an external agent this would be a trivial computational task requiring only to count, e.g. the number of 1s in the initial state. With no external agent using only local interactions between the elements the computation of the state with the highest initial density is far from trivial. This situation has been studied intensely in complexity theory (see, e.g. Mitchell et al. 1994; Crutchfield and Mitchell 1995; Wolfram 2002). A simple example of such a density classification is the *majority rule*, where the update rule maps an element to the state, which holds the majority in the neighbourhood under consideration. In order to get one step closer to the biological phenomenon we consider the case of two spatial dimensions. Figure 2 summarizes the update rule, which retains some features of the majority rule with the additional condition of shuffling undecided neighbourhoods. There, the update rule operates on the number of 1s in the nine-element neighbourhood for this automaton in two spatial dimensions. Update rules involving only the number of, e.g. 1s and not their distribution in the neighbourhood are called *totalistic cellular automata* in CA theory. Figure 3 shows spatial snapshots at different time points starting from an initial density of 1s, p_1, close to 0.5. The corresponding time course of the density is displayed in the bottom part of Fig. 3. The capacity of this automaton to classify initial densities by converging to a specific state is summarized in Fig. 4. There, densities $\rho(t)$ for different initial densities, p_1, are compared. It is seen that, while highly biased densities are rapidly classified by this automaton, the difficult cases, where $p_1 \approx 0.5$, can lead to long transients (and also to an occasional mis-classification). Peak et al. (2004) evaluate such cellular automata in terms of the event size distributions and other statistical parameters, and compare these quantities with the stomatal patterns. Both, in theory and experiment, they focus on the long transients.

This approach is conceptually different from the usual framework of theoretical biology, where specific mathematical models are formulated and then analyzed with methods from nonlinear dynamics. Here the mathematical description focuses on analogies. A quantitative analysis then relies on

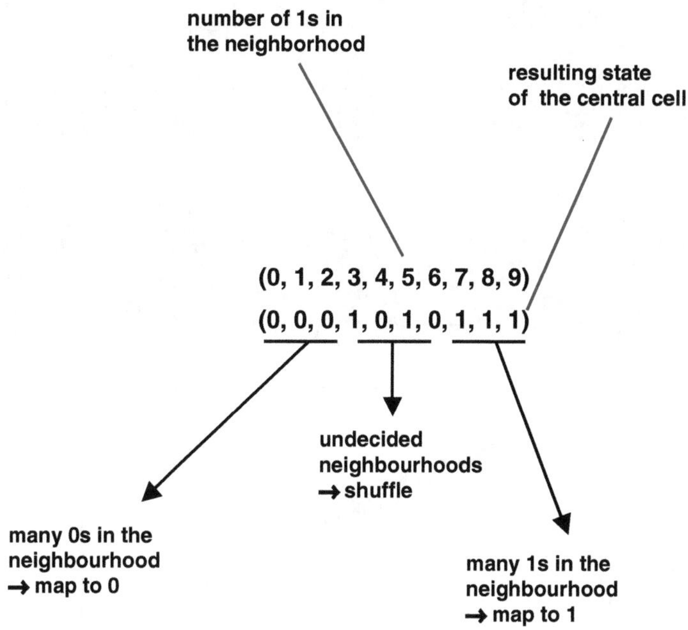

Fig. 2. Summary of the update rule for a totalistic CA suitable for density classification. The first line of digits in parentheses gives the number of 1s in a spatial (nine-element) neighbourhood, while the second line lists the corresponding state of the central cell in the next time step

comparing statistical properties of the real-life data with those from the theoretical counterpart. For stomatal dynamics, Peak et al. (2004) analyse event sizes, the scaling of transients and the distribution of inter-peak intervals in the average opening state (see also West et al. 2005). One can view this study as an example of a new trend in theoretical biology on its way to an understanding at a system-wide level: Ever larger systems and, on the other hand, an increasing importance of stochastic contributions (usually coming from few-molecule configurations beyond the usual average concentrations far beyond the molecular level) require models with abstract, discrete dynamics and the use of stochastic processes to implement the effect of stochasticity on this abstract level (see also Bornholdt, 2005).

5 Ci signalling network

Lateral diffusion of CO_2 is clearly demonstrated in homobaric leaves and, depending on the magnitude of concentration gradients for diffusion, also in functionally heterobaric leaves. Observations of the spatiotemporal

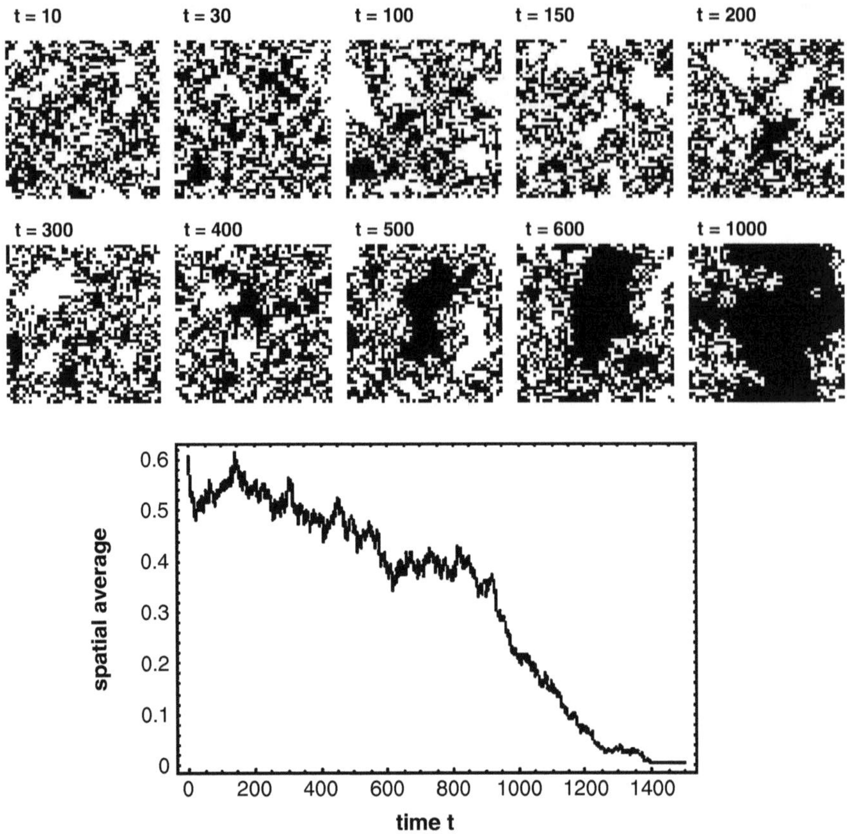

Fig. 3. Spatial snapshots of the CA from Fig. 2 at different time points (top part) and time course of the density ρ(t) of 1s (i.e. of the spatial average over the lattice at each time point)

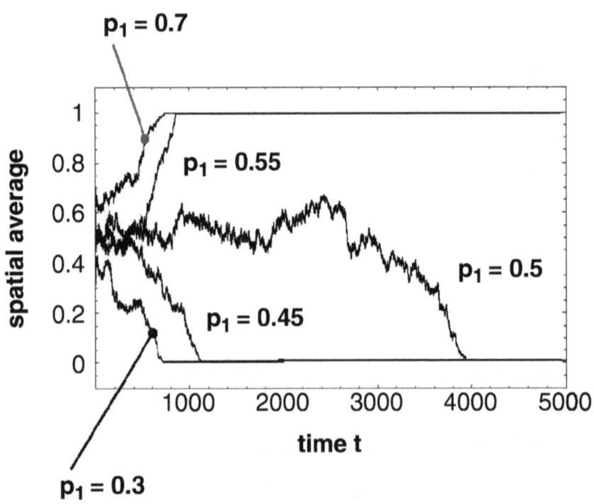

Fig. 4. Density ρ(t) as a function of time t for different values of the initial density of 1s, p_1. The cellular automaton is the same as in Figs 2 and 3

dynamics of synchronization/desynchronization of heterogeneous photosynthetic activity in leaves can be explained by Ci signalling. Summarizing Ci signalling (Fig. 5), we can consider Ci as a central pool and in addition a carbohydrate pool, and when we deal with CAM plants also an organic acid pool. The Ci pool is emptied by CO_2 fixation via the carboxylase activity of RubisCO and via PEPC and by loss of CO_2 via open stomata in respiration, and there is also sometimes a minimal loss of CO_2 to the atmosphere during phase III of CAM notwithstanding closed stomata because the inward-outward directed concentration gradient is so large in phase III. The Ci pool is filled by CO_2 uptake from the atmosphere via the stomata, by respiration, by photorespiration, and in CAM by organic acid decarboxylation. The mechanisms of Ci signalling to the other two pools can be standard based on substrate kinetics of the carboxylating enzymes. Ci affects the oxygenation activity of RubisCO via the CO_2/O_2 ratio and the respective substrate

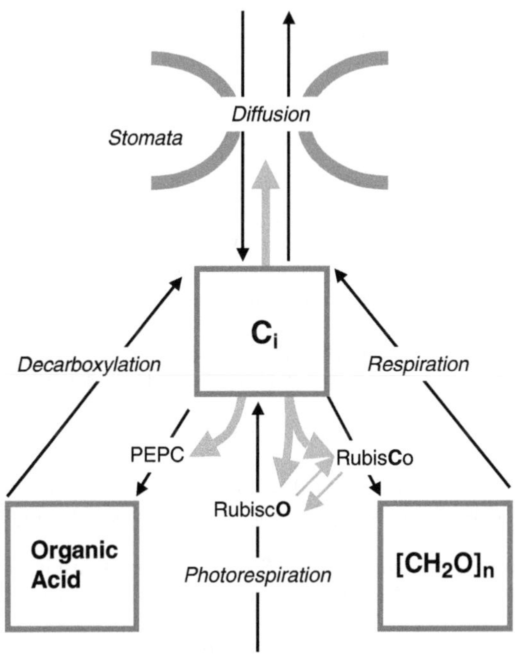

Fig. 5. Scheme of a Ci-signalling network. Pools are in quadratic boxes, i.e. Ci, carbohydrate ($[CH_2O]_n$) and organic acid. (We are writing organic acid and not just malate because although *K. daigremontiana* is only accumulating malate during the dark period of CAM, *C. minor* is accumulating both malate and citrate (Lüttge 2006). Thin black arrows show influx and efflux of the pools. Broad grey arrows indicate signalling pathways. Rubis**CO** = oxygenase and Rubis**CO** = carboxylase activity of ribulose-bis-phosphate carboxylase/oxygenase, PEPC = phospho*enol*pyruvate carboxylase

affinities. The mechanism of Ci signalling to the stomata is unexplained, as outlined in section 2.

Clearly, there is a substantial feedback between the spatiotemporal patterning of photosynthetic activity and stomatal dynamics. The successful analysis by Peak et al. (2004), however, suggests that one should theoretically approach these entangled spatiotemporal dynamics with similar mathematical tools. Thus, in the vein of their study novel mathematical modelling approaches are required for further progress of understanding these networks.

References

Assmann SM (1999) The cellular basis of guard cell sensing of rising CO_2. Plant Cell Environ 22:629–637

Badger MR, Hanson D, Price GD (2002) Evolution and diversity of CO_2 concentrating mechanisms in cyanobacteria. Funct Plant Biol 29:161–173

Beyschlag W, Eckstsein J (1997) Stomatal patchiness. Prog Bot 59:283–298

Birkenhead K, Willmer CM (1986) Some biochemical characteristics of guard cell and mesophyll cell protoplasts from *Commelina communis* L. J Exp Bot 37:119–128

Bohn A, Geist A, Rascher U, Lüttge U (2001) Responses to different external light rhythms by the circadian rhythm of Crassulacean acid metabolism in *Kalanchoe daigremontiana*. Plant Cell Environment 84:811–820

Bornholdt S (2005) Less is more in modeling large genetic networks. Science 310: 449–451

Brearley J, Venis MA, Blatt MR (1997) The effect of elevated CO_2 concentrations on K^+ and anion channels of *Vicia faba* L. guard cells. Planta 203:145–154

Cardon ZG, Berry J (1992) Effects of O_2 and CO_2 concentration on the steady-state fluorescence yield of single guard cell pairs in intact leaf discs of *Tradescantia albiflora*. Evidence for Rubisco-mediated CO_2 fixation and photorespiration in guard cells. Plant Physiol 99:1238–1244

Cramer MD, Gao ZF, Lips H (1999) The influence of dissolved inorganic carbon in the rhizosphere on carbon and nitrogen metabolism in salinity-treated tomato plants. New Phytol 142:441–450

Crutchfield JP, Mitchell M (1995) The evolution of emergent computation. Proc Natl Acad Sci USA 92:10742–10746

Drechsler Z, Sharkia R, Cabantchik ZT, Beer S (1993) Bicarbonate uptake in the marine macroalga *Ulva* sp. is inhihibited by classical probes of anion exchange by red blood cells. Planta 191:34–40

Drechsler Z, Sharkia R, Cabantchik ZT, Beer S (1994) The relationship of arginine groups to photosynthetic HCO_3^- uptake in *Ulva* sp. mediated by a putative anion exchanger. Planta 194:250–255

Duarte HM (2006) Chronobiologie von *Clusia minor*: circadianer Rhythmus in einer Pflanze mit C_3/CAM-intermediärem photosynthetischen Verhalten. Dr. rer. nat. Dissertation, Darmstadt

Duarte HM, Lüttge U (2006a) Gas-exchange, photorespiration and spatiotemporal dynamics of relative quantum use efficiency in leaves of the C_3-photosynthesis/crassulacean acid metabolism-intermediate species *Clusia minor* L. (Clusiaceae) in both modes of photosynthesis. Trees (in press)

Duarte HM, Lüttge U (2006b) Endogenous circadian rhythmicity of gas exchange, quantum use efficiency and heterogeneous spatial patterns in leaves of the C_3-photosysnthesis/crassulacean acid metabolism (CAM) intermediate *Clusia minor* L. (Clusiaceae) in both modes of photosynthesis. Plant Biol (in press)
Duarte HM, Jakovljevic I, Kaiser F, Lüttge U (2005) Lateral diffusion of CO_2 in leaves of the crassulacean acid metabolism plant *Kalanchoë daigremontiana* Hamet et Perrier. Planta 220:809–816
Frechilla S, Talbott LD, Zeiger E (2002) The CO_2 response of *Vicia* guard cells acclimates to growth environment. J Exp Bot 53:545–550
Gotov K, Taylor S, Zeiger E (1988) Photosynthetic carbon fixation in guard cell protoplasts of *Vicia faba*. Evidence from radiolabel experiments. Plant Physiol 86:700–705
Hanba YT, Miyazawa S-I, Terashima I (1999) The influence of leaf thickness on the CO_2 transfer conductance and leaf stable carbon isotope ratio for some evergreen tree species in Japanese warm-temperate forests. Funct Ecol 13:632–639
Hanba YT, Shibasaka M, Hahashi Y, Hayakawa T., Kasamo K, Terashima I, Katsuhara M (2004) Overexpression of the barley aquaporin HvPIP2;1 increases internal CO_2 conductance and CO_2 assimilation in the leaves of transgenic rice plants. Plant Cell Physiol 45:521–529
Hanstein SM (2002) CO_2-triggered chloride release from guard cells in intact fava bean leaves. Kinetics of the onset of stomatal closure. Plant Physiol 130:940–950
Hanstein SM, Felle HH (2002) CO_2-triggered chloride release from guard cells in intact fava bean leaves. Kinetics of the onset of stomatal closure. Plant Physiol 130:940–950
Heber U (2002) Irrungen, Wirrungen? The Mehler reaction in relation to cyclic electron transport in C_3 plants. Photosynth Res 73:223–231
Heber U, Bukhov NG, Shuvalov VA, Kobayashi Y, Lange OL (2001) Protection of the photosynthetic apparatus against damage by excessive illumination in homoiohydric leaves and poikilohydric mosses and lichens. J Exp Bot 52:1999–2006
Hedrich R, Marten I (1993) Malate-induced feedback regulation of plasma membrane anion channels could provide a CO_2 sensor to guard cells. EMBO J 12:897–901
Hedrich R, Marten I, Lohse G, Dietrich P, Winter H, Lohaus G, Heldt H-W (1994) Malate-sensitive anion channels enable guard cells to sense changes in the ambient CO_2 concentration. Plant J 6:741–748
Hütt MT, Neff R (2001) Quantification of spatiotemporal phenomena by means of cellular automata techniques. Physica A 289:498–516
Huxman TE, Monson RK (2003) Stomatal responses of C_3, C_3-C_4 and C_4 *Flaveria* species to light and intercellular CO_2 concentration: implications for the evolution of stomatal behaviour. Plant Cell Environ 26:313–322
Jahnke S, Krewitt M (2002) Atmospheric CO_2 concentration may directly affect leaf respiration measurement in tobacco, but not respiration itself. Plant Cell Environ 25:641–651
Jahnke S, Pieruschka R (2006) Air pressure in leaf chambers – a neglected issue in gas exchange measurements. J Exp Bot (in press)
Jewer PC, Neales TF, Incoll LD (1985) Stomatal responses to carbon dioxide of isolated epidermis from a C_3 plant, the *Argenteum* mutant of *Pisum sativum* L. and a crassulacean-acid-metabolism plant *Kalanchoë daigremontiana* Hamet et Perr. Planta 164:495–500
Kaldenhoff R (2005) Besides water: Functions of plant membrane intrinsic proteins and quaporins. Progr Bot 67:206–218
Klockenbring T, Meinhard M, Schnabl H (1998) The stomatal phosphoenolpyruvate carboxylase—a potential target for selective proteolysis during stomatal closure? J Plant Physiol 152:222–229
Lawson T, Oxborough K, Morison JTL, Baker NR (2003) The responses of guard cell and mesophyll cell photosynthesis to CO_2, O_2, light, and water stress in a range of species are similar. J Exp Bot 54:1743–1752

Leegood RC (2002) C_4 photosynthesis: principles of CO_2 concentration and prospects for its introduction into C_3 plants. J Exp Bot 53:581–590

Leymarie J, Lascève G, Vavasseur A (1998) Interaction of stomatal responses to ABA and CO_2 in *Arabidopsis thaliana.* Austral J Plant Physiol 25:785–791

Lüttge U (2002) CO_2-concentrating: consequences in crassulacean acid metabolism. J Exp Bot 53:2131–2142

Lüttge U (2006) Photosynthetic flexibility and ecophysiological plasticity: questions and lessons from *Clusia,* the only CAM tree, in the neotropics. New Phytol 171:7–25

Lüttge U, Beck F (1992) Endogenous rhythms and chaos in crassulacean acid metabolism. Planta 188:28–38

Mansfield TA, Hetherington AM, Atkinson CJ (1990) Some current aspects of stomatal physiology. Annu Rev Plant Physiol Plant Mol Biol 41:55–75

Maxwell K, Caemmerer S von, Evans JR (1997) Is low internal conductance to CO_2 diffusion a consequence of succulence in plants with crassulacean acid metabolism. Aust J Plant Physiol 24:777–786

Mimura T, Müller R, Kaiser WM, Shimmen T, Dietz K-J (1993) ATP-dependent carbon transport in perfused *Chara* cells. Plant Cell Environ 16:653–661

Mitchell M, Crutchfield JP, Hraber PT (1994) Evolving cellular automata to perform computations: Mechanisms and impediments. Physica D 75:361–391

Morison JLL, Gallouëtt E, Lawson T, Cornic G, Heerbin R, Baker NR (2005) Lateral diffusion of CO_2 in leaves is not sufficient to support photosynthesis. Plant Physiol 139:254–266

Mott KA (1990) Sensing of atmospheric CO_2 by plants. Plant Cell Environ 13:731 -737

Neger FW (1912) Spaltöffnungsschluβ und künstliche Turgorsteigerung. Ber Dt Bot Ges 30:179–194

Neger FW (1918) Die Wegsamkeit der Laubblätter für Gase. Flora 111:152–161

Olsen RL, Pratt RB, Gump P, Kemper A, Tallman G (2002) Red light activates a chloroplast-dependent ion uptake mechanism for stomatal opening under reduced CO_2 concentrations in *Vicia* spp. New Phytol 153:497–508

Osmond CB (1978) Crassulacean acid metabolism: A curiosity in context. Annu Rev Plant Physiol 29:379–414

Osmond CB, Grace CE (1995) Perspectives on photoinhibition and photorespiration in the field: quintessential inefficiencies of the light and dark reactions of photosynthesis? J Exp Bot 46:1351–1362

Outlaw WH (1989) Critical examination of the quantitative evidence for and against photosynthetic CO_2 fixation by guard cells. Physiol Plantarum 77:275–281

Palmqvist K, Ögren E, Lernmark U (1994) The CO_2-concentrating mechanism is absent in the green alga *Coccomyxa*: a comparative study of photosynthetic CO_2 and light responses of *Coccomyxa, Chlamydomonas reinhardtii* and barley protoplasts. Plant Cell Environ 17:65–72

Parkhurst DF (1994) Diffusion of CO_2 and other gases inside leaves. New Phytol 126:449–479

Peak D, West JD, Messinger SM, Mott KA (2004) Evidence for complex, collective dynamics and emergent, distributed computation in plants. Proc Natl Acad Sci USA 101:918–922

Pieruschka R, Schurr U, Jahnke S (2005) Lateral gas diffusion inside leaves. J. Exp Bot 56:857–864

Pieruschka R, Schurr U, Jensen M, Wolff WF, Jahnke S (2006) Lateral diffusion of CO_2 from shaded to illuminated leaf parts affects photosynthesis inside homobaric leaves. New Phytol 169:779–787

Portis AR (1992) Regulation of ribulose-1.5-bisphosphate carboxylase/oxygenase activity. Annu Rev Plant Physiol Plant Mol Biol 43:415–437

Price GD, Caemmerer S von, Evans JR, Yu J-W, Lloyd J, Oja V, Kell P, Harrison K, Gallagher A, Badger MR (1994) Specific reduction of chloroplast carbonic anhydrase activity by antisense RNA in transgenic tobacco plants has a minor effect on photosynthetic CO_2 assimilation. Planta 193:331–340

Price GD, Maeda S-I, Omata T, Badger MR (2002) Modes of active inorganic carbon uptake in the cyanobacterium, *Synechococcus* sp.PCC7942. Funct Plant Biol 29:131–149
Rascher U, Blasius B, Beck F, Lüttge U (1998) Temperature profiles for the expression of endogenous rhythmicity and arrythmicity of CO_2 exchange in the CAM plant *Kalanchoë daigremontiana* can be shifted by slow temperature changes. Planta 207:76–82
Rascher U, Hütt M-T, Siebke K, Osmond B, Beck F, Lüttge U (2001) Spatiotemporal variation of metabolism, in a plant circadian rhythm: the biological clock as an assembly of coupled individual oscillators. Proc Natl Acad Sci USA 98:11801–11805
Rascher U, Lüttge U (2002) High-resolution chlorophyll fluorescence imaging serves as a non-invasive indicator to monitor the spatio-temporal variations of metabolism during the day-night cycle and during the endogenous rhythm in continuous light in the CAM plant *Kalanchoë daigremontiana.* Plant Biol 4:671–681
Raschke K, Patzke J, Daley PF, Berry JA (1990) Spatial and temporal heterogeneities of photosynthesis detected through analysis of chlorophyll-fluorescence images of leaves. In Baltscheffsky M (ed) Current research in photosynthesis. Kluwer, Boston, pp 573–578
Reckmann U, Scheibe R, Raschke K (1990) Rubisco activity in guard cells compared with the solute requirement for stomatal opening. Plant Physiol 92:246–253
Rother T, Acker G, Scheibe R (1988) Immunogold localization of chloroplast protein in spinach leaf mesophyll, edpidermis, and guard cells. Botanica Acta 101:311–320
Schnabl H (1981) The compartmentation of carboxylating and decarboxylating enzymes in guard cell protoplasts. Planta 152:307–313
Schroeder JI, Allen GJ, Hugouvieux V, Kwak JM, Waner D (2001) Guard cell signal transduction. Annu Rev Plant Physiol Plant Mol Biol 52:627–658
Schulz M, Hunte C, Schnabl H (1992) Multiple forms of phosphoenolpyruvate carboxylase in mesophyll, epidermal and guard cells of *Vicia faba.* Physiol Plant 86:315–321
Sharkia R, Beer S, Cabantchik ZI (1994) A membrane-located polypeptide of *Ulva* sp. which may be involved in HCO_3^- uptake is recognized by antibodies raised against the human red-blood-cell anion-exchange protein. Planta 194:247–249
Shimazaki K-I, Zeiger E (1985) Cyclic and noncyclic photophosphorylation in isolated guard cell chloroplasts from *Vicia faba* L. Plant Physiol 78:211–214
Tallman G (2004) Are diurnal patterns of stomatal movement the result of alternating metabolism of endogenous guard cell ABA and accumulation of ABA delivered to the apoplast around guard cells by transpiration? J Exp Bot 55:1963–1976
Terashima I (1992) Anatomy of non-uniform leaf photosynthesis. Photos Res 31:195–212
Terashima I, Miyazawa S-I, Hanba YT (2001) Why are sun leaves thicker than shade leaves?—Consideration based on analyses of CO_2 diffusion in the leaf. J Plant Res 114:93–105
Turing AM (1952) The chemical basis of morphogenesis. Philos Trans R Soc 237:37–72
Uehlein N, Lovisolo C, Siefrietz F, Kaldenhoff R (2003) The tobacco aquaporin NtAQP1 is a membrane CO_2 pore with physiological function. Nature 425:734–737
Vaughn KC, Vaughn MA (1988) Is ribulose bisphosphate carboxylase present in guard cell chloroplasts? Physiol Plant 74:409–413
Vavasseur A, Raghavendra AS (2005) Guard cell metabolism and CO_2 sensing. New Phytol 165:665–682
Warren CR (2004) The photosynthetic limitation posed by internal conductance to CO_2 movement is increased by nutrient supply. J Exp Bot 55:2313–2321
West JD, Peak D, Peterson JQ, Mott KA (2005) Dynamics of stomatal patchiness for a single surface of *Xanthium strumarium* L. leaves observed with fluorescence and thermal images. Plant Cell Environ 28:633–641
Wolfram S (2002) A new kind of science. Wolfram Media Publishing, IL, USA
Zeiger E, Talbott LD, Frechilla S, Srivastava A, Zhy J (2002) The guard cell chloroplast: a perspective for the twenty-first century. New Phytol 153:415–424
Zhu J, Talbott LD, Jui X, Zeiger E (1998) The stomatal response to CO_2 is linked to changes in guard cell zeaxanthin. Plant Cell Environ 21:813–820

Ulrich Lüttge
Institut für Botanik
Technische Universität Darmstadt
Schnittspahnstrasse 3–5
64287 Darmstadt
Germany
Tel.: +49-6151-163700
Fax: +49-6151-164630
e-mail: luettge@bio.tu-darmstadt.de

Marc-Thorsten Hütt
Computational Systems Biology
International University Bremen
School of Engineering and Science
Campus Ring 1
28759 Bremen
Tel.: +49-421200-3238
Fax.: +49-421200-3249
e-mail: m.huett@iu-bremen.de

Plant Haemoglobins, Nitrate and Nitric Oxide: Old Players, New Games

Yoshinari Ohwaki and Werner M. Kaiser

1 Introduction

In higher plants, three distinct classes of haemoglobins have been found which are denominated as symbiotic (symHb), non-symbiotic (nsHb) and truncated haemoglobins (trHb). Symbiotic haemoglobin has a well characterized function in nitrogen fixation, in providing an adequate supply of oxygen to the terminal oxidases of the symbiotic bacteria in the nodule, and at the same time preventing inactivation of the nitrogenase enzyme which is sensitive to oxygen. Much less is so far known about two other plant haemoglobin forms, the so-called "non-symbiotic haemoglobin" and "truncated haemoglobin", which will be considered in more detail below.

Expression of haemoglobin in non-nodulated plants was first reported in *Trema tomentosa* (Bogusz et al. 1988), and root-specific expression of the promoters of *Trema* haemoglobin was achieved in transgenic tobacco (Bogusz et al. 1990). The finding of functional haemoglobin in non-nodulating plants suggests that all plants may have haemoglobin genes, and implies that haemoglobins have a function in normal cells. Meanwhile, nsHbs have been reported to exist in a range of plants, including monocots (Taylor et al. 1994) and dicots (Hunt et al. 2001), and are now considered ubiquitous in the plant kingdom. The function of nsHb was first suggested to be related to oxygen transport (Bogusz et al. 1988; Andersson et al. 1996) or to represent an oxygen sensor to change metabolism from the oxidative to the fermentative pathway (Appleby et al. 1998). Recent work on the properties of haemoglobin and functional analysis of transgenic plants, however, has opened new aspects. One that will be considered in some detail here is that nsHb may have physiological relevance with respect to nitric oxide (NO) metabolism. NO is a small, gaseous free radical molecule that has been extensively studied as an important signalling compound in mammals, covering a wide range of regulatory functions in the cardiovascular, nervous and immunological system (Alderton et al. 2001). Accumulated knowledge on NO in plants suggests that it plays also important roles in the control of

Progress in Botany, Vol. 68

growth and development including germination, cell differentiation, flowering and senescence (Beligni and Lamattina 2000; Corpas et al. 2004; He et al. 2004; Guo and Crawford 2005) as well as in defence responses to biotic and abiotic stresses (Neill et al. 2002; Romero-Puertas et al. 2004). Thus it is extremely important to find out how the activity of NO is controlled in vivo.

Basically, control may be exerted by modulation of production or modulation of consumption, or both. In plants, NO may be formed either by nitrite- or by L-arginine-dependent reactions (Fig. 1). To date, the larger group of the nitrite-dependent NO sources includes nitrate reductase (NR), plasma-membrane bound nitrite:NO-reductase (NI-NOR) and mitochondrial electron transport. The L-arginine-dependent group presently consists

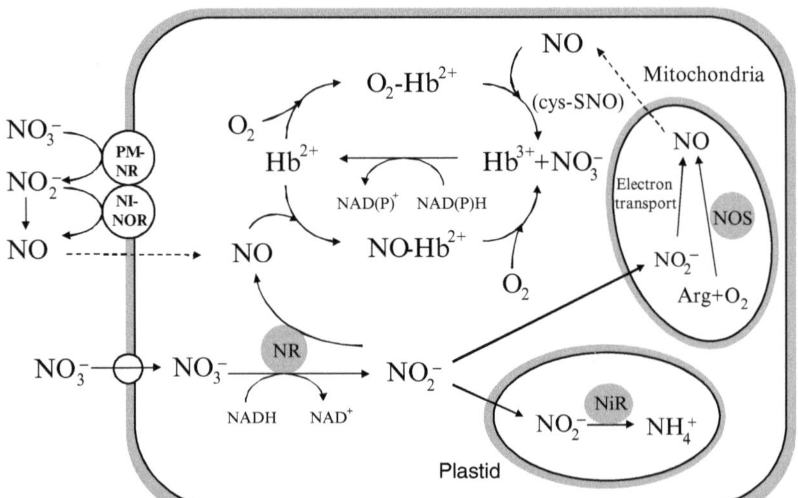

Fig. 1. Schematic presentation of the NO generating reactions, and putative haemoglobin(Hb)-mediated NO scavenging in plant cells (abbreviations used in the figure are explained in this legend): In the root apoplast, nitrate-dependent NO production is catalysed by plasmalemma-bound nitrate reductase (*PM-NR*) and a closely associated nitrite:NO reductase (*NI-NOR*). NO might be also produced at physiological nitrite concentrations via non-enzymatic reduction of nitrite, but only if the pH in the apoplast would drop significantly below pH 5. NO synthesis in the cytosol is catalysed by either nitrate reductase (NR), which reduces nitrite to NO, or by NO synthase (NOS; AtNOS1) in the mitochondria, which uses L-arginine (Arg) as substrate. Mitochondrial electron transport also reduces nitrite to NO but only under hypoxia/anoxia. The reaction catalysed by non-symbiotic haemoglobin (nsHb) may convert NO into nitrate at the expense of O_2 and NAD(P)H. The mechanism for this nsHb-mediated reaction has not been fully established yet. One possible route is the dioxygenase reaction, where haem-bound oxygen (O_2-Hb^{2+}) reacts with NO to yield nitrate. *S*-nitrosylation of haemoglobin (cys-SNO) is proposed to be involved in this reaction. Another possible route might be a denitrosylase reaction, where haem-bound NO (NO-Hb^{2+}) reacts with oxygen to form nitrate. The methaemoglobin (Hb^{3+}) produced by this reaction can be reduced by NAD(P)H to haemoglobin (Hb^{2+})

of only one enzyme, NO synthase (NOS). A gene for NOS has been recently identified in *Arabidopsis* (AtNOS1), which has no sequence similarity to any mammalian isoform. Nevertheless, it appears to oxidize L-arginine to L-citrulline and NO by using O_2 and NADPH, in a way similar to mammalian eNOS and nNOS (Guo et al. 2003).

As pointed out recently, NO production through the nitrite-dependent pathways can be regulated by variety of environmental and developmental cues (Meyer and Stöhr 2002; Tischner et al. 2004; Gupta et al. 2005; Planchet et al. 2005, 2006). The biological half life of NO is probably short, and there are many potential NO quenchers, e.g. thiols or tocopherols and haem proteins such as nsHbs. This rapid NO scavenging may be as important for its signalling function as its production.

Here, we describe the main physiological and biochemical functions reported for nsHb in plants, but with a strong focus on its interaction with NO in nitrate assimilation, flooding (or anoxia-) tolerance and in plant-microbe interactions.

2 Properties of nsHbs

2.1 Gene and protein expression

A number of nsHb genes have been isolated from a variety of plants. Their coding sequences are highly similar and are interrupted by three introns located at identical positions (Arredondo-Peter et al. 1997; Aréchaga-Ocampo et al. 2001; Hunt et al. 2001). For example, the deduced amino acid sequence of barley nsHb shares 71% identity with a non-legume haemoglobin gene, a further 16% of the residues being conservative replacements (Taylor et al. 1994). The sequence of nsHb is clearly distinct from the symbiotic class of proteins, e.g. the predicted amino acid sequence of nsHb in soybean is only 57–58% identical to symHb (Andersson et al. 1996).

The molecular phylogenetic analysis of nsHb revealed that two distinct classes of nsHb exist in dicots, namely class 1 and class 2 nsHb. Monocots appear to have only class 1 nsHb (Hunt et al. 2001). Although class 1 and class 2 nsHb genes contain the conserved distal histidine (E7), proximal histidine (F8), proline (C2) and phenylalanine (cd1) residues (Hunt et al. 2001), they share only limited sequence similarities and showed distinct expression profiles. *AHB1* (class 1) and *AHB2* (class 2) nsHb genes from *Arabidopsis* share only 69% identity at the nucleotide level (Trevaskis et al. 1997), and the class 2 nsHb gene sequences in canola (*Brassica napus*) and cotton (*Gossypium hirsutum*) are only 55–60% identical to a class 1 nsHb gene in *Arabidopsis* and 60–70% identical to the symHbs of legumes and *Casuarina* (Hunt et al. 2001). The class 1 nsHb gene is induced by low oxygen condition, by nitrate treatment (see below) and by high sucrose levels (Trevaskis et al. 1997). In contrast, class 2 nsHb gene expression is not affected by hypoxia, but is induced by low temperature (Trevaskis et al. 1997) and cytokinin treatment (Hunt et al. 2001). More recently, it has been reported that the expression of class 1 nsHb from *Lotus japonicus* is induced by low temperature and by the plant hormones ABA and cytokinin, as well as by anoxia (Shimoda et al. 2005). The promoter of class 1 nsHb (*OsNSHB2*) in rice is significantly up-regulated by cytokinin (Ross et al. 2004). These results suggest that the functions of the two classes of nsHbs may overlap (Ross et al. 2004).

2.2 Cellular and subcellular localization of Hbs

Expression of class 1 nsHb has been frequently observed in germinating seedlings (Duff et al. 1998; Hunt et al. 2001; Lira-Ruan et al. 2001; Ross et al. 2001), but that was not obvious in teosinte (*Zea mays* ssp. *parviglumis*; Aréchaga-Ocampo et al. 2001). NsHbs exist in different cell types of developing seedlings including the aleurone, scutellum, root cap cells and differentiating sclerenchyma and tracheary elements. The specific cellular localization of the nsHb protein indicates specific developmental functions in plants (Ross et al. 2001). In contrast to class 1 nsHbs, expression of class 2 nsHbs has been found in the bolt stem, rosette leaves and roots of mature flowering plants, but not in young plants (Hunt et al. 2001). In *Arabidopsis*, class 2 nsHb is activated during embryogenesis and seed maturation, suggesting a function in seed development (Hunt et al. 2001). This agrees with the finding that a class 2 nsHb gene is specifically expressed under conditions inducing somatic embryogenesis in *Cichorium* (Hendriks et al. 1998). NsHbs appear to be located in the cytosol, because the DNA sequence of nsHbs does not have a transit signal peptide (Taylor et al. 1994; Arredondo-Peter et al. 1997; Trevaskis et al. 1997). However, the alfalfa MHb1 protein (Seregélyes et al. 2000) and the cotton GhHb1 protein (Qu et al. 2005) were predominantly present in the nucleus, and only a small amount appeared in the cytoplasm. In yeast (*Saccharomyces cerevisiae*), a flavohaemoglobin (flavoHb) encoded by a single nuclear gene *YHB1* lacking a predicted mitochondrial signal peptide was localized in both, the mitochondrial matrix and the cytosol (Cassanova et al. 2005).

A new haemoglobin having a central domain similar to trHb of microorganisms was isolated from *Arabidopsis* (Watts et al. 2001), wheat (Larsen

2003) and *Medicago truncatula* (Vieweg et al. 2005). Many trHbs display amino acid sequences 20–40 residues shorter than non-vertebrate haemoglobin, and possess an alternative folding pattern with a 2-on-2 sandwich of α-helices, distinct from other haemoglobins having 3-on-3 structures (Wittenberg et al. 2002). TrHbs are distributed in eubacteria, cyanobacteria, unicellular eukaryotes and in higher plants, and form a phylogenetically distinct family separate from other haemoglobins (Wittenberg et al. 2002; Milani et al. 2005). In plants, trHbs are constitutively expressed throughout the plant (Watts et al. 2001; Larsen 2003), but expression in the roots appears to be about four times higher than in shoots (Watts et al. 2001). These patterns are similar to class 1 nsHb, but trHb showed no induction by hypoxia (Watts et al. 2001; Larsen 2003).

2.3 Oxygen affinity

Recombinant class 1 nsHb protein has a remarkably high affinity for O_2, which is mainly due to an extraordinary low oxygen dissociation constant (Arredondo-Peter et al. 1997; Duff et al. 1997; Trevaskis et al. 1997; Hargrove et al. 2000). This much lower O_2 dissociation constant of nsHbs compared to cytochrome oxidase makes it unlikely that these proteins supply O_2 for the terminal oxidases in mitochondria (Trevaskis et al. 1997). Moreover, the extraordinary slow dissociation of oxygen would suggest that nsHb is not involved in oxygen transport (Arredondo-Peter et al. 1997; Trevaskis et al. 1997). Class 2 nsHb proteins have a much lower affinity for O_2 than class 1 nsHbs, which are similar to symbiotic haemoglobin (Trevaskis et al. 1997). It was proposed that symbiotic haemoglobin genes in legumes and *Casuarina* in evolution arose from class 2 genes (Trevaskis et al. 1997; Hunt et al. 2001). The affinity of trHbs for O_2 is higher than that of vertebrate haemoglobin and myoglobins but lower than that of cytochrome oxidase from plants (Watts et al. 2001).

2.4 Structural aspects

In any haem protein, the structure of the haem pocket plays a crucial role in controlling protein function (for review, see Gow et al. 2005). The haem of ferrous and ferric forms of the nsHb is a low spin 6-coordinate, which differs markedly from traditional pentacoordinate haemoglobins (Arredondo-Peter et al. 1997; Duff et al. 1997; Goodman and Hargrove 2001; Perazzolli et al. 2004; Sáenz-Rivera et al. 2004). Hexacoordinate haemoglobins have been found in bacteria, protozoa and animals as well as in plants, suggesting that 6-coordinate haemoglobins are ubiquitous in living organism (Kundu et al. 2003). The distal histidine in the haem pocket forms a strong hydrogen bonding with bound O_2, which contributes to the slow dissociation constant (Arredondo-Peter et al. 1997; Duff et al. 1997; Das et al. 1999). The 6-coordinate conformation of nsHb may suggest a role in electron transfer to haem iron and bound ligand (Duff et al. 1997; Weiland et al. 2004).

2.5 Reaction of nsHbs with NO

It has been well documented that haemoglobins react with NO. FlavoHbs that are found in a variety of prokaryotic and eukaryotic microorganisms possess NO dioxygenase (Gardner et al. 1998; Hausladen et al. 1998) or denitrosylase activities (Hausladen et al. 2001) to generate nitrate aerobically, and a slower NO reductase activity to produce nitrous oxide (N_2O) under anaerobic conditions (Hausladen et al. 1998; Kim et al. 1999). The flavoHb consist of a haemoglobin domain and a flavin-containing reductase domain, which is distinct from nsHb which does not posses a flavin domain (Gardner 2005). The reduction of ferric haem iron back to the ferrous form is critical for maximal catalytic turnover of the enzyme. The C-terminal reductase domain binds NAD(P)H and transfers electrons to ferric haem iron via bound FAD. By metabolizing NO, flavoHbs may prevent NO-mediated modification of thiol and metal centres of other critical enzymes and regulatory proteins. In addition, it may decrease the formation of the highly reactive peroxynitrite, thereby preventing the irreversible oxidation and nitration of proteins or lipids (for review, see Frey and Kallio 2003; Gardner 2005). When the haem and flavin domains were separately expressed in *Escherichia coli*, that haem domain in flavoHb could confer resistance to growing cells from NO donors, but failed to provide protection from NO to the respiratory chain (Hernández-Urzúa et al. 2003).

While most bacteria species express flavoHbs, some bacteria such as *Viteroscilla* sp. and *Campylobacter jejuni* express single-domain haemoglobins, which do not possess flavin domains (Frey and Kallio 2003). Expression of these single-domain haemoglobins also conferred resistance to the toxicity of NO, indicating that the C-terminal reductase domain is not obligatory for a NO degrading activity in vivo (Frey et al. 2002; Elvers et al. 2004). It was suggested that an external reductase system may reduce the haem iron to the biochemically active ferrous form in single-domain haemoglobins.

An NO metabolizing activity has also been described for class 1 nsHb (Fig. 1). NO consumption by class 1 nsHb exhibits characteristics of NO dioxygenase activity. The reaction proceeds aerobically (Igamberdiev et al. 2004) and either NADH or NADPH can be used as electron donors, with the NADH dependent rate being about 2–2.5 times faster than the NADPH rate (Igamberdiev et al. 2004; Perazzolli et al. 2004; Seregélyes et al. 2004). The activity was sensitive to diphenylene iodonium, an inhibitor of flavoproteins (Igamberdiev et al. 2004; Seregélyes et al. 2004), and nitrate was a reaction product (Igamberdiev et al. 2004; Perazzolli et al. 2004).

In *Arabidopsis*, AHb1 possesses a redox-active cysteine residue within the distal haem pocket, which is not present in flavoHbs (Gow et al. 2005). A cysteine-dependent mechanism of haemoglobin to metabolize NO has been reported for the parasitic nematode *Ascaris* (Minning et al. 1999). The cysteine residue near the ligand-binding site in *Ascaris* haemoglobin binds NO to give *S*-nitrosocysteine, which interacts with dioxygen bound to haem to form nitrate. Perazzolli et al. (2004) found that *S*-nitrosylated AHb1 is endogenously produced during the course of NO consumption, and suggested that AHb1 can scavenge NO through a similar cysteine-dependent mechanism as in *Ascaris*. On the other hand, Igamberdiev et al. (2005b) demonstrated that the mutation of the single cystein residue in barley nsHb has no effects on the NO metabolizing activity.

As for the single-domain haemoglobin in bacteria, NO consumption by nsHbs requires interaction with other molecules possessing methaemoglobin reduction activity (Seregélyes et al. 2004), and methaemoglobin reductase activity has indeed been reported in alfalfa root extracts (Igamberdiev et al. 2004). It was also suggested that ascorbate can accomplish this reduction (Igamberdiev et al. 2005b, c).

3 NsHb and NO in nitrate assimilation

3.1 Nitrate assimilation as a source for NO

NO emission from plants and cell suspensions is common when nitrate is the N-source, but not with ammonium (Wildt et al. 1997; Planchet et al. 2005). NO emission from cells or leaves during nitrate reduction is usually much higher under anoxia than in air (see below). This is at least partly due to a limitation by the tissue nitrite concentration (Morot-Gaudry-Talarmain et al. 2002; Rockel et al. 2002; Planchet et al. 2005), and because nitrate is a competitive inhibitor for NO formation, the nitrate/nitrite ratio has been suggested to be an important factor determining the rate of NO production in plants (Rockel et al. 2002; Vanin et al. 2004).

Originally, it was thought that nitrite- and NAD(P)H-dependent NO formation is exclusively catalysed by nitrate reductase (NR) itself (Dean and Harper 1988; Yamasaki and Sakihama 2000; Rockel et al. 2002). NO emission from tobacco leaves was completely suppressed by the application of tungstate, which, when replacing molybdate, blocks NR activity (Planchet et al. 2005). As NR-dependent NO formation was competitively inhibited by nitrate (Ki=50 μM), nitrite and nitrate appear reduced on NR at the same site (Kaiser et al. 2002). The posttranslational modulation of NR activity affected the NO production from nitrite in vitro and in vivo (Rockel et al. 2002), and tobacco plants with NR mutated in the regulatory Ser 521 had a light-dark-pattern of NO emission opposite to the wild type (Lea et al. 2004). However, NR (soluble or PM-bound) is not the only source for nitrite-derived NO. Even completely NR-free tobacco suspension cells produced NO when nitrite was added externally (Planchet et al. 2005). Inhibitors of mitochondrial electron transport blocked this NO formation (Modolo et al. 2005; Planchet et al. 2005), and it is now clear that purified mitochondria from roots, but probably not from leaves, are able to reduce nitrite to NO (Gupta et al. 2005). Other MoCo enzymes such as xanthine oxidase or aldehyde oxidase probably do not contribute to NO production from nitrite in plants (Modolo et al. 2005; Planchet et al. 2005), although these proteins can potentially produce NO from nitrate and nitrite (Harrisson 2002). In tobacco roots, NI-NOR-catalysed NO formation was described (Stöhr et al. 2001). This root-specific protein is associated with a root-specific, succinate-dependent PM-bound nitrate reductase (PM-NR), and reduces nitrite to NO with reduced cytochrome *c* as an electron donor, but not with NADH. NO produced by this enzyme in the apoplast may enter the cell, and was suggested to be involved in regulating nitrate assimilation (Stöhr and Ullrich 2002).

As mentioned above, plants may also produce NO independent of nitrite by NO-synthase. However, under normal conditions, the frequently used NOS inhibitors had no effect on NO production (Rockel et al. 2002; Sakihama et al. 2002; Vanin et al. 2004; Planchet et al. 2005).

3.2 A matter of speculation: Effects of NO on nitrate assimilation and related processes

While NO is undoubtedly a by-product of nitrate assimilation in higher plants, it is not clear whether it has a function in that overall process. We will only briefly discuss some (rather speculative) ideas on this aspect. Nitrite has been suggested to be involved in the nitrate assimilation as a regulatory signal (Loqué et al. 2003), and NO could act as an additional signal to control nitrate uptake and assimilation (Meyer and Stöhr 2002). In analogy to protein phosphorylation/dephosphorylation, NO may influence the nitrate assimilation pathway by transcriptional and post-transcriptional modulation of enzymes. Nitrite reductase-deficient transgenic tobacco with a continuous overproduction of NO showed increased protein tyrosine nitration, suggesting that high NO production resulted in increased peroxynitrite production (Morot-Gaudry-Talarmain et al. 2002). Peroxynitrite is known to be produced by the reaction of NO with superoxide (Arteel et al. 1999), and it was suggested that NR produces both NO and superoxide which subsequently generate peroxynitrite (Yamasaki and Sakihama 2000). The formation of peroxynitrite may contribute to the toxic effects of NO, possibly through inhibition of signal transduction by tyrosine nitration (Morot-Gaudry-Talarmain et al. 2002).

Enzyme activity may be regulated by NO through *S*-nitrosylation of cysteine residues. In animals, *S*-nitrosylation of cysteine appears involved in signal transduction (Stamler et al. 2001; Foster and Stamler 2004) and the level of *S*-nitrosylation is also regulated by enzymatic denitrosylation (Liu et al. 2001), which is analogous to regulation by protein phosphorylation/dephosphorylation. Denitrosylation activity has been reported for peroxidases (Abu-Soud and Hazen 2000) and glutathion-dependent formaldehyde dehydrogenases (Liu et al. 2001). These enzymes were also found in plants (Sakamoto et al. 2002; Díaz et al. 2003), suggesting that similar reactions are involved in signal transduction of plants. NO-mediated *S*-nitrosylation was suggested to regulate K^+ channels in guard cells of *Vicia faba* (Sokolovski and Blatt 2004). Glutamine synthetase (GS) was also found to be a target for *S*-nitrosylation when *Arabidopsis* leaves were exposed to NO gas (Lindermayr et al. 2005). GS catalyses the ATP-dependent condensation of ammonium ions with glutamate to produce glutamine (Hodges 2002). Glutamine itself is involved in the regulation of nitrate assimilatory enzymes, including nitrate reductase, nitrite reductase and GS (Sivasankar et al. 1997; Oliveira and Coruzzi 1999). Although the consequence of *S*-nitrosylation of GS on its activity has not been examined to date, this might represent an interesting feed-back loop in nitrate reduction and nitrogen assimilation.

Mammalian cytosolic and mitchondrial aconitase are sensitive to inactivation by NO (Gardner et al. 1997). Aconitase contains an Fe-S cluster which catalyses the reversible isomerization of citrate to isocitrate. In addition, the cytosolic isoform of aconitase functions as RNA binding protein that regulates uptake, sequestration, and utilization of iron in animals (for review, see Hentze and Kühn 1996). The RNA-binding property of cytosolic aconitase can be enhanced by NO with a concomitant loss of aconitase activity. In tobacco, it was demonstrated that the activity of both, cytosolic and mitochondrial aconitase was inhibited by NO (Navarre et al. 2000). Mitochondrial aconitase is part of the Krebs cycle, but the role of cytosolic aconitase is not well understood in plants. Isocitrate produced by catalytic reaction of aconitase is further converted by isocitrate dehydrogenase to 2-oxoglutarate, which is the substrate for the GS-GOGAT pathway (Hodges 2002). Both mitochondrial and cytosolic aconitases have been suggested to produce isocitrate (Gálvez et al. 1999; Abiko et al. 2005). In a tomato mutant having reduced levels of cytosolic and mitochondrial aconitase protein and activity, the levels of 2-oxoglutarate and glutamine were markedly reduced (Carrari et al. 2003). Thus, the inhibition of aconitase activity by NO may influence amino acid metabolism by limiting the supply of carbon skeletons for GS-GOGAT pathway as well as suppressing ATP production in mitochondria.

NO produced during nitrate reduction may not only affect the Krebs cycle, but may also inhibit mitochondrial electron transport and oxidative phosphorylation. There is evidence that NO inhibits the activity of the cytochrome pathway but not of the alternative oxidase (AOX) (Millar and Day 1996; Yamasaki et al. 2001; Huang et al. 2002; Zottini et al. 2002). Expression of the AOX gene is induced by NO (Huang et al. 2002). These changes in electron flow may reduce generation of reactive oxygen species (ROS), but result in lower ATP synthesis rates (Yamasaki et al. 2001). Thus NO-derived modification of mitochondrial functions may represent another feed-back loop to nitrate assimilation and the GS/GOGAT reactions.

3.3 Role of nsHbs in nitrate assimilation

Many genes, not only of the nitrate assimilation pathway, are rapidly induced by nitrate (Wang R et al. 2000, 2003; Wang YH et al. 2001). The coordinated changes of gene expression triggered by nitrate probably serve to prevent energy loss and accumulation of toxic metabolites. The most nitrate-responsive genes are those involved in nitrite reduction, but in addition class 1 nsHbs (*AHB1*) were among the most strongly nitrate-induced genes in *Arabidopsis* found by microarray analysis (Wang R et al. 2000). Using a nitrate reductase-null *Arabidopsis* mutant, Wang R et al. (2004) further demonstrated that the nitrate response of *AHB1* as well as of other key nitrate assimilatory genes is independent of nitrate reduction, and that nitrate itself is a signal. In rice cell cultures, the class 1 nsHb gene (*ORYsa GLB1a*) was rapidly induced by nitrate, nitrite and by NO without de novo protein synthesis in the cytoplasm (Ohwaki et al. 2005). In bacteria, flavoHb and a single-domain haemoglobin possessing NO detoxification activity were induced by nitrite or NO (Elvers et al. 2004; LaCelle et al. 1996; Poole et al. 1996). In addition to nitrate, nitrite and NO, plants may use cytokinin to induce nsHbs. Ross et al. (2004) found that a cytokinin-regulated transcription factor, ARR1, activates the *OsNSHB2* encoding class 1 nsHb in rice, and cytokinin significantly activates the rice *OsNSHB2* promoter. Cytokinin has also been reported to be involved in long distance communication of the nitrogen status in plants (for review, see Forde 2002; Sakakibara 2003). Cytokinin treatment rapidly stimulates NO release in tobacco cell cultures (Tun et al. 2001).

The rapid, nitrate-dependent induction of class-1 nsHb genes suggests a general role of nsHbs in nitrate assimilation and related processes. Moreover, the fact that nsHbs can be induced by nitrite and NO as well (Fig. 2; Sakamoto et al. 2004; Ohwaki et al. 2005; Shimoda et al. 2005) implies that nsHbs are involved in nitrite and NO detoxification (Igamberdiev et al. 2004; Perazzolli et al. 2004; Sregélyes et al. 2004). Our own work has shown that leaves of rice plants overproducing class 1 nsHb emitted less NO than WT when they were grown with nitrate (Ohwaki and Kaiser, unpublished). On the other hand, in a nsHb underexpressing line, aconitase activity was lower than in a nsHb overexpressing line (Igamberdiev et al. 2005c).

A catalytic activity of nsHbs in nitrite metabolism has also been described. In *Arabidopsis*, recombinant class 1 nsHb (AtGLB1), class 2 nsHb (AtGLB2) and trHb (AtGLB3) exhibited peroxidase-like activity to oxidize NO_2^- (Sakamoto et al. 2004). Among the three types of haemoglobins, class

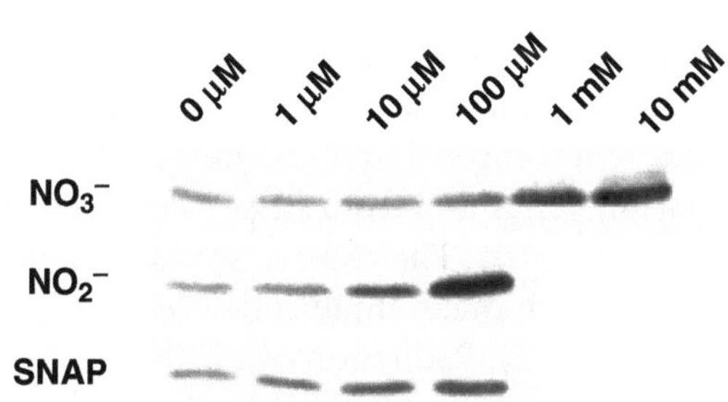

Fig. 2. Synthesis and accumulation of the class 1 nsHb protein in rice cells are induced by nitrate (NO_3^-), nitrite (NO_2^-) and nitric oxide (NO). Total cellular proteins, extracted from the cells treated at various concentrations of NO_3^-, NO_2^-, or with an NO donor (SNAP) were subjected to SDS-PAGE for Western blotting. The antibody for rice Hb1 was a gift from Dr. G. Sarath, University of Nebraska, Lincoln

1 nsHb was most effective in utilizing NO_2^- to generate reactive nitrogen species (RNS). Although the physiological relevance of these reactions is not yet clear, cellular detoxification of NO_2^- or involvement in cellular signalling via tyrosine nitration has been suggested. It was postulated that the peroxidase-like activity of nsHbs may be involved in the conversion of nitrite to nitrogen oxide (NO_2) and in the formation of a presently unidentified form of nitrogen in plants (Morikawa et al. 2005). Obviously, more intense transgenic approaches are required to unravel the consequences of nsHb expression on the entire process of nitrate assimilation.

4 NsHb and NO in anoxia tolerance

4.1 NO production under anoxia

As mentioned above, NO production by plant cells is much higher under anoxia than in air. For example, in tobacco or maize suspension cells or in alfalfa root cultures, NO production was hardly detectable under aerobic conditions, whereas rates were at least 2 orders of magnitude higher under hypoxic or anoxic conditions (Dordas et al. 2003, 2004; Planchet et al. 2005). In tobacco leaves grown with nitrate, NO emission under anoxia increased up to 1000-fold over air-levels, reaching 100 nmol g^{-1} FW h^{-1} (Planchet et al. 2005). As in air, NO formation under anoxia was directly or indirectly NR dependent (Planchet et al. 2005). Enhanced emission of NO under anoxia was associated with nitrite accumulation in leaves (Planchet et al. 2005), which is apparently due to two effects: i) activation of NR triggered by cellular acidification (Kaiser and Brendle-Behnisch 1995), and ii) inhibition (probably also by cytosolic acidification) of plastidic nitrite reduction (Botrel et al. 1996). While NR is undoubtedly able to reduce nitrite to NO, in air as well as under anoxia, mitochondrial electron transport is able to reduce nitrite to NO only under anoxia, but not in air (Gupta et al. 2005). This is apparently due to a competition of nitrite with O_2. Interestingly, only root mitochondria or mitochondria from suspension cells were able to produce NO, not leaf mitochondria. The reasons are as yet unknown. Based on inhibitor studies it was shown that both terminal oxidases participate in NO production (Gupta et al. 2005). Reduction of nitrite to NO by mitochondrial electron transport has also been reported for green algae (Tischner et al. 2004) and animals (Kozlov et al. 1999). In a recent report, it was indicated that plant mitochondria also contain NOS (*AtNOS1*) activity (Guo and Crawford 2005). However, involvement of NOS in NO formation under anoxia is improbable, because the reaction requires molecular oxygen.

4.2 Role of nsHbs in anoxia tolerance

One feature common to all class 1 nsHbs is their enhanced expression under anoxia. The induction of class 1 nsHb genes under anaerobic condition in plants was first reported for isolated barley aleurone layers and roots (Taylor et al. 1994). NsHb mRNA was not detectable in aleurone layers of barley at 20% and 10% O_2, was hardly visible at 5% O_2 and became easily detectable at 2% and 0% O_2. In roots of *Arabidopsis*, accumulation of the GLB1 (AHB1) protein was induced below 5% O_2, and was further increased in 0.1% O_2 (Hunt et al. 2002). The response of nsHb expression to oxygen limitation was observed after 1h of incubation under N_2, and required lower O_2 concentration than alcohol dehydrogenase (ADH) or lactate dehydrogenase (LDH) induction (Taylor et al. 1994). The induction of class 1 nsHbs under anoxia has also been reported in cotton *GhHb1* (Qu et al. 2005), *Lotus japonicus LjHb1* (Shimoda et al. 2005), wheat *TaHb1* (Larsen 2003), *Arabidopsis AtGLB1* (Yang et al. 2005) and alfalfa *Mhb1* (Seregélyes et al. 2000). The level of nsHb protein was increased under flooded conditions in roots of barley (Taylor et al. 1994) and in rice (Lira-Ruan et al. 2001). This pattern suggests a role for nsHb in hypoxia tolerance, and indeed enhanced nsHb expression improved the survival of hypoxia of *Arabidopsis* (Hunt et al. 2002).

In the aleuron layer of barley, not only anoxia, but also respiratory inhibitors including CO, cyanide, and antimycine A, as well as uncouplers of oxidative phosphorylation and ATP synthesis strongly elevated mRNA levels of nsHb (Nie and Hill 1997). The conclusion was that the expression of nsHb is not directly influenced by O_2 but by the availability of ATP. It was suggested that cytosolic Ca^{2+} is involved in anoxia-induced nsHb gene expression (Nie et al. 2005). Recently, it was demonstrated that expression of class 1 nsHb genes is also rapidly up-regulated by NO in *L. japonicus* plants (Shimoda et al. 2005) and in rice cells (Ohwaki et al. 2005). In bacteria, flavoHbs are also induced by anoxia, and this appears directly mediated by NO, involving nitrosylation of the O_2-response regulator FNR (Cruz-Ramos et al. 2002).

The role of nsHb in hypoxia tolerance has been examined with transgenic plants over- or underexpressing haemoglobin (Table 1). Maize cells overexpressing barley nsHb showed hardly any change of the energy status when they were grown under nitrogen for 12 h. In contrast, WT cells and cells underexpressing nsHb had much lower ATP levels (Sowa et al. 1998). Similarly, alfalfa root cultures overexpressing barley haemoglobin maintained root growth and ATP levels when incubated in 3% O_2, whereas growth and ATP levels of WT and underexpressing cultures were impaired (Dordas et al. 2003). High-affinity ligand binding is essential for class 1 nsHb

Table 1. Growth and stress response of transgenics to haemoglobin over-or-underexpression. *OE* overexpression, *UE* underexpression

Plants	Transgene	Type	Effects	References
Alfalfa root culture	Barley class 1 nsHb	OE	Hypoxia: high root growth, unaltered ATP levels, high ascorbate levels, low NO emission, high NO degradation activity, only little cell disintegration High activity of aconitase under normoxic and hypoxic condition	Dordas et al. (2003), Igamberdiev et al. (2004, 2005c)
		UE	Hypoxia: low root growth, declined ATP levels, high NO emission, low NO degradation activity, strong cell disintegration	
Arabidopsis	*Arabidopsis* class 1 nsHb	OE	Hypoxia: high survival and growth rate, no enhanced expression of glycolytic enzymes Normoxia: early enhanced growth	Hunt et al. (2002)
Arabidopsis	*Arabidopsis* class 1 nsHb	OE	Hypoxia: high growth, low NO emission Pathogen infection: unaltered NO accumulation and hypersensitive cell death	Perazzolli et al. (2004)
		UE	Hypoxia: low growth and high NO emission	

Arabidopsis	*Arabidopsis* class 1 nsHb	OE	Hypoxia: high survival rate, low H_2O_2 levels and ADH activity. Tolerance to H_2O_2 treatment	Yang et al. (2005)
		UE	Hypoxia: low survival rate, high H_2O_2 levels and ADH activity. Susceptible to H_2O_2 treatment	
Arabidopsis	*E. coli* flavoHb (*hmp*)	OE	Pathogen infection: Low NO emission, H_2O_2 and SA levels, and *Pal* expression, and diminished hypersensitive response (HR)	Zeier et al. (2004)
Maize cells	Barley class 1 nsHb	OE	Hypoxia: unaltered ATP levels, low ADH activity, low NO emission	Sowa et al. (1998), Dordas et al. (2004), Manac'h-Little et al. (2005)
		UE	Hypoxia: low ATP levels, high NO emission High ethylene formation under normoxic and hypoxic condition	
Tobacco plants	Alfalfa class 1 nsHb	OE	Pathogen infection: high NO degradation activity. Less necrosis, high ROS and SA production, elevated *PR-1a* expression	Seregélyes et al. (2003, 2004)

to improve survival of hypoxia. Replacement of a histidine residue at position E7 with leucine resulted in a lower affinity for oxygen than the WT protein, and transgenic *Arabidopsis* plants overexpressing leucine-substituted nsHb protein did not show improved survival under anoxia (Hunt et al. 2002). It was suggested that haemoglobin helps to maintain the energy status of cells in a low oxygen environment by promoting the glycolytic flux through NADH re-oxidation (Sowa et al. 1998). NADH re-oxidation appeared correlated with NO degradation by nsHb, and the overall rate was comparable to ADH activity. Thus, it was suggested that the NADH oxidation by nsHb could substitute for ADH activity and prevent formation of ethanol (Igamberdiev et al. 2004). Positive effects on hypoxia tolerance of *Arabidopsis* were observed when the *GLB1 (AtGLB1)* gene was overexpressed (Hunt et al. 2002; Perazzolli et al. 2004; Yang et al. 2005). Effects on glycolysis were less clear. An *Arabidopsis* line overexpressing nsHb had lower ADH activity than the WT or than an underexpressing line under hypoxia (Yang et al. 2005), which is consistent with other findings (Sowa et al. 1998; Dordas et al. 2003). On the other hand, it was postulated that changes in glycolytic flux may not be a direct consequence of nsHb overexpression, because expression of genes for glycolytic enzymes was not changed in nsHb overexpressing lines (Hunt et al. 2002).

A function of haemoglobin in hypoxia tolerance has been recently correlated with its NO consuming activity. The amount of NO accumulating in nsHb underexpressing alfalfa root cultures was 2.5-fold higher than that in the nsHb overexpressing line (Dordas et al. 2003), and NO levels under anoxia were inversely related to the nsHbs content of maize cell lines (Dordas et al. 2004). In *Arabidopsis*, overexpression of nsHb reduced NO emission from plants under hypoxic stress, whereas lines underexpressing nsHb showed enhanced emission of NO (Perazzolli et al. 2004). Extracts from nsHb overexpressing root cultures possessed twice the NO conversion rate than the extracts from WT or from lines underexpressing nsHb (Igamberdiev et al. 2004). All this is good evidence that nsHbs modulate the NO level in plant cells and contribute to survival under anoxia. Igamberdiev et al. (2005a) therefore proposed the "Hb/NO cycle" as a NAD(P)H- and Hb-dependent NO scavenging system, operative under hypoxic conditions, that would contribute to maintain cellular ATP levels. It is important to be aware, however, that this Hb/NO cycle will work only under hypoxic condition but not under anoxia, since NO scavenging by the Hb/NO cycle requires at least low levels of oxygen (Fig. 1).

Another function for nsHbs might be protection of plants from oxidative stress (Table 1). Overexpression of nsHbs improved the antioxidant system (Igamberdiev et al. 2005c, Yang et al. 2005). In nsHb underexpressing

Arabidopsis, the H_2O_2 content after hypoxic stress was higher than in WT plants, and the survival rate upon re-oxygenation after hypoxia was lower. In contrast, nsHb overexpressing lines had lower H_2O_2 content and higher survival rate than WT (Yang et al. 2005). These results are consistent with the finding that in plants challenged with avirulent *Pseudomonas syringae*, levels of H_2O_2 at the infection site were lower in plants overexpressing bacterial NO dioxygenase than in WT (Zeier et al. 2004). Biochemical studies indicated that class 1 nsHb possesses a peroxidase-like activity (Sakamoto et al. 2004). On the other hand, no effect of nsHb expression was found on H_2O_2 levels in alfalfa root cultures (Igamberdiev et al. 2005c).

Lysigenous aerenchyma formation in stems and roots is one form of programmed cell death (PCD) and represents an important adaptation to hypoxia. It provides an internal pathway facilitating gas diffusion between shoots and roots and improves oxygen supply to the roots (for review, see Drew et al. 2000). It is usually related to higher ethylene production, increased activity of 1-amino-cyclopropane-1-carboxylic acid (ACC) synthase, ACC oxidase, and cellulase (He et al. 1996). Nitric oxide and ethylene are thought to act antagonistically. Indeed, roots of nsHb-underexpressing alfalfa (which have higher NO levels) showed symptoms of cellular digration characteristic of PCD (Dordas et al. 2003). Maize cells underexpressing nsHb produced more ethylene than WT or nsHb overexpressors, which was associated with higher ACC oxidase activity than in WT cells or nsHb overexpressing cells (Manac'h-Little et al. 2005). These observations are certainly suggestive for an interaction of NO with ethylene formation and with ethylene controlled processes in plants.

5 NsHb and NO in plant-microbe interactions

A wealth of evidence supports the notion that plants produce NO upon challenge by pathogens, and that NO acts as a signal triggering defence reactions. Nevertheless, the sources of NO and the factors controlling NO levels upon infection are far from being understood, and even the above basic role of NO as such is not absolutely sure (Planchet et al. 2006). NOS-like activity was enhanced in resistant tobacco after infection with tobacco mosaic virus (Durner et al. 1998). NO production was drastically enhanced in soybean cells (Delledonne et al. 1998) or in tobacco plants (Mur et al. 2005) inoculated with an avirulent strain of *P. syringae*, and as this increase was inhibited by NOS inhibitors, NOS appeared to be the major NO source. Similarly, a transient NO production by epidermal tobacco cells was triggered by the fungal elicitor cryptogein, and this NO production was sensitive to NOS inhibitors (Foissner et al. 2000). More recently, it was demonstrated that lipopolysaccharides (LPS) as cell-surface components of Gram-negative bacteria could induce a NO burst in *Arabidopsis* suspension cells (Zeidler et al. 2004), which was again suppressed by NOS inhibitors. Using AtNOS1 mutants, the LPS-induced NO burst was shown to originate from AtNOS1 (Zeidler et al. 2004). On the other hand, Clarke et al. (2000) demonstrated that the stimulation of NO formation from *Arabidopsis* suspension cultures in response to an avirulent strain of *P. syringae* was insensitive to the application of NOS inhibitors. Modolo et al. (2005) indicated that although NOS activity increased substantially in *Arabidopsis* leaves inoculated with *P. syringae*, NO formation was dependent on nitrite and on mitochondrial electron transport rather than on L-arginine.

Cell death associated with the hypersensitive response (HR) is a typical reaction against infection by avirulent pathogens. Exogenously applied NO donors induced cell death in *Arabidopsis* cell suspensions (Clarke et al. 2000), and NO appeared integral to elicitation of cell death in tobacco infected by an avirulent strain of *P. syringae* (Mur et al. 2005). Delledonne et al. (1998) reported that NO donors potentiated the induction of cell death by ROS in soybean cell suspensions, and NOS inhibitors suppressed the HR caused by avirulent *P. syringae*. The HR was activated by interaction of NO with H_2O_2 generated from O_2^- by superoxide dismutase, and it was suggested that the HR requires a balance between NO and ROS (Delledonne et al. 2001). However, using an NO overproducing tobacco transformant a balanced coproduction of NO and ROS was not confirmed as being obligatory for the HR (Planchet et al. 2006), and other work suggested that NO itself is a sufficient signal to induce cell death via cGMP mediated pathway in *Arabidopsis* cell suspensions (Clarke et al. 2000). But occasionally it has been questioned that NO in general is an obligatory part of the signalling chain leading to the HR (Planchet et al. 2006).

Like NO production, haemoglobin expression is also responsive to pathogens. A rapid and transient induction of *GhHb1*, a class 1 nsHb gene in roots of cotton was observed following infection of *Verticillium dahliae*, a soil-borne fungus causing Verticillium wilt of cotton (Qu et al. 2005). Although the signals involved in the induction of nsHb in response to pathogen infection have not been examined to date, Ross et al. (2004) found a W-box seemingly involved in plant defence signalling, in the promoters of rice and *Arabidopsis* nsHbs. It might well be that NO or H_2O_2 produced by plants upon infection with pathogens are actually the triggers of nsHb expression (Ohwaki et al. 2005; Qu et al. 2005; Shimoda et al. 2005).

Transgenic tobacco overexpressing alfalfa nsHbs produced less necroses than WT plants when an NO donor, or avirulent *P. syringae*, or Tobacco Necrosis Virus were infiltrated into the leaves (Seregélyes et al. 2003, 2004). These results suggest that the above described NO-scavenging activity of nsHb can affect the HR in response to pathogen attack. Moreover, Yang et al. (2005) showed that leaves or protoplasts isolated from nsHb overexpressing *Arabidopsis* had lower rates of cell death when treated with H_2O_2 than leaves or protoplasts from WT or lines underexpressing nsHbs. However, as for NO (see above), the role of nsHbs in the HR is not unequivocal (Table 1). Overexpression of nsHb (AHb1) in *Arabidopsis* did not affect the NO-mediated HR in response to avirulent *P. syringae*, which may perhaps be attributed to the slow kinetics of NO detoxification by nsHbs (Perazzolli et al. 2004). These authors also suggested that AHb1 does not interfere with the above described "NO burst" in the HR. *Arabidopsis* plants overexpressing the *E. coli hmp* gene encoding a NO dioxygenase showed reduced NO emission at the site of infection with avirulent *P. syringae* compared with WT, and development of the HR was diminished (Zeier et al. 2004). Tobacco overproducing nsHb had elevated concentrations of ROS and SA and higher expression of PR-1a than WT after inoculation with avirulent *P. syringae* (Seregélyes et al. 2003, 2004). In contrast, levels of H_2O_2 and SA in plants expressing bacterial *hmp* genes after infection with avirulent *P. syringae* were lower than in WT, and induction of *Pal* and *PR-1* genes was suppressed in *hmp* plants (Zeier et al. 2004).

NsHbs may also be involved in symbiotic interactions between plants and microbes. Shimoda et al. (2005) showed that the expression of class 1 nsHb (*LjHb1*) in *L. japonicus* was enhanced at an early stage of infection with symbiotic rhizobium (*Mesorhizobium loti*), and it was suggested that nsHb are involved in controlling the early stage of the legume-rhizobium symbiosis by modulating NO levels. It was also indicated that *LjHb1* from *L. japonicus* (Shimoda et al. 2005; Uchiumi et al. 2002), and *MtTrHb1* and *MtTrHb2* encoding proteins homologous to trHbs from *Medicago truncatula* were expressed in root nodules (Vieweg et al. 2005). They suggested that these haemoglobins might be involved in the detoxification of NO produced in the nodules in order to maintain the activity of nitrogen fixation. In plant–mycorrhiza interactions, the expression of haemoglobin was repressed in the roots of *L. japonicus* and *Medicago sativa* colonized by *Glomus* sp., suggesting that mycorrhizal fungi control the expression of the nsHb genes during the establishment of symbiosis with plants (Uchiumi et al. 2002). Also, *MtTrHb2* was expressed in mycorrhizal roots of *M. truncatula* colonized by the mycorrhizal fungus *G. intraradices*, and may act as a scavenger for NO in arbusculated cells (Vieweg et al. 2005).

6 Conclusions

Enhanced NO emission from nsHb-underexpressing lines of various transgenic plants (Table 1) points to a potentially important role of nsHbs in NO scavenging, thereby contributing to a control of NO levels in vivo. High NO production affects many processes including energy production, antioxidant metabolism and biosynthesis of hormones (Table 1). The finding that nsHb proteins possess NO dioxygenase activity similar to bacterial flavoHbs supports the idea of an NO scavenging function of nsHbs. Most of the catalytic activity presented in nsHbs has been examined with recombinant protein. The native protein in vivo, however, can interact with other molecules (Goodman and Hargrove 2001), and these interactions may be important for the function of nsHbs. To date, no organism appears to posses both, hexacoordinate haemoglobin and flavoHbs, which may indicate that the two Hb types have a similar function in NO scavenging (Kundu et al. 2003).

Whether the reaction of nsHb with NO fulfils as yet other purposes than simply scavenging a toxic by-product is not clear yet. It seems likely that the NO deoxygenase reaction of nsHbs at the same time enhances O_2 and NAD(P)H consumption. This may serve as an alternate pathway for fermentation under hypoxic stress (Igamberdiev and Hill 2004). Although it has not yet been demonstrated in plants, the O_2 consumption associated with NO degradation of nsHbs might also help to avoid oxygen toxicity, as suggested for the parasitic nematode *Ascaris* (Minning et al. 1999). These reactions may be specifically important during transients from hypoxia/anoxia back to normoxia. Indeed, one of the key features of nsHb-overexpressing plants is that they have an enhanced tolerance to hypoxic stresses (Table 1). It seems likely that this tolerance is not only important for flooding survival, but also critical for germination. NsHb expression is increased during seed germination, and overexpression of nsHb improved early growth in normal atmospheric conditions, possibly through enhanced tolerance of a localized, transient hypoxia during germination (Duff et al. 1998; Hunt et al. 2001, 2002; Lira-Ruan et al. 2001; Ross et al. 2001).

In plant/pathogen interactions, regulation of NO levels through interaction with nsHbs may affect NO toxicity for both, pathogens and hosts cells, and may also interact with NO signalling during the HR. If a basic action of nsHbs is to decrease NO levels, anti-pathogenic effects including induction of the HR should be weakened. Available data (see Table 1) are partly controversial, probably due to insufficient methods for determining rates of NO production and NO concentrations in tissues and cells. Obviously, more quantitative approaches for NO determination are required to fully elucidate the concerted action of nsHb and NO in plants.

Acknowledgements

Y.O. was recipient of fellowships by OECD Co-operative Research Programme: Biological Resource Management for Sustainable Agriculture Systems. W.K. was supported by the DFG, SFB 567 and Ka 456-15/1-3.

References

Abiko T, Obara M, Ushioda A, Hayakawa T, Hodges M, Yamaya T (2005) Localization of NAD-isocitrate dehydrogenase and glutamate dehydrogenase in rice roots: candidates for providing carbon skeletons to NADH-glutamate synthase. Plant Cell Physiol 46:1724–1734

Abu-Soud HM, Hazen SL (2000) Nitric oxide is a physiological substrate for mammalian peroxidases. J Biol Chem 275:37524–37532

Alderton WK, Cooper CE, Knowles RG (2001) Nitric oxide synthase: structure, function and inhibition. Biochem. J. 357:593–615

Andersson CR, Jensen EO, Llewellyn DJ, Dennis ES, Peacock WJ (1996) A new hemoglobin gene from soybean: a role for hemoglobin in all plants. Proc Natl Acad Sci USA 93:5682–5687

Appleby CA, Bogusz D, Dennis ES, Peacock WJ (1998) A role for haemoglobin in all plant roots? Plant Cell Environment 11:359–367

Aréchaga-Ocampo E, Saenz-Rivera J, Sarath G, Klucas RV, Arredondo-Peter R (2001) Cloning and expression analysis of hemoglobin genes from maize (*Zea mays* ssp. *mays*) and teosinte (*Zea mays* ssp. *parviglumis*). Biochim Biophys Acta 1522:1–8

Arredondo-Peter R, Hargrove MS, Sarath G, Moran JF, Lohrman J, Olson JS, Klucas RV (1997) Rice hemoglobins. Gene cloning, analysis, and O_2-binding kinetics of a recombinant protein synthesized in *Escherichia coli*. Plant Physiol 15:1259–1266

Arteel GE, Briviba K, Sies H (1999) Protection against peroxynitrite. FEBS Lett 445:226-230

Beligni MV, Lamattina L (2000) Nitric oxide stimulates seed germination and de-etiolation, and inhibits hypocotyl elongation, three light-inducible responses in plants. Planta 210:215–221

Bogusz D, Appleby CA, Landsmann J, Dennis ES, Trinick MJ, Peacock WJ (1988) Functioning haemoglobin genes in non-nodulating plants. Nature 331:178–180

Bogusz D, Llewellyn DJ, Craig S, Dennis ES, Appleby CA, Peacock WJ (1990) Nonlegume hemoglobin genes retain organ-specific expression in heterologous transgenic plants. Plant Cell 2:633–641

Botrel A, Magné C, Kaiser WM (1996) Nitrate reduction, nitrite reduction and ammonium assimilation in barley roots in response to anoxia. Plant Physiol Biochem 34:645–652

Carrari F, Nunes-Nesi A, Gibon Y, Lytovchenko A, Loureiro ME, Fernie AR (2003) Reduced expression of aconitase results in an enhanced rate of photosynthesis and marked shifts in carbon partitioning in illuminated leaves of wild species tomato. Plant Physiol 133:1322–1335

Cassanova N, O'Brien KM, Stahl BT, McClure T, Poyton RO (2005) Yeast flavohemoglobin, a nitric oxide oxidoreductase, is located in both the cytosol and the mitochondrial matrix: effects of respiration, anoxia, and the mitochondrial genome on its intracellular level and distribution. J Biol Chem 280:7645–7653

Clarke A, Desikan R, Hurst RD, Hancock JT, Neill SJ (2000) NO way back: nitric oxide and programmed cell death in *Arabidopsis thaliana* suspension cultures. Plant J 24:667–677

Corpas FJ, Barroso JB, Carreras A, Quirós M, León AM, Romero-Puertas MC, Esteban FJ, Valderrama R, Palma JM, Sandalio LM, Gómez M, del Río LA (2004) Cellular and subcellular localization of endogenous nitric oxide in young and senescent pea plants. Plant Physiol 136:2722–2733

Cruz-Ramos H, Crack J, Wu G, Hughes MN, Scott C, Thomson AJ, Green J, Poole RK (2002) NO sensing by FNR: regulation of the *Escherichia coli* NO-detoxifying flavohaemoglobin, Hmp. EMBO J 21:3235–3244

Das TK, Lee HC, Duff SMG, Hill RD, Peisach J, Rousseau DL, Wittenberg BA, Wittenberg JB (1999) The heme environment in barley hemoglobin. J Biol Chem 274:4207–4212

Dean JV, Harper JE (1988) The conversion of nitrite to nitrogen oxide(s) by the constitutive NAD(P)H-nitrate reductase enzyme from soybean. Plant Physiol 88:389–395

Delledonne M, Xia Y, Dixon RA, Lamb C (1998) Nitric oxide functions as a signal in plant disease resistance. Nature 394:585–588

Delledonne M, Zeier J, Marocco A, Lamb C (2001) Signal interactions between nitric oxide and reactive oxygen intermediates in the plant hypersensitive disease resistance response. Proc Natl Acad Sci USA 98:13454–13459

Díaz M, Achkor H, Titarenko E, Martínez MC (2003) The gene encoding glutathione-dependent formaldehyde dehydrogenase/GSNO reductase is responsive to wounding, jasmonic acid and salicylic acid. FEBS Lett 543:136–139

Dordas C, Hasinoff BB, Igamberdiev AU, Manac'h N, Rivoal J, Hill RD (2003) Expression of a stress-induced hemoglobin affects NO levels produced by alfalfa root cultures under hypoxic stress. Plant J 35:763–770

Dordas C, Hasinoff BB, Rivoal J, Hill RD (2004) Class-1 hemoglobins, nitrate and NO levels in anoxic maize cell-suspension cultures. Planta 219:66–72

Drew MC, He CJ, Morgan PW (2000) Programmed cell death and aerenchyma formation in roots. Trends Plant Sci 5:123–127

Duff SMG, Guy PA, Nie X, Durnin DC, Hill RD (1998) Haemoglobin expression in germinating barley. Seed Sci Res 8:431–436

Duff SMG, Wittenberg JB, Hill RD (1997) Expression, purification, and properties of recombinant barley (*Hordeum* sp.) hemoglobin. Optical spectra and reactions with gaseous ligands. J Biol Chem 272:16746–16752

Durner J, Wendehenne D, Klessig DF (1998) Defense gene induction in tobacco by nitric oxide, cyclic GMP, and cyclic ADP-ribose. Proc Natl Acad Sci USA 95:10328–10333

Elvers KT, Wu G, Gilberthorpe NJ, Poole RK, Park SF (2004) Role of an inducible single-domain hemoglobin in mediating resistance to nitric oxide and nitrosative stress in *Campylobacter jejuni* and *Campylobacter coli*. J Bacteriol 186:5332–5341

Foissner I, Wendehenne D, Langebartels C, Durner J (2000) In vivo imaging of an elicitor-induced nitric oxide burst in tobacco. Plant J 23:817–824

Forde BG (2002) The role of long-distance signalling in plant responses to nitrate and other nutrients. J Exp Bot 53:39–43

Foster MW, Stamler JS (2004) New insights into protein *S*-nitrosylation. Mitochondria as a model system. J Biol Chem 279:25891–25897

Frey AD, Kallio PT (2003) Bacterial hemoglobins and flavohemoglobins: versatile proteins and their impact on microbiology and biotechnology. FEMS Microbiol Rev 27:525–545

Frey AD, Farrés J, Bollinger CJ, Kallio PT (2002) Bacterial hemoglobins and flavohemoglobins for alleviation of nitrosative stress in *Escherichia coli*. Appl Environ Microbiol 68:4835–4840

Gálvez S, Lancien M, Hodges M (1999) Are isocitrate dehydrogenases and 2-oxoglutarate involved in the regulation of glutamate synthesis? Trends Plant Sci 4:484–490

Gardner PR (2005) Nitric oxide dioxygenase function and mechanism of flavohemoglobin, hemoglobin, myoglobin and their associated reductases. J Inorg Biochem 99:247–266

Gardner PR, Costantino G, Szabó C, Salzman AL (1997) Nitric oxide sensitivity of the aconitases. J Biol Chem 272:25071–25076

Gardner PR, Gardner AM, Martin LA, Salzman AL (1998) Nitric oxide dioxygenase: an enzymic function for flavohemoglobin. Proc Natl Acad Sci USA 95:10378–10383

Goodman MD, Hargrove MS (2001) Quaternary structure of rice nonsymbiotic hemoglobin. J Biol Chem 276:6834–6839
Gow AJ, Payson AP, Bonaventura J (2005) Invertebrate hemoglobins and nitric oxide: how heme pocket structure controls reactivity. J Inorg Biochem 99:903–911
Guo FQ, Crawford NM (2005) *Arabidopsis* nitric oxide synthase1 is targeted to mitochondria and protects against oxidative damage and dark-induced senescence. Plant Cell 17:3436–3450
Guo FQ, Okamoto M, Crawford NM (2003) Identification of a plant nitric oxide synthase gene involved in hormonal signaling. Science 302:100–103
Gupta KJ, Stoimenova M, Kaiser WM (2005) In higher plants, only root mitochondria, but not leaf mitochondria reduce nitrite to NO, in vitro and in situ. J Exp Bot 56:2601–2609
Hargrove MS, Brucker EA, Stec B, Sarath G, Arredondo-Peter R, Klucas RV, Olson JS, Phillips GN, Jr. (2000) Crystal structure of a nonsymbiotic plant hemoglobin. Structure 8:1005–1014
Harrison R (2002) Structure and function of xanthine oxidoreductase: where are we now? Free Radic Biol Med 33:774–797
Hausladen A, Gow AJ, Stamler JS (1998) Nitrosative stress: metabolic pathway involving the flavohemoglobin. Proc Natl Acad Sci USA 95:14100–14105
Hausladen A, Gow A, Stamler JS (2001) Flavohemoglobin denitrosylase catalyzes the reaction of a nitroxyl equivalent with molecular oxygen. Proc Natl Acad Sci USA 98:10108–10112
He C, Finlayson SA, Drew MC, Jordan WR, Morgan PW (1996) Ethylene biosynthesis during aerenchyma formation in roots of maize subjected to mechanical impedance and hypoxia. Plant Physiol 112:1679–1685
He Y, Tang RH, Hao Y, Stevens RD, Cook CW, Ahn SM, Jing L, Yang Z, Chen L, Guo F, Fiorani F, Jackson RB, Crawford NM, Pei ZM (2004) Nitric oxide represses the *Arabidopsis* floral transition. Science 305:1968–1971
Hendriks T, Scheer I, Quillet MC, Randoux B, Delbreil B, Vasseur J, Hilbert JL (1998) A non-symbiotic hemoglobin gene is expressed during somatic embryogenesis in *Cichorium*. Biochim Biophys Acta 1443:193–197
Hentze MW, Kühn LC (1996) Molecular control of vertebrate iron metabolism: mRNA-based regulatory circuits operated by iron, nitric oxide, and oxidative stress. Proc Natl Acad Sci USA 93:8175–8182
Hernández-Urzúa E, Mills CE, White GP, Contreras-Zentella ML, Escamilla E, Vasudevan SG, Membrillo-Hernández J, Poole RK (2003) Flavohemoglobin Hmp, but not its individual domains, confers protection from respiratory inhibition by nitric oxide in *Escherichia coli*. J Biol Chem 278:34975–34982
Hodges M (2002) Enzyme redundancy and the importance of 2-oxoglutarate in plant ammonium assimilation. J Exp Bot 53:905–916
Huang X, von Rad U, Durner J (2002) Nitric oxide induces transcriptional activation of the nitric oxide-tolerant alternative oxidase in *Arabidopsis* suspension cells. Planta 215:914–923
Hunt PW, Watts RA, Trevaskis B, Llewelyn DJ, Burnell J, Dennis ES, Peacock WJ (2001) Expression and evolution of functionally distinct haemoglobin genes in plants. Plant Mol Biol 47:677–692
Hunt PW, Klok EJ, Trevaskis B, Watts RA, Ellis MH, Peacock WJ, Dennis ES (2002) Increased level of hemoglobin 1 enhances survival of hypoxic stress and promotes early growth in *Arabidopsis thaliana*. Proc Natl Acad Sci USA 99:17197–17202
Igamberdiev AU, Hill RD (2004) Nitrate, NO and haemoglobin in plant adaptation to hypoxia: an alternative to classic fermentation pathways. J Exp Bot 55:2473–2482
Igamberdiev AU, Seregélyes C, Manac'h N, Hill RD (2004) NADH-dependent metabolism of nitric oxide in alfalfa root cultures expressing barley hemoglobin. Planta 219:95–102
Igamberdiev AU, Baron K, Manac'h-Little N, Stoimenova M, Hill RD (2005a) The haemoglobin/nitric oxide cycle: involvement in flooding stress and effects on hormone signalling. Ann Bot 96:557–564

Igamberdiev AU, Bykova NV, Hill RD (2005b) Nitric oxide scavenging by barley hemoglobin is facilitated by a monodehydroascorbate reductase-mediated ascorbate reduction of methemoglobin. Planta (published on line Dec 8):1–8

Igamberdiev AU, Stoimenova M, Seregélyes C, Hill RD (2005c) Class-1 hemoglobin and antioxidant metabolism in alfalfa roots. Planta (published on line Nov 12):1–6

Kaiser WM, Brendle-Behnisch E (1995) Acid-base-modulation of nitrate reductase in leaf tissues. Planta 196:1–6

Kaiser WM, Weiner H, Kandlbinder A, Tsai CB, Rockel P, Sonoda M, Planchet E (2002) Modulation of nitrate reductase: some new insights, an unusual case and a potentially important side reaction. J Exp Bot 53:875–882

Kim SO, Orii Y, Lloyd D, Hughes MN, Poole RK (1999) Anoxic function for the *Escherichia coli* flavohaemoglobin (Hmp): reversible binding of nitric oxide and reduction to nitrous oxide. FEBS Lett 445:389–394

Kozlov AV, Staniek K, Nohl H (1999) Nitrite reductase activity is a novel function of mammalian mitochondria. FEBS Lett 454:127–130

Kundu S, Trent JT, 3rd, Hargrove MS (2003) Plants, humans and hemoglobins. Trends Plant Sci 8: 387–393

LaCelle M, Kumano M, Kurita K, Yamane K, Zuber P, Nakano MM (1996) Oxygen-controlled regulation of the flavohemoglobin gene in *Bacillus subtilis.* J Bacteriol 178:3803–3808

Larsen K (2003) Molecular cloning and characterization of cDNAs encoding hemoglobin from wheat (*Triticum aestivum*) and potato (*Solanum tuberosum*). Biochim Biophys Acta 1621:299–305

Lea US, Ten Hoopen F, Provan F, Kaiser WM, Meyer C, Lillo C (2004) Mutation of the regulatory phosphorylation site of tobacco nitrate reductase results in high nitrite excretion and NO emission from leaf and root tissue. Planta 219:59–65

Lindermayr C, Saalbach G, Durner J (2005) Proteomic identification of *S*-nitrosylated proteins in Arabidopsis. Plant Physiol 137:921–930

Lira-Ruan V, Sarath G, Klucas RV, Arredondo-Peter R (2001) Synthesis of hemoglobins in rice (*Oryza sativa* var. Jackson) plants growing in normal and stress conditions. Plant Sci 161:279–287

Liu L, Hausladen A, Zeng M, Que L, Heitman J, Stamler JS (2001) A metabolic enzyme for *S*-nitrosothiol conserved from bacteria to humans. Nature 410:490–494

Loqué D, Tillard P, Gojon A, Lepetit M (2003) Gene expression of the NO_3^- transporter NRT1.1 and the nitrate reductase NIA1 is repressed in *Arabidopsis* roots by NO_2^-, the product of NO_3^- reduction. Plant Physiol 132:958–967

Manac'h-Little N, Igamberdiev AU, Hill RD (2005) Hemoglobin expression affects ethylene production in maize cell cultures. Plant Physiol Biochem 43:485–489

Meyer C, Stöhr C (2002) Soluble and plasma membrane-bound enzymes involved in nitrate and nitrite metabolism. In: Foyer CH, Noctor G (eds) Photosynthetic nitrogen assimilation and associated carbon and respiratory metabolism. Kluwer Academic Publishers, Dordrecht, pp 49–62

Milani M, Pesce A, Nardini M, Ouellet H, Ouellet Y, Dewilde S, Bocedi A, Ascenzi P, Guertin M, Moens L, Friedman JM, Wittenberg JB, Bolognesi M (2005) Structural bases for heme binding and diatomic ligand recognition in truncated hemoglobins. J Inorg Biochem 99:97–109

Millar AH, Day DA (1996) Nitric oxide inhibits the cytochrome oxidase but not the alternative oxidase of plant mitochondria. FEBS Lett 398:155–158

Minning DM, Gow AJ, Bonaventura J, Braun R, Dewhirst M, Goldberg DE, Stamler JS (1999) *Ascaris* haemoglobin is a nitric oxide-activated "deoxygenase". Nature 401: 497–502

Modolo LV, Augusto O, Almeida IMG, Magalhaes JR, Salgado I (2005) Nitrite as the major source of nitric oxide production by *Arabidopsis thaliana* in response to *Pseudomonas syringae.* FEBS Lett 579:3814–3820

Morikawa H, Takahashi M, Sakamoto A, Ueda-Hashimoto M, Matsubara T, Miyawaki K, Kawamura Y, Hirata T, Suzuki H (2005) Novel metabolism of nitrogen in plants. Z Naturforsch 60:265–271

Morot-Gaudry-Talarmain Y, Rockel P, Moureaux T, Quilleré I, Leydecker MT, Kaiser WM, Morot-Gaudry JF (2002) Nitrite accumulation and nitric oxide emission in relation to cellular signaling in nitrite reductase antisense tobacco. Planta 215:708–715

Mur LAJ, Santosa IE, Laarhoven LJJ, Holton NJ, Harren FJM, Smith AR (2005) Laser photoacoustic detection allows in planta detection of nitric oxide in tobacco following challenge with avirulent and virulent *Pseudomonas syringae* Pathovars. Plant Physiol 138:1247–1258

Navarre DA, Wendehenne D, Durner J, Noad R, Klessig DF (2000) Nitric oxide modulates the activity of tobacco aconitase. Plant Physiol 122:573–582

Neill SJ, Desikan R, Clarke A, Hurst RD, Hancock JT (2002) Hydrogen peroxide and nitric oxide as signalling molecules in plants. J Exp Bot 53:1237–1247

Nie X, Hill RD (1997) Mitochondrial respiration and hemoglobin gene expression in barley aleurone tissue. Plant Physiol 114:835–840

Nie X, Durnin DC, Igamberdiev AU, Hill RD (2005) Cytosolic calcium is involved in the regulation of barley hemoglobin gene expression. Planta (published on line Sep 22):1–8

Ohwaki Y, Kawagishi-Kobayashi M, Wakasa K, Fujihara S, Yoneyama T (2005) Induction of class-1 non-symbiotic hemoglobin genes by nitrate, nitrite and nitric oxide in cultured rice cells. Plant Cell Physiol 46:324–331

Oliveira IC, Coruzzi GM (1999) Carbon and amino acids reciprocally modulate the expression of glutamine synthetase in *Arabidopsis*. Plant Physiol 121:301–310

Perazzolli M, Dominici P, Romero-Puertas MC, Zago E, Zeier J, Sonoda M, Lamb C, Delledonne M (2004) *Arabidopsis* nonsymbiotic hemoglobin AHb1 modulates nitric oxide bioactivity. Plant Cell 16:2785–2794

Planchet E, Gupta KJ, Sonoda M, Kaiser WM (2005) Nitric oxide emission from tobacco leaves and cell suspensions: rate limiting factors and evidence for the involvement of mitochondrial electron transport. Plant J 41:732–743

Planchet E, Sonoda M, Zeier J, Kaiser WM (2006) Nitric oxide (NO) as an intermediate in the cryptogein-induced hypersensitive response—a critical re-evaluation. Plant Cell Environ 29:59–69

Poole RK, Anjum MF, Membrillo-Hernández J, Kim SO, Hughes MN, Stewart V (1996) Nitric oxide, nitrite, and Fnr regulation of *hmp* (flavohemoglobin) gene expression in *Escherichia coli* K-12. J Bacteriol 178:5487–5492

Qu ZL, Wang HY, Xia GX (2005) *GhHb1*: a nonsymbiotic hemoglobin gene of cotton responsive to infection by *Verticillium dahliae*. Biochim Biophys Acta 1730:103–113

Rockel P, Strube F, Rockel A, Wildt J, Kaiser WM (2002) Regulation of nitric oxide (NO) production by plant nitrate reductase in vivo and in vitro. J Exp Bot 53:103–110

Romero-Puertas MC, Perazzolli M, Zago ED, Delledonne M (2004) Nitric oxide signalling functions in plant-pathogen interactions. Cell Microbiol 6:795–803

Ross EJH, Shearman L, Mathiesen M, Zhou YJ, Arredondo-Peter R, Sarath G, Klucas RV (2001) Nonsymbiotic hemoglobins in rice are synthesized during germination and in differentiating cell types. Protoplasma 218:125–133

Ross EJH, Stone JM, Elowsky CG, Arredondo-Peter R, Klucas RV, Sarath G (2004) Activation of the *Oryza sativa* non-symbiotic haemoglobin-2 promoter by the cytokinin-regulated transcription factor, ARR1. J Exp Bot 55:1721–1731

Sáenz-Rivera J, Sarath G, Arredondo-Peter R (2004) Modeling the tertiary structure of a maize (*Zea mays* ssp. *mays*) non-symbiotic hemoglobin. Plant Physiol Biochem 42:891–897

Sakakibara H (2003) Nitrate-specific and cytokinin-mediated nitrogen signaling pathways in plants. J Plant Res 116:253–257

Sakamoto A, Ueda M, Morikawa H (2002) *Arabidopsis* glutathione-dependent formaldehyde dehydrogenase is an S-nitrosoglutathione reductase. FEBS Lett 515:20–24

Sakamoto A, Sakurao S, Fukunaga K, Matsubara T, Ueda-Hashimoto M, Tsukamoto S, Takahashi M, Morikawa H (2004) Three distinct *Arabidopsis* hemoglobins exhibit peroxidase-like activity and differentially mediate nitrite-dependent protein nitration. FEBS Lett 572:27–32

Sakihama Y, Nakamura S, Yamasaki H (2002) Nitric oxide production mediated by nitrate reductase in the green alga *Chlamydomonas reinhardtii*: an alternative NO production pathway in photosynthetic organisms. Plant Cell Physiol 43:290–297

Seregélyes C, Mustárdy L, Ayaydin F, Sass L, Kovács L, Endre G, Lukacs N, Kovács I, Vass I, Kiss GB, Horváth GV, Dudits D (2000) Nuclear localization of a hypoxia-inducible novel non-symbiotic hemoglobin in cultured alfalfa cells. FEBS Lett 482:125–130

Seregélyes C, Barna B, Hennig J, Konopka D, Pasternak TP, Lukács N, Fehér A, Horváth GV, Dudits D (2003) Phytoglobins can interfere with nitric oxide functions during plant growth and pathogenic responses: a transgenic approach. Plant Sci 165:541–550

Seregélyes C, Igamberdiev AU, Maassen A, Hennig J, Dudits D, Hill RD (2004) NO-degradation by alfalfa class 1 hemoglobin (Mhb1): a possible link to *PR-1a* gene expression in Mhb1-overproducing tobacco plants. FEBS Lett 571:61–66

Shimoda Y, Nagata M, Suzuki A, Abe M, Sato S, Kato T, Tabata S, Higashi S, Uchiumi T (2005) Symbiotic rhizobium and nitric oxide induce gene expression of non-symbiotic hemoglobin in *Lotus japonicus*. Plant Cell Physiol 46:99–107

Sivasankar S, Rothstein S, Oaks A (1997) Regulation of the accumulation and reduction of nitrate by nitrogen and carbon metabolites in maize seedlings. Plant Physiol 114:583–589

Sokolovski S, Blatt MR (2004) Nitric oxide block of outward-rectifying K^+ channels indicates direct control by protein nitrosylation in guard cells. Plant Physiol 136:4275–4284

Sowa AW, Duff SMG, Guy PA, Hill RD (1998) Altering hemoglobin levels change energy status in maize cells under hypoxia. Proc Natl Acad Sci USA 95:10317–10321

Stamler JS, Lamas S, Fang FC (2001) Nitrosylation: the prototypic redox-based signaling mechanism. Cell 106:675–683

Stöhr C, Ullrich WR (2002) Generation and possible roles of NO in plant roots and their apoplastic space. J Exp Bot 53:2293–2303

Stöhr C, Strube F, Marx G, Ullrich WR, Rockel P (2001) A plasma membrane-bound enzyme of tobacco roots catalyses the formation of nitric oxide from nitrite. Planta 212:835–841

Taylor ER, Nie XZ, MacGregor AW, Hill RD (1994) A cereal haemoglobin gene is expressed in seed and root tissues under anaerobic conditions. Plant Mol Biol 24:853–862

Tischner R, Planchet E, Kaiser WM (2004) Mitochondrial electron transport as a source for nitric oxide in the unicellular green alga *Chlorella sorokiniana*. FEBS Lett 576: 151–155

Trevaskis B, Watts RA, Andersson CR, Llewellyn DJ, Hargrove MS, Olson JS, Dennis ES, Peacock WJ (1997) Two hemoglobin genes in *Arabidopsis thaliana*: the evolutionary origins of leghemoglobins. Proc Natl Acad Sci USA 94:12230–12234

Tun NN, Holk A, Scherer GFE (2001) Rapid increase of NO release in plant cell cultures induced by cytokinin. FEBS Lett 509: 174–176

Uchiumi T, Shimoda Y, Tsuruta T, Mukoyoshi Y, Suzuki A, Senoo K, Sato S, Kato T, Tabata S, Higashi S, Abe M (2002) Expression of symbiotic and nonsymbiotic globin genes responding to microsymbionts on *Lotus japonicus*. Plant Cell Physiol 43:1351–1358

Vanin AF, Svistunenko DA, Mikoyan VD, Serezhenkov VA, Fryer MJ, Baker NR, Cooper CE (2004) Endogenous superoxide production and the nitrite/nitrate ratio control the concentration of bioavailable free nitric oxide in leaves. J Biol Chem 279:24100–24107

Vieweg MF, Hohnjec N, Küster H (2005) Two genes encoding different truncated hemoglobins are regulated during root nodule and arbuscular mycorrhiza symbioses of *Medicago truncatula*. Planta 220:757–766

Wang R, Guegler K, LaBrie ST, Crawford NM (2000) Genomic analysis of a nutrient response in Arabidopsis reveals diverse expression patterns and novel metabolic and potential regulatory genes induced by nitrate. Plant Cell 12:1491–1509

Wang R, Okamoto M, Xing X, Crawford NM (2003) Microarray analysis of the nitrate response in Arabidopsis roots and shoots reveals over 1,000 rapidly responding genes and new linkages to glucose, trehalose-6-phosphate, iron, and sulfate metabolism. Plant Physiol 132:556–567

Wang R, Tischner R, Gutiérrez RA, Hoffman M, Xing X, Chen M, Coruzzi G, Crawford NM (2004) Genomic analysis of the nitrate response using a nitrate reductase-null mutant of Arabidopsis. Plant Physiol 136: 2512–2522

Wang YH, Garvin DF, Kochian LV (2001) Nitrate-induced genes in tomato roots. Array analysis reveals novel genes that may play a role in nitrogen nutrition. Plant Physiol 127:345–359

Watts RA, Hunt PW, Hvitved AN, Hargrove MS, Peacock WJ, Dennis ES (2001) A hemoglobin from plants homologous to truncated hemoglobins of microorganisms. Proc Natl Acad Sci USA 98:10119–10124

Weiland TR, Kundu S, Trent JT, 3rd, Hoy JA, Hargrove MS (2004) Bis-histidyl hexacoordination in hemoglobins facilitates heme reduction kinetics. J Am Chem Soc 126:11930–11935

Wildt J, Kley D, Rockel A, Rockel P, Segschneider HJ (1997) Emission of NO from several higher plant species. J Geophysical Res 102:5919–5927

Wittenberg JB, Bolognesi M, Wittenberg BA, Guertin M (2002) Truncated hemoglobins: a new family of hemoglobins widely distributed in bacteria, unicellular eukaryotes, and plants. J Biol Chem 277:871–874

Yamasaki H, Sakihama Y (2000) Simultaneous production of nitric oxide and peroxynitrite by plant nitrate reductase: in vitro evidence for the NR-dependent formation of active nitrogen species. FEBS Lett 468:89–92

Yamasaki H, Shimoji H, Ohshiro Y, Sakihama Y (2001) Inhibitory effects of nitric oxide on oxidative phosphorylation in plant mitochondria. Nitric Oxide 5:261–270

Yang LX, Wang RY, Ren F, Liu J, Cheng J, Lu YT (2005) AtGLB1 enhances the tolerance of *Arabidopsis* to hydrogen peroxide stress. Plant Cell Physiol 46:1309–1316

Zeidler D, Zähringer U, Gerber I, Dubery I, Hartung T, Bors W, Hutzler P, Durner J (2004) Innate immunity in *Arabidopsis thaliana*: Lipopolysaccharides activate nitric oxide synthase (NOS) and induce defense genes. Proc Natl Acad Sci USA 101:15811–15816

Zeier J, Delledonne M, Mishina T, Severi E, Sonoda M, Lamb C (2004) Genetic elucidation of nitric oxide signaling in incompatible plant-pathogen interactions. Plant Physiol 136:2875–2886

Zottini M, Formentin E, Scattolin M, Carimi F, Lo Schiavo F, Terzi M (2002) Nitric oxide affects plant mitochondrial functionality in vivo. FEBS Lett 515:75–78

Yoshinari Ohwaki
National Agricultural Research Center
Tsukuba, Ibaraki 305-8666 Japan
e-mail: ohwaki@affrc.go.jp

Werner M. Kaiser
Julius-von-Sachs-Institute for Biosciences
University of Würzburg
Julius-von-Sachs-Platz 2
D-97082 Würzburg
e-mail: kaiser@botanik.uni-wuerzburg.de

Living in Day-Night Cycles—Specific Diel Leaf Growth Patterns and the Circadian Control of Photomorphogenesis

Shizue Matsubara and Achim Walter

1 Introduction

For photoautotrophic organisms, light is an essential source of energy. Throughout the entire life cycle, plants continuously monitor this energy source and decide the timing of important developmental switches during their life cycle. Since light is highly variable, plants have evolved sensitive and sophisticated systems to "measure" and "interpret" light signals and at the same time acquired high plasticity in coping with light fluctuations. The spatial and temporal heterogeneity in the light environment induces numerous responses mainly in the aerial parts of plants (Björkman 1981): state transition and thermal energy dissipation in the thylakoid membranes of chloroplasts (Demmig-Adams and Adams 1992; Allen and Nilsson 1997; Niyogi 1999; Wollman 2001), ontogenic modification in sun and shade leaves of some plant species (Sack et al. 2003; Terashima et al. 2006) or adjustment of canopy architecture (Terashima and Hikosaka 1995; Evans and Poorter 2001; Frak et al. 2002; Niinemets et al. 2004). Such phenotypic plasticity is vital for plants to achieve efficient photosynthesis, resource allocation and biomass production under the ever-changing light environment typical in many habitats (Schurr et al. 2006).

Responses to changing light regimes, such as formation of sun and shade leaves or adjustment of canopy structure, take place in the course of growth and have long been studied in terms of photomorphogenesis. Most prominent acclimation processes occur when plants develop in the dark or in shade, leading to "etiolation" or more generally termed "shade avoidance syndrome", characterized by morphological and developmental responses such as hypocotyl and petiole elongation, reduction in leaf area or acceleration of flowering. Changes in organ morphology and developmental transitions are among the most dramatic and visible phenotypes that originate from photomorphogenesis-related gene expression. In response to a multitude of external signals, proper outputs are elicited and are coordinated

Progress in Botany, Vol. 68

with endogenous growth and developmental programmes through dynamic regulatory networks (Walter and Schurr 2005; Schurr et al. 2006). Unravelling intrinsic regulatory networks that cause phenotypic variations when triggered by certain external stimuli and elucidating the mechanisms by which these phenotypic variations are suppressed under "normal" conditions are central focuses in plant developmental biology. Towards this goal, it is fundamental to analyse the processes causing a specific phenotype at time scales and spatial resolutions that are relevant for the dynamics of the signal transduction system triggered by internal and/or external cues.

Leaf growth is highly dynamic. It varies spatially across the lamina (Avery 1933; Maksymowych 1973) and shows temporal oscillations of different frequencies. Significant progress in visualization and quantification of these dynamic growth processes has been made by using digital image sequence processing (DISP) approaches (Schmundt et al. 1998; Walter and Schurr 2005). A hallmark of leaf growth dynamics is the pronounced day-night rhythmicity that shows distinct time courses in different species and is maintained for several days even if plants are kept under continuous illumination (Price et al. 2001; Walter and Schurr 2005). This diel rhythmicity of leaf growth contrasts with the rather stable activity of root growth under the same constant conditions (Walter et al. 2002b; Walter and Schurr 2005; Schurr et al. 2006). Recently, the molecular mechanisms giving rise to diel growth patterns have been studied in leaves of *Populus deltoides* Bartr. ex. Marsh in an attempt to clarify the endogenous control of leaf growth dynamics (Matsubara et al. 2006).

The periodicity of roughly 24 h (i.e. circadian) and its persistency under continuous illumination suggest the control of leaf growth by the endogenous clock by which diverse organisms anticipate the day-night cycle on Earth and synchronize internal events with it. Many metabolic, physiological and developmental processes in plants are controlled by circadian clocks (Somers 1999; McClung 2001; Staiger 2002; Lüttge 2003; Schultz and Kay 2003; Millar 2004; Misquitta and Herrin 2005; Mittag et al. 2005). Resonating the timing of these circadian-regulated events with the external light-dark cycles increases competitive advantage and reproductive fitness of photosynthetic organisms (Ouyang et al. 1998; Green et al. 2002; Dodd et al. 2005). Plants regularly adjust (or "entrain") their endogenous clocks to the period of environmental cycles by using light as the primary cue (Somers et al 1998a; Somers 1999; McClung 2001; Frankhauser and Staiger 2002; Quail 2002a,b; Más et al. 2003a; Millar 2004). This light-signal input to circadian clocks engages photosensory proteins (photoreceptors).

The photoreceptors provide information about the intensity, colour, directionality and diurnal duration of light to induce specific responses in growth and development of plants, including photomorphogenesis (Neff et al. 2000; Smith 2000; Quail 2002a,b; Liscum et al. 2003; Schultz and Kay 2003). Our understanding of the molecular mechanisms of light-signal transduction in the control of de-etiolation, hypocotyl elongation and flowering has been greatly advanced in the last years (Schaffer et al. 1998; Dowson-Day and Millar 1999; Morelli and Ruberti 2000; Hayama and Coupland 2003; Más et al. 2003a; Franklin and Whitelam 2005; Vandenbussche et al. 2005). By comparison, the molecular mechanisms of light acclimation in leaf growth and canopy development remain elusive even though marked progress has been achieved for understanding the control of individual processes involved, such as cell division and cell elongation (Van Volkenburgh 1999; Cosgrove 2000; den Boer and Murray 2000; Stals and Inzé 2001; Menges et al. 2002; Tsukaya 2002, 2003; Beemster et al. 2005).

In this review, we focus on the molecular mechanisms of endogenous (circadian clocks) and environmental control of diel leaf growth dynamics, wherein light is regarded as a key environmental factor. It has recently been shown that rapid cell elongation in shade avoidance, an output from light signal transduction, is gated by the endogenous clock in *Arabidopsis*, becoming arrested at subjective dawn and most pronounced at subjective dusk (Salter et al. 2003). Hence, we discuss the regulation of diel leaf growth cycles in the context of shade avoidance responses or plant photomorphogenesis, which is currently considered as a general model for the interactions of development with environment (Smith 1995, 2000; Pigliucci 1998; Pepper et al. 2002).

2 Endogenous patterns of leaf growth

Plant leaves display a variety of forms and structures. Yet, the basic construction of leaves is rather conservative. A typical leaf of dicotyledonous plants consists of a lamina (blade), which is supported by a network of veins, and a petiole (or petiolule for compound leaves). Most monocotyledonous plants, on the other hand, have no petiole and linear leaves with veins running in parallel along the leaf length direction. The leaf lamina usually contains several layers of palisade and spongy parenchyma cells where the bulk of chloroplasts performs photosynthesis. Above and below these photosynthetic cells, there is a layer of epidermal cells that are devoid of chloroplasts and covered with cuticles on the side of the leaf surface. Stomatal guard cells are distributed between the epidermal cells on both sides in some plants or on one side in others. Veins and petioles are part of the transport system, containing vascular tissues with xylem, phloem and sclerenchyma, but also confer mechanical support for leaves, spanning the lamina (veins) or holding the lamina in a certain position or angle (petioles). Leaf growth is thus an embodiment of growth activities of all these cells that differ in their physiological functions and physical as well as metabolic properties. This should be kept in mind when analysing the spatial and temporal dynamics of leaf growth.

2.1 Diel growth patterns in different species and light environments

Day-night growth rhythms have been found in all species thus far investigated, both in monocots and dicots (Table 1). Since leaf growth in monocots is practically restricted to one dimension (i.e. leaf length) and rapidly growing regions are "hidden" under the sheath, classical methods using rulers and linear variable displacement transducers (LVDT) are usually employed to analyse leaf growth in monocots although they do not provide any spatial information on growth, and in the case of using rulers, temporal resolution is limited. For analyses in dicot leaves, for which growth needs to be calculated as an increase in area (i.e. in two dimensions), methods allowing high temporal and spatial resolution have become available only in recent times (Schmundt et al. 1998). Data obtained by one- or two-dimensional analysis methods have shown diel patters in different dicot species (Table 1). In general, the majority of monocot leaves grow in the presence of light whereas there is no such general tendency among dicot leaves.

Distinct diel growth patterns have been measured in different species by applying the DISP method (Fig. 1). Leaf growth rate of *Ricinus communis* L. (Fig. 1A) peaked at dawn to early morning with a steep post-midnight increase and a decrease in the afternoon (Walter et al. 2002a). This type of temporal pattern has also been found in leaves of *Nicotiana tabacum* L. (Walter and Schurr 2000, 2005). These two plants share similar spatial distribution patterns of growth rate with a clear gradient from basal (high growth rate) to apical (low growth rate) regions of the lamina (Schmundt et al. 1998; Schurr et al. 2000; Walter et al. 2002a; Walter and Schurr 2000, 2005).

A contrasting diel course has been observed in *P. deltoides* (Walter et al. 2005; Matsubara et al. 2006). It was characterized by a post-midnight decrease and an increase during the day (Fig. 1B). *Glycine max* L. exhibited a comparable post-midnight decrease towards dawn, but thereafter the growth rate in this species continued to increase from morning until midnight (Ainsworth et al. 2005), whereas it reached a peak at around dusk in *P. deltoides* (Walter et al. 2005; Matsubara et al. 2006). Similarity in leaf growth characteristics of *P. deltoides* and *G. max* is found not only in the

Table 1. Variation in diel leaf growth patterns. *LVDT* linear variable displacement transducer; *DISP* digital image sequence processing; *CAM* crassulacean acid metabolism

Genera	Growth rate				Methods	References
	Dawn-morning	Day	Dusk	Night		
Monocotyledons						
Zea		Max		Min	LVDT	Watts 1974
		Max		Min	Ruler	Acevedo et al. 1979
Triticum					LVDT	Christ 1978
Sorghum	Min		Max		Ruler	Acevedo et al. 1979
Oryza		Max		Min	LVDT	Seneweera et al. 1995
Festuca			Min	Max	LVDT	Durand et al. 1995
Dicotyledons						
Helianthus	Min			Max	Ruler	Boyer 1968
Glycine		Min		Max	Ruler	Bunce 1977
	Min			Max	DISP	Ainsworth et al. 2005
Phaseolus	Min	Max			LVDT	Davies and Van Volkenburgh 1983
Vitis		Max		Min	LVDT	Shackel et al. 1987
Lycopersicon	Min		Max		LVDT	Price et al. 2001
Betula		Max		Min	LVDT	Taylor and Davies 1985

Acer		Min		Max	LVDT	Taylor and Davies 1985
Ricinus (Fig. 1A)	Max		Min		DISP	Schmundt et al. 1998
Nicotiana	Max		Min		DISP	Walter and Schurr 2000
Populus (Fig. 1B)	Min		Max		DISP	Walter et al. 2005 Matsubara et al. 2006
Gymnosperms						
Chamaecyparis (Fig. 1C)	Min		Max		DISP	Lai et al. 2005
CAM plants						
Mesembryanthemum (C_3)	Max		Min		DISP	Gouws et al. 2005
Opuntia (C_4; Fig. 1D)		Max		Min	DISP	
Kalanchoe (C_4)		Max		Min	DISP	

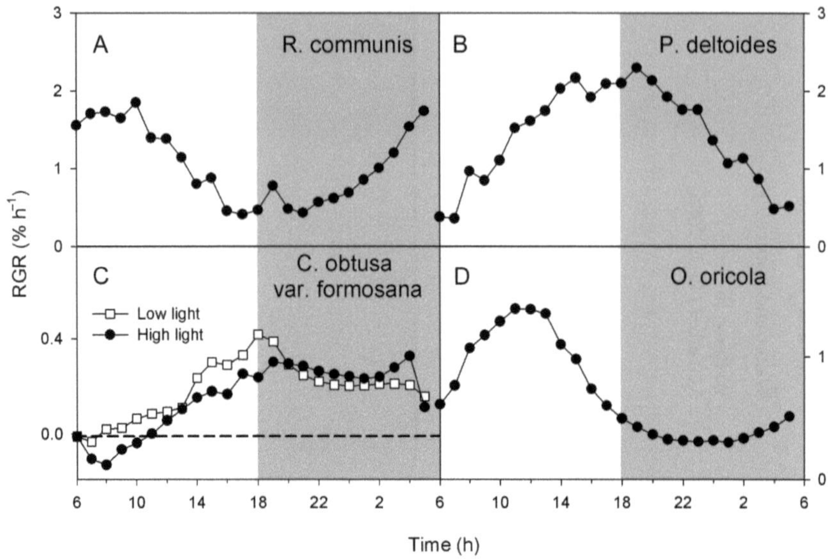

Fig. 1. The variability of diel leaf growth cycles among different species. Depicted are 1-h mean values of leaf relative growth rate (RGR) of **A** *Ricinus communis* ($n = 14$), **B** *Populus deltoides* ($n = 25$), **C** *Chamaecyparis obtusa* ($n = 38$, grown in high light intensity; $n = 31$ grown in low light intensity) and **D** cladodes of *Opuntia oricola* ($n = 4$)

timing of growth but also in the spatial distribution. In these species, the spatial patterns of leaf growth lack a base-tip gradient across the growing lamina (Ainsworth et al. 2005; Walter et al. 2005; Matsubara et al. 2006), again differing from *N. tabacum* and *R. communis.*

Diel growth cycles have been detected even in diminutive leaves of gymnosperm seedlings (Lai et al. 2005) and succulent leaves and cladodes of crassulacean acid metabolism (CAM) plants (Gouws et al. 2005). *Chamaecyparis obtusa* Sieb. et Zucc. var. *formosana* (Hayata) Rehde, which is adapted to shady conditions of forest understorey in foggy montane areas of Taiwan, showed a growth pattern with a gradual increase during the day followed by a plateau at night, and an abrupt decrease at dawn (Fig. 1C; Lai et al. 2005). In addition to a clear base-tip gradient across the very small leaves, it has been revealed that leaves of *C. obtusa* continually shrink for several hours in the morning when exposed to relatively high irradiance (Fig. 1C). Presumably this is indicative of intolerance of this species to increased light-induced transpiration and the resulting decline in leaf water potential (Lai et al. 2005). By contrast, leaves of *C. formosensis* Matsum, occurring in open gaps, did not undergo such shrinking but were less flexible than *C. obtusa* in adjusting the growth activities to sudden transitions in light environment (Lai et al. 2005).

Leaf and cladode growth in four species of CAM plants, *Kalanchoë beharensis* Drake et Castillo, *Opuntia oricola* Philbrick, *O. phaeacantha* Engelm. and *O. engelmanii* Salm-Dyck, was centred at around midday when these plants were growing in dry soils and performing CAM (Fig. 1D; Gouws et al. 2005). The timing of this maximal growth coincided with the phase III of CAM, in which CO_2 fixation via ribulose-1,5-bisphosphate-carboxylase-oxygenase (Rubisco) takes place without opening stomata by using CO_2 released from malic acid that was formed by phosphoenolpyruvate-carboxylase and stored in the vacuole during the previous night. It has been pointed out that the cellular status prevailing the phase III of CAM, i.e. high carbon availability, high turgor and low cytoplasmic pH, would offer a favourable condition for growth (Gouws et al. 2005). Furthermore, under well-watered conditions in which operation of C_3 metabolism supposedly increases, cladodes of *O. oricola* and *O. phaeacantha* displayed different diel growth patterns compared with the ones observed under the dry/CAM conditions (Gouws et al. 2005), indicating a link between the metabolic and growth cycles in CAM plants.

2.2 Circadian clock, gene expression and diel metabolic activities

Marked diurnal changes occur in the concentrations of metabolites in leaves, including the major substrates for growth such as carbohydrates and amino acids (e.g. Matt et al. 1998; Urbanczyk-Wochniak et al. 2005; Walter and Schurr 2005). Leaf carbohydrate concentrations are typically higher during the day than during the night, with excursions of starch being the most prominent (e.g. Matt et al. 1998; Urbanczyk-Wochniak et al. 2005; Walter et al. 2005; Walter and Schurr 2005; Matsubara et al. 2006). These changes reflect the day-night transitions between the autotrophic and heterotrophic state (Fig. 2) which repeat in photosynthetic tissue every day (also with some modifications like CAM). Not surprisingly, the expression of many genes involved in photosynthesis is controlled by the endogenous clock and/or light (Millar and Kay 1996; Harmer et al. 2000; Schaffer et al. 2001; Urbanczyk-Wochniak et al. 2005). Such transcriptional regulation of photosynthetic genes would have obvious advantage as gene expression in two compartments, cell nucleus and chloroplast, can be coordinated to support efficient operation of photosynthesis (Dodd et al. 2005; Misquitta and Herrin 2005).

According to a current model (Wang et al. 1997; Shaffer et al. 1998; Somers et al. 1998b; Wang and Tobin 1998; Green and Tobin 1999; Alabadí et al. 2001, 2002; Mizoguchi et al. 2002), the circadian clock in *Arabidopsis* is composed of a negative feedback loop between the pseudo response

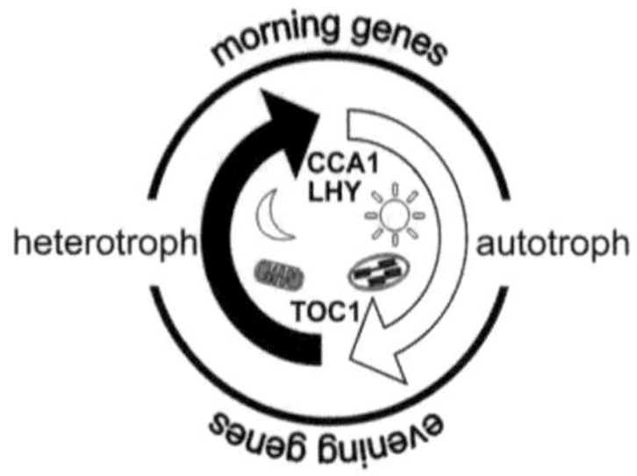

Fig. 2. Diel cycles in leaves. In the course of day-night cycles, two different sources alternately provide energy in leaves: during the day energy is directly delivered by photosynthesis while nocturnal processes are dependent on energy from starch mobilization and mitochondrial respiration. Many genes involved in these pathways are up-regulated in the corresponding phases. At the transitions between day and night, genes in specific pathways are up-regulated in anticipation of the upcoming change in the environment. This rhythmic gene expression persists under continuous light or continuous dark, indicating that it is controlled by an endogenous clock, a self-sustaining oscillatory system. The current model of such an oscillator in *Arabidopsis* consists of a negative feedback loop between CCA1 (circadian clock associated 1), LHY (late elongated hypocotyl) and TOC1 (timing of chlorophyll *a*/*b*-binding proteins CAB expression 1)

regulator TOC1 (timing of chlorophyll *a*/*b*-binding proteins CAB expression 1) and two transcription factors related to MYB (myleoblastosis), CCA1 (circadian clock associated 1) and LHY (late elongated hypocotyl). The two transcription factors CCA1 and LHY, which activate gene expression in the clock-output pathway during the day (e.g. light-harvesting antenna proteins, also known as CAB), simultaneously repress *TOC1*, a positive element for *CCA1*/*LHY*. As the levels of CCA1 and LHY decrease in the evening, repression of *TOC1* is relieved, which in turn leads to the accumulation of CCA1 and LHY towards the morning.

It has been estimated that ca. 6% of 8600 genes on *Arabidopsis* oligonucleotide arrays (Harmer et al. 2000) or 2% of 7800 clones of expressed sequenced tags (Schaffer et al. 2001) could be regulated by the circadian clock. Genes in the same metabolic pathways are often found to be expressed in clusters at distinctive times of a day. In accordance with the carbon metabolic cycles in leaves (Fig. 2), genes involved in photosynthesis (e.g. genes encoding proteins in photosystem I and II or light-harvesting antenna I and II, chlorophyll *a* oxygenase) were expressed mostly during the day as opposed to the expression patterns of genes in starch mobilization (e.g. β-amylase, putative fructose-bisphosphate aldolase, sucrose-phosphate synthase homolog) concentrated at night (Harmer et al. 2000; Schaffer et al. 2001; Smith et al. 2004). Many genes in the phenylpropanoid pathway (e.g. phenylalanine ammonia lyase, chalcone synthase, flavonol synthase, glutathione S-transferase) are presumably "morning genes", the transcripts of which increase during the night to reach the maximum at around dawn (Borevitz et al. 2000; Harmer et al. 2000; Rogers et al. 2005). This expression pattern has been associated with preparation of UV- and photo-protective compounds early in the day (Harmer et al. 2000). Coordinated expression of a group of genes in sugar allocation and storage (e.g. genes encoding sugar transporters, glycolytic enzymes, oxidative pentose phosphate enzymes and galactinol synthase), was observed near the end of the day ("evening genes"), suggesting an important role of circadian clocks in temporal coordination of assimilate trafficking (Harmer et al. 2000).

This global picture of circadian gene expression could conform to the day-night metabolic cycles (Fig. 2) in leaves of many species and will interact with the diel growth patterns (Table 1, Fig. 1A–C). Growth processes and mechanisms, endogenously shaping different diel patterns, need to be compatible with, and may possibly be constrained by, the timetable of these circadian programmes.

2.3 Gating of cell cycle and diel growth processes

The transcriptional control of diel leaf growth pattern has been investigated in *P. deltoides* (Matsubara et al. 2006), a species having a post-midnight decline in growth rate (Table 1, Fig. 1B). For this study, POP2 microarrays containing 24,912 clones of expressed sequence tags obtained from poplar (Sterky et al. 2004) were used. By comparing transcripts from leaf samples collected at well-defined growth rates that were determined by an online growth analysis tool, concerted down-regulation of a number of ribosomal protein (RP) genes was revealed in poplar leaves concomitant with the nocturnal growth deceleration (Matsubara et al. 2006). In parallel with the down-regulation of RP genes, the transcript levels of histone *H2B*, an S-phase gene (Breyne et al. 2002; Menges et al. 2002), were also rapidly reduced in leaves of *P. deltoides* after midnight (Matsubara et al. 2006). Ribosome biogenesis is mainly controlled at the transcriptional and translational level in yeast and mammalian cells, respectively (Jorgensen et al. 2004). In plants, there seems to be a control at the transcriptional level, similar to the situation in yeast, as co-expression of RP genes has been reported for plants, especially in meristematic tissues, by different transcriptomic studies (Sterky et al. 2004; Schmid et al 2005; Matsubara et al. 2006). It has been suggested that biosynthesis of ribosomes may be a critical factor for the G1/S transition in the mitotic cell cycle (Cuadrado et al. 1985).

Circadian rhythmicity of cell-division frequency has been observed in shoot meristems, but apparently not in root meristems, with the maxima occurring at night in both long-(*Spinacia oleracea* L, *Papaver somniferum* L.) and short-day plants (*Perilla ocymoides* L, *G. max*) (Bünning 1952). Similar circadian oscillations in reproduction rate have been documented for other photosynthetic organisms, namely cyanobacteria *Synechococcus* PCC 7942 (Mori et al. 1996) or unicellular algae *Chlamydomonas reinhardtii* Dangeard (Goto and Johnson 1995) and *Euglena gracilis* Klebs (Hagiwara et al. 2002; Bolige et al. 2005a,b). It has been demonstrated that cell division activities of these unicellular photoautotrophs are temporally restricted (or "gated") by the circadian clock such that the population size grows during subjective night. While questions about the molecular mechanisms of the cell-cycle gating are largely unsolved, the study with *Euglena* (Bolige et al. 2005a) has shown that the gating happens by forbidding cell-cycle transitions (closing the gate), and not by permitting them (opening the gate), at primarily G2/M but also G1/S and S/G2. Further, it seems that *Euglena* cells require photoinduction to become committed to progress to the next phase in the cell cycle (Hagiwara et al. 2002). In other words, they need to be illuminated at least for a short period prior to the cell-cycle transitions. The efficiency of such photoinduction was found to be lowest at subjective dawn and highest at subjective dusk, leading in the latter case to a rapid increase in the population size, especially the number of cells in the G1 phase, in the dark period immediately after dusk (Hagiwara et al. 2002).

The nocturnal preference of cell division has been associated with a strategy to avoid UV-sensitive DNA replication during the day, known as the "escape from light" hypothesis (Paietta 1982; Pittendrigh 1993). Lately, this hypothesis has been modified to include the selective advantage of "resistance to light" (Bolige et al. 2005b) to explain the observed circadian variation in the survival rate of *Chlamydomonas* and *Euglena* under UV radiation (Nikaido and Johnson 2000; Bolige et al. 2005b). For the evolution and survival of photoautotrophic organisms, preparation of photoprotective molecules in advance of daily light- and UV-exposure may have been as vital as programming radiation-sensitive reactions for the night (Bolige et al. 2005b). Circadian clocks may have evolved and given a selective advantage to the photosynthetic organisms (Ouyang et al. 1998; Green et al. 2002; Dodd et al. 2005), facilitating temporal organization as well as synchronization of photosynthesis, photoprotection and cell division with regular cycles of the light environment.

3 Linking light environment with diel growth dynamics

3.1 Shade avoidance, signalling networks and the role of the circadian clock

Along with the regular day-night cycle, plants are subjected to variations in their light environments throughout their life cycles. Shade avoidance (as well as etiolation of seedlings) is one of the light-induced growth syndromes in which organ morphology and structure, and in a longer term also developmental programmes (e.g. early flowering in *Arabidopsis*), are dramatically changed. It is often manifested by enhanced elongation of stem-like organs (including petioles) concomitant with suppression of leaf-blade expansion and shoot branching (Aphalo et al. 1999; Vandenbussche et al. 2005). These changes in growth enable plants to position their leaves in exposed locations, continuously adjust their canopy structure to the light environment throughout the vegetative growth season and compete for light with their neighbours in dense vegetation (Aphalo et al. 1999; Vandenbussche et al. 2005). Many angiosperms, ranging from small herbs to large trees, employ the shade-avoidance strategy (Smith 2000).

Growth and developmental responses to changes in light environments are released by the photosensory signal transduction network involving three major families of photoreceptors: phytochromes (Phy) for perception of wavelengths in red (R) and far-red (FR) regions and cryptochromes (Cry) and phototropins for blue and UV-A regions. Combined activities of these photoreceptors provide multifaceted information about surrounding light environments to generate differential responses (Ahmad and Cashmore 1996; Neff et al. 2000; Smith 2000; Quail 2002a,b; Liscum et al. 2003; Schultz and Kay 2003). Shade signature in the light environment is detected by plants predominantly as decrease in R:FR ratio that reflects depletion of light in blue and R regions by chlorophylls and carotenoids of nearby leaves and vegetation and/or enrichment of FR by reflection and scattering from them (Morelli and Ruberti 2000; Smith 2000; Franklin and Whitelam 2005; Vandenbussche et al. 2005). Decline in R:FR, albeit to a lesser extent, occurs also outside the shade at dawn and dusk (so-called end-of-day FR). The R- and FR-sensing photoreceptor Phy, a homodimeric chromoprotein having a linear tetrapyrrole (phytochromobilin), plays central roles in shade avoidance as well as responses to end-of-day FR. Three major Phy proteins in angiosperms are PhyA, PhyB and PhyC while PhyD and PhyE have so far been found only in dicot plants (Franklin and Whitelam 2005).

The striking feature of Phy is the reversible photochromism. In the dark, Phy is synthesized in the R-absorbing form (Pr, absorption maximum approximately 665 nm) and accumulates in the cytoplasm. Upon light absorption, it changes the conformation to the FR-absorbing form (Pfr, absorption maximum approximately 730 nm), becomes biologically "activated" and translocates from the cytoplasm into the nucleus (Nagatani 2004). The photoconversion and nucleo-cytoplasmic translocation are both reversible. In reality, the absorption spectra of Pr and Pfr largely overlap below 700 nm, which implies that a mixture of the two forms exists under almost all light conditions (Smith 2000; Franklin and Whitelam 2005). Hence, the base of the shade avoidance syndrome is the equilibrium between the populations of Phy proteins in the Pr- and Pfr-form (Smith 2000). This photoreversibility and reciprocity (dependence on the total number of photons absorbed irrespective of the duration of exposure) are the characteristics of classic Phy-responses, including shade avoidance (Neff et al. 2000).

3.2 Light signalling pathways

One of the genes that are strongly induced under low R:FR is the *Arabidopsis* homeobox-leucine-zipper transcription factor *ATHB-2* (Fig. 3), also known as homeobox-leucine-zipper protein 4 (*HAT4*), which is a negative regulator of gene expression (Steindler et al. 1999). Rapid and marked up-regulation of *ATHB-2* occurs under low R:FR, which is readily reversible by a subsequent high-R:FR treatment (Carabelli et al. 1996). It has been demonstrated that shade-avoidance responses can be mimicked by increasing ATHB-2 levels (Steindler et al. 1999). Elevated ATHB-2 inhibits cotyledon expansion by restricting elongation of epidermal and mesophyll cells. In hypocotyl, overexpression of *ATHB-2* results in enhanced longitudinal expansion in epidermal and cortical cells while it inhibits proliferative thickening of the vascular system. These anatomical changes in hypocotyl are also found in roots (Steindler et al. 1999). In contrast, transgenic lines with reduced levels of ATHB-2 display reverse phenotypes characterized by short hypocotyl and larger cotyledons. Based on these observations and the requirement of auxin for the expression of shade-avoidance responses, a role of ATHB-2 for auxin lateral flow in hypocotyl and root has been postulated (Steindler et al. 1999; Morelli and Ruberti 2000).

Partial phenocopy of the *ATHB-2* overexpressor, i.e. enhanced elongation and reduced thickening in hypocotyl and root (Oyama et al. 1997), can be found in the *Arabidopsis* photomorphogenic mutant *hy5* (long hypocotyl 5)

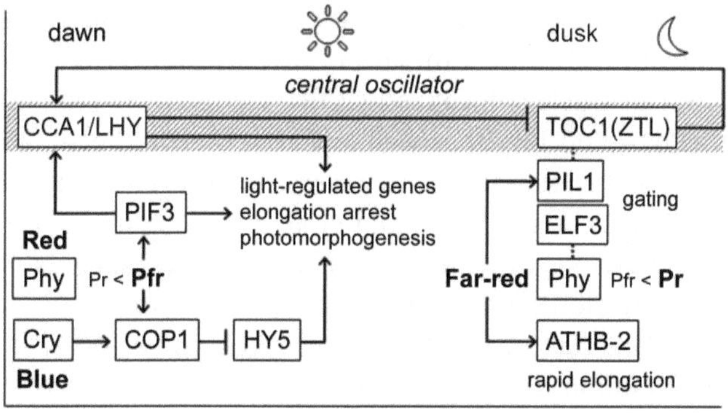

Fig. 3. Simplified schematic model of the light and circadian regulation of shade-avoidance responses. When activated by red light, phytochrome (Phy; Pfr) interacts with the transcription factor PIF3 (phytochrome interacting factor 3) in the nucleus. PIF3 binds to G-box, a DNA-sequence motif present in a range of light-regulated promoters. The two key genes in the central oscillator of the circadian clock, *CCA1* (circadian clock associated 1) and *LHY* (late elongated hypocotyl), possess G box and are thus expressed by Phy-PIF activity. Subsequently, CCA1 and LHY, themselves transcription factors, induce expression of various morning genes (e.g. chlorophyll *a*/*b*-binding proteins *CAB*) while repressing expression of *TOC1* (timing of CAB expression 1), the third component of the circadian oscillator, which positively regulates expression of *CCA1* and *LHY*. Blue light detected by cryptochrome (Cry) and red light producing Pfr enhance accumulation of another transcription factor HY5 (long hypocotyl 5) by interacting with COP1 (constitutively photomorphogenic 1), an E3 ubiquitin ligase which targets HY5 for degradation. HY5 also promotes expression of light-regulated genes having G box in promoter regions. Far-red light, on the other hand, de-activates Pfr and forms Pr. Genes such as *PIL1* (PIF3-like 1) and *ATHB-2* (also known as homeobox-leucine-zipper protein *HAT4*) are rapidly up-regulated under the light environments with low red:far-red ratios, leading to the acceleration of elongation growth, most pronouncedly at the end of the day. This circadian gating of rapid elongation growth presumably requires PIL1 and TOC1 at the same time. Transcriptional repression by CCA1/LHY and degradation by ZTL (zeitlupe) determines TOC1 abundance. In parallel, ELF3 (early flowering 3) negatively regulates light-signal input to the clock in the evening, preventing the downstream events of Pfr

lacking functional HY5 protein (Fig. 3), a bZIP transcription factor that interacts with light-regulated promoters containing G-box (Chattopadhyay et al. 1998). The abundance of HY5, which is directly correlated with the expression of photomorphogenic phenotypes (Osterlund et al. 2000), is controlled by COP1 (constitutively photomorphogenic 1; Fig. 3), a RING-finger protein which functions as an E3 ubiquitin-protein ligase (Schwechheimer et al. 2001) and regulates the expression of a large number of transcription factors in *Arabidopsis* (Ma et al. 2002). A line of evidence suggests that COP1 is a repressor of light-regulated gene expression and directly interacts with HY5 in the nucleus to target its degradation via the 26S proteasome (Osterlund et al. 2000; Ma et al. 2002). Illumination reduces the nuclear abundance of COP1, thereby reducing the rate of HY5 protein degradation and hence allowing HY5 accumulation in the nucleus (Osterlund et al. 2000). The level of nuclear-localized COP1 seems to be regulated by light through Phy (PhyA and PhyB) as well as Cry (Osterlund et al. 2000; Fig. 3). In addition to the depolarization of plasma membrane mediated by blue light receptors (Parks et al. 1998; Folta et al. 2003), light also inhibits elongation through activities of transcriptional regulators, such as HY5, and at the same time promotes thickening in hypocotyl and root. It has been reported that AXR2 (auxin resistant 2; also known as IAA7, indole acetic acid 7) and SLR (solitary root, also known as IAA14), negative regulators of auxin signalling, are among the target genes of HY5, suggesting modulation of auxin sensitivity by the light (Cluis et al. 2004).

Another transcription factor interacting with Phy is the basic helix-loop-helix transcription factor PIF3 (phytochrome interacting factor 3; Fig. 3) that regulates genes having the light-regulated G-box in promoter regions (Ni et al. 1999; Martinez-García et al. 2000). The PIF3-antisense plants of *Arabidopsis* (Halliday et al. 1999) show morphological phenotypes resembling *hy5* mutants (Oyama et al. 1997), ATHB-2 overexpressors (Steindler et al. 1999) or plants grown under low R:FR. Conversely, mutants containing increased levels of PIF3 have a short hypocotyl and larger cotyledons (Halliday et al. 1999), as has been observed in ATHB 2-antisense plants (Steindler et al. 1999). The strong similarity in the phenotypes of these mutants and transgenic plants indicates that these pathways may share the same output components in the downstream region, some of which are presumably involved in auxin signalling (Steindler et al. 1999; Morelli and Ruberti 2000; Tanaka et al. 2002; Cluis et al. 2004). In fact, low R:FR induces up-regulation of a number of genes associated with auxin signalling as revealed by a recent microarray study on shade avoidance (Devlin et al. 2003).

3.3 Cross-talk with circadian clock

A connection between light signalling and the endogenous clock is provided by direct binding of Phy (Pfr) to PIF3 in the nucleus, which leads to gene expression of the two key components CCA1 and LHY in the central clock oscillator (Fig. 3), both containing G-box motifs in the promoter regions (Ni et al. 1999; Martinez-García et al. 2000). Because CCA1 and LHY are themselves MYB-related transcription factors controlling the circadian gene expression of various proteins, such as CAB or TOC1 (Wang et al. 1997; Schaffer et al. 1998; Green and Tobin 1999; Harmer et al. 2000; Albadí et al. 2001), the Phy-PIF3 interaction could be one of the mechanisms by which light entrains the circadian clock at dawn (Somers et al. 1998a; Frankhauser and Staiger 2002; Quail 2002a,b).

Recently, it has been shown that reduced levels of TOC1, which positively regulates *CCA1* and *LHY* expression in the negative feedback loop of the circadian oscillator (Alabadí et al. 2001; Fig. 3), cause arrhythmia in the circadian gene expression under R light while maintaining rhythmicity under blue and white light albeit with shorter periods (Más et al. 2003a). Because the R-dependent induction of *CCA1* and *LHY* is impaired in the absence of functional TOC1 and TOC1 can interact with PIF3, TOC1-mediated modulation of PIF3 binding to *CCA1*/*LHY* promoters has been postulated (Más et al. 2003a). The intimate cross-talk between the circadian oscillator and light-signalling pathways is further underpinned by ZTL (zeitlupe; Fig. 3), a protein that facilitates the proteasome-dependent degradation of TOC1 (Más et al. 2003b) and can interact with PhyB and Cry1 (Jarillo et al. 2001). Moreover, evidence has been presented that the expression of Phy and Cry is in turn regulated by the circadian clock (Kozma-Bognár et al. 1999; Harmer et al. 2000; Tóth et al. 2001), meaning these photoreceptors are input as well as output components of the clock (Harmer et al. 2000). The fact that many, although not all (e.g. Somers et al. 1998b), of the circadian-clock mutants and transgenic plants exhibit elongated (or shortened) hypocotyl and/or early-flowering phenotypes (e.g. Schaffer et al. 1998; Dowson-Day and Millar 1999; Más et al. 2003a; Kim et al. 2005) highlights the convergence of light- and clock-controlled output pathways in growth and development.

3.4 Circadian gating of shade avoidance

Both natural elongation growth of hypocotyl (Dowson-Day and Millar 1999) and rapid elongation in shade avoidance (Salter et al. 2003) have been shown to be gated by the endogenous clock in *Arabidopsis*, with an arrest at dawn and acceleration at dusk. A function in circadian gating of light signal input has been proposed for ELF3 (early flowering 3; Fig. 3). Unlike wild-type plants, in which *CAB* expression is not responsive to light pulses during subjective night (Millar and Kay 1996) and elongation growth of hypocotyl always stops at subjective dawn (Dowson-Day and Millar 1999), ELF3 mutants show acute activation of *CAB* upon illumination even during subjective night (McWatters et al. 2000) and no sign of elongation arrest at subjective dawn when kept in the light (Dowson-Day and Millar 1999). Conversely, overexpression of ELF3 results in diminished responsiveness to light stimuli during subjective night (Covington et al. 2001). Thus, it has been suggested that ELF3, having the maximal expression in the evening, is a negative regulator of light input to the endogenous clock, facilitating the gating of light signal input to prevent the expression of light-inducible genes (e.g. *CAB*) in the evening (Covington et al. 2001). This means that activation of genes involved in inhibition of hypocotyl elongation in the morning (Fig. 3) may also be repressed around the end of the day by ELF3 abundance even if there is still some light. Although the exact mode of ELF3 action in the nucleus is not yet known, *in vitro* interaction of ELF3 and PhyB indicates a potential role of ELF3 in restraining PhyB from light signal input (Liu et al. 2001).

More recently, circadian gating of *PIL1* and *PIL2* (PIF3-like 1 and 2; Fig. 3, only PIL1 is shown) has been implicated with the gating of shade-avoidance responses (Salter et al. 2003). Transcript levels of these genes, encoding basic helix-loop-helix transcription factors with high protein sequence similarity to PIF3, increase in response to low R:FR, with the induction of PIL1 being more rapid and strong compared with that of PIL2 (Salter et al. 2003). Requirement of PIL1 for shade-avoidance responses has been established by the demonstration that acceleration of hypocotyl elongation by low R:FR is abolished in PIL1 mutants (Salter et al. 2003). Because *PIL1* expression reaches the maximum at subjective dawn while maximal enhancement of hypocotyl elongation by low R:FR occurs at subjective dusk, it has been proposed that PIL1-dependent shade-avoidance responses may be limited by TOC1 (Salter et al. 2003), which accumulates at dusk and interacts with PIL1 (Makino et al. 2002). The finding that shade avoidance is gated by the circadian clock such that it coincides with the natural elongation rhythm of hypocotyl (Dowson-Day and Millar 1999; Salter et al. 2003; Franklin and Whitelam 2005) supports the notion that rapid shade-avoidance responses of *Arabidopsis* seedlings engage part of the endogenous controlling pathways for the hypocotyl elongation. Because the phenotypes of PIL1 mutants do not differ from wild-type plants under prolonged low R:FR, involvement of PIL2, which responds to low R:FR more slowly than PIL1, has been postulated for longer-term shade-avoidance responses, such as petiole elongation and flowering (Salter et al. 2003).

4 Leaf growth dynamics in dicot plants and mechanisms of shade avoidance

In hypocotyl and seedlings of *Arabidopsis*, the mechanisms controlling endogenous elongation growth and rapid elongation in shade avoidance have at least several operational traits in common: signalling by photoreceptors for light input, interactions of important transcriptional regulators with the clock oscillator, and output patterns of gating. Observations that end-of-day FR can mimic effects on growth in shade avoidance syndrome (Franklin and Whitelam 2005) also suggest an overlap between the regulatory mechanisms of shade avoidance (long-term low R:FR) and the key signal transduction pathways of daily growth programmes, including end-of-day FR (short-pulse low R:FR). Differential roles suggested for PIL1 and PIL2 in controlling the short- and long-term responses to low R:FR, respectively, may be a way to adjust the regulation of endogenous daily growth programmes to the environment (Salter et al. 2003).

Control mechanisms of elongation growth and shade avoidance in hypocotyl and seedling growth are not identical. However, there are good indications of coupling of the signal transduction pathways. Shade-avoiding and shade-tolerant species differ in their growth responses to increasing light. The shade-avoiding species *Betula pendula* Roth and *Phaseolus vulgaris* L. increase cell wall extensibility upon illumination with white light (Van Volkenburgh and Cleland 1981; Taylor and Davies 1985, 1986). Both species show the same pattern of diel growth activity. In contrast, the shade-tolerant *Acer pseudoplatanus* L. did not show increased cell wall extensibility upon illumination with white light and a diel leaf growth pattern differing from that of the shade-avoiding species (Taylor and Davies 1985, 1986).

The diel phasing of leaf growth activity and of light-signal response differs in shade-avoiding and shade-tolerant plants. Therefore, it can be postulated that the photomorphogenic pathways linking light-signal input with auxin signalling, an important factor for cell wall extension, may be differently regulated in these plants, resulting in contrasting effects of light on leaf growth. Our knowledge about the molecular control of photomorphogenic responses in *Arabidopsis* has been greatly increased to allow us to pinpoint some of the central regulators (e.g. HY5, PIF3, ATHB-2) and their phases of action in day-night cycles determined by the circadian oscillators (CCA1, LHY, TOC1) and gate keepers (e.g. ELF3, PIL1). Nevertheless, information on the activities and regulation of their orthologous genes in other species is very limited.

External trigger (light) and internal regulatory systems (circadian clock) interact in controlling growth dynamics in aerial parts of plants (Fig. 4). Two contrasting diel growth patterns in leaves (laminae) of C_3 dicot plants are represented by the two growth curves of Fig. 4 (solid line, increasing at midnight, compare Fig. 1A; dashed line, decreasing at midnight, compare Fig. 1B). On top of the shaded area that is common to both growth patterns, plants like *N. tabacum* and *R. communis* (Schmundt et al. 1998; Schurr et al. 2000) show increased growth activities from dawn to midday (solid line,

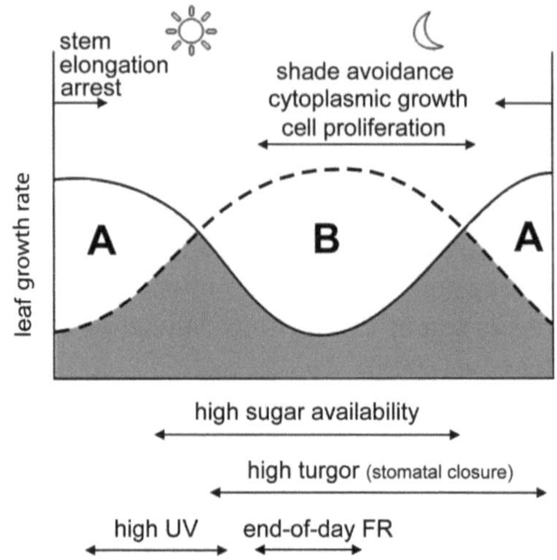

Fig. 4. Comparison of two diel growth patterns in C_3 dicot leaves in the context of several endogenous and environmental factors impinging on them. *Solid line*, growth rhythm with maximal growth rate at dawn; *dashed line*, growth rhythm with maximal growth rate at dusk. *Shaded area* is common to both types of growth. Areas *A* and *B* represent differential up-regulation of growth by the two types

areas A) while others like *P. deltoides* and *G. max* (Ainsworth et al. 2005; Walter et al. 2005; Matsubara et al. 2006) rely more heavily on the period from dusk to midnight (dashed line, area B).

Even with the limited number of factors shown here, it is clear that the up-regulation of growth activities in A and B would engage different control mechanisms to coordinate with the diel cycles of other cellular processes (Fig. 2) as well as the environment. For example, the negative effect of stomatal opening on leaf growth manifested in *C. obtusa* in the morning (Fig. 1C; Lai et al. 2005) needs to be overcome in type-A plants while relative importance of such problems would be minor in type-B plants. Circadian gating of cell cycle (Goto and Johnson 1995; Mori et al. 1996; Hagiwara et al. 2002; Bolige et al. 2005a) could restrict the contribution of cell proliferation to the overall leaf growth more strongly in type-A plants compared with type-B plants (Matsubara et al. 2006). Also, gated growth rhythms in stem-like organs (Lecharny et al. 1985; Dowson-Day and Millar 1999) may cause resource competition in type-B plants around the end of the day whereas the peak growth activities in leaf blades and stems/petioles are probably separated in type-A plants. Systematic screening of leaf growth in a range of dicot species is needed to examine if these diel growth patterns are associated with certain life strategies, such as shade tolerance.

5 Perspective

Given the small number of species, for which diel growth patterns in leaves are well characterized, and also the lack of the genetic information for many species except for the few model plants (mostly *Arabidopsis*), it is difficult to draw a picture of molecular networks controlling distinct endogenous leaf growth patterns. Even though fundamental processes for growth control are conserved in most plants, extrapolation of knowledge and models obtained from *Arabidopsis* to other plants needs caution because the same processes can be regulated differently. For studying growth in particular, one should always keep in mind a possible influence by the life strategy of *Arabidopsis* plants (formation of rosette, long-day plants with a tendency to accelerate flowering under unfavourable conditions). Nevertheless, the emerging techniques for growth analysis facilitating elucidation of dynamic processes, both above- and belowground (Walter and Schurr 2005; Schurr et al. 2006), can now be applied to diverse species to extend our knowledge in plant growth mechanisms. Furthermore, the ongoing effort in genome sequencing in a variety of plants will soon enable comparable growth studies in those plants at all levels, from gene to community, to explore "plant growth diversity". Together with better understanding of phenotypic plasticity within species, uncovering interspecific differences in such plasticity would be beneficial for a broad spectrum of fields, including technical transfer to agriculture and green biotechnology as well as improvement of ecosystem models for environmental research.

Acknowledgements

We thank Uli Schurr for his comments on this manuscript and for his continuing interest in and support for the elucidation of plant growth dynamics.

References

Acevedo E, Fereres E, Hsiao TC, Henderson DW (1979) Diurnal growth trends, water potential and osmotic adjustment of maize and *Sorghum* leaves in the field. Plant Physiol 64:476–480

Ahmad M, Cashmore AR (1996) Seeing blue: the discovery of cryptochrome. Plant Mol Biol 30:851–861

Ainsworth EA, Walter A, Schurr U (2005) *Glycine max* leaflets lack a base-tip gradient in growth rate. J Plant Res 118:343–346

Alabadí D, Oyama T, Yanovsky MJ, Harmon FG, Más P, Kay SA (2001) Reciprocal regulation between TOC1 and LHY/CCA1 within the *Arabidopsis* circadian clock. Science 293:880–883

Alabadí D, Yanovsky MJ, Más P, Harmer SL, Kay SA (2002) Critical role for CCA1 and LHY in maintaining circadian rhythmicity in *Arabidopsis.* Curr Biol 12:757–761

Allen JF, Nilsson A (1997) Redox signalling and the structural basis of regulation of photosynthesis by protein phosphorylation. Physiol Plant 100:863–868

Aphalo PJ, Ballaré CL, Scopel AL (1999) Plant-plant signalling, the shade-avoidance response and competition. J Exp Bot 50:1629–1634

Avery GS (1933) Structure and development of tobacco leaves. Am J Bot 20:565–592

Beemster GTS, De Veylder L, Vercruysse S, West G, Rombaut D, Van Hummelen P, Galichet A, Gruissem W, Inzé D, Vuylsteke M (2005) Genome-wide analysis of gene expression profiles associated with cell cycle transition in growing organs of *Arabidopsis.* Plant Physiol 138:734–743

Björkman O (1981) Responses to different quantum flux densities. In: Lange OL, Nobel PS, Osmond CB, Ziegler H (eds) Encyclopaedia of plant physiology, vol 12A. Physiological plant ecology I. Responses to the physical environment. Springer-Verlag, Berlin, Heidelberg, New York, pp 96–117

Bolige A, Hagiwara S, Zhang Y, Goto K (2005a) Circadian G2 arrest as related to circadian gating of cell proliferation growth in *Euglena.* Plant Cell Physiol 46:931–936

Bolige A, Kiyota M, Goto K (2005b) Circadian rhythms of resistance to UV-C and UV-B radiation in *Euglena* as related to "escape from light" and "resistance to light". J Photochem Photobiol B: Biol 81:43–54

Borevitz JO, Xia Y, Blount J, Dixon RA, Lamb C (2000) Activation tagging identifies a conserved MYB regulator of phenylpropanoid biosynthesis. Plant Cell 12:2383–2394

Boyer JS (1968) Relationship of water potential to growth of leaves. Plant Physiol 43:1056–1062

Breyne P, Dreesen R, Vandepoele K, De Veylder L, Van Breusegem F, Callewaert L, Rombauts S, Raes J, Cannoot B, Engler G, Inzé D, Zabeau M (2002) Transcriptome analysis during cell division in plants. Proc Natl Acad Sci USA 99:14825–14830

Bunce JA (1977) Leaf elongation in relation to leaf water potential in soybean. J Exp Bot 28:156–161

Bünning E (1952) Über den Tagesrhythmus der Mitosehäufigkeit in Pflanzen. Z Bot 40:193–199

Carabelli M, Morelli G, Whitelam G, Ruberti I (1996) Twilight-zone and canopy shade induction of the *Athb-2* homeobox gene in green plants. Proc Natl Acad Sci USA 93: 3530–3535

Chattopadhyay S, Ang L-H, Puente P, Deng X-W, Wei N (1998) *Arabidopsis* bZIP protein HY5 directly interacts with light-responsive promoters in mediating light control of gene expression. Plant Cell 10: 673–683

Christ RA (1978) The elongation rate of wheat leaves. I. The elongation rates during night and day. J Exp Bot 29:603–610

Cluis CP, Mouchel CF, Hardtke CS (2004) The *Arabidopsis* transcription factor HY5 integrates light and hormone signaling pathways. Plant J 38:332–347

Cosgrove DJ (2000) Expansive growth of plant cell walls. Plant Physiol Biochem 38:109–124

Covington MF, Panda S, Liu XL, Strayer CA, Wagner DR, Kay SA (2001) ELF3 modulates resetting of the circadian clock in *Arabidopsis.* Plant Cell 13:1305–1315

Cuadrado A, Navarrete MH, Canovas JL (1985) The effect of partial protein synthesis inhibition on cell proliferation in higher plants. J Cell Sci 76:97–104

Davies WJ, van Volkenburgh E (1983) The influence of water deficit on the factors controling the daily pattern of growth of *Phaseolus* trifoliates. J Exp Bot 34:987–999

Demmig-Adams B, Adams WW III (1992) Photoprotection and other responses of plants to high light stress. Annu Rev Plant Physiol Plant Mol Biol 43:599–626 den Boer BGW, Murray JAH (2000) Triggering the cell cycle in plants. Trends Cell Biol 10:245–250

Devlin PF, Yanovsky MJ, Kay SA (2003) A genomic analysis of the shade avoidance response in *Arabidopsis*. Plant Physiol 133:1617–1629

Dodd AN, Salathia N, Hall A, Kévei E, Tóth R, Nagy F, Hibberd JM, Millar AJ, Webb AAR (2005) Plant circadian clocks increase photosynthesis, growth, survival, and competitive advantage. Science 309:630–633

Dowson-Day MJ, Millar AJ (1999) Circadian dysfunction causes aberrant hypocotyl elongation patterns in *Arabidopsis*. Plant J 17:63–71

Durand JL, Onillon B, Schnyder H, Rademacher I (1995) Drought effects on cellular and spatial parameters of leaf growth in tall fescue. J Exp Bot 46:1147–1155

Evans JR, Poorter H (2001) Photosynthetic acclimation of plants to growth irradiance: the relative importance of specific leaf area and nitrogen partitioning in maximizing carbon gain. Plant Cell Environ 24:755–767

Folta KM, Lieg EJ, Durham T, Spalding EP (2003) Primary inhibition of hypocotyl growth and phototropism depend differently on phototropin-mediated increases in cytoplasmic calcium induced by blue light. Plant Physiol 133:1464–1470

Frak E, Le Roux X, Millard P, Adam B, Dreyer E, Escuit C, Sinoquet H, Vandame M, Variet-Grancher C (2002) Spatial distribution of leaf nitrogen and photosynthetic capacity within the foliage of individual trees: disentangling the effects of local light quality, leaf irradiance, and transpiration. J Exp Bot 53:2207–2216

Frankhauser C, Staiger D (2002) Photoreceptors in *Arabidopsis* thaliana: light perception, signal transduction and entrainment of the endogenous clock. Planta 216:1–16

Franklin KA, Whitelam GC (2005) Phytochromes and shade-avoidance responses in plants. Ann Bot 96:169–175

Goto K, Johnson CH (1995) Is the cell division cycle gated by a circadian clock? The case of *Chlamydomonas reinhardtii*. J Cell Biol 129:1061–1069

Gouws LM, Osmond CB, Schurr U, Walter A (2005) Distinctive diel growth cycles in leaves and cladodes of CAM plants: differences from C_3 plants and putative interactions with substrate availability, turgor and cytoplasmic pH. Funct Plant Biol 32:421–428

Green RM, Tingay S, Wang ZY, Tobin EM (2002) Circadian rhythms confer a higher level of fitness to *Arabidopsis* plants. Plant Physiol 129:576–584

Green RM, Tobin EM (1999) Loss of the circadian clock-associated protein 1 in *Arabidopsis* results in altered clock-regulated gene expression. Proc Natl Acad Sci USA 96: 4176–4179

Hagiwara S, Bolige A, Zhang Y, Takahashi M, Yamagishi A, Goto K (2002) Circadian gating of photoinduction of commitment to cell-cycle transitions in relation to photoperiodic control of cell reproduction in *Euglena*. Photochem Photobiol 76:105–115

Halliday KJ, Hudson M, Ni M, Quail PH (1999) *poc1*: an *Arabidopsis* mutant perturbed in phytochrome signaling because of a T DNA insertion in the promoter of *PIF3*, a gene encoding a phytochrome-interacting bHLH protein. Proc Natl Acad Sci USA 96:5832–5837

Harmer SL, Hogenesch JB, Straume M, Chang H-S, Han B, Zhu T, Wang X, Kreps JA, Kay SA (2000) Orchestrated transcription of key pathways in *Arabidopsis* by the circadian clock. Science 290:2110–2113

Hayama R, Coupland G (2003) Shedding light on the circadian clock and the photoperiodic control of flowering. Curr Opin Plant Biol 6:13–19

Jarillo JA, Capel J, Tang RH, Yang HQ, Alonso JM, Ecker JR, Cashmore AR (2001) An *Arabidopsis* circadian clock component interacts with both CRY1 and phyB. Nature 410:487–490

Jorgensen P, Tyers M, Warner JR (2004) Forging the factory: ribosome synthesis and growth control in budding yeast. In: Hall MN, Raff M, Thomas G (eds) Cell growth: control of cell size. Cold Spring Harbor Laboratory Press, New York, pp 329–370

Kim W-Y, Hicks KA, Somers DE (2005) Independent roles for EARLY FLOWERING 3 and ZEITLUPE in the control of circadian timing, hypocotyl length, and flowering time. Plant Physiol 139:1557–1569

Kozma-Bognár L, Hall A, Ádam É, Thain SC, Nagy F, Millar AJ (1999) The circadian clock controls the expression pattern of the circadian input photoreceptor, phytochrome B. Proc Natl Acad Sci USA 96:14652–14657

Lai I-L, Scharr H, Chavarría-Krauser A, Küsters R, Wu J-T, Chou C-H, Schurr U, Walter A (2005) Leaf growth dynamics of two congener gymnosperm tree species reflect the heterogeneity of light intensities given in their natural ecological niche. Plant Cell Environ 28:1496–1505

Lecharny A, Schwall M, Wagner E (1985) Stem extension rate in light-grown plants. Plant Physiol 79:625–629

Liscum E, Hodgson DW, Campbell TJ (2003) Blue light signaling through the cryptochromes and phototropins. So that's what the blues is all about. Plant Physiol 133:1429–1436

Liu XL, Covington MF, Frankhauser C, Chory J, Wagner DR (2001) *ELF3* encodes a circadian clock-regulated nuclear protein that functions in an *Arabidopsis PHYB* signal transduction pathway. Plant Cell 13:1293–1304

Lüttge U (2003) Circadian rhythmicity: Is the "Biological Clock" hardware or software? Prog Bot 64:277–319

Ma L, Gao Y, Qu L, Chen Z, Li J, Zhao H, Deng XW (2002) Genomic evidence for COP1 as a repressor of light-regulated gene expression and development in *Arabidopsis.* Plant Cell 14:2383–2398

Makino S, Matsushika A, Kojima M, Yamashino T, Mizuno T (2002) The APRR1/TOC1 quintet implicated in circadian rhythms of *Arabidopsis thaliana*: characterization with APRR1-overexpressing plants. Plant Cell Physiol 43:58–59

Maksymowych R (1973) Analysis of leaf development. Cambridge University Press, Cambridge

Martínez-García JF, Huq E, Quail PH (2000) Direct targeting of light signals to a promoter element-bound transcription factor. Science 288:859–863

Más P, Alabadí D, Yanovsky MJ, Oyama T, Kay SA (2003a) Dual Role of TOC1 in the control of circadian and photomorphogenic responses in *Arabidopsis.* Plant Cell 15:223–236

Más P, Kim WY, Somers DE, Kay SA (2003b) Targeted degradation of TOC1 by ZTL modulates circadian function in *Arabidopsis* thaliana. Nature 426:567–570

Matsubara S, Hurry V, Druart N, Benedict C, Janzik I, Chavarría-Krauser A, Walter A, Schurr U (2006) Nocturnal changes in leaf growth of *Populus deltoids* are controlled by cytoplasmic growth. Planta 223:1315–1328

Matt P, Schurr U, Klein D, Krapp A, Stitt M (1998) Growth of tobacco in short-day conditions leads to high starch, low sugars, altered diurnal changes in the *Nia* transcript and low nitrate reductase activity, and inhibition of amino acid synthesis. Planta 207:27–41

McClung CR (2001) Circadian rhythms in plants. Annu Rev Plant Physiol Plant Mol Biol 52:139–162

McWatters HG, Bastow RM, Hall A, Millar AJ (2000) The *ELF3 zeitnehmer* regulates light signalling to the circadian clock. Nature 408:716–720

Menges M, Hennig L, Gruissem W, Murray JAH (2002) Cell cycle-regulated gene expression in *Arabidopsis.* J Biol Chem 277:41987–42002

Millar AJ (2004) Input signals to the plant circadian clock. J Exp Bot 55:277–283

Millar AJ, Kay SA (1996) Integration of circadian and phototransduction pathways in the network controlling CAB gene transcription in *Arabidopsis.* Proc Natl Acad Sci USA 93:15491–15496

Misquitta RW, Herrin DL (2005) Circadian regulation of chloroplast gene transcription: A review. Plant Tissue Cult 15:83–101

Mittag M, Kiaulehn S, Johnson CH (2005) The circadian clock in *Chlamydomonas reinhardtii.* What is it for? What is it similar to? Plant Physiol 137:399–409

Mizoguchi T, Wheatley K, Hanzawa Y, Wright L, Mizoguchi M, Song HR, Carré IA, Coupland G (2002) *LHY* and *CCA1* are partially redundant genes required to maintain circadian rhythms in *Arabidopsis.* Dev Cell 2:629–641

Morelli G, Ruberti I (2000) Shade avoidance responses. Driving auxin along lateral routes. Plant Physiol 122:621–626

Mori T, Binder B, Johnson CH (1996) Circadian gating of cell division in cyanobacteria growing with average doubling times of less than 24 hours. Proc Natl Acad Sci USA 93:10183–10188

Nagatani A (2004) Light-regulated nuclear localization of phytochromes. Curr Opin Plant Biol 7:708–711

Neff MM, Frankhauser C, Chory J (2000) Light: an indicator of time and place. Genes Dev 14:257–271

Ni M, Tepperman JM, Quail PH (1999) Binding of phytochrome B to its nuclear signalling partner PIF3 is reversibly induced by light. Nature 400:781–784

Niinemets Ü, Kull O, Tenhunen JD (2004) Within canopy variation in the rate of development of photosynthetic capacity is proportional to integrated quantum flux density in temperate deciduous trees. Plant Cell Environ 27:293–313

Nikaido SS, Johnson CH (2000) Daily and circadian variation in survival from ultraviolet radiation in *Chlamydomonas reinhardtii.* Photochem Photobiol 71:758–765

Niyogi KK (1999) Photoprotection revisited: genetic and molecular approaches. Annu Rev Plant Physiol Plant Mol Biol 50:333–359

Osterlund MT, Hardtke CS, Wei N, Deng XW (2000) Target destabilization of HY5 during light-regulated development of *Arabidopsis.* Nature 405:462–466

Ouyang Y, Andersson CR, Kondo T, Golden SS, Johnson CH (1998) Resonating circadian clocks enhance fitness in cyanobacteria. Proc Natl Acad Sci USA 95:8660–8664

Oyama T, Shimura Y, Okada K (1997) The *Arabidopsis* HY5 gene encodes a bZIP protein that regulates stimulus-induced development of root and hypocotyl. Genes Dev 11:2983–2995

Paietta J (1982) Photooxidation and the evolution of circadian rhythmicity. J Theor Biol 97:77–82

Parks BM, Cho MH, Spalding EP (1998) Two genetically separable phases of growth inhibition induced by blue light in *Arabidopsis* seedlings. Plant Physiol 118:609–615

Pepper AE, Corbett RW, Kang N (2002) Natural variation in *Arabidopsis* seedling photomorphogenesis reveals a likely role for *TED1* in phytochrome signalling. Plant Cell Environ 25:591–600

Pigliucci M (1998) Developmental phenotypic plasticity: where internal programming meets the external environment. Curr Opin Plant Biol 1:87–91

Pittendrigh CS (1993) Temporal organization: reflections of a Darwinian clock-watcher. Annu Rev Physiol 55:17–54

Price LE, Bacon MA, Young PC, Davies WJ (2001) High-resolution analysis of tomato leaf elongation: the application of novel time-series analysis techniques. J Exp Bot 52:1925–1932

Quail PH (2002a) Photosensory perception and signalling in plant cells: new paradigms? Curr Opin Cell Biol 14:180–188

Quail PH (2002b) Phytochrome photosensory signalling networks. Nat Rev Mol Cell Biol 3:85–93

Rogers LA, Dubos C, Cullis IF, Surman C, Poole M, Willment J, Mansfield SD, Campbell MM (2005) Light, the circadian clock, and sugar perception in the control of lignin biosynthesis. J Exp Bot 416:1651–1663

Sack L, Cowan PD, Jaikumar N, Holbrook NM (2003) The "hydrology" of leaves: co-ordination of structure and function in temperate woody species. Plant Cell Environ 26:1343–1356

Salter MG, Franklin KA, Whitelam GC (2003) Gating of the rapid shade-avoidance response by the circadian clock in plants. Nature 426:680–683

Schaffer R, Landgraf J, Accerbi M, Simon V, Larson M, Wisman E (2001) Microarray analysis of diurnal and circadian-regulated genes in *Arabidopsis*. Plant Cell 13:113–123

Schaffer R, Ramsay N, Samach A, Corden S, Putterill J, Carré IA, Coupland G (1998) The *late elongated hypocotyl* mutation of *Arabidopsis* disrupts circadian rhythms and the photoperiodic control of flowering. Cell 93:1219–1229

Schmid M, Davison TS, Henz SR, Pape UJ, Demar M, Vingron M, Schölkopf B, Weigel D, Lohmann JU (2005) A gene expression map of *Arabidopsis thaliana* development. Nat Genet 37:501–506

Schmundt D, Stitt M, Jähne B, Schurr U (1998) Quantitative analysis of the local rates of growth of dicot leaves at a high temporal and spatial resolution, using image sequence analysis. Plant J 16:505–514

Schultz TF, Kay SA (2003) Circadian clocks in daily and seasonal control of development. Science 301:326–328

Schurr U, Heckenberger U, Herdel K, Walter A, Feil R (2000) Leaf development in *Ricinus communis* during drought stress: dynamics of growth processes, of cellular structure and of sink-source transition. J Exp Bot 51: 1515–1529.

Schurr U, Walter A, Rascher U (2006) Functional dynamics of plant growth and photosynthesis—from steady-state to dynamics—from homogeneity to heterogeneity. Plant Cell Environ 29:340–352

Schwechheimer C, Serino G, Callis J, Crosby WL, Lyapina S, Deshaies RJ, Gray WM, Estelle M, Deng X-W (2001) Interactions of the COP9 signalosome with the E3 ubiquitin ligase SCF^{TIR1} in mediating auxin response. Science 292:1379–1382

Seneweera SP, Basra AS, Barlow EW, Conroy JP (1995) Diurnal regulation of leaf blade elongation in rice by CO_2. Plant Physiol 108:1471–1477

Shackel KA, Matthews MA, Morrison JC (1987) Dynamic relation between expansion and cellular turgor in growing grape (*Vitis vinifera* L.) leaves. Plant Physiol 84:1166–1171

Smith H (1995) Physiological and ecological function within the phytochrome family. Annu Rev Plant Physiol Plant Mol Biol 46:289–315

Smith H (2000) Phytochromes and light signal perception by plants—an emerging synthesis. Nature 407:585–591

Smith SM, Fulton DC, Chia T, Thorneycroft D, Chapple A, Dunstan H, Hylton C, Zeeman SC, Smith AM (2004) Diurnal changes in the transcriotome encoding enzymes of starch metabolism provide evidence for both transcriptional and posttranscriptional regulation of starch metabolism in *Arabidopsis* leaves. Plant Physiol 136:1–13

Somers DE (1999) The physiology and molecular bases of the plant circadian clock. Plant Physiol 121:9–19

Somers DE, Devlin PF, Kay SA (1998a) Phytochromes and cryptochromes in the entrainment of the *Arabidopsis* circadian clock. Science 282:1488–1490

Somers DE, Webb AAR, Pearson M, Kay SA (1998b) The short-period mutant, toc1-1, alters circadian clock regulation of multiple outputs throughout development in *Arabidopsis thaliana*. Development 125:485–494

Staiger D (2002) Circadian rhythms in *Arabidopsis*: time for nuclear proteins. Planta 214:334–344

Stals H, Inzé D (2001) When plant cells decide to divide. Trends Plant Sci 6: 359–364

Steindler C, Matteucci A, Sessa G, Weimar T, Ohgishi M, Aoyama T, Morelli G, Ruberti I (1999) Shade avoidance responses are mediated by the ATHB-2 HD-Zip protein, a negative regulator of gene expression. Development 126:4235–4245

Sterky F, Bhalerao RR, Unneberg P, Segerman B, Nilsson P, Brunner AM, Charbonnel-Campaa L, Jonsson-Lindvall J, Tandre K, Strauss SH, Sundberg B, Gustafsson P, Uhlén M, Bhalerao RP, Nilsson O, Sandberg G, Karlsson J, Lundeberg J, Jansson S (2004) A Populus EST resource for plant functional genomics. Proc Natl Acad Sci USA 101:13951–13956

Tanaka S-I, Nakamura S, Mochizuki N, Nagatani A (2002) Phytochrome in cotyledons regulates the expression of genes in the hypocotyl through auxin-dependent and –independent pathways. Plant Cell Physiol 43:1171–1181

Taylor G, Davies WJ (1985) The control of leaf growth of *Betula* and *Acer* by photoenvironment. New Phytol 101:259–268

Taylor G, Davies WJ (1986) Leaf growth of *Betula* and *Acer* in simulated shade light. Oecologia 69:589–593

Terashima I, Hikosaka K (1995) Comparative ecophysiology of leaf and canopy photosynthesis. Plant Cell Environ 18:1111–1128

Terashima I, Hanba YT, Tazoe Y, Vyas P, Yano S (2006) Irradiance and phenotype: comparative eco-development of sun and shade leaves in relation to photosynthetic CO_2 diffusion. J Exp Bot 57:343–354

Tóth R, Kevei É, Hall A, Millar AJ, Nagy F, Kozma-Bognár L (2001) Circadian clock-regulated expression of phytochrome and cryptochrome genes in *Arabidopsis.* Plant Physiol 127:1607–1616

Tsukaya H (2002) The leaf index: heteroblasty, natural variation, and the genetic control of polar processes of leaf expansion. Plant Cell Physiol 43:372–378

Tsukaya H (2003) Organ shape and size: a lesson from studies of leaf morphogenesis. Curr Opin Plant Biol 6:57–62

Urbanczyk-Wochniak E, Baxter C, Kolbe A, Kopka J, Sweetlove LJ, Fernie AR (2005) Profiling of diurnal patterns of metabolite and transcript abundance in potato (*Solanum tuberosum*) leaves. Planta 221:891–903

Vandenbussche F, Pierik R, Millenaar FF, Voesenek LACJ, Van Der Straeten D (2005) Reaching out of the shade. Curr Opin Plant Biol 8:462–468

Van Volkenburgh E (1999) Leaf expansion—an integrating plant behaviour. Plant Cell Environ 22:1463-1473

Van Volkenburgh E, Cleland RE (1981) Control of light-induced bean leaf expansion: role of osmotic potential, wall yield stress, and hydraulic conductivity. Planta 153: 572–577

Walter A, Schurr U (2000) Spatio-temporal variation of leaf growth, development and function. In: Marshall B, Roberts JA (eds) Leaf development and canopy growth. Sheffield Academic Press, Sheffield, pp 96–117

Walter A, Schurr U (2005) Dynamics of leaf and root growth: endogenous control versus environmental impact. Ann Bot 95:891–900

Walter A, Feil R, Schurr U (2002a) Restriction of nyctinastic movements and application of tensile forces to leaves affects diurnal patterns of expansion growth. Funct Plant Biol 29:1247–1258

Walter A, Spies H, Terjung S, Küsters R, Kirchgeaner N, Schurr U (2002b) Spatio-temporal dynamics of expansion growth in roots: automatic quantification of diurnal course and temperature response by digital image sequence processing. J Exp Bot 53:689–698

Walter A, Christ MM, Barron-Gafford GA, Grieve KA, Murthy R, Rascher U (2005) The effect of elevated CO_2 on diel leaf growth cycle, leaf carbohydrate content and canopy growth performance of *Populus deltoids.* Global Change Biol 11:1207–1219

Wang Z-Y, Tobin EM (1998) Constitutive expression of the *Circadian clock associated 1* (*CCA1*) gene disrupts circadian rhythms and suppresses its own expression. Cell 93:1207–1217

Wang Z-Y, Kenigsbuch F, Sun L, Harel E, Ong MS, Tobin EM (1997) A myb-related transcription factor is involved in the phytochrome regulation of an *Arabidopsis Lhcb* gene. Plant Cell 9:491–507

Watts WR (1974) Leaf extension in *Zea mays.* J Exp Bot 25:1085–1096

Wollman F-A (2001) State transitions reveal the dynamics and flexibility of the photosynthetic apparatus. EMBO J 20:3623–3630

Shizue Matsubara,
Achim Walter
Institute of Chemistry and Dynamics of the Geosphere ICG-3: Phytosphere
Research Center Jülich
52425 Jülich
Germany

Corresponding author
Shizue Matsubara
Institute of Chemistry and Dynamics of the Geosphere ICG-3: Phytosphere
Research Center Jülich
52425 Jülich
Germany
Tel: +49 2461 618690
Fax: +49 2461 612492
e-mail: s.matsubara@fz-juelich.de

Ecology

Competitive Networks, Indirect Interactions, and Allelopathy: A Microbial Viewpoint on Plant Communities

Ragan M. Callaway and Timothy G. Howard

"... the more numerous and complex the competitive networks on any substratum, the more slowly will the available space be dominated by any single competitor."

Buss and Jackson, 1979

1 Introduction

The ecological processes that promote coexistence among species requiring the same resources have been sought for decades. The most powerful mechanisms that promote diversity appear to be those that disrupt community equilibrium, such as herbivory or abiotic disturbance, but these mechanisms are not omnipresent, therefore generating interest in mechanisms that operate when communities are at equilibrium. Understanding the relative importance of mechanisms that promote coexistence is not trivial, because coexistence is elemental to biological diversity. Conceptual models for coexistence under equilibrium conditions are generally based on niche theory (Tilman 2000). In other words, competing species can coexist if they do not use exactly the same resources at exactly the same time, or require precisely the same conditions to maximize growth and reproduction. Ecologists have convincingly demonstrated spatial niche partitioning (Parrish and Bazzaz 1976; Cody 1986), niches based on resource monopolization and renewal and resource ratios (Connell 1971; MacArthur 1972; Tilman 1982), and temporal niche partitioning (Chesson 1983; Chesson and Huntly 1993). Recently, Clark et al. (2004) demonstrated that niche-based theory may not be obligatory; reporting that "massive stochasticity" in fecundity and recruitment among tree species appeared to promote coexistence independently of niche differences.

Niche-based coexistence functions by reducing the probability of competitive exclusion; species avoid competition when they occupy different niches. Competition has long been considered one of the more important forces structuring communities (Watt 1947; Harper 1977). In 1940, Griggs

Progress in Botany, Vol. 68

concluded that "[t]he explanation of rarity must, therefore, lie in an evaluation of the competitive competency of species". Despite this honoured and hoary perspective, and the general paradigm that competition works against coexistence, other evidence indicates that competition also has the potential to have counterintuitive and coexistence-promoting effects in communities. Whether or not competition, or interference, promotes or inhibits coexistence depends how *groups* of species interact. If groups of species compete in non-hierarchical *networks*, then resource monopolization by a single species decreases, and therefore networks maintain higher diversity than would be predicted from pairwise interactions or competitive hierarchies. If species interact through competitive *hierarchies*, this can lead to exclusion in the absence of non-equilibrium mechanisms. The reason for this difference is that networks of competing species cause indirect interactions which can have powerful positive effects on coexistence. Indirect interactions among competitors occur when interactions between two species are caused or altered by simultaneous interactions with additional species (Buss and Jackson 1979; Lawlor 1979; Stone and Roberts 1991; Miller 1994; Wooton 1994) or through cumulative, "diffuse" effects that occur when numerous species have different kinds of direct effects that act on a single species (Davidson 1980; Wilson and Keddy 1986; Vandermeer 1980).

Indirect interactions involving consumers are based on solid theory (Andrewartha and Birch 1954; MacArthur 1972), straightforward mechanisms, and have been thoroughly examined empirically (e.g. Paine 1966; Wooton 1992; Pennings 1996); but indirect interactions among competitors remain somewhat mysterious. Indirect interactions are produced by something analogous to an alliance—either two species join together to elicit a stronger synergistic negative effect, or one species ameliorates the negative direct effect of another neighbour, resulting in a facilitative or positive effect. The latter case has been described as "an enemy of my enemy is my friend" and is the most common indirect interaction described in the literature.

Indirect interactions among competitors cannot occur when competitive interactions within a community are hierarchical in nature (transitive patterns), in other words, when all species in a community can be ranked in linear order of competitive ability and all species higher in competitive rank outcompete *all* species lower in competitive rank. However, when species compete intensely against some neighbouring species, but less intensely against others without forming clear hierarchies, the resulting networks or intransitive patterns can create indirect effects (Karlson and Jackson 1981). For example, Levine (1976) modelled the effects of adding a third competitor in a system with two competing species and found that the additional species could change the cumulative effect of one species from competitive to facilitative because of how it suppressed a shared competitor. The third species may have competitive effects on both other species, but as long as the third species indirectly relieves the competitive pressure on a neighbour more than it directly hurts that same neighbour, indirect facilitation can occur.

Indirect interactions among competitors are exceptionally difficult to study empirically, simply because of the logistical and interpretive problems of manipulating more than one species in an ecologically meaningful way, but a number of experimental studies have demonstrated strong indirect interactions among competing plant species (Miller 1994; Li and Wilson 1998; Levine 1999; Callaway and Pennings 2000).

Networks may be mechanistically based on resource competition but, because plant species compete for very similar resources, resource competition may be limited as a driver of indirect interactions. Therefore, communities that are highly structured by resource competition may be more likely to demonstrate non-species-specific competitive hierarchies, in which some species are simply better at acquiring resources than others. Furthermore, if there is only one predominantly limiting resource in communities in a system then hierarchies are to be expected; if species A outcompetes species B for water, and species B outcompetes species C for water, it is hard to see how species C might outcompete species A for water. But if plants in a community are competing simultaneously for *different* resources, and competitive abilities for different resources are not linked, species-specific interactions may develop (see Tilman 1982, 1988).

Allelopathy is an alternative mechanism to resource competition. In contrast to the limited suite of resources for which plants can compete, plants are known to produce over 100,000 different low-molecular-mass natural biochemicals, many of which are released into the surrounding environments and appear to be species-specific (Flores 1999). Far more are likely to be discovered. This creates the potential for highly species-specific biochemical, or allelopathic, interactions among plants that therefore have the potential to create highly complex interactive networks. The central objective of this paper is to consider allelopathy as a possible mechanism for the establishment of non-hierarchical networks in plant communities which may promote coexistence in equilibrium conditions.

Here we try to synthesize 1) the evidence for hierarchical and non-hierarchical interactions in communities, 2) the evidence for indirect interactions among plants, 3) the potential for allelopathy to mediate indirect effects, 4) new insight about interspecific variation in tolerance to the allelopathic effects of species, and 5) novel discoveries in microbial ecology suggesting that balanced interplay between resource competition and allelopathy establishes non-hierarchical relationships and promote coexistence. To be clear, the connections are tenuous, but enough pieces of the puzzle are in place to make a synthesis of these ideas timely.

2 Do hierarchies or networks characterize competitive interactions in plant communities?

There are examples of both hierarchies and networks in the literature. Even so, whether or not plant communities show hierarchical transitivity or non-hierarchical intransitivity has received considerable attention (Keddy and Shipley 1989; Herben and Krahulec 1990; Silvertown and Dale 1991; Grace et al. 1993; Shipley 1993; Connolly 1997). If competitive relationships among species in a community are transitive, simple plant traits can be used to predict competitive ability (Herben and Krahulec 1990) and indirect interactions are not likely to strongly affect coexistence. If competitive relationships are non-transitive, predicting competitive ability from specific traits is more difficult because many different traits have the potential to establish competitive dominance, but indirect interactions are much more likely to develop.

Of course some communities may be hierarchical and others non-hierarchical, but the evidence suggests that most plant communities are hierarchical and transitive in nature. Good examples include Goldsmith (1978), Mitchley and Grubb (1986), Wilson and Keddy (1986), Keddy and Shipley (1989) and Keddy et al. (2002). Shipley (1993) analysed ten published matrices of replacement series experiments and found nine cases of complete transitivity. Goldberg (1997) reviewed studies using additive experiments and found mostly transitive rankings among species for both competitive effect and competitive response.

Transitive rankings may be the most common pattern in the literature, but this may be affected by methodological bias. Rankings in competitive ability has been determined most often with replacement series, additive experimental designs (Snaydon 1991), or simply one on one trials. The replacement series design maintains a constant total density and varies relative proportions of two competing species in a series of replicates (de Wit 1961; Trenbath 1974; Harper 1977). Additive experimental designs keep focal species at a constant density (often a single individual) while varying the density of a competitor. Most, but not all, studies of competitive rankings compare all possible pairwise comparisons using the methods described above, and then use these pairwise interactions to assign competitive rankings. As noted, rankings developed in this way tend to show strong transitive hierarchies that are inconsistent with the non-transitive networks necessary to produce indirect interactions. Interestingly, a number of experimental field removals (or perturbation experiments) of different species from intact communities contrast strikingly with conclusions based on pairwise rankings, and instead demonstrate complex networks of interacting species.

For example, Fowler (1981) removed individual species from old field communities in North Carolina, USA and then measured the responses of the remaining species. From her published results, we ranked the responses (regardless of statistical significance) of the manipulated species to the removal of each other and found no evidence for a competitive hierarchy. For example, removal of the numerical dominant, *Plantago lanceolata*, caused large increases in *Cynodon dactylon* and *Paspalum leava* and had no affect on *Poa pratensis*, suggesting that *Plantago* was a strong competitor against *Cynodon* and *Paspalum* but a weak competitor against *Poa*. If Fowler's community was hierarchical, then *Cynodon* and *Paspalum* would also be expected to be even weaker competitors against *Poa*. This was not the case; *Paspalum* showed an intermediate competitive ranking versus *Poa*, and *Cynodon* strongly suppressed *Poa*. Other species showed similar non-transitive relationships. Importantly, Fowler's non-transitive networks corresponded with documentation of indirect interactions. Two species responded to the effects of three or more neighbours in ways that were not predictable from pairwise interactions. For example, *Plantago* suppressed winter annuals, but only in the presence of *Rumex acetosella*. The results from another field perturbation experiment by Allen and Forman (1976) also show strong non-hierarchical relationships.

Miller (1994) conducted a perturbation experiment by selectively removing species from experimental plots and comparing the growth of remaining species to their growth without competitors. Miller's results were more hierarchical than Fowler's (*Ambrosia artemisiifolia* was the competitive dominant and had strong direct suppressive effects on all other species), but interactions among several species were non-transitive and strong indirect effects were demonstrated. Overall, Miller found that the indirect effects from his experiments were larger than predicted by theoretical studies and argued that such strong indirect effects should be the rule rather than the exception in plant communities.

As noted by Silander and Antonovics (1982), experimental perturbation studies have the advantage of testing specific biotic interactions in natural conditions. They performed perturbation experiments on a set of adjacent coastal plant communities. They found that species interactions structure these plant communities, but the interactions were "specific or diffuse, reciprocal or non-reciprocal and may vary in different environments". In other words, they found little evidence for strict competitive hierarchies. Similarly, Von Holle and Simberloff (2004) manipulated invaded communities and came to the conclusion that "positive, indirect interactions may be more important in structuring this community than negative, direct interactions between functional group members".

Other insight into competitive rankings has come from recent studies showing that simple competitive hierarchies do not fall out as clearly when they are compared across life history stages. For example, Suding and Goldberg (2001) quantified competitive response ability for survival and growth for three herbaceous prairie species. They found that disturbance shifted a "distinct competitive hierarchy" for growth to "competitive equivalence" for survival. The competitive rankings for survival in gaps were opposite those in undisturbed communities, where strong competitive hierarchies were detected. Similarly, Howard and Goldberg (2001) found strong competitive hierarchies based on size, but no hierarchies based on survival. Such variation in hierarchy position across lifestages may promote coexistence.

Herben et al. (2001) suggested an intriguing explanation for the differences we have described between pairwise experiments in homogeneous environments and perturbation experiments in the field. They argued that competition is highly asymmetrical when mineral nutrition and water are sufficient and only competition for light exists, allowing species to form clear competitive hierarchies (see Keddy and Shipley 1989; Gaudet and Keddy 1988, 1995; Goldberg and Landa 1991). Under such circumstances, competitive success can be determined by one resource that is spatially homogeneous, and Herben et al. (2001) suggest that size is the most important trait in homogeneous conditions. In contrast, competitive interactions in more heterogeneous conditions are likely to be determined by more than one trait or more than one resource and cannot be easily expressed by a single quantity. Plant size is much less important. Therefore interactions may involve several traits, become more species-specific and create non-hierarchical networks of competitive interactions.

In sum, it would appear that either competitive hierarchies are common and competitive networks are relatively rare, or particular methodological approaches to measuring competitive relationships have skewed the current perspective. Replacement series and additive design experiments seem to generally demonstrate hierarchies, while perturbation experiments often show networks. It is not clear why this difference might exist, but perturbation experiments measure the responses of species while they are still competing with a suite of other species, creating a situation that is not easy to interpret. A final problem, suggesting an irreducibly complex relationship between the niche and competitive hierarchy models for community structure, is that few tests of competitive relationships have considered abiotic conditionality. Plant species compete quite differently in different abiotic conditions (Pennings and Callaway 1992; Callaway et al. 1996), often reversing competitive dominance. For example, Novoplansky and Goldberg (2001) found that providing water to plants in pulses substantially altered competitive hierarchies (but see Keddy et al. 1994). This suggests that slight changes in water, fertilizer, or light might create entirely different hierarchies, a result that may support coexistence if these resources vary over time.

The verdict still appears to be out on the general occurrence and the importance of non-transitive competitive relationships in plant communities, but the occurrence of indirect interactions (see the next section) and demonstrated examples non-transitive competitive relationships in the field indicate that non-transitive networks exist in plant communities.

3 Do indirect interactions occur among competing plants?

Yes, but there have been few empirical studies, and therefore we do not know how important indirect interactions are relative to other processes. In this section, we present several empirical, experimental studies that demonstrate indirect interactions in plant communities.

As noted above, Miller (1994) used five competing old-field species to develop a model of plant interactions designed to estimate direct and indirect effects occurring between all possible pairs of species in a community. The model assumed "competitor equivalence"—all species have the same per-gram competitive effect on a focal species regardless of their identity, therefore the yields of different associate species could be added for a net effect. Competitor equivalence is controversial, but Miller integrated his model with an elegant 2-year field experiment in which he selectively removed species from experimental plots and compared the growth of remaining species to their potential growth without competitors. In this way, he quantified direct and indirect effects among six old-field plant species (two species were substituted for each other between the years) over 2 years.

Miller found that direct and indirect effects were common and strong, but interactions among the five species in the second year occurred in an unpredictable network and interaction strengths were not equivalent among species. Instead, interactions sorted themselves with direct negative effects among particular species being balanced by positive indirect effects. Total effects were always neutral to negative. The strongest example of offsetting direct and indirect effects was for *Ambrosia* and *Agropyron*. The proportional direct effect of *Ambrosia* on the potential growth of *Agropyron* was –41%, yet the proportional indirect effect was +43%, apparently because *Ambrosia* also highly suppressed other competitors.

Miller's results supported a model developed by Case (1991), which suggested that native plant species may be completely displaced by exotics in pairwise interactions, but can coexist with exotics in diverse communities, apparently because of the greater number of indirect interactions.

In another example, large tussocks of *Carex nudata* directly facilitate other plant species by protecting them from flooding disturbance. In addition to this direct facilitative effect, *Carex* has indirect facilitative effects on the species growing within its root mats and tussocks. Levine (1999) manipulated the presence of *Carex* and *Mimulus guttatus* in a factorial design and found that thinning the leaves of *Carex* resulted in much bigger *Mimulus*, indicating that the tussocks strongly suppressed *Mimulus*. Dense *Carex* also suppressed the liverwort *Conocephalum conicum*. However, Levine found that the effect of *Carex* on *Conocephalum* was reversed in the presence of *Mimulus*. In other words, when *Conocephalum* had to cope with a very competitive neighbor, *Mimulus*, it helped to be buried inside dense tussocks of a weak competitor, *Carex*. Without the exceptionally competitive neighbour, these same *Carex* tussocks had negative effects.

Li and Wilson (1998) tested whether the presence of conspecifics enhanced the growth and survivorship of *Symphoricarpos occidentalis* seedlings either with or without the perennial grass *Bromus inermus*. After two growing seasons seedling survivorship was reduced in the presence of *Bromus*. For *Symphoricarpos*, conspecific neighbours significantly decreased growth rates when no grasses were present, but conspecific neighbours increased growth rates when growing in stands of *Bromus*.

In the upper zones of coastal salt marshes in southern California, two dominant perennial species have strikingly opposite effects on most co-occurring winter annual species. *Arthrocnemum subterminale*, a succulent sub-shrub, facilitates two annual species, *Parapholis incurva* and *Hutchisinia procumbens* (Callaway 1994); whereas *Monanthecloe littoralis*, a clonal grass, appears to eliminate virtually all annuals by establishing a thick vegetative mat. Pennings and Callaway (2000) tested the hypothesis that *Monanthechloe* might directly outcompete annual species, but that the positive effect of *Arthrocnemum* might buffer annual species from the full negative effect of *Monanthechloe*. Over a 13-year period they found that increasing abundance of *Monanthechloe* correlated with decreases in the abundance of all other species, but *Arthrocnemum* was correlated with much weaker negative effects of *Monanthechloe*. In field experiments, in patches of *Monanthechloe*, several of the annual species survived only when *Arthrocnemum* was present. The most striking indirect effects in the factorial field experiment occurred for *Spergularia* and *Limonium*. In the absence of *Monanthechloe*, *Spergularia* was much more common when *Arthrocnemum* was removed, indicating a competitive effect of *Arthrocnemum*. But when the dominant competitor *Monanthechloe* was present, *Arthrocnemum* strongly facilitated *Spergularia*.

Limonium was indifferent to the removal of *Arthrocnemum* when *Monanthechloe* was absent, but in the presence of *Monanthechloe*, *Limonium* recruits were only found in plots in which *Arthrocnemum* was also present.

4 Allelopathy and indirect interactions among plants

In the previous section, we presented experimental field studies that demonstrated indirect effects in plant communities. None of these studies suggested mechanisms, with the exception of possible competition for light in the *Carex-Mimulus-Conocephalum* interactions, by which non-hierarchical networks were established. In this section, we present studies, admittedly with much weaker causal certainty, which suggest possible links between allelopathy and indirect effects.

In grasslands of the northern Rocky Mountains invaded by *Centaurea maculosa*, *Lupinus sericeus* appears to play a role much like *Arthrocnemum* does in California salt marshes. By integrating ecological, physiological, biochemical signal transduction, and genomic approaches to study of root exudates from *Centaurea*, Bais et al. (2003) were able to isolate a chemical, (±)-catechin, with phytotoxic properties produced by *Centaurea* roots. While certainly not the only factor contributing to the success of *Centaurea*, many studies suggest that (±)-catechin enhances the ability of the invader to competitively exclude North American plant species (Ridenour and Callaway 2001; Bais et al. 2003; Weir et al. 2003; Perry et al. 2005a,b). However, not all North American species are highly susceptible to *Centaurea*. Weir et al. (2006) found that field plots in areas invaded by *Centaurea* containing *Lupinus* were much more likely to also contain native grasses. Furthermore, native grasses had much greater cover in plots with *Lupinus* than without *Lupinus*. In transplant experiments, the native grasses *Festuca idahoensis* and *Pseudoroegneria spicata* grew larger when next to *Lupinus* than when far from *Lupinus*, indicating a facilitative effect of *Lupinus*. However, this facilitative effect occurred only in dense stands of *Centaurea*. In the absence of *Centaurea*, the effect of *Lupinus* was competitive, indicating that the facilitative effect may have been indirect. Importantly for the general thesis of this paper, the mechanism for the indirect effect may be chemically mediated. When *Lupinus* was experimentally exposed to (±)-catechin it increased exudation of organic acids, primarily oxalic acid, from its roots. Mechanistically, the oxalate exuded from the roots of *Lupinus* blocks generation of reactive oxygen species in other plant species that are susceptible (±)-catechin and reduces oxidative damage that is commonly generated in response to (±)-catechin. This was shown by adding oxalate to substrate used to grow native grasses and *Arabidopsis thaliana* in vitro, which alleviated the phytotoxic effects of (±)-catechin. Simply put, pairwise comparisons indicate competitive interactions among *Lupinus* and grasses, but when *Centaurea*, *Lupinus*, and grasses are considered as a network of interacting species, allelopathically mediated indirect interactions become apparent.

Similar indirect facilitative and interfering interactions may occur among *Quercus agrifolia*, *Pholistima auritum*, and annual grasses in California woodlands. Parker and Muller (1982) found that *P. auritum*, often occurs in virtually "pure stands" directly beneath the canopies of some *Q. agrifolia* individuals, with a strong shift to dominance by annual grasses at the outside edges of the tree canopies. Why *Pholistima* can be so strongly associated with *Q. agrifolia* is not known, but the absence of annual grass species in the understory is clearly not due to the direct effects of the oaks. In fact, if *Pholistima* is not present some of these grass species perform far better under oak canopies than in the open grassland (Parker and Muller 1982, Callaway et al. 1991). Parker and Muller found that litter and leachates from *Pholistima* were highly inhibitory to understory grasses. In controlled conditions fresh *Pholistima* litter reduced the germination of *Bromus diandrus* and *Avena fatua* from 96% and 93%, respectively, to zero. In field experiments, fresh *Pholistima* litter reduced *Bromus* germination by 73% and *Avena* by 96%. However, when experiments were conducted with *Pholistima* litter that had been leached (litter was placed in running deionized water for 48 h) at least 92% of seeds germinated in every treatment. The abundance of *Pholistima* under some oaks, and its rarity in the open, indicates that the oaks are directly facilitating *Pholistima*. However, by facilitating *Pholistima* a situation is created in which the allelopathic effects of *Pholistima* inhibit annual grasses. In other words, oaks have powerful negative indirect effects on grasses.

All true epiphytes benefit directly from their hosts. However, interactions among epiphyte species may create linked indirect interactions among autotrophic organisms that are somewhat unique. In mixed evergreen-deciduous forests of the southeastern USA, the vascular epiphytes *Tillandsia usneoides* and *Polypodium polypodioides* are much more common on some host species than on others (Callaway et al. 2001). Furthermore, transplant experiments showed that the growth rates of *Tillandsia* strands were higher on the host species on which they occurred most frequently in nature. Not only do host trees have species-specific direct effects, they also harbour unique communities of non-vascular epiphytes which also correlated highly with the relative abundance of the vascular epiphytes. In experiments in which the foliose *Parmotrema* lichen species were removed from branch segments of *Q. virginiana* (a preferred host of *Tillandsia* on which *Parmotrema* was abundant) growth rates of *Tillandsia* were 19.8% lower than on branches for which *Parmotrema* was not removed. This suggests that *Q. virginiana* may indirectly facilitate *Tillandsia* by directly facilitating *Parmotrema*. Furthermore, *Tillandsia* seedlings that were watered with extracts from *Cryptothecia rubrocincta*, a lichen species common on poor *Tillandsia* host tree species, had significantly lower growth and survival than did those watered with extracts from *Parmotrema*, *Pyxine caesiopruinosa*, "green algae", or rainwater. This suggests that the poor hosts may indirectly suppress *Tillandsia* through their facilitative effects on *Cryptothecia*. Although far from conclusive, these results suggest that different epiphytic lichen species occurring on different host tree species have the potential to indirectly affect, both in positive and negative ways, the distribution and abundance of vascular epiphytes, perhaps through their biochemical effects.

An interesting parallel has been described in intertidal marine communities. Of the competitive rankings conducted in intertidal communities, most are hierarchical, or transitive, suggesting that competitive networks in these systems may be the exception (Buss and Jackson 1979). However, one study of coral reef communities has showed non-transitive networks, and these appeared to be organized by variability in biochemical effects and responses among species (Jackson and Buss 1975). They subjected co-occurring coral reef species to homogenates made from 11 of these species and found that five of nine sponge species and one of two ascidian species exhibited species-specific allelochemical effects. Furthermore, some allelopathic effects were non-transitive in nature suggesting they could function as drivers of indirect interactions and maintain coexistence in the absence of predation and disturbance. To my knowledge, this is the only research in any system that has explicitly connected species-specific allelopathic interactions to the conceptual area of indirect interactions.

5 Do hierarchies in allelopathic tolerance exist?

In the previous sections, we presented evidence for indirect interactions among competing species in plant communities, some evidence for non-transitive networks in plant communities, and for the potential of allelopathy to act as a mechanism causing indirect interactions. As discussed above, this begs the question of whether or not allelopathic effects might drive non-transitive relationships among species and indirect effects. We know of no studies in which the allelopathic effects of each species in a community network have been tested against all other species; in other words, there is no evidence that non-hierarchical, non-transitive networks in *plant* communities can be based on allelopathy. We do know, however, that competitively dominant species can have strong allelopathic effects on some species but weak or no allelopathic effects on other species. Strongly suppressing some species but not others creates the potential for the intransitive conditions necessary for indirect effects—some species must compete well against some neighbouring species, but poorly against others.

Ortega and Pearson (2005) found that the abundance of *C. maculosa* correlated with dramatic declines in some native species, but not all. The negative effects of *C. maculosa* on native taxa varied in strength by functional group and species, with the strongest effects on numerically dominant native species in non-invaded communities. Interestingly, the numerically dominant species tended to be the tallest, and therefore plant size was not a good indicator of competitive ability against *C. maculosa.* This conflicts with the majority of competitive rankings in the literature, which indicate a strong correlation between competitive ability and either height or mass (Keddy 1992).

Perry et al. (2005b) tested the tolerances of 23 grassland species to (±)-catechin, (the suspected allelopathic chemical exuded from *C. maculosa* roots) and found that this too varied widely in sensitivity, at least in controlled laboratory incubations. Inhibition of seedling root elongation was the strongest and most common effect of (±)-catechin treatment. At concentrations of 1 mg/ml (±)-catechin, a concentration similar to that frequently observed in the fields (Bais et al. 2003; Perry et al. 2005b), native seedling root lengths were reduced by more than 40% for 11 species, whereas three species did not respond to (±)-catechin. Based on EC50 (experimentally derived concentrations at which root lengths were reduced by 50%), seven species were "highly sensitive" with EC50s ranging from 0.43 to 0.71 mg/ml, eight species were "sensitive" with EC50s ranging from 1.64 to 2.13 mg/ml, four species were "resistant" with EC50s ranging from 3.28 to 3.45 mg/ml, and two species "more resistant" with EC50s greater than 4 mg/ml. The three species that did not respond to (±)-catechin were considered "highly resistant".

Variation in tolerance to (±)-catechin has also been documented in the field. Thelen et al. (2005) measured the response of eight native herbaceous species to experimental injections of (±)-catechin into rhizospheres. The proportional declines of these species, relative to controls, ranged from +6.9% (not significant) to –72.2% (highly significant). Similar responses were observed in another year with declines ranging from +16% (not significant) to –89.9% (highly significant). Species responded differently to (±)-catechin from year to year, but across all species the response in year 1 was correlated to the response in year 2 ($r = 0.88$, $P = 0.049$).

Only four species were tested in both field and laboratory trials, and for these species response to (±)-catechin tended to correlate ($r = 0.90$, $P = 0.098$).

There was not enough overlap in the species used by Ortega and Pearson (2005) and the species used by Perry et al. (2005b) to correlate sensitivity to (±)-catechin with declines in the field, but considered together, these studies show substantial variation in the competitive and allelopathic effects of *C. maculosa*. Other studies have shown more consistent allelopathic effects among target species (Ahmed and Wardle 1994), while yet others show variation in tolerance similar to what we describe here (Halligan 1976). Interestingly, the interspecific variation in response to allelopathic bioassays described by Halligan was highly correlated with field distributions.

Pinus sylvestris trees are highly variable in the composition and concentrations of monoterpenes (Iason et al. 2005). They found that the chemical diversity of monoterpenes of individual trees was significantly associated with the species richness of understory communities. Similar relationships have been demonstrated among specific *Thymus vulgaris* "chemotypes" and ecotypes of associated species (Linhart and Thompson 1999; Ehlers and Thompson 2004). These results indicate that that chemical diversity of dominant species, apparently through species-specific effects, may have powerful effects on the diversity of plant communities.

Not much allelopathy would be required to shift competitive hierarchies to competitive networks. Allelopathic interactions among all species are certainly not necessary. All it would take is one species "acting out of order" to create non-transitive processes. As we will see in the next section, simple networks can develop from balanced relationships between resource competition and allelopathy.

6 What can we learn from microbes?

Allelopathic interactions can be important in plant communities (for examples see Webb et al. 1967; Williamson 1990; Nilsson 1994; Czarnota et al. 2001; Bertin et al. 2003; Hierro and Callaway 2003). And as described above, the allelopathic effects of a species or a plant-produced chemical on other species can vary widely and may cause indirect interactions among plants (Parker and Muller 1982; Callaway et al. 2001; Weir et al. 2006). To our knowledge there is no direct evidence for competitive networks in plant communities being produced by allelopathic interactions among plants, but there is strong evidence for allelopathically caused indirect effects among microbes, suggesting fascinating potential scenarios for plant communities.

In a paper titled "Chemical warfare between microbes promotes biodiversity", Czárán et al. (2002) used a spatially explicit theoretical game model to explore the potential for biochemical antibiotic interactions (equivalent to allelopathic interactions among plants) within microbial communities to drive indirect interactions and maintain diversity. Microorganisms commonly excrete antibiotic chemicals that inhibit or kill conspecifics or other species, but these chemicals are highly variable in their effects because resistance appears to evolve quickly (Reeves 1972; Berdy 1974; Starmer et al. 1987). Examples include antimicrobial toxins produced by bacteria such as colicins from *Escherichia coli* and nisins from lactic acid bacteria (James et al. 1991; Riley and Gordon 1992). Czárán et al. (2002) concluded that the excretion of antibiotic compounds and the resource competitive effects caused by the associated metabolic costs produced non-transitive networks of interactions analogous to the rock-scissors-paper game in which each of these responses defeats one other response, and is defeated in turn by one other response. In other words, modelled variation in allelopathic effects and responses appeared to drives a cyclical non-hierarchical among species in which A eliminates B, B eliminates C, but C eliminates A. They argued such a non-hierarchical network is likely to maintain diversity even in systems at equilibrium.

Kirkup and Riley (2004) investigated narrow-spectrum biochemical antibiotics produced by particular genotypes of *Escherichia coli*, and which are active against other conspecific genotypes. These antibiotic chemicals are called colicins. Despite strong toxic effects, genotypes that produce colicin coexist with sensitive and resistant strains in natural systems. Laboratory experiments indicated that coexistence among genotypes is promoted by the allelopathic effects of colicin because of non-transitive interactions (Kerr et al. 2002). In these interactive networks, genotypes that produce colicins kill sensitive genotypes, sensitive genotypes outcompete resistant genotypes, but resistant genotypes outcompete colicin producing genotypes. In pairwise competition, competitive exclusion is always the case. Kirkup and Riley established different experimental combinations of *E. coli* genotypes in the colons of mice and monitored changes in bacterial population dynamics in faecal pellets. Their in vivo experiments corroborated theoretical predictions and laboratory results. Because of non-transitive pairwise interactions, the toxic effects of colicins promote genotypic diversity.

7 Conclusions

Indirect effects within groups of interacting plant species may be one of ecology's most overlooked phenomena, and one that could transform our understanding of the mechanisms that maintain coexistence and diversity. Ecologists have tried for decades to understand competition in the context of *avoiding* competition, through niche partitioning and variation in particular resource requirements and uptake rates, and in the context of non-equilibrium processes such as herbivory and non-biotic disturbance. New perspectives on allelopathy in general (Baldwin 2003; Fitter 2003), the potential for highly species-specific allelopathic interactions among members of a community (Perry et al. 2005b; Weir et al. 2006), and insight from allelopathically caused indirect interactions among microbes (Czárán et al. 2002; Kirkup and Riley 2004) suggest that the effects of allelopathy on community organization may be important. The demands of sorting through species-specific allelopathic effects and indirect interactions among groups of plant species are daunting, but if such interactions contribute in novel ways to coexistence and biological diversity, the rewards will be great.

Acknowledgements

Funding was provided by the Aldo Leopold Wilderness Center, the USDA, DoD SERDP, the National Science Foundation, Civilian Research and Development Foundation, and the University of Michigan Rackham School of Graduate Studies. We thank Deborah E. Goldberg for shaping our perspectives on plant competition, competitive hierarchies, and the organization of communities.

References

Ahmed M, Wardle DA (1994) Allelopathic potential of vegetative and flowering ragwort (*Senecio jabobaea* L.) plants against associated pasture species. Plant Soil 164:61–68
Allen EB, Forman RT (1976) Plant species removals and old-field community structure and stability. Ecology 57:1233–1243
Andrewartha HG, Birch LC (1954) The distribution and abundance of animals. The University of Chicago Press, Chicago, Illinois
Bais HP, Vepachedu R, Gilroy S, Callaway RM, Vivanco JM 2003 Allelopathy and exotic plants: from genes to invasion. Science 301:1377–1380
Baldwin IT (2003) At last, evidence of weapons of mass destruction. Science STKE, p 42
Berdy J (1974) Recent developments of antibiotic research and classification of antibiotics according to chemical structure. Adv Appl Microbiol 18:309–406
Bertin C, Yang CX, Weston LA (2003) The role of root exudates and allelochemicals in the rhizosphere. Plant Soil 256:67–83
Buss LW, Jackson JBC (1979) Competitive networks: nontransitive competitive relationships in cryptic coral reef environments. Am Nat 113:223–234
Callaway RM (1994) Facilitative and interfering effects of *Arthrocnemum subterminale* on winter annuals in California salt marsh. Ecology 75:681–686
Callaway RM, Nadkarni NM, Mahall BE. (1991) Facilitating and interfering effects of *Quercus douglasii* in central California. Ecology 72:1484–1499

Callaway RM, DeLucia EH, Moore D, Nowak R., Schlesinger WD (1996) Competition and facilitation: contrasting effects of *Artemisia tridentata* on *Pinus ponderosa* versus *P. monophylla*. Ecology 77:2130–2141

Callaway RM, Pennings SC (2000) Facilitation may buffer competitive effects: indirect and diffuse interactions among salt marsh plants. Am Nat 156:416–424

Callaway RM, Reinhart KO, Tucker SC, Pennings SC (2001) Effects of epiphytic lichens on host preference of the vascular epiphyte *Tillandsia usneoides*. *Oikos* 94:433–441

Case TJ (1991) Invasion resistance, species build-up and community collapse in metapopulation models with interspecies competition. Biol J Linnean Soc 42:239–266

Chesson PL (1983) Coexistence of competitors in a stochastic environment: the storage effect. Lecture Notes in Biomathematics 52:188–198

Chesson PL, Huntly N (1993) Temporal hierarchies of variation and the maintenance of diversity. Plant Species Biol 8:195–206

Clark JS, LaDeau S Ibanez I (2004) Fecundity of trees and the colonization competition hypothesis. Ecol Monogr 74:415–442

Cody ML (1986) Structural niches in plant communities. In: Diamond J, Case TJ (eds) Community ecology. Harper and Row, New York, pp 381–405

Connell JH (1971) On the role of natural enemies in preventing competitive exclusion in some marine animals and in rain forest trees. In: Den Boer PJ, Gradwell G (eds) Dynamics of populations. Wageningen, pp 298–312

Connolly J (1997) Substitutive experiments and the evidence for competitive hierarchies in plant communities. Oikos 80:179–182

Czárán TL, Hoekstra RF, Pagie L (2002) Chemical warfare between microbes promotes biodiversity. Proc Natl Acad Sci 99:786–790

Czarnota MA, Paul RN, Dayan FE, Nimbal CI, Weston LA (2001) Mode of action, localization of production, chemical nature, and activity of sorgoleone: a potent PSII inhibitor in *Sorghum* spp. root exudates. Weed Technology 15:813–825

Davidson DW (1980) Some consequences of diffuse competition in a desert ant community. Am Nat 116:92–105

de Wit CT (1961) Space relationship within populations of one or more species. Soc Exp Biol Symp 15:314–329

Ehlers BK, Thompson J (2004) Do co-occurring plant species adapt to one another? The response of *Bromus erectus* to the presence of different *Thymus vulgaris* chemotypes. Oecologia 141:512–518

Fitter A (2003) Making allelopathy respectable. Science 301:1337–1338

Flores HE (1999) "Radicle" biochemistry: the biology of root-specific metabolism. Trends Plant Sci 4:220–226

Fowler N (1981) Competition and coexistence in a North Carolina grassland: II. The effects of the experimental removal of species. J Ecol 69:843–854

Gaudet CL, Keddy PA (1995) Competitive performance and species distribution in shoreline plant communities: a comparative approach. Ecology 76: 280–291

Gaudet CL, Keddy PA (1988) A comparative approach to predicting competitive ability from plant traits. Nature 334:242–243

Goldberg DE (1997) Competitive ability: definitions, contingency and correlated traits. In: Silvertown J, Franco M, Harper MJL (eds) Plant life histories: ecology, phylogeny and evolution, 1st edn. The Royal Society, Cambridge, pp 283–306

Goldberg DE, Landa K (1991) Competitive effect and response: hierarchies and correlated traits in the early stages of competition. J Ecol 79:1013–1030

Goldsmith FB (1978) Interaction (competitive) studies as a step towards the synthesis of sea-cliff vegetation. J Ecol 66: 921–931

Grace JB, Gutenspergen GR, Keough J (1993) The examination of a competition matrix for transitivity and intransitive loops. Oikos 68:91–98

Griggs RF (1940) The ecology of rare plants. Bull Torrey Bot Club 67:575–594

Halligan JP (1976) Toxicity of *Artemisia californica* to four associated herb species. Am Mid Nat 95:406–421
Harper JL (1977) Population biology of plants. Academic Press, London
Hierro JL, Callaway RM (2003) Allelopathy and exotic plant invasion. Plant Soil 256:25–39
Herben T, Krahulec F (1990). Competitive hierarchies, reversals of rank order and the de Wit approach: are they compatible? Oikos 58:254–256
Herben T, Hara T, Hadincová V, Kranhulec F, Pecháckova S, Skálová H, Suziki J (2001) Neighborhood effects and genetic structure in a clonal grass: the role of the spatial structure of the environment. Plant Spec Biol 16:1–11
Howard TG, Goldberg DE (2001) Competitive response hierarchies for germination, growth, and survival and their influence on abundance. Ecology 82:979–990
Iason GR. Lennon JJ, Pakeman RJ, Thoss V, Beaton JK, Sim DA Elston DA. (2005) Does chemical composition of individual Scots pine trees determine the biodiversity of their associated ground vegetation? Ecol Lett 8:364–369
Jackson JBC, Buss LW (1975) Allelopathy and spatial competition among coral reef invertebrates. Proc Natl Acad Sci 72:5160–5163
James R, Lazdunski C, Pattus F (1991) Bacteriocins, microcins, and antibiotics. Springer, New York
Karlson RH, Jackson JBC (1981) Competitive networks and community structure: a simulation study. Ecology 62:670–678
Keddy PA (1992) Assembly and response rules: two goals for predictive community ecology. J Veg Sci 3:157–164
Keddy PA, Shipley B (1989) Competitive hierarchies in herbaceous plant communities. Oikos 54:234–241
Keddy PA, Twolan-Strutt L, Wisheu IC (1994) Competitive effect and response rankings in 20 wetland plants: Are they consistent across three environments? J Ecol 82:635–643
Keddy PA, Nielsen K, Weiher E, Lawson R (2002) Relative competitive performance of 63 species of terrestrial herbaceous plants. J Veg Sci 13:5–16
Kerr B, Riley MA, Feldman M, Bohannan B (2002) Local dispersal and interaction promote coexistence in a real life game of rock–paper–scissors. Nature 418:171–174
Kirkup BC, Riley MA (2004) Antibiotic-mediated antagonism leads to a bacterial game of rock–paper–scissors in vivo. Nature 428:412–414
Lawlor LR (1979) Direct and indirect effects of n-species competition. Oecologia 45:355–364
Levine JM (1999) Indirect facilitation: evidence and predictions from a riparian community. Ecology 80:1762–1769
Levine SH (1976) Competitive interactions in ecosystems. Am Nat 110:903–910
Li X, Wilson SD. (1998) Facilitation among woody plants establishing in an old field. Ecology 79:2694–2705
Linhart YB, Thompson JD (1999) Thyme is of the essence: biochemical polymorphism and multi-species deterrence. Evol Ecol Res 1:151–171
MacArthur RH (1972) Geographical ecology. Harper and Row, New York
Miller TE (1994) Direct and indirect species interactions in an early old-field plant community. Am Nat 143:1007–1025
Mitchley J, Grubb PJ (1986) Control of relative abundance of perennials in chalk grassland in southern England I. Constancy of rank order and results of pot- and field-experiments on the role of interference. J Ecol 74:1139–1166
Nilsson MC (1994). Separation of allelopathy and resource competition by the boreal dwarf shrub *Empetrum hermaphroditum* Hagerup. Oecologia 98:1–7
Novoplansky A, Goldberg DE (2001) Effects of water pulsing on individual performance and competitive hierarchies in plants, J Veg Sci 12:199–208
Ortega YK, Pearson DE (2005) Weak vs. strong invaders of natural plant communities: assessing invasibility and impact. Ecol Appl 15:651–661
Paine RT (1966) Food web complexity and species diversity. Am Nat.100:65–75

Parker VT, Muller CH (1982) Vegetational and environmental changes beneath isolated live oak trees (*Quercus agrifolia*) in a California annual grassland. Am Midl Nat 107: 69–81

Parrish JAD, Bazzaz FA (1976) Underground niche separation in successional plants. Ecology 57:1281–1288

Pennings SC (1996) Indirect interactions on coral reefs. In: Birkeland C (ed) Life and death of coral reefs. Chapman and Hall, New York, pp 249–272

Pennings S, Callaway RM (1992) Salt marsh plant zonation: the importance and intensity of competition and physical factors. Ecology 73:681–690

Perry LG, Thelen GC, Ridenour WM, Weir TL, Callaway RM, Paschke MW, Vivanco JM (2005a) Dual role for an allelochemical: (±)-catechin from *Centaurea maculosa* root exudates regulates conspecific seedling establishment. J Ecol 93:1126–1135

Perry LG, Johnson C, Alford ER, Vivanco JM, Paschke MW. (2005b) Screening of grassland plants for restoration after spotted knapweed invasion. Restoration Ecol 13:725–735

Reeves P (1972) The bacteriocins. Springer, New York

Ridenour WM, Callaway RM (2001) The relative importance of allelopathy in interference: the effects of an invasive weed on a native bunchgrass. Oecologia 126:444–450

Riley MA, Gordon DM (1999) The ecological role of bacteriocins in bacterial competition. Trends Microbiol 7:129–133

Shipley B (1993) A null model for competitive hierarchies in competition matrices. Ecology 74: 1693–1699

Silander JA, Antonovics J (1982) Analysis of interspecific interactions in a coastal plant community—a perturbation approach. Nature 298:557–560

Silvertown J, Dale P (1991) Competitive hierarchies and the structure of herbaceous plant-communities. Oikos 61:441–444

Snaydon RW (1991) Replacement or additive designs for competition studies? J Appl Ecol 28:930–946

Starmer WT, Ganter PF, Aberdeen V, Lachance MA, Phaff HJ (1987) The ecological role of killer yeasts in natural communities of yeasts. Can J Microbiol 33:783–796

Stone L, Roberts A (1991) Conditions for a species to gain an advantage from the presence of competitors. Ecology 72:1964–1972

Suding KN, Goldberg DE (2001) Do disturbances alter competitive hierarchies? Mechanisms of change following gap creation. Ecology 82:2133–2149

Thelen, GC, Vivanco JM, Newingham B, Good W, Bais HP, Landres P, Caesar HA, Callaway RM 2005 Insect herbivory stimulates allelopathic exudation by an invasive plant and the suppression of natives. Ecology Letters 8:209–217

Tilman D (1982) Resource competition and community structure. Princeton University Press, Princeton, N.J.

Tilman D (1988) Plant strategies and the dynamics and structure of plant communities. Princeton University Press, Princeton, N.J.

Tilman D (2000) Causes, consequences and ethics of biodiversity. Nature 405:208–211

Trenbath BR (1974) Biomass productivity of mixtures. Adv Agron 26:177–210

Vandermeer J (1980) Saguaros and nurse trees: a new hypothesis to account for population fluctuations. Southwest Nat 25:357–360

Von Holle B, Simberloff D (2004) Testing Fox's assembly rule: does plant invasion depend on recipient community structure? Oikos 105:551–563

Watt AS (1947) Pattern and process in the plant community. J Ecol 35:1–22

Webb LJ, Tracey JG, Haydock KP (1967) A factor toxic to seedlings of the same species associated with living roots of the non-gregarious subtropical rain forest tree *Grevillea robusta*. J Appl Ecol 4:13–25

Weir TL, Bais HP, Vivanco JM (2003) Intraspecific and interspecific interactions mediated by a phytotoxin, (±)-catechin, secreted by the roots of *Centaurea maculosa* (Spotted knapweed). J Chem Ecol 29:2379–2393

Weir TL, Bais HP, Stull VJ, Callaway RM, Thelen GC, Ridenour WM, Bhamidi S, Stermitz FR, Vivanco JM. (2006) Oxalate contributes to the resistance of *Gaillardia grandiflora* and *Lupinus sericeus* to a phytotoxin produced by *Centaurea maculosa*. Planta (in press)
Williamson GB (1990) Allelopathy, Koch's postulates, and the neck riddle. In: Grace JB, Tilman D (eds) Perspectives on plant competition. Academic Press, San Diego, pp 143–162
Wilson SD, Keddy PA (1986) Measuring diffuse competition along an environmental gradient: results from a shoreline plant community. Am Nat 127:862–869
Wooton JT (1992) Indirect effects, prey susceptibility, and habitat selection: impacts of birds on limpets and algae. Ecology 73:981–991
Wooton JT (1994) The nature and consequences of indirect effects in ecological communities. Annu Rev Ecol Syst 25:443–466

Ragan M. Callaway
Division of Biological Sciences
The University of Montana
Missoula, MT
USA 59812
Phone: 406-243-5077
e-mail: ray.callaway@mso.umt.edu

Timothy G. Howard
The New York Natural Heritage Program
Albany, NY
USA 12233-4757

Quaternary Palaeoecology: Isotopes as Valuable Aids in Palaeoecological Research

Burkhard Frenzel

1 Use of various isotope species as indicators of food consumed in the past

When dating organic material by the ^{14}C-method, in general other isotope species are registered as well. Particularly $\delta^{13}C$ values, but also $\delta^{15}N$, $\delta^{12}C$ and $\delta^{18}O$ may become very meaningful, because they can shed some light on the physiological groups of plants involved, such as C_3 or C_4 plants, on various types of food preferred in the past by animals and humans, on former temperature or moisture conditions etc., provided comprehensive observations or experiments have already been performed for correctly interpreting these additional data. Since these ways of interpretation undoubtedly have a lot of pitfalls, some critical comments will be found in the following paragraphs. Nevertheless, progress in correctly interpreting the palaeoecological meaning of the fossil material analysed depends on this particular method of research.

In Japan, people of the Jomon culture, which began at approximately 16,540 cal BP (Nakamura et al. 2001), had already produced pottery. The question arose as to which food sources these people had used, because this would help to better understand the means and intensity of previous human impact on the environment. To answer this question, Yoneda et al. (2002) analysed at first the $\delta^{13}C$ and $\delta^{15}N$-ratios of C_3 and C_4 plants thriving there, of terrestrial herbivores, marine shellfish, marine fishes, and marine mammals. From isotope determinations from fossil human skeletons of the Jomon period, it could then be shown that the people at that time had preferred C_3 plants and terrestrial mammals, such as *Cervus nippon* and *Sus leucomystax*, together with marine fish and marine mammals. Comparable investigations were performed by Cook et al. (2001) and by Bonsall et al. (2004) on the Mesolithic and Neolithic sites of Lepenski Vir at the Iron Gates Gorge, western Romania, and of its surroundings. During Late Mesolithic times, people there had evidently used more than 58% fish protein. Thus, the impact of humans on the terrestrial vegetation should

Progress in Botany, Vol. 68

have been rather small at that time. Yet during Early Neolithic times, diets changed remarkably: Now the proteins consumed originated about 50–80% from domesticated animals, whereas in Roman times and during the Middle Ages a strong proportion of C_4 plants, possibly millet, was found in the consumed food. Thus, since Neolithic times forest clearance has been taking place.

Quite similar research has been done by Arneborg et al. (1999) on the diets of the Greenland Vikings in south-southwestern Greenland. In about 1000 AD, these people had consumed largely terrestrial food, yet the share of this diet decreased constantly so that marine food dominated during the middle of the fifteenth century. It is thought that these changes in diet might have strongly contributed to the dying out of the Vikings there. Yet this may be questioned, in view of repeatedly found reports on vigorous Eskimo attacks at that time.

The $\delta^{13}C/\delta^{12}C$-ratio in animal tissues is sometimes used for reconstructing the food preferences. This procedure seems to need more contributing research, since Panarello and Fernández (2002) have shown that various hair types from primary consumers of the Altiplano of Argentina differ in their $\delta^{13}C$-values, and that there were considerable differences between the $\delta^{13}C$ signatures of carnivores and the herbivores they had eaten. On the other hand, Shen et al. (2001) state that, in analysing the $\delta^{13}C$-values of soil organic matter in the Ding hu shan Biosphere Reserve, Guan dong Province, China, it would be possible to differentiate, not only between the formerly existing C_3 and C_4 plants, but also between coniferous and broadleaved plants. I wonder whether this is indeed possible.

The well known fact that the $\delta^{13}C$-ratios in wood cellulose can be used for reconstructing past temperatures is confirmed by the analysis of last glacial *Chamaecyparis* trees in Japan by Takahashi et al. (2001). Andrews et al. (1998) use the $\delta^{13}C$ and $\delta^{18}O$-values in pedogenetic carbonate-concretions of the Thar Desert for reconstructing the history of the monsoon system in Upper Quaternary times. Nakamura et al. (1998) used the C-isotope signatures of dissolved inorganic material in water samples of Lake Biwa, Central Japan, to analyse the intensity of seasonal mixing of the water masses. Kacanski et al. (2001) have used the $\delta^{13}C$- and $\delta^{18}O$-ratios in carbonates of a speleothem, which was growing for more than 2000 years in a cave of eastern Serbia, for studying times and intensities of climatic changes there. Regrettably, the intensity of former forest clearances on top of the cave has not been analysed. It is astonishing to note that this holds true for nearly all investigations of speleothems, although the water-budgets in caves are strongly influenced by the vegetation growing on top.

2 General difficulties in choosing appropriate samples for obtaining reliable ^{14}C age-data

In a very comprehensive investigation, Walker et al. (2001) have given valuable recommendations for selecting appropriate samples for radiocarbon dating: AMS-^{14}C dating done on material of the same stratigraphical horizon will in general give comparable age data. Nevertheless, a critically performed sample selection is always necessary. Moreover, late-glacial "bulk-sediments" often yield quite aberrant age data. The same holds for humine acids, and dating of the humus fraction by the AMS-^{14}C technique generally reveal age data, which are significantly higher than those of the plant-remains from the same stratigraphical horizon. Moreover, coleoptera can give quite aberrant age data. In addition, it seems that organic carbon in opal-rich marine sediments can hardly be correctly dated by the ^{14}C-technique, because modern C is frequently and intensively fixed by diatomite opal (Zheng et al. 2002). To avoid these difficulties, it is often recommended to rely on data obtained from charcoal instead on those from bulk sediments. However, Sveinbjörnsdóttir et al. (2004) and Gavin (2001) could show that by doing so, considerable differences in age data occur, between charcoal from the inner parts of tree trunks and charcoal from the twigs of the same tree. Similar effects might have played a role in a study of Kennett et al. (2002), where differences in age data were found when marine mussels and terrestrial charcoal of the same geological exposure in southern Peru were compared. Another way to avoid these difficulties is to rely on AMS-^{14}C age data of single pollen grains. However, even that is not always free of pitfalls. For instance, Vandergoes and Prior (2003) stated that very scrupulously prepared pollen grains obtained from Upper Pleistocene sediments from South Westland, New Zealand, always revealed much higher AMS-^{14}C age data than the surrounding bulk sediment. It is assumed that the sediment had been influenced by humus substances or by some other organic material, but it has not been checked whether the dated pollen grains might have originated from much older sediments, i.e. that they were redeposited here, as repeatedly observed. Mensing and Southon (1999), analysing single pollen grains by AMS radiocarbon for dating marine sediments off the coast of northwestern North America, found that by using this method, these sediments appeared to be much older than in previous datings, where other methods were applied. For explanation, they assume contamination by older pollen. On the other hand, they observed, that single pollen grains on top of the Mount Mazama volcanic ash were much older than the ash. Here again, redeposition of the pollen dated grains needs to be considered. Thus if single pollen grains are to be used as objects for AMS dating, a very scrupulously performed stratigraphic procedure will always be indispensable. Hajdas et al. (1998) found that AMS-^{14}C age-data of terrestrial macrofossils, deposited in Lago Grande di Monticchio, Central Italy, generally revealed much lower ^{14}C-ages than expected by counting the ages of "warves". On the other hand, both, the "warve"-ages and the AMS-^{14}C ages were much less than the original assumptions based on the calculated ages of the sediments which are more than 22,000 years BP old. Again the question is how the correctness of the datings might be reliably checked. Evidently, a comprehensively and scrupulously done stratigraphical analysis of the studied sediment-sequence seems to be the conditio sine qua non. Another possibility for avoiding these difficulties in the dating of very old sediments would be to work in deep-lying laboratories, in which the background of cosmic radiation might be minimized (Plastino et al. 2001).

Another problem is the method used to pretreat the samples to be dated. Van Strydonck et al. (2001) found, e.g. when analysing a mesolithic site in dune fields of the Netherlands, that acid-base-acid pretreatment produced soluble material, the ^{14}C ages of which were always up to more than 3400 years older compared with the insoluble material. It is therefore recommended to rely on the insoluble residue only. Other difficulties are caused by frost action or by swelling and shrinking of the sediments. A good example for this is given by Kovda et al. (2001), when studying the Gilgai microrelief of vertisols in the Stavropol uplands on the northern flank of the Caucasus Mountains. Gilgais are approximately 50 cm high small hills, surrounded by depressions. The distance between neighbouring gilgais is ca. 3–5 m. The humic acids of the depressions gave ages of about 70±45 BP, whereas the small "hills" had ages of 5 610±180 BP and with increasing depths there repeatedly occurred inverse age data. Similar findings are described by Vasil'chuk et al. (2001) on the Jamal Peninsula, Northwest Siberia. Here, within the permafrost region, the ^{14}C age-data of peat sequences were repeatedly found to be inversed. This was explained by "allochthonous re-deposition of peat" as water-levels of near-by ponds were held to have quickly changed. We observed the same phenomenon within the permafrost region of Central Tibet, where evidently sliding of peat-layers caused by short melting periods of the uppermost peat-layers had repeatedly occurred. Thus, Solow (2003) is right in criticizing the very often practised procedure of interpolating age data in geological profiles between a lowermost and an uppermost age-date, assuming that the sediment in between was accumulated at a constant velocity. This should first be proven!

Dating "organic" material on potsherds is also problematic if it cannot be proven that the dated material had been put into the former pots by humans, because organic material can also have originated in the sediments used for preparing the former pots (Anderson et al. 2005).

Other difficulties frequently arise in lake or cave-sediments. They are caused by "dead" radiocarbon, which originates from older sediments (Genty et al. 1999; Gibert et al. 1999). On the other hand, it turns out that ostracode shells have to be intensively leached before being ^{14}C-dated, because they can be affected by being old or too young C (Hajdas et al. 2004). The effect of "dead" radiocarbon interfering with the true ages of archaeological material was studied intensively by Fischer and Heinemeier (2003) when investigating prehistoric food residues on pottery found in Seeland, northern Denmark. Here even modern fish-bones reveal ages of about 100–500 years, and it turned out that the fossil material on the inner sides of archaeological potsherds is about 100–300 years older than the charcoal on the outer side of the same potsherd. The authors are evidently right in warning against a too uncritical acceptance of AMS-^{14}C age-data as reliable data. In general, the capability of the AMS-^{14}C-technique to date very small samples is of utmost importance for dating material of the past, when only very small samples of organic material are available. However, there are difficulties, which frequently have not been taken into consideration. In this respect, a paper by Megens et al. (1998) is worth mentioning, since it could be shown in the Ems-Dollart estuary that carbohydrates and proteins of organic material here are much younger than the bulk samples, whereas lipids are much older and have lower $\delta^{13}C$-values. Of course, studying the influence of various types of groundwater, even much more negative differences can be found (Boaretto et al. 1998; Pazdur et al. 1999).

Studying the ^{14}C-ages of animal remains, it becomes important to know on which material the animals had fed. Good examples for this necessity can be found in several papers devoted to the radiocarbon ages of *Rattus exulans*, which is held to have been brought to New Guinea and other islands of this region by humans. Yet ^{14}C age data seemed to show that this rat had come to New Guinea long before humans. In between, it could be shown (Beavan and Sparks 1998; Beavan-Athfield et al. 2001; Higham et al. 2004) that these rats fed on a wealth of different types of food, growing in water and on land so that their ^{14}C-ages could become very high due to the incorporation of much "dead" carbon in their diet. Moreover, the various tissues of modern rats of this species can have quite different ^{14}C-ages (Beavan-Athfield et al. 2001), differing from 2487±50 years BP (liver) to 2946±80 years BP (coat) so that there are no reliable age-data of the introduction of this animal species to the isles.

Repeatedly land snails have been used for ^{14}C-dating of sediment layers. Yet Goodfriend et al. (1999) have shown that shells of these snails (here *Rhabdotus dealbatus, R. alternatus* in Texas) can give quite divergent AMS-^{14}C ages within the same shell. So the apical part can be 1200 years older than the basal part, and age differences between various individuals of the two species, found at the same site and within the same layer can differ by up to 700±180 years. This list of negative effects can be extended by observations of Girbal et al. (2001) on oxalate biodeposits of lichens on stone surfaces. Here it was shown that the obtained age-data were not reliable at all. Here, it may be recommended, to proceed as described by Geyh and Eitel (1998). who tried to radiometrically date calcretes of Namibia and compared U/Th, OSL, TL, and ^{14}C age-data with each other, because the results of just one method may be misleading.

Thus, it turns out that parallel to the profound improvement of dating techniques, which is so important, there arise new and sometimes unexpected interpretation difficulties of the obtained results. Their solution evidently requires a much better and much more subtle knowledge of the processes involved and will become a challenge for future research. For avoiding at least some of these difficulties, Lowe and Walker (2000) recommend using a good quality protocol for working in the field as well as in the laboratory.

3 Problems in dating palaeopedological processes

Pessenda et al. (2001) compared at identical depths ^{14}C-ages of the soil organic matter with those of the humine-fraction and of charcoal in eight Brazilian soil profiles. It turned out that the age-data of the humine-fraction were always higher than those of the soil organic matter, but within the uppermost 50–100 cm they were almost identical to those of charcoal. However, at greater depths the ^{14}C-ages of charcoal were much higher. Astonishingly, the authors did not take into consideration that this charcoal from greater depths should date from the time when the sediment had been accumulated, but not from that of modern pedogenetic processes. Combining ^{14}C-datings of total soil organic matter, of humic-acids and of charcoal with analyses of the $\delta^{13}C$-contents, Pessenda et al. (1998) investigated the history of C_3 and C_4 plants within the southern Brazilian Amazon region (Rondonia), where presently savannah and forest vegetation co-occur. After having studied the $\delta^{13}C$-values of modern C_3 and C_4 plants from the same location, it was concluded that during the entire Holocene period, forest plants had always dominated, while during the Upper Holocene C_4 plants spread into some regions. This is assumed to indicate a drier climate during the Middle Holocene, although I still think the dating of this event was not convincingly done, because the authors extrapolate from a very poor age-data base.

The ages of fossil soils below funeral burrows may become very interesting for dating archaeological facts. Yet Alexandrovsky and Chichagova

(1998) found that these fossil soils may give much lower ages than would be expected by their archaeological contents. The age differences can amount to 2500–5500 years, but it has to be mentioned that only the ^{14}C-ages of easily dislocated humine acids had been investigated. On the other hand, Becker-Heidmann et al. (2002) found quite aberrant age values compared to what would be expected in vertisols of Israel and near Hyderabad in India. Here ^{14}C and $\delta^{13}C$ had always been analysed on the same samples and at vertical distances of only 2 cm. The problems of dating samples out of vertisols, where age differences of up to 14,000 years can be found, have already been mentioned. In this connection, it is interesting to be informed by Römkens et al. (1998) that in arable land of the Netherlands there occurs a strong vertical displacement of soil organic matter of various sizes, which causes quite remarkable age differences. Thus, by dating fossil soils, which had been influenced by former land-use, one should evidently try to get independent information about the formerly practised type of land-use, which may sometimes be rather difficult.

4 Regional patterns of ^{14}C-reservoir ages in oceans, lakes and rivers

For a correct dating of past events by means of the ^{14}C-method, one should know whether the activities measured have been caused by radioactive decay only, or whether additional influences have to be taken into consideration as well. The first question is whether ^{14}C after its production in the high atmosphere has been rapidly and homogeneously mixed into the lower layers of the troposphere, so that everywhere the starting point for incorporating this isotope into the biosphere is exactly the same. It could be shown (Knox and McFadgen 2001; Hogg et al. 2002; McCormac et al. 2004; Sakamoto et al. 2003) that the ages of modern samples of the Southern Hemisphere are generally a few decades higher than those of samples of simultaneously grown organisms from the Northern Hemisphere. This has been known for a long time. However, this age difference differs occasionally (Hogg et al. 2002; McCormac et al. 2002, 2004; Sakamoto et al. 2003). Even within the Northern Hemisphere the offset in ages differs regionally from time to time (Knox and McFadgen 2001: between −20 to +60 years), evidently depending on the atmosphere's circulation patterns. Obviously, the Tunguska meteorite impact did not influence the ^{14}C production of that time (1908 AD.; Yonenobu and Takenaka 1998). Nevertheless, the ^{14}C production has repeatedly oscillated in the past (Voelker et al. 2000). Phases of higher ^{14}C-concentrations, combined with a reduced geomagnetic field intensity, have been the "Mono Lake event"

(33,500–34,500 cal BP), and the "Laschamp event" (40,300–41,700 cal BP). Another remarkable phase of increased ^{14}C-production was the time of about 38,000 cal BP. Each of these phases was characterized by a rapid increase in the $\Delta^{14}C$-values. The first phase seems to have coincided with a remarkable melt-water spike in the North Atlantic, which had caused an increased age of the planktonic reservoir. These changes in the ocean's water-ventilation can cause fluctuations in the $\Delta^{14}C^{atm}$ of up to 25%, combined with considerable changes in the surface water reservoir ages (R) of several decennia. For ^{14}C-changes caused by the "Maunder Minimum" in the sun's activity, see Damon et al. (1999). A much shorter periodicity of about 200–230 years in atmospheric radiation, which influences the quality of ^{14}C-datings during Upper Quaternary times was found by Goslar et al. (2001; see also Frenzel 1998). The melt-water spikes mentioned are assumed to have influenced the top-to-bottom age-variations, which are generally thought to be of up to 500 years in the Pacific Ocean and of ca. 1000 years in the Atlantic Ocean (Stocker and Wright 1998). Thus, the last mentioned authors point out that it might be dangerous to assume constant reservoir ages in the run of time, when dealing with ages of marine organisms. Adkins et al. (2002) give interesting data about the velocity of the vertical water circulation in the Pacific and in the Atlantic Oceans in this connection. This vertical water circulation was also investigated by Nydal (2000), and Reimer and Reimer (2001) produced a databank of regional reservoir effect data, i.e. of correction factors which have to be taken into consideration, for the comparison of marine and terrestrial ^{14}C age-data (see below).

For understanding the following paragraphs it needs to be mentioned that the mean marine correction of ^{14}C-values of the global oceans in comparison to those of terrestrial sites, is called (R). The differences are caused by the incorporation of old or "dead" ^{14}C due to the vertical turn-over of waters. R is approximately 400 years, i.e. in general the age values of marine samples should be ca. 400 years higher than those of contemporarily formed terrestrial samples. However, due to the circulation-patterns of water, major regional deviations from this generalised value may occur. They are internationally indicated by the symbol (ΔR). For reservoir offset models to calibrate radiocarbon data, see Jones and Nicholls (2001).

According to investigations at 26 localities of the Mediterranean and the Black Sea, Siani et al. (2000) have shown that the mean ΔR seems to be 390 ± 85 years in the Mediterranean Sea, 415 ± 90 years in the Western Black Sea and 440 ± 40 years in the Red Sea. Reimer and McCormac (2002) stress that in the "pre-bomb-time" (this means before the time of nuclear weapon experiments, here for the period 1835–1950 AD), the regionally strongly divergent ΔR-value in the Mediterranean Sea has amounted to about 300 years;

however, it may be that at 6000 cal BP, completely different values had existed. This suggestion is important, since it is a warning against a too easily performed extrapolation from modern conditions into the past. In the Lagoon of Venice ΔR amounts to about 1200–1300 years (Zoppi et al. 2001), and in the Aegean Sea, Isle of Youra, this value seems to be about 515±22 years, but there are also some deviations from this value (Facorellis et al. 1998).

Sometimes there exist difficulties at the coasts of the eastern Mediterranean Sea when calibrated AMS-^{14}C age data are compared with ages given in archaeological or, sometimes, written reports: Mostly these types of data coincide quite well in their numerical values (e.g. Görsdorf et al. 1998; Boaretto et al. 2005). However, very frequently the AMS-^{14}C ages are 200–300 years higher than archaeological data would suggest (e.g. Bruins and van der Plicht 1998, 2001; Segal et al. 1998; Bonani et al. 2001; Burton and Levy 2001; Bourke et al. 2004). Keenan (2002) has speculated that these age differences were caused by a strong Holocene upwelling of deep waters of the eastern Mediterranean Sea, by which old or even "dead" ^{14}C came into surface ocean waters, where it could then influence terrestrial organic material growing in the downwind direction by degassing its ^{14}C contents. Regrettably, Keenan has not tried to calculate the amounts of ^{14}C needed to arrive at this effect, and, as already stated, the age difference to higher values does not hold true for all of the meanwhile available data. It needs to be mentioned here, that penetration of water from the sea into the aquifers of coastal regions of the Near East has been repeatedly found (Bruce et al. 2001; Sivan et al. 2001, 2004; Weinstein et al. 2001; Yechieli et al. 2001). In these investigations it could be shown that there exist strong regional differences in the amounts of the penetrating water and of its vertical movement in sediments near to the coast, and groundwaters can move upwards here quite rapidly (Yechieli et al. 2001). Thus, at present one should agree with Avner and Carmi (2001) that much more research work is needed here.

Southon et al. (2002) have produced a very interesting map of modern ΔR values of the entire Indian Ocean: At present, upwelling is obviously very intensive in the western part and in the Arabian Sea (Duffa et al. 2001), there causing high ^{14}C-ages of modern corals, whereas it seems to be negligible in the South China Sea and in the southwestern Pacific Ocean. At the mouth of the river Mekong (Dang et al. 2004) ΔR amounts to −74±39 years, only. Roughly the same holds for the eastern Indian Ocean (Hua et al. 2004) and the coasts of Kenya (Grumet et al. 2002), while the importance of upwelling increases at the eastern coast of Africa above 3–4°S.

It has just been stated that the ΔR value of surface waters of the southwestern Pacific Ocean is only very small (Duffa et al. 2001). Yet according to the ocean currents, it increases off the isle of Taiwan (Ishigaki Island) to

290±40/455±45 years (Hideshima et al. 2001; Ryukyu Islands, see Morimoto et al. 2004), i.e. to about the same values as found at the eastern coasts of Honshu and Hokkaido (Yoneda et al. 2001). At the southern Kurile Islands, however, it increases to 711±46 years (Kuzmin et al. 2001). These last-mentioned values are very important for archaeology, because the prehistoric cultures here had strongly used marine food. Thus without the knowledge of these ΔR-values, quite misleading age data of terrestrial cultures and of the then occurring changes in the terrestrial vegetation can be obtained (Kuzmin et al. 2001).

Meanwhile, the problem has become very important, when certain West-Pacific isles, including Australia, have been settled by humans. For answering this question, again the ΔR-values at the coasts of these islands become significant, since the diet of former people was consistently strongly depended on marine food, which of course influences the terrestrial archaeological data, if human bones and rests of the former diets are being dated. Key et al. (2002), Hughen et al. (2004a) and Leboucher et al. (1999) have contributed many very interesting data from various parts of the Pacific Ocean, most of all of its southern part. By comparing ^{14}C-data with U/Th-data, Hughen et al. (2004b) even contribute to a better understanding of changing ΔR-values in the run of time of the Holocene. Other investigations show here again, strong changes in ΔR-values at the regional scale, evidently depending on the marine circulation patterns (Samoa: Phelan 1999; Petchey 2001; New Ireland: Petchey et al. 2004; the Marquesas: Paterne et al. 2004; Hawaii: Druffel et al. 2001). Comparable observations were made on the coast of Peru (Owen 2002; Fontugne et al. 2004). It should be mentioned that both of these papers contain an analysis of the changing ΔR-values during the entire Holocene. It becomes evident that depending on changes in the ocean currents, these values show consistently strong changes, which, of course, also contributes to a better understanding of climate history.

Meanwhile local differences in the ΔR-values of marine material similar to the coasts of the Pacific Ocean have also been found at the coasts of the Atlantic Ocean. This led Cordero et al. (2003), quite correctly to me, to the statement that evidently the global sea average reservoir effect is of very minor importance. However, from the coasts of the Isla Livingston, Antarctica, to the Brazilian coasts the ΔR-values differ from about 1900±140 years to 220±20 or even 8±17 years (Eastoe et al. 2002; Cordero et al. 2003; Angulo et al. 2005). Off the coast of Venezuela (Cariaco Basin), this value has been found to be ca. 312 years (Guilderson et al. 2005). Thus, the Late Glacial chronology of the Cariaco Basin would be very reliable, if there were not data from the age analysis of "warves" at the end of the Allerød-interstadial showing an offset of about 600 years, compared with tree-ring data, which at the beginning of the Younger Dryas cold period had soon changed to about 400 years, only (Kromer et al. 2004; see below). ΔR-values of various parts of the North Atlantic Ocean are given by Ascough et al. (2004) and by Sveinbjörnsdóttir et al. (1998).

Further difficulties occur in terrestrial sites, caused by solution of carbonaceous rocks, by pedogenesis, by variable volumes of sediment transport in running water, and by biochemical processes, such as photosynthesis, respiration and fermentation. In arid to semi-arid regions such processes can strongly influence the $\Delta^{14}C$ values, as shown by Stein et al. (2004), who investigated sediments of Lake Lisan, the predecessor of the Dead Sea. During full-glacial times of the Last Glaciation, at about 15,000 cal BP, the ΔR-values were extremely high here, but they decreased, when climate ameliorated and became moister during interstadial and Holocene times. This was confirmed by van der Borg et al. (2004), also studying the $\Delta^{14}C$ values of Lake Lisan and additionally those of Lake Suigetsu in Japan. It could be shown that these values differed strongly (up to 32,000 cal BP) from those of the Cariaco Basin, the Iceland Sea and a Bahama stalagmite. This becomes very important for the establishment of a calibration of ^{14}C-data versus calendar years (see section 5 below). These difficulties are strongly enlarged, if ground-water exerts a strong influence (Gallagher et al. 2000; Geyh 2000; Goslar et al. 2000a,b; Gibert et al. 2002). Yet this disturbing influence was not constant over time, as could be shown by Geyh et al. (1998) at the Schleinsee, southern Germany, the Proscansko Lake in the Plitvice National Park, Slovenia, and the Laguna Leja on the Chilean Altiplano. To avoid these obstacles to accurate ^{14}C-dating, the authors recommend either dating from the same sediment-layer organic material and lake marls, or, better, organic, short-living terrestrial material. These changing $\Delta^{14}C$-values also strongly affect the radiocarbon dating of sediments of Lake Kinneret (i.e. Sea of Galilee: Stiller et al. 2001). Here, from ca. 8000 to 6000 BP, this reservoir effect had amounted to about 1000 years. Between ca. 2500 and 1000 BP it increased to 1800 years and afterwards it decreased again.

It has already been stated in section 1, to what extent freshwater reservoir effects had influenced the correct dating of Mesolithic human bones at the Golden Gate site in westernmost Romania via the food consumed by prehistoric humans (Cook et al. 2001). Other negative influences in freshwater lakes can occur through disturbances caused by the boring technique, by bioturbation or by periodic or aperiodic redepositioning of already sedimented material (Moreton et al. 2004). Thus, the authors recommend caution when interpreting such data material, particularly statements about effects of climate change, as are repeatedly being made at present, should be considered with care. This warning seems to be justified for the discussion of the ^{14}C ages of surface sediments at the bottom of Lake Baikal (Piotrowska et al. 2004). Here, these sediments were dated at three sites by the AMS-^{14}C technique, performed on single pollen grains. It turned out that they gave ages of between 1250 and 2000 years. The reasons for this are not clear, and the authors conclude "that this could result from specific properties of pollen grains"; however, a possible hard-water effect has not been analysed.

From all the above facts, it can be concluded that the use of excellent physical and chemical techniques in dating fossil material needs always to be combined with a painstaking geological and palaeobiological analysis of the studied material. This has already to be done in the field but also later in the laboratory, where all the obtained results must be critically interpreted, under consideration of the wealth of available literature.

5 Problem of constructing a generally accepted calibration curve for translating ^{14}C age-data into those of calendar years

It is a well known fact that radiocarbon age data cannot be directly compared with astronomical years. To do so, one has to fix these steadily changing age differences, i.e. to determine ^{14}C-ages of samples whose exact ages are known. The best way for doing so is the ^{14}C-datation of exactly dated tree-rings. The Hohenheim series, the longest series existing at present, goes back to 12,460 BP (Friedrich et al. 2004, see also Kromer et al. 1998), after misleading sequences caused by the impact of cockchafers (*Melolontha melolontha*, *M. hippocastani*) have been eliminated. Yet it is well known that not all tree-rings are annually formed features (e.g. Biondi and Fessenden 1999). This is a serious warning against a too carelessly made comparison of "annually" layered structures with ^{14}C-datations. Generally great care should be taken when doing so; at any rate, it must at first be proven that these structures are really annually formed features. Examples for this are the so-called "warves", which were originally assumed to be annually layered sediments in lakes in front of glaciers. But now, each finely layered sediment in water is called "warve" (e.g. Hughen et al.1998, 2004a,b) without having proven that these layers were truly a product of an annual rhythm. Even the typical warves can still pose serious dating problems (Wohlfarth and Possnert 2000).

Great success has been achieved by Reimer et al. (2002, 2004), due to an intensive international cooperation of several research groups (see also Scott et al. 2003). The INTCAL04 ^{14}C-calibration curve goes now back to 26,000 cal BP. Nevertheless, within this span of time, at least within the older parts, there do exist phases in which the dating accuracy is not as reliable as it is in the younger parts (see also Blackwell et al. 2006). For improving this situation and for extending the calibration curve farther back into the past, very intensive and multi-facetted research work needs to be done. The main problems are the extreme changes in the ^{14}C-content of the atmosphere between approximately 30,000 and 45,000 BP, which have already been mentioned in section 4.

The general way to overcome these difficulties is to compare ^{14}C age-data with independently obtained data by employing other methods. All studies agree that the critical time is the above mentioned period (Geyh and Schlüchter 1998; Jöris and Weninger 1998; Kitagawa and van der Plicht 1998, 2000; Voelker et al. 1998; Goslar et al. 2000a,b; Stein et al. 2000; Yokoyama et al. 2000; Cutler et al. 2004; Chiu et al. 2005; Fairbanks et al. 2005). However, the differences in the observed age data and in their tendencies are great. Based on the cooperation of various research groups, always analysing paired $^{230}Th/^{234}U/^{238}U$ and ^{14}C-dates in pristine corals, Fairbanks et al. (2005) stress that the ^{14}C-calibration curve can now be extended back to 50,000 BP. This is doubted by Reimer et al. (2004) and van der Plicht (2004), because the age differences between the various attempts are too great (see also van Andel 2005). Therefore, these new curves should rather be termed "comparison curves" instead of "calibration curves" (Reimer et al. 2006). Independent of this terminological problem, the attempts to produce reliable calibrations of old ^{14}C data need to be intensified. Bird et al. (1999) state, that a wet oxidation, stepped-combustion procedure of "old" charcoal might provide reliable ^{14}C-ages of charcoal up to at least 50,000 BP.

6 Conclusion

Despite all the difficulties mentioned above, it would be wrong to assume that all the attempts to date past events are erroneous or misleading. I think that just the opposite is true: obviously, our present dating techniques are much better and much more efficient than our understanding of the different processes acting in nature. Therefore, we have to considerably improve our knowledge and our understanding of all the different processes which have played a role in the past. For doing so, much time and own efforts are necessary instead of following a "publish or perish" strategy and to incorporate steadily growing groups of scientists into the preparation of a single paper. On the other hand, I think serious attempts should be made to considerably improve physical dating techniques besides the ^{14}C-technique: Many important processes were already active at times which exceed the dating capacity of the ^{14}C-technique and they may even turn out to be very important for understanding what is presently going on in nature and for evaluating what might happen in the future.

References

Adkins JF, Griffin S, Kashgarian M, Cheng M, Druffel ERM, Boyle EA, Edwards RL, Shen ChCh (2002) Radiocarbon dating of deep sea corals. Radiocarbon 44:567–580

Alexandrovsky AL, Chichagova OA (1998) The ^{14}C age of humic substances in paleosols. Radiocarbon 40:991–997

Anderson A, Chappell J, Clark G, Phear S (2005) Comparative radiocarbon dating of lignite-pottery, and charcoal samples from Babeldaob Island, Republic of Palau. Radiocarbon 47:1–9

Andrews JE, Singhvi AK, Kailath AJ, Kuhn R, Dennis PF, Tandon SK, Dhir RP (1998) Do stable isotope data from calcrete record Late Pleistocene monsoonal climate variation in the Thar Desert of India? Quat Res 50:240–251

Angulo RJ, de Souza MC, Reimer PJ, Sasaoka SK (2005) Reservoir effect of the southern and southeastern Brazilian coast. Radiocarbon 47:67–73

Arneborg J, Heinemeier J, Lynnerup N, Nielsen HL, Rud N, Sveinbjörnsdóttir ÁE (1999) Change of diet of the Greenland Vikings determined from stable carbon isotope analysis and ^{14}C of their bones. Radiocarbon 41:157–168

Ascough PhL, Cook GT, Dugmore AJ, Barber J, Higney E, Scott EM (2004) Holocene variations in the Scottish marine reservoir effect. Radiocarbon 46:611–620

Avner U, Carmi I (2001) Settlement patterns in the southern Levant deserts during the 6th–3rd millennia BC: a revision based on ^{14}C dating. Radiocarbon 43:1203–1216

Beavan NR, Sparks RJ (1998) Factors influencing ^{14}C ages of the Pacific rat *Rattus exulans*. Radiocarbon 40:601–613

Beavan-Athfield NR, McFadgen BG, Sparks RJ (2001) Environmental influences on dietary carbon and ^{14}C ages in modern rats and other species. Radiocarbon 43:7–14

Becker-Heidmann P, Andresen O, Kalmar D, Scharpenseel H-W, Yaalon DH (2002) Carbon dynamics in vertisols as revealed by high-resolution sampling. Radiocarbon 44:63–73

Biondi F, Fessenden JE (1998) Radiocarbon analysis of *Pinus lagunae* tree rings: implications for tropical dendrochronology. Radiocarbon 41:241–249

Bird MI, Ayliffe LK, Fifield LK, Turney CSM, Cresswell RG, Barrows TT, David B (1999) Radiocarbon dating of "old" charcoal using a wet oxidation, stepped-combustion procedure. Radiocarbon 41:127–140

Blackwell PG, Buck CE, Reimer PJ (2006) Important features of the new radiocarbon calibration curves. Quat Sci Rev (in press)

Boaretto E, Thorling L, Sveinbjörnsdóttir ÁE, Yechieli Y, Heinemeier J (1998) Study of the effect of fossil organic carbon on ^{14}C in groundwater from Hvinningdal, Denmark. Radiocarbon 40:915–920

Boaretto E, Jull AJT, Gilboa A, Sharon I (2005) Dating the Iron Age I/II transition in Israel: first intercomparison results. Radiocarbon 47:39–55

Bonani G, Haas H, Hawass Z, Lehner M, Nakhla Sh, Nolan J, Wenke R, Wölfli W (2001) Radiocarbon data of Old and Middle Kingdom monuments in Egypt. Radiocarbon 43:1297–1320

Bonsall C, Cook GT, Hedges REM, Higham TFG, Pickard C, Radovanovic I (2004) Radiocarbon and stable isotope evidence of dietary change from the Mesolithic to the Middle Ages in the Iron Gates: new results from Lepenski Vir. Radiocarbon 46:293–300

Bourke St, Zoppi U, Meadows J, Hua Qu, Gibbins S (2004) The end of the Chalcolithic Period in the South Jordan Valley: new ^{14}C determinations from Teleilat Ghassul, Jordan. Radiocarbon 46:315–323

Bruce D, Friedman GM, Kaufman A, Yechieli Y (2001) Spatial variations of radiocarbon in the coastal aquifer of Israel—indicators of open and closed systems. Radiocarbon 43:783–791

Bruins HJ, van der Plicht J (1998) Early Bronze Jericho: high precision ^{14}C dates of short-lived palaeobotanic remains. Radiocarbon 40:621–628

Bruins HJ, van der Plicht J (2001) Radiocarbon challenges archaeo-historical time frameworks in the Near East: the Early Bronze Age of Jericho in relation to Egypt. Radiocarbon 43:1321–1332

Burton M, Levy ThE (2001) The Chalcolithic radiocarbon record and its use in southern Levantine archaeology. Radiocarbon 43:1223–1246

Chiu TzCh, Fairbanks RG, Mortlock RA, Bloom AL (2005) Extending the radiocarbon calibration beyond 26,000 years before present using fossil corals. Quat Sci Rev (in press)

Cook GT, Bonsall C, Hedges REM, McSweeney K, Boronean V, Pettitt PB (2001) A freshwater diet-derived ^{14}C reservoir effect at the Stone Age sites in the Iron Gates Gorge. Radiocarbon 43:453–460

Cordero RR, Panarello H, Lanzelotti S, Dubois CMF (2003) Radiocarbon age offsets between living organisms from the marine and continental reservoir in coastal localities of Patagonia (Argentina). Radiocarbon 45:9–15

Cutler KB, Grey SG, Burr GS, Edwards RL, Taylor FW, Cabioch G, Beck JW, Cheng H, Moore J (2004) Radiocarbon calibration and comparison to 50 kyr BP with paired ^{14}C and ^{230} Th dating of corals from Vanuatu and Papua New Guinea. Radiocarbon 46:1127–1160

Damon PE, Eastoe ChJ, Mikheeva IB (1999) The Maunder Minimum: an interlaboratory comparison of $\Delta^{14}C$ from AD 1688 to AD 1710. Radiocarbon 41:47–50

Dang Ph, Mitsuguchi T, Kitagawa H, Shibata Y, Kobayashi T (2004) Marine reservoir correction in the south of Vietnam estimated from an annually-banded coral. Radiocarbon 46:657–660

Druffel ERM, Griffin S, Guilderson TP, Kashgarian M, Southon J, Schrag DP (2001) Changes of subtropical North Pacific radiocarbon and correlation with climate variability. Radiocarbon 43:15–25

Duffa K, Bushan R, Somayajulu BLK (2001) ΔR correction values for the northern Indian Ocean. Radiocarbon 43:483–488

Eastoe CJ, Fish S, Fish P, Gaspar MD, Long A (2002) Reservoir corrections for marine samples from the South Atlantic coast, Santa Catarina State, Brazil. Radiocarbon 44:145–148

Facorellis Y, Maniatis Y, Kromer B (1998) Apparent 14C ages of marine mollusk shells from a Greek island: calculations of the marine reservoir effect in the Aegean Sea. Radiocarbon 40:963–973

Fairbanks RG, Mortlock RA, Chiu TzCh, Cao L, Kaplan A, Guilderson TP, Fairbanks TW, Bloom AL, Grootes PM, Nadeau M-J (2005) Radiocarbon calibration curve spanning 0 to 50,000 years BP based on paired $^{230}Th/^{234}U/^{238}U$ and ^{14}C dates on pristine corals. Quat Sci Rev (in press)

Fischer A, Heinemeier J (2003) Freshwater reservoir effect in ^{14}C dates of food residue on pottery. Radiocarbon 45:449–466

Fontugne M, Carré M, Bentaleb I, Julien M, Lavallée D (2004) Radiocarbon reservoir age variations in the South Peruvian upwelling during the Holocene. Radiocarbon 46:531–537

Frenzel B (1998) Klimavariabilität während der Eiszeit. Annalen der Meteorologie 36:75–93

Friedrich M, Remmele S, Kromer B, Hoffmann J, Spurk M, Kaiser KF, Orcel C, Küppers M (2004) The 12,460-year Hohenheim oak and pine tree-ring chronology from Central Europe—a unique annual record for radiocarbon calibration and paleoenvironment reconstructions. Radiocarbon 46:1111–1122

Gallagher D, McGee EJ, Kalin RM, Mitchell PI (2000) Performance of models for radiocarbon dating of groundwater: an appraisal using selected Irish aquifers. Radiocarbon 42:235-248

Gavin DG (2001) Estimation of inbuilt age in radiocarbon ages of soil charcoal for fire history studies. Radiocarbon 43:27–44

Genty D, Massault M, Gilmour M, Baker A, Verheyden S, Kepens E (1999) Calculation of past dead carbon proportion and variability by the comparison of AMS ^{14}C and TIMS U/Th ages on two Holocene stalagmites. Radiocarbon 41:251–270

Geyh MA (2000) An overview of ^{14}C analysis in the study of groundwater. Radiocarbon 42:99–114

Geyh MA, Eitel B (1998) Radiometric dating of young and old calcretes. Radiocarbon 49:795–802

Geyh MA, Schlüchter C (1998) Calibration of the ^{14}C time scale beyond 22 000 BP. Radiocarbon 40:475–482
Geyh MA, Schotterer U, Grosjean M (1998) Temporal changes of the ^{14}C reservoir effect in lakes. Radiocarbon 40:921–931
Gibert E, Travi Y, Massault M, Chernet T, Barbecot F, Laggoun-Defarge F (1999) Comparing carbonate and organic AMS-^{14}C ages in Lake Abiyata sediments (Ethiopia): hydrochemistry and paleoenvironmental implications. Radiocarbon 41:271–286
Gibert E, Travi Y, Massault M, Tiercelin J-J, Chernet T (2002) AMS-14C chronology of a lacustrine sequence from Lake Langano (main Ethiopian rift): correction and validation steps in relation with volcanism, lake water and carbon balances. Radiocarbon 44:65–92
Girbal J, Prada JL, Rocabayera R, Argemí M (2001) Dating of biodeposits of oxalates at the Arc de Berà in Tarragona, Spain. Radiocarbon 43:637–645
Görsdorf J, Dreyer G, Hartung U (1998) New ^{14}C dating of the Archaic royal necropolis Umm el-Qaab at Abydos (Egypt). Radiocarbon 40:641–647
Goodfriend GA, Ellis GL, Toolin LJ (1999) Radiocarbon age anomalies in land snail shells from Texas: ontogenetic, individual, and geographic patterns of variation. Radiocarbon 41:149–156
Goslar T, Arnold M, Tisnérat-Laborde N, Hatté Chr, Paterne M, Ralska-Jasiewiczowa M (2000a) Radiocarbon calibration by means of warves versus ^{14}C ages of terrestrial macrofossils from Lake Gosciaz and Lake Perespilno, Poland. Radiocarbon 42:335–348
Goslar T, Hercman H, Pazdur A (2000b) Comparison of U-series and radiocarbon dates of speleothems. Radiocarbon 42:403–414
Goslar T, Tisnérat-Laborde N, Paterne M (2001) Searching solar periodicities in the late-glacial record of atmospheric radiocarbon. Radiocarbon 43:339–344
Grumet NS, Guilderson ThP, Dunbar RB (2002) Pre-bomb radiocarbon variability inferred from a Kenyan coral record. Radiocarbon 44:581–590
Guilderson ThP, Cole JE, Southon JR (2005) Pre-bomb $\Delta^{14}C$ variability and Suess effect in Cariaco Basin surface waters as recorded in hermatypic corals. Radiocarbon 47:57–65
Hajdas I, Bonani G, Zolitschka B, Brauer A, Negendank J (1998) ^{14}C ages of terrestrial macrofossils from Lago Grande di Monticchio (Italy). Radiocarbon 40:803–807
Hajdas I, Bonani G, Herrgesell Zimmermann S, Mendelson M, Hemming S (2004) ^{14}C ages of ostracods from Pleistocene lake sediments of the Western Great Basin, USA—results of progressive acid leaching. Radiocarbon 46:189–200
Hideshima Sh, Matsumoto E, Abe O, Kitagawa H (2001) Northwest Pacific marine reservoir correction estimated from annually banded coral from Ishigaki Island, southern Japan, Radiocarbon 43:473–476
Higham TFG, Hedges REM, Anderson AJ, Ramsey CB, Fankhauser B (2004) Problems associated with the AMS dating of small bone samples: the question of the arrival of Polynesian rats to New Zealand. Radiocarbon 46:207–218
Hogg AG, McCormac EG, Higham TFG, Reimer PJ, Baillie MGL, Palmer JG (2002) High-precision radiocarbon measurements of contemporaneous tree-ring dated wood from the British Isles and New Zealand AD 1850–1950. Radiocarbon 44:633–640
Hua Q, Woodroffe CD, Barbetti M, Smithers SG, Zoppi U, Fink D (2004) Marine reservoir correction for the Cocos (Keeling) Islands, Indian Ocean. Radiocarbon 46:603–610
Hughen KA, Overpeck JT, Lehman SJ, Kashgarian M, Southon IR, Peterson LC (1998) A new ^{14}C calibration data set for the last deglaciation based on marine warves. Radiocarbon 40:483–494
Hughen KA, Baillie MGL, Bard E, Beck JW, Bertrand ChJH, Blackwell PG, Buck CE, Burr GS, Cutler KB, Damon PE, Edwards RL, Fairbanks RG, Friedrich M, Guilderson ThP, Kromer B, McCormac G, Manning St, Ramsey ChB, Reimer PJ, Reimer RW, Remmele S., Southon JR, Stuiver M, Talamo S, Taylor FW, van der Plicht J, Weyhenmeyer CE (2004a) MARINE04 marine radiocarbon age calibration, 0–26 cal kyr BP. Radiocarbon 46:1059–1086

Hughen KA, Southon JR, Bertrand ChJH, Frantz B, Zermeno P (2004b) Cariaco Basin calibration update: revisions to calendar and ^{14}C chronologies for core PL07-58PC. Radiocarbon 46:1161–1187

Jöris O, Weninger B (1998) Extension of the ^{14}C calibration curve to ca. 40,000 cal BC by synchronizing Greenland $^{18}O/^{16}O$ ice core records and North Atlantic foraminifera profiles: a comparison with U/Th coral data. Radiocarbon 40:495–504

Jones M, Nicholls G (2001) Reservoir offset models for radiocarbon calibration. Radiocarbon 43:119–124

Kacanski A, Carmi I, Shemesh A, Kronfeld J, Yam R, Flexer A (2001) Late Holocene climatic change in the Balkans: speleothem isotopic data from Serbia. Radiocarbon 43:647–658

Keenan DJ (2002) Why early-historical radiocarbon dates downwind from the Mediterranean are too early. Radiocarbon 44:225–237

Kennett DJ, Ingram BL, Southon JR, Wise K (2002) Differences in ^{14}C age between stratigraphically associated charcoal and marine shell from the Archaic Period sites of kilometer 4, southern Peru: old wood or old water? Radiocarbon 44:53–58

Key RM, Quay PD, Schlosser P, McNichol AP, von Reden KF, Schneider RJ, Elder KL, Stuiver M, Ostlund HG (2002) WOCE radiocarbon IV: Pacific Ocean results; P10, P13N, P14C, P18, P19 and S4P. Radiocarbon 44:239–392

Kitagawa H, van der Plicht J (1998) A 40 000-year warve chronology from Lake Suigetsu, Japan, extension of the ^{14}C calibration curve. Radiocarbon 40:505–515

Kitagawa H, van der Plicht J (2000) Atmospheric radiocarbon calibration beyond 11,900 cal BP from Lake Suigetsu laminated sediments. Radiocarbon 42:369–380

Knox FB, McFadgen BG (2001) Least-squares fitting smooth curves to decadal radiocarbon calibration data from AD 1145 to AD 1945. Radiocarbon 43:87–116

Kovda I, Lynn W, Williams D, Chichagova O (2001) Radiocarbon age of vertisols and its interpretation using data on Gilgai complex in the North Caucasus. Radiocarbon 43:603–609

Kromer B, Spurk M, Remmele S, Barbetti M, Toniello V (1998) Segments of atmospheric ^{14}C change as derived from Late Glacial and Early Holocene floating tree-ring series. Radiocarbon 40:351–358

Kromer B, Friedrich M, Hughen KA, Kaiser F, Remmele S, Schaub M, Talamo S (2004) Late Glacial ^{14}C ages from a floating, 1382-ring pine chronology. Radiocarbon 46:1203–1205

Kuzmin YaV, Burr GS, Jull AJT (2001) Radiocarbon reservoir correction ages in the Peter the Great Gulf, Sea of Japan, and eastern coast of the Kunashir, southern Kuriles (northwestern Pacific). Radiocarbon 43:477–484

Leboucher V, Orr J, Jean-Baptiste Ph, Arnold M, Monfray P, Tisnerat-Laborde N, Poisson A, Duplessy J-C (1999) Oceanic radiocarbon between Antarctica and South Africa along WOCE Section 16 at 30°E. Radiocarbon 41:51–73

Lowe JJ, Walker MJC (2000) Radiocarbon dating the Last Glacial–Interglacial transition (ca. 14–9 ^{14}C ka BP) in terrestrial and marine records: the need for new quality assurance protocols. Radiocarbon 42:53–68

McCormac FG, Reimer PJ, Hogg AG, Higham TFG, Baillie MGL, Palmer J, Stuiver M (2002) Calibration of the radiocarbon time scale for the Southern Hemisphere: AD 1850–950

McCormac FG, Hogg AG, Blackwell PG, Buck CE, Higham TFG, Reimer PJ (2004) SHCAL04 Southern Hemisphere calibration, 0–11,0 cal kyr BP. Radiocarbon 46:1087–1092

Megens L, van der Plicht J, de Leeuw JW (1998) Molecular, radioactive and stable carbon isotope characterization of estuarine particulate organic matter. Radiocarbon 40:985–990

Mensing SA, Southon JR (1999) A simple method to separate pollen for AMS radiocarbon dating and its application to lacustrine and marine sediments. Radiocarbon 41:1–8

Moreton StG, Rosqvist GC, Davies SJ, Bentley MJ (2004) Radiocarbon reservoir ages from freshwater lakes, South Georgia, Sub-Antarctic: modern analogues from particulate organic matter and surface sediments. Radiocarbon 46:621–626

Morimoto M, Kitagawa H, Shibata Y, Kayanne H (2004) Seasonal radiocarbon variations of surface sea water recorded in a coral from Kikai Island, subtropical northwestern Pacific. Radiocarbon 46:643–648

Nakamura T, Kojima S, Ohta T, Oda H, Ikeda A, Okuno M, Yokota KI, Mizutani Y, Kretschmer W (1998) Isotopic analysis and cycling of dissolved inorganic carbon at Lake Biwa, Central Japan. Radiocarbon 40:933–944

Nakamura T, Taniguchi Y, Isuji S, Oda H (2001) Radiocarbon dating of charred residues on the earliest pottery in Japan. Radicarbon 43:1129–1138

Nydal R (2000) Radiocarbon in the ocean. Radiocarbon 42:81–98

Owen BD (2002) Marine carbon reservoir age estimates for the far south coast of Peru. Radiocarbon 44:701–708

Panarello HO, Fernández J (2002) Stable carbon isotope measurements on hair from wild animals from Altiplano of Jujuy, Argentina. Radiocarbon 44:709–710

Paterne M, Ayliffe LK, Arnold M, Cabioch G, Tisnérat-Laborde N, Hatté Chr, Douville E, Bard E. (2004) Paired ^{14}C and $^{230}Th/U$ dating of surface corals from the Marquesas and Vanuatu (sub-equatorial Pacific) in the 3000 to 15,000 cal yr interval. Radiocarbon 46:551–566

Pazdur A, Goslar T, Pawlyta M, Hercman H, Gradzinski M (1999) Variations of isotopic composition of carbon in the karst environment from southern Poland, present and past. Radiocarbon 41:81–97

Pessenda LCR, Gouveia SEM, Aravena R, Gomes BM, Boulet R, Ribeiro AS (1998) ^{14}C dating and stable carbon isotopes in soil organic matter in forest-savanna boundary areas in the southern Brazilian Amazon region. Radiocarbon 40:1013–1022

Pessenda LCR, Gouveia SEM Aravena R (2001) Radiocarbon dating of total soil organic matter and humin fraction and its comparison with ^{14}C ages of fossil charcoal. Radiocarbon 43:595–601

Petchey FJ (2001) Radiocarbon determinations from the Mulifauna Lapita site, Upolu, Western Samoa. Radiocarbon 43:63–68

Petchey F, Phelan M, White JP (2004) New ΔR values for the southwest Pacific Ocean. Radiocarbon 46:1005–1014

Phelan MB (1999) A ΔR correction value for Samoa from known-age-marine shells. Radiocarbon 41:99–101

Piotrowska N, Bluszcz A, Demske D, Granoszewski W, Heumann G (2004) Extraction and AMS radiocarbon dating of pollen from Lake Baikal sediments. Radiocarbon 46:181–187

Plastino W, Kaihola L, Bartolomei P, Bella F (2001) Cosmic background reduction in the radiocarbon measurements by liquid scintillation spectrometry at the underground laboratory of Gran Sasso. Radiocarbon 43:157–161

Reimer PJ, McCormac FG (2002) Marine radiocarbon reservoir corrections for the Mediterranean and Aegean Seas. Radiocarbon 44:150–166

Reimer PJ, Reimer RW (2001) A marine reservoir correction database and on-line interface. Radiocarbon 43:461–463

Reimer PJ, Hughen KA, Guilderson ThP, McCormac G, Baillie MGL, Bard C, Barratt Ph, Beck JW, Buck CE, Damon PE, Friedrich M, Kromer B, Ramsey ChrB, Reimer RW, Remmele S, Southon JR, Stuiver M, van der Plicht J (2002) Preliminary report of the first workshop of the Intcal04 radiocarbon calibration/comparison working group. Radiocarbon 44:653–661

Reimer PJ, Baillie MGL, Bard E, Bayliss A, Beck JW, Bertrand ChJH, Blackwell PG, Buck CE, Burr GS, Cutler KB, Damon PE, Edwards RL, Fairbanks RG, Friedrich M, Guilderson ThP, Hogg AG, Hughen KA, Kromer B, McCormac G, Manning St, Ramsay ChB, Reimer RW, Remmele S, Southon JR, Stuiver M, Talamo S, Taylor FW, van der Plicht J, Weyhenmeyer CE (2004) INTCAL04 terrestrial radiocarbon age calibration, 0–26 cal kyr BP. Radiocarbon 46:1029–1058

Reimer PJ, Baillie MGL, Bard E, Beck JW, Blackwell PG, Buck CE, Burr, GS, Friedrich M, Guilderson ThP, Hughen KA, Kromer B, McCormac G, Manning St, Reimer RW, Southon JR, van der Plicht J, Weyhenmeyer CE (2006) Comment on "Radiocarbon calibration curve spanning 0 to 50,000 years BP based on paired $^{230}Th/^{234}U/^{238}U$ and ^{14}C dates on pristine corals" by Fairbanks RG, Mortlock RA, Chiu T.-C, Cao L, Kaplan A, Guilderson TP, Fairbanks TW, Bloom AL, Grootes PM, Nadeau M-J, and "Extending the radiocarbon calibration beyond 26.000 years Before Present using fossil corals" by Chiu T-C, Fairbanks RG, Mortlock RA and Bloom AL. Quat Sci Rev (in press)

Römkens PFAM Hassink J, van der Plicht J (1998) Soil organic ^{14}C dynamics: effects of pasture installation on arable land. Radiocarbon 40:1023–1031

Sakamoto M, Imamura M, van der Plicht J, Mitsutani T, Sahara M (2003) Radiocarbon calibration for Japanese wood samples. Radiocarbon 45:81–89

Scott EM (2003) The Fourth International Radiocarbon Intercomparison (FIRI). Radiocarbon 45:135–408

Segal D, Carmi I, Gal Z, Smithline H, Shalem D (1998) Dating a Chalcolithic burial cave in Peqi'in, Upper Galilee, Israel. Radiocarbon 40:707–712

Shen ChD, Yi WX, Sun YM, Xing ChP, Yang Y, Yuan Ch, Li Zh, Peng ShL, An ZhSh, Liu TSh (2001) Distribution of ^{14}C and ^{13}C in forest soils of the Dinghushan Biosphere Reserve. Radiocarbon 43:671–678

Siani G, Paterne M, Arnold M, Bard E, Métivier B, Tisnerat N, Bassinot F (2000) Radiocarbon reservoir ages in the Mediterranean Sea and Black Sea. Radiocarbon 42:271–280

Sivan O, Herut B, Yechieli Y Lazar B (2001) Radiocarbon dating of porewater—correction for diffusion and diagenetic processes. Radiocarbon 43:765–771

Sivan O, Lazar B, Boaretto E, Yechieli Y, Herut B (2004) Radiocarbon in porewater of continental shelf sediments (southeast Mediterranean). Radiocarbon 46:633–642

Solow AR (2003) Characterizing the error in the estimated age-depth relationship. Radiocarbon 45:501–506

Southon J, Kashgarian M, Fontugne M, Métivier B, Yim W-WS (2002) Marine reservoir corrections for the Indian Ocean and Southeast Asia. Radiocarbon 44:167–180

Stein M, Goldstein StL, Schramm A (2000) Radiocarbon calibration beyond the dendrochronology range. Radiocarbon 42:415–422

Stein M, Migowski Cl, Bookman R, Lazar B (2004) Temporal changes in radiocarbon reservoir ages in the Dead Sea-Lake Lisan system. Radiocarbon 46:649–655

Stiller M, Kaufman A, Carmi I, Mintz G (2001) Calibration of lacustrine sediment ages using the relationship between ^{14}C levels in lake waters and in the atmosphere: the case of Lake Kinneret. Radiocarbon 43:821–830

Stocker TF, Wright DG (1998) The effect of a succession of ocean ventilation changes on ^{14}C. Radiocarbon 40:359–366

Sveinbjörnsdóttir ÁE, Heinemeier J, Kristensen P, Rud N, Geirsdóttir Á, Hardardóttir J (1998) ^{14}C AMS dating of Icelandic lake sediments. Radiocarbon 40:865–872

Sveinbjörnsdóttir ÁE, Heinemeier J, Gudmundsson G (2004) ^{14}C dating of the settlement of Iceland. Radiocarbon 46:387–394

Takahashi HA, Yonenobu H, Nakamura T, Wada H (2001) Seasonal fluctuations of stable carbon isotopic composition in Japanese cypress tree rings from the Last Glacial period—possibility of paleoenvironment reconstruction. Radiocarbon 43:433–438

Van Andel TjH (2005) The ownership of time: approved ^{14}C calibration or freedom of choice? Antiquity 79:944–948

Van der Borg K, Stein M, de Jong AFM, Waldmann N, Goldstein SL (2004) Near-zero $\Delta^{14}C$ values at 32 kyr cal BP observed in the high-resolution ^{14}C record from U-Th dated sediment at Lake Lisan. Radiocarbon 46:785–795

Vandergoes MJ, Prior ChrA (2003) AMS dating of pollen concentrations—a methodological study of Late Quaternary sediments from South Westland, New Zealand. Radiocarbon 45:479–491

Van der Plicht J, Beck JW, Bard E, Baillie MGL, Buck CE, Friedrich M, Guilderson TP,Hughen KA, Kromer B, McCormac FG, Ramsey CB, reamer PJ, Reimer RW, Remmele S, Richards DA, Southon JR, Stuiver M, Weyhenmeyer CE (2004) NOTCAL04—comparison / calibration ^{14}C records 26–50 cal kyr BP. Radiocarbon 46:1225–1238

Van Strydonck MJY, Crombé Ph, Maes A (2001) The site of Verrebroek "Dok" and its contribution to the absolute dating of the Mesolithic in the Low Countries. Radiocarbon 43:997–1005

Vasil'chuk YuK, Jungner H., Vasil'chuk AC (2001) ^{14}C dating of peat and $\delta^{18}O$–δD in ground ice from Northwest Siberia. Radiocarbon 43:527–540
Voelker AHL, Sarnthein M, Grootes PM, Erlenkeuser H, Laj C, Mazaud A, Nadeau M-J, Schleicher M (1998) Correlation of marine ^{14}C-ages from the Nordic Seas with the GISP2 isotope record: implications for the ^{14}C calibration beyond 25 ka BP. Radiocarbon 40:517–534
Voelker AHL, Grootes PM, Nadeau M-J, Sarnthein M (2000) Radiocarbon levels in the Icelandic Sea from 25–53 kyr and their links to the Earth's magnetic field intensity. Radiocarbon 42:437–452
Walker MJC, Bryant C, Coope GR, Harkness DD, Lowe JJ, Scott EM (2001) Towards a radiocarbon chronology of the Late-Glacial: sample selection strategies. Radiocarbon 43:1007–1019
Weinstein-Evron M, Vogel JC, Kronfeld J (2001) Further attempts at dating the palynological sequence of the Hula L07 core, Upper Jordan Valley, Israel. Radiocarbon 43:561–570
Wohlfarth B, Possnert G (2000) AMS radiocarbon measurements from the Swedish varved clays. Radiocarbon 42:323–333
Yechieli Y, Sivan O, Lazar B, Vengosh A, Ronen D, Herut B (2001) Radiocarbon in seawater intruding into the Israeli Mediterranean coastal aquifer. Radiocarbon 43:773–781
Yokoyama Y, Esat TM, Lambeck K, Fifield LK (2000) Last Ice Age millennial scale climate changes recorded in Huon Peninsula corals. Radiocarbon 42:383–401
Yoneda M, Hirota M, Uchida M, Uzawa K, Tanaka A, Shibata Y, Morita M (2001) Marine radiocarbon reservoir effect in the western North Pacific observed in archaeological fauna. Radiocarbon 43:465–471
Yoneda M, Hirota M, Uchida M, Tanaka A, Shibata Y, Morita M, Akazawa T (2002) Radiocarbon and stable isotope analysis on the earliest Jomon skeletons from the Tochibara rockshelter, Nagano, Japan. Radiocarbon 44:549–557
Yonenobu H, Takenaka Ch (1998) The Tunguska event as recorded in a tree trunk. Radiocarbon 40:367–371
Zheng Y, Anderson RF, Froelich PhN, Beck W, McNichol AP, Guilderson Th (2002) Challenges in radiocarbon dating organic carbon in opal-rich marine sediments. Radiocarbon 44:123–136
Zoppi O, Albani A, Ammerman AJ, Hua Q, Lawson EM, Sernandrei-Barbero R (2001) Preliminary estimates of the reservoir age in the Lagoon of Venice. Radiocarbon 43:489–494

Professor Dr. Dr.h.c. Burkhard Frenzel
Institut für Botanik – 210
Universität Hohenheim
D-70593 Stuttgart
Germany

Competition for Resources in Trees: Physiological Versus Morphological Plasticity

Thorsten E.E. Grams and Christian P. Andersen

Abstract

In this review, we examine two hypotheses related to resource acquisition in trees. The first hypothesis states that when competition is size-asymmetrical, then allocation changes leading to morphological shifts are more important than physiological shifts in obtaining limited resources. For example, the most effective response to competition for light (unidirectional resource supply) is a change in architectural arrangement of the biomass, i.e. a morphological response, both to maximize light exposure but also to shade neighbouring plants. A second hypothesis states that when competition is size-symmetrical, physiological adjustments are more important than morphological adjustments in obtaining limited resources. One example is below-ground competition for nutrients, which is generally considered size-symmetric, since resource supply is multi-directional. In this case, architectural arrangement is not important since all biomass is equally effective in accessing resources, and therefore physiological adjustments can lead to greater uptake rates than morphological shifts. However, below ground, resource distribution is often patchy, which may lead to conditions analogous to unidirectional resource supply. If resource supply is patchy below ground, then morphological adjustments may be more effective than physiological adjustments in capturing resources, similar to conditions when resource supply is unidirectional (and hence competition is size-asymmetrical). Despite this possibility, there is little direct evidence to suggest that competition below-ground is size-asymmetric.

Above ground, the literature supports the importance of morphological shifts in response to the primary resource in question, i.e. light. Physiological responses may be essential for tree survival in the shade but hardly relate to their competitive success. Below ground, however, the lack of comprehensive tree studies coupled with complexities associated with the simultaneous limitation of more than one resource makes it difficult to rank the relative importance of physiological vs. morphological plasticity. We conclude that

Progress in Botany, Vol. 68

below ground both morphological and physiological adjustments in response to competition for resources are important in trees.

1 Introduction

For decades, scientists have debated plant strategies for acquiring limited resources both above and below ground (Grime 1977, 1979, Tilman 1985, 1988). Our understanding of resource acquisition and plant competition comes primarily from studies with non-woody species, and several recent reviews have summarized major findings (Poorter and Navas 2003; Craine 2005). Trees differ considerably from annual and herbaceous plants, both in overall growth strategy and allocation of resources for growth, maintenance and reproduction. In addition, as trees age, increased biomass is required for structural and architectural purposes relative to physiologically active tissues involved in resource acquisition. These factors make trees somewhat unique in their ability to respond to resource limitation, which has led to increased interest in studying competitive strategies in trees (e.g. Küppers 1989; Grams et al. 2002; Matyssek et al. 2005; Balandier et al. 2006).

Recently, debate has focused on differences in resource supply and the relative importance of morphological shifts compared with physiological shifts in response to resource limitation (Schwinning and Weiner 1998). Herbaceous plants exhibit a range of plasticity in response to resource limitation, including changes in allocation leading to altered morphology and changes in physiological adjustments leading to increased efficiency of resource uptake. Even though it is clear that most trees are able to respond to their environment in one or more ways to improve resource capture and chances of survival, the extent and relative importance of morphological versus physiological plasticity is not well understood.

1.1 Size-asymmetric versus size-symmetric competition

Schwinning (1996) and Schwinning and Weiner (1998) discussed the concepts of size-symmetrical and -asymmetrical competition, based on the supply of resource in question. If resource supply was unidirectional, as in the case of light, then competition for light was size-asymmetrical. In size-asymmetrical competition, resource capture is disproportionate to size, i.e. biomass involved in capturing a particular resource. It is currently under debate what measure best represents the "biomass involved" (Anten and Hirose 1998). We believe that "biomass involved" must include both the harvesting organ and the structural biomass needed to support it. Thus, in the example of competition for light, resource captured should be related to the sum of leaf and branch biomass (cf. Grams et al. 2002; Kozovits et al. 2005b). Not only do larger or higher canopies capture a disproportionate share of the contested resource (i.e. light), but they also reduce light availability to neighbours, effectively reducing the growth of their competitor. Therefore, to generate size-asymmetric competition, the resource in question must be "pre-emptable" (Schwinning and Weiner 1998).

Conversely, resources that are multidirectional are thought to result in size-symmetric competition. In size-symmetric competition, resource capture is proportional to the biomass invested to capture the resource. For example, nitrate supply in soils is considered multidirectional due to its high diffusivity in the liquid phase of the soil (Tinker and Nye 2000). Thus, competition for nitrate is considered more size-symmetric, and proportionally to their size, larger root systems are able to acquire more resource than smaller root systems. In this latter example, there is no nutrient 'shading' effect as is the case above ground with light, and resource capture is not thought to be "pre-emptable". However, it is currently under debate if below-ground competition is always size-symmetric, since soil heterogeneity and low diffusivity of ions such as phosphate could make below-ground competition for resources size-asymmetric. This topic will be discussed more thoroughly in section 2.7 below.

In this review, we provide an overview of morphological and physiological plasticity in response to resource limitation, drawing primarily from the tree literature. However, the majority of competition studies have been conducted with non-woody species, so these studies will be included to the extent that they shed light on competitive strategies and identify important areas for future research. Two hypotheses are stated below to help frame the discussion and hopefully stimulate additional research in competition and resource capture in trees.

1.2 Hypotheses

Two hypotheses related to resource supply in relation to physiological and morphological plasticity will serve to organize the discussion. The first hypothesis states that when competition is size-asymmetrical, then allocation changes leading to morphological shifts are more important than physiological adjustments in obtaining limited resources (Fig. 1). The second hypothesis states that when competition is size-symmetrical, physiological adjustments are more important than morphological adjustments in obtaining limited resources. It is important to note, however, that in most cases woody plants

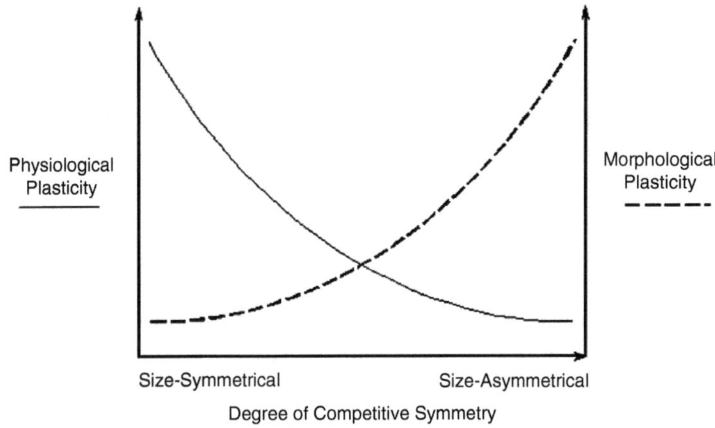

Fig. 1. Hypothetical importance of physiological versus morphological plasticity based on resource competitive symmetry in trees

will exhibit both morphological and physiological shifts in response to resource limitation. In this review we examine the literature in support of these hypotheses, focusing on woody plant species.

2 Background

2.1 Above-ground competition in response to resource limitation

Plants compete above ground for energy (i.e. light) as the essential resource for photosynthetic C fixation. In a strict sense, C itself is another above-ground resource plants compete for, similar to the uptake of N below ground, which is also energy dependent. In contrast to light, supply of C is considered multi-directional. In principle, there are gradients in CO_2 concentration in forests, which develop and change with space and time. For example, concentration of CO_2 is lower during daytime compared with nighttime and is higher close to the ground (Broadmeadow et al. 1992; Buchmann et al. 1996), but gradients are rather small, due to the high diffusivity and turbulent mixture of CO_2 in air. Therefore C, as a resource with multi-directional supply, is not considered pre-emptable, resulting in size-symmetrical competition. Following our hypotheses posed above (see section 1.2), physiological mechanisms should dominate the competition for C provided energy is not limiting resource capture. In contrast, due to its strong unidirectional supply by the sun, light is considered pre-emptable, resulting in size-asymmetric competition. Thus, morphological responses should be crucial in competition for light.

2.2 Above-ground morphological plasticity

The life-form "tree" is an evolutionary response to competition for light and the strategy of height growth in response to light competition. In particular in dense (forest) stands, height growth is a dominant strategy in competition for light (Weiner and Thomas 1992; Weiner and Fishman 1994; Berntson and Wayne 2000), which requires increased allocation of resources to the stem (Anten and Hirose 1998; Vanninen and Mäkelä 2005). In general, tree crowns grow towards gaps or space with high light availability where competition for light is low. Vigorously growing parts of crowns are supported by the tree at the expense of investments in crown regions with low incident light and subsequent diminished growth (Canham 1988; Young and Hubbell 1991). Responses to gaps and neighbors via structural crown interaction are crucial for species composition within communities (Tremmel and Bazzaz 1995; Umeki 1997; Suzuki 2002). Plants detect their neighbours by changes in light quality e.g. via the phytochrome system, and they may alter crown architecture before light availability becomes reduced (Ballaré et al. 1987, 1988). Due to their high morphological plasticity, in particular tree canopies of broadleaved species are often displaced and rarely positioned directly above their stem bases (Muth and Bazzaz 2002).

However, competitiveness of woody plants in above-ground resource capture cannot be reduced to height or canopy growth. Thorough understanding of competitive interactions among trees requires the study of spatial arrangements of canopies and plastic responses in canopy volume (Muth and Bazzaz 2003). In dense canopies of juvenile beech and spruce trees, competitiveness of individuals was indicated by biomass-related above-ground space occupation (Grams et al. 2002; Kozovits et al. 2005b), underlining the importance of space occupation above ground when competition is for light. When grown in mixture with spruce, juvenile beech plants displayed smaller crown volumes per unit of shoot biomass compared with beech monoculture. Conversely, in mixture with beech, spruce had greater space occupation per unit of shoot biomass (Grams et al. 2002; Kozovits et al. 2005b). The decline in above-ground space occupation (crown volume per biomass invested) in beech in mixed culture resulted from size-independent changes in the architectural arrangement of the canopy: lower investments into foliage per unit of shoot biomass in mixed compared with monoculture (Kozovits et al. 2005b). This confirmed earlier findings by Küppers (1985) and Schulze et al. (1986) that in a dense canopy, high competitive ability was related to increased above-ground space occupation, i.e. casting shade towards neighbours, at low biomass investments. These results support our hypothesis that in the case of size-asymmetric competition (e.g. for light), responses in morphology are crucial in obtaining limited resources.

Occupying above-ground space not only casts shade on the competitor, but also is the basis for effective capture of light energy. Anten and Hirose (1998) used the amount of absorbed light per above-ground biomass as a quantitative measure for the degree of size-asymmetric above-ground competition. By this they confirmed the general assumption (Weiner and Thomas 1986) of size-asymmetric competition for light. However, in multi-specific herbaceous stands with similar LAI but relatively low leaf area ratio (LAR) of the dominant species, above-ground competition might also be size-symmetric (Hirose and Werger 1995; Anten and Hirose 1999).

The amount of light absorbed by a leaf is not only determined by the incident PPFD at the leaf lamina, but also by the proportion of photons absorbed by the leaf. When a leaf is growing under high irradiance, it typically develops a thicker (multi-layer) palisade parenchyma. Thus, less leaf area is constructed per unit biomass, i.e. specific leaf area (SLA) is lower (Carpenter and Smith 1981; Witkowski and Lamont 1991), which increases absorption of incident light . This plastic response in leaf morphology contributes to higher overall light capture of the crown. Thus in part, above-ground competition for energy (i.e. light) is also determined on the leaf level. Conversely, when an individual leaf is growing under low irradiance, leaves develop with higher SLA, which increases leaf area per unit of biomass. In addition, leaves are placed horizontally to maximize light interception (in contrast to more vertically oriented leaves in the sun crown of a tree; Givnish 1988). Together with avoidance of self-shading (e.g. by reduction in height growth), this increases the amount of absorbed light under low irradiance (cf. Messier et al. 1999). Typically, plasticity in leaf morphology is greater in broadleaf species compared with conifers (Niinemets and Kull 1995; Bartelink 1997; Grote and Reiter 2004; Stancioiu and O'Hara 2006) and in shade tolerant compared with shade-intolerant trees (Messier et al. 1999).

2.3 Above-ground physiological plasticity

Plants exhibit also a great range of physiological plasticity in responses to their light environment. Here, we concentrate on physiological aspects of light absorption and efficient use of absorbed light for C fixation. We will not discuss physiological responses to excess of light such as photoinhibition or induction of xanthophyll cycle (for reviews on these topics see Demmig-Adams and Adams 1992; Choudhury and Behera 2001; Latowski et al. 2004).

Light dependence of trees varies greatly with genotype and species. Typically, pioneer species are more light demanding and are more sensitive to shading by neighbours than late successional species, which is sensed via shifts in light quality by the phytochrome system (Ballaré et al. 1990; Ellenberg 1996; Gilbert et al. 2001).

Competition for light also involves the absorption of incident light, which is not only a question of low SLA (i.e. thick palisade parenchyma) but also depends on amount and type of pigments in the leaf. Under low irradiance plants increase chlorophyll concentration (per unit leaf mass) and, in particular of chlorophyll b (decreased chlorophyll a to b ratio; Lichtenthaler et al. 1981; Hoflacher and Bauer 1982; Grassi and Bagnaresi 2001), which reflects an increase in the proportion of chlorophyll in the light-harvesting complexes and a decrease in photosystem II complexes (Sims and Pearcy 1994). These physiological responses to low irradiance (e.g. due to self or neighbour shading) enable leaves to absorb a higher proportion of incident PPFD. Thus, shade leaves invest pigments and leaf mass (i.e. high SLA) to maximize light absorption via a large leaf area, which is not critical for sunlit leaves due to high resource availability in terms of light (Givnish 1988).

Under low irradiance, limited energy availability is the main constraint for C gain. However as stated above, carbon itself has a multi-directional supply and therefore is considered not pre-emptable. Thus, if light energy is not limiting, C gain is postulated to be dominated by physiological mechanisms (see section 1.2).

Effective CO_2 uptake of plants depends on both stomatal aperture (regulation of CO_2 diffusion into the leaf) and the rate of biochemical C fixation via the carboxylation reaction of ribulose-1, 5bis-phosphate carboxylase/oxygenase (rubisco). In addition to the substomatal CO_2 concentration, stomatal aperture largely depends on the water status of the plant, i.e. availability of the below-ground resource water. In addition to CO_2 supply, biochemical parameters of the CO_2-fixing enzyme rubisco, such as maximum velocity of C fixation (V_{cmax}) and photosynthetic capacity, determine CO_2 uptake of plants. Increase of V_{cmax} with light availability not only depends on morphological plasticity of a leaf (i.e. increased thickness of palisade parenchyma, see above) but might also be affected by physiological plasticity at the cell level. Beside morphological adjustments, higher partitioning of leaf nitrogen to rubisco and bioenergetics will further increase photosynthetic capacity (e.g. Grassi and Bagnaresi 2001). In general, a lot of emphasis is given to these biochemical and physiological processes due to their essential nature for life on earth. However, rubisco biochemistry and C gain rarely explain the competitive success of trees (Grams et al. 2002; Kozovits et al. 2005a,b; Reiter et al. 2005).

With respect to above-ground competition, neither the highest energy efficiency in C uptake nor the best physiological adjustment to low irradiance will be effective in compensating for a reduction in C gain by shade cast from a neighbour via morphological aspects of interplant competition, at least for trees competing in dense stands. Thus, when plants compete for light, their most effective strategies are related to morphological adjustments.

2.4 Below-ground competition in response to resource limitation

Roots and associated mycorrhizal symbionts exhibit both physiological and morphological plasticity in response to inadequate supply of nutrients and nutrient patchiness (Friend et al. 1990; Robinson and van Vuuren 1998; Fitter et al. 2000). Compared with above-ground competition, which is primarily for light, below-ground competition occurs for a suite of nutrients as well as water. Typically, it is thought that plants adjust to conditions of resource imbalance by allocating more carbon to the tissues that acquire the most strongly limiting resource (Chapin et al. 1987; Gleeson and Tilman 1992), which in the case of below-ground competition leads to increased root biomass relative to shoot biomass. In many plants, carbohydrates serve as signalling molecules responsible for shifting allocation between sources (leaves) and sinks (roots, non-photosynthesizing tissues) in response to resource limitation (Koch 1996; Farrar and Jones 2000; Andersen 2003). Hexoses and hexokinases appear to be the signalling molecules that act at the gene level to up-regulate growth, storage and other metabolic processes in tissues (Koch 1996). Although the mechanisms underlying root plasticity are variable and complex, it appears that carbohydrate signalling may provide a rapid means for plants to respond both physiologically and morphologically to stress and resource limitations (Andersen 2003).

It is important to consider the soil resource in question when characterizing plasticity of tree response to resource limitation. Roots come in contact with soil resources through the process of diffusion, mass flow and by root interception through root growth (Casper and Jackson 1997). Certain nutrients are relatively immobile and move primarily by diffusion, such as P, K, Ca and ammonium, while other nutrients move as solvents in the water by (mass) flow, such as nitrate. As a result, the volume of soil exploited by roots can be 100–1000 times greater for mobile compared with immobile nutrients (Fitter et al. 2002). In the field, supply of N, P and K to plant roots usually occurs through both diffusion and mass flow, while direct root interception accounts for less than 10% of nutrients taken up by the root (Casper and Jackson 1997). Nutrient mobility and chemical interactions in soil lead to resource patchiness, which influences the effectiveness of physiological versus morphological plasticity (Fitter et al. 2002; Hodge 2006).

Since supply of many nutrients to roots occurs through mass flow, plant water use can strongly influence nutrient supply and uptake (Casper and Jackson 1997). As a result, above-ground competition may influence nutrient uptake not only by reducing carbohydrate availability for root growth and nutrient uptake, but also by influencing movement of soil nutrients to the root surface through changes in transpiration and plant water use.

2.5 Below-ground morphological plasticity

One of the most well documented responses to soil nutrient and water limitation is increased carbon allocation to roots and increased root growth (Clarkson 1985; Cannell et al. 1988; Farrar and Jones 2000). This response is often accompanied by increases in root-shoot ratio (Ingestad and Agren 1991; Ericsson 1995), showing morphological plasticity at the whole-plant level. Ericsson (1995) found that shifts in allocation to roots varied depending on the mineral nutrient that was limiting, with root growth favoured when N, P and S were deficient. Morphological plasticity has also been observed within root systems in response to localized nutrient enrichment (Hodge 2006). Philipson and Coutts (1977) found that roots of Sitka spruce grow preferentially in high N versus low N soil. In Douglas-fir stands, fine root production was found to be inversely correlated with N availability (Grier et al. 1985).

Friend et al. (1990) conducted split-root experiments with Douglas-fir seedlings in microcosms and demonstrated that the degree of morphological plasticity in response to N limitation was related to the overall N status of the plant. Lateral roots from nutrient deficient and sufficient plants were placed in either high or low N micro-environments in side chambers. Root proliferation in response to high N micro-environment was twice as great in N deficient plants compared with N sufficient plants. These results suggest that trees can effectively respond to nutrient patchiness in soil via plasticity at the individual root level. In addition, they demonstrate that plasticity is not only a function of root environment, but also dependent on whole-plant nutrition.

Roots also show morphological plasticity through increased specific root length (length per unit of biomass), which effectively increases the ratio of surface area to volume (Aerts et al. 1991; Casper and Jackson 1997). Morphological plasticity in specific root length is seen in response to nutrient patchiness and the form of nutrients available (Pregitzer et al. 1993; Robinson 1994). Root proliferation into areas of nutrient enrichment often results in higher density and smaller diameter roots (Casper and Jackson 1997). Bilbrough and Caldwell (1995) found that specific root length of *Agropyron desertorum* showed increased specific root length in N-enriched patches. Woolfolk and Friend (2003) used split root systems of *Populus deltoides* to examine the influence of different ratios of NH_4:NO_3 on root growth and development. Specific root length varied with nutrient supply, with seedlings showing the highest specific root length when 80% of the N added was in the form of NO_3.

To the extent that mycorrhizal hyphae are considered extensions of the root system, increased colonization in response to nutrient deficiency also could be considered morphological plasticity. Although the mechanisms are not well understood, it appears that plants may regulate the degree of colonization through changes in carbon allocation and root exudation (Graham et al. 1981; Smith and Read 1997). Nutrient deficiency often increases the extent of mycorrhizal development on root systems, which increases soil exploration and nutrient uptake (Smith and Read 1997; Wallenda et al. 2000). In addition, mycorrhizal hyphae can exploit nutrient patches through hyphal proliferation, although roots may be more responsive than hyphae to nutrient patches (Hodge 2006). The ability of mycorrhizal root systems to rapidly access inorganic and organic forms of N may increase their competitiveness with soil microbial communities, allowing plants to exert stronger control over N cycling (Chapman et al. 2006).

Root and mycorrhizal turnover rate is another example of morphological plasticity in plants. King et al. (2002) found that mycorrhizal colonization increased root longevity in *Pinus taeda*. In *Picea abies*, mycorrhizal longevity was related to branching order, and N addition altered branching density and therefore mycorrhizal longevity. Johnson et al. (2000) found that roots born in the fall and winter had longer median life-spans than roots born in spring or summer and that elevated CO_2 and N fertilization both affected root life-span. Root longevity and turnover contribute significantly to soil C and therefore are critical determinants controlling N and C cycling in ecosystems (Pregitzer et al. 1995; Zak et al. 2000).

It is clear that morphological plasticity is widespread and an important competitive strategy in both trees and herbaceous plants in response to resource limitation below ground. Since soil nutrients vary significantly in their distribution and mobility in soil, it is difficult to make generalizations about the relative importance of morphological plasticity, particularly in trees.

2.6 Below-ground physiological plasticity

Roots and their mycorrhizal symbionts exhibit also a wide range of physiological adjustments that increase resource acquisition. One important physiological response to nutrient limitation appears to be increased root exudation

(Walker et al. 2003; Jones et al. 2004). Since rhizosphere and microbial communities are generally C-limited (Zak et al. 1994), increased root exudation of carbon compounds is thought to stimulate rhizosphere microbial activity, possibly leading to increased nutrient cycling and availability. Root exudates are important as chemical attractants and repellents; they can regulate soil microbial communities and in some cases communicate with other roots in the vicinity (Walker et al. 2003). Some exudates appear to be chemical messengers that regulate root competition among different plant species (Callaway and Aschehoug 2000). Other compounds such as flavonoids can stimulate Rhizobium genes responsible for nodulation in N-fixing species (Peters et al. 1986). By stimulating and controlling rhizosphere organisms, plants can favour conditions of increased nutrient turnover and availability for plant uptake, as well as influence the extent of symbiotic associations.

The extent to which the plant actively controls exudation to carry-out these important rhizosphere processes is not well understood. There is some evidence that ATP-binding cassette transporters (ABC transporters) are widespread and may be involved in secretion of root exudates (Walker et al. 2003). Since plants have the ability to synthesize and secrete thousands of compounds that may play a role in rhizosphere dynamics, the potential for plant control of root competition for resources through physiological plasticity is high.

Physiological adjustments such as proton pumps and enzyme systems that lead to increased root uptake provide a means to respond to increased resource availability (Jackson et al. 1990; Aerts 1999). It is thought that physiological plasticity may be more important for rapidly versus. slowly diffusing nutrients (Aerts 1999), consistent with our hypotheses (Fig. 1). For example, P is relatively immobile in soils and therefore plants would benefit little from physiological adjustments without increased root growth into P rich areas (i.e. morphological adjustment). Conversely, the kinetics of uptake may be more important than root architectural arrangement for ions that move readily and homogeneously through the soil via diffusion and mass flow (Fitter et al. 2002). In general, this supports our hypotheses since competition for mobile, evenly distributed nutrients should be size-symmetric, while immobile nutrients tend to be distributed more heterogeneously, leading to size-asymmetric competition (Schwinning 1996).

Physiological plasticity in mycorrhizal symbionts is extremely important, since mycorrhizae are thought to be responsible for most of the nutrient uptake that occurs in forested ecosystems (Wallenda et al. 2000). Recently, it has become apparent that certain mycorrhizae contain proteolytic enzymes that may play an important role in organic nutrient uptake (Smith and Read 1997; Read and Perez-Moreno 2003). It appears that ectomycorrhizal fungi exhibiting proteolytic capabilities are highest in low N areas (Lilleskov et al.

2002). The ability to utilize organic forms of soil nutrients may be especially important in soils where mineralization is slow or inhibited, such as low temperature soils. Whether or not these proteolytic enzyme systems are substrate inducible (e.g. present only when substrate is available) or constitutive is poorly understood. Nonetheless, the ability of mycorrhizal roots to access organic forms of N increases their competitiveness with microbial communities and may increase plant control of nutrient cycling in ecosystems (Chapman et al. 2006; Hodge 2006).

2.7 Evidence for size-asymmetric competition below ground

It is generally thought that below-ground competition is size-symmetric due to the multidirectional supply of nutrients in the soil solution. Nutrient ions such as nitrate and K move both by diffusion and mass flow, while others such as phosphate adsorb strongly to surfaces (Casper and Jackson 1997). Tightly absorbed ions are rather immobile (Tinker and Nye 2000), and therefore are distributed heterogeneously in the soil. In general, when soil resources are heterogeneously distributed, the competitive situation is largely similar to above-ground competition for light, because heterogeneously distributed resources are "pre-emptable". Thus, theoretically competition for certain soil resources below ground should be size-asymmetric (Weiner 1990; Blair 2001).

Many experimental tests to confirm the hypothesis that below-ground competition is size-asymmetric have been unsuccessful (Casper and Cahill 1996; Weiner et al. 1997; Cahill and Casper 2000; Blair 2001). Blair (2001) conducted a pot experiment using *Ipomoea tricolor* and heterogeneous nutrient patches and found that competition below-ground was size-symmetric. Similarly, Cahill and Casper (2000) found that below-ground competition between several grasses and *Amaranthus retroflexus* was proportional to neighbour root biomass in a field study, supporting size-symmetry of below-ground competition.

Other studies have shown that competition below ground can be size-asymmetric (Leuschner et al. 2001; Rajaniemi 2003). Rajaniemi (2003) found size-asymmetric competition to occur between *Bromus* plants of different sizes in an unhomogenized field soil. Fransen et al. (2001) found size-symmetric competition in homogeneous soil and size-asymmetric competition in heterogeneous soil. They also found that physiological plasticity was more important than morphological plasticity over the long-term as competition became more size-symmetric, consistent with hypotheses presented in Fig. 1. Canham et al. (2004) examined the relative importance of shading versus crowding in a mixed stand of western hemlock (*Tsuga heterophylla)* and western red cedar (*Thuja plicata*) in British Columbia and found that for both species, growth declined more steeply as a function of crowding than shading. In addition, there was very little effect of crowding by red cedars on radial growth of hemlock, but hemlock had a strong effect on the growth of red cedars. Crowding was assumed to include both below-ground competition and above-ground inhibition of crown development. Although the mechanisms behind the size-asymmetry in interspecific competition were not determined, the authors concluded that it likely resulted both from below-ground competition and hemlocks negative influence on forest floor nutrient dynamics.

A number of studies have shown that resource capture is not proportional to biomass production below ground, providing support for size-asymmetric competition below ground. In a sagebrush steppe, tussock grass and sagebrush took up similar amounts of phosphate even though the tussock grass had significantly more roots in nutrient patches (Caldwell et al. 1991, 1996; Casper and Jackson 1997). These and other studies demonstrate that resource capture is not always proportional to biomass investment in the patches, consistent with size-asymmetry. This could result from physiological adjustments, but it also may have resulted from intraspecific root competition.

Size-asymmetric competition below ground also may result from root interference or allelopathy (Cahill and Casper 2000). In this case, roots of certain plants may be able to restrict the growth of competitor plants, resulting in a situation similar to light limitation above ground by shading.

Collectively, there is evidence that both symmetric and asymmetric competition for soil resources occurs, and in some cases there is support for both hypotheses 1 and 2. However, the degree to which patchy nutrient supply and competition for rather immobile resources results in size-asymmetry in below-ground competition is currently under debate. Likewise, it remains unclear if physiological or morphological aspects dominate the competition for highly mobile resources, such as nitrate. Unlike competition above ground, which is primarily for light, below-ground competition may occur simultaneously for several nutrients, each varying in their supply and patchiness. Compared with above-ground resources, we know far less about resource distribution and heterogeneity below ground, which greatly hampers our ability to study plant competition for below-ground resources.

3 Competition at high versus low resource availability

There is still debate regarding underlying mechanisms of response to competition in species adapted to high versus low resource availability (e.g. nutrient rich versus nutrient poor environments; Aerts 1999).

Rates of root and mycorrhizal turnover are affected by resource availability (Pregitzer et al. 1995; Eissenstat et al. 2000; Johnson et al. 2001; Majdi et al. 2001). Root turnover is often greater in nutrient rich than nutrient poor soils (Aber et al. 1985; Pregitzer et al. 1993, 1995; Johnson et al. 2000), although this is not always the case (Burton et al. 2000). In addition, other factors such as soil temperature may interact with nutrient availability in affecting root longevity (King et al. 1999). Burton et al. (2000) suggest that plasticity in root longevity in northern hardwood forests may be regulated by carbohydrate supply from the shoot; when N supply is high, root metabolic activity and sink strength are high, providing carbon to meet theses metabolic demands. As N availability decreases, decreased root sink strength and hence carbohydrate supply leads to shorter root lifespan. However, it is also possible that plasticity in root turnover is species dependent, with species adapted to nutrient poor sites favouring nutrient retention (and slow turnover) versus rapid growth and high turnover (Aerts 1999). At high levels of nutrient availability, competition above ground for light becomes more important, and traits of species from fertile environments lead to a high rate of ecosystem N cycling. Thus, slower growing and nutrient conserving species may get out-competed due to both competition for light and high rates of nutrient turn-over (Aerts 1999). Wilson and Tilman (1991, 1993) examined the effect of N-addition to transplants of grass species and found that competition shifted from primarily below ground in unfertilized plots to both below- and above ground in fertilized plots. In addition, the intensity of below-ground competition was greater at low than high light (Wilson and Tilman, 1993). The overall intensity of total above- plus below-ground competition did not change with fertilization, but rather the intensities of above- and below-ground competition were negatively correlated.

According to the theory of Grime (1979), increased productivity in resource-rich environments should result in increased density-dependent competition and mortality. Above ground, for example, increased irradiance enhanced asymmetry competition (Bazzaz and Harper 1976; Weiner 1985). However, high resource availability of a multi-directional supplied resource (e.g. high CO_2 concentration) might reduce size-asymmetry of above-ground competition compared with a situation with low resource availability (Wayne and Bazzaz 1997). In regenerating yellow birch (*Betula alleghaniensis*) stands grown under elevated CO_2, size-asymmetry of above-ground competition was reduced (Wayne and Bazzaz 1997). Likewise other studies confirmed that growth of light limited woody plants is more stimulated by elevated CO_2 than that of dominant or light-saturated individuals (Würth et al. 1998; Hättenschwiler and Körner 2000; Granados and Körner 2002). Increase in above-ground size-symmetry of competition can result from both morphological and physiological responses to increase CO_2 availability: (1) elevated

CO_2 increases the light availability of shaded plants by a morphological response of dominant plant, in that they developed a reduced leaf area per unit biomass (i.e. lower LAR) under higher CO_2 availability (Wayne and Bazzaz 1997). Hence, light availability of shaded plants was increased by higher light penetration through the canopy. Likewise, Hirose and Werger (1995) and Anten and Hirose (1999) found size-symmetry of above-ground competition to be increased in multi-specific herbaceous stands with relatively low LAR of the dominant species; (2) a direct physiological effect of elevated CO_2 increasing the light-use efficiency of woody plants for growth, i.e. c. 50% higher growth under elevated CO_2 (Bazzaz and Miao 1993, Wayne and Bazzaz 1997). Similarly, Hättenschwiler and Körner (2000) found increased growth of understory seedlings under elevated CO_2 when growing in a light environment close to their CO_2 light compensation point. Apparently, under these conditions elevated CO_2 increased seedling carbon balance.

4 Synthesis

4.1 Morphological versus physiological responses

In reality, plants exhibit a range of plasticity in response to competition for resources, and in many cases both physiological and morphological changes result from competition for resources (Fig. 1). Nonetheless, there is much debate regarding the relative importance of physiological versus morphological adjustments (e.g. Schwinning and Weiner 1998; Weiner 2004). Küppers (1989) emphasized that in woody plants analysis of above-ground competition should pursue "an integrated view of carbon gain, increment of biomass and its architectural arrangement in space..., especially in situations where crowns compete for space and light". In other words, integration of biomass increment and spatial arrangement together with the flux of resources along this structure (e.g. C gain, N uptake) allows for a mechanistic understanding of (woody) plant competitiveness. Thus following Küppers (1989), both morphological and physiological aspects of resource competition have to be considered. Along this line, Schwinning (1996) decomposed the relative rate of resource uptake (S) into three terms: *i* a resource availability factor (R), *ii* an allometry or morphology factor (A) and *iii* a resource capture factor (C):

$$S = R \cdot A \cdot C \qquad (1)$$

The relative rate of resource uptake (S) therefore depends on the availability of the resource in question (R) and the degree of both morphological (A) and physiological (C) plasticity of individual species. In more detail, Schwinning (1996) formulated the above concept using the following equation:

$$\frac{1}{B_X} \cdot \frac{dU_X}{dt} = \frac{1}{V_X} \cdot \frac{dI_X}{dt} \times \frac{V_X}{B_X} \times \left(\frac{dU_X}{dt} \Big/ \frac{dI_X}{dt} \right) \qquad (2)$$

where B_X is the biomass involved in the uptake of resource X. In our understanding the term "biomass involved" includes structural biomass, such as branches, to place the directly in the resource uptake involved biomass, such as leaves. The time derivative of U_X is the rate of resource uptake and the time derivate of I_X is the rate of resource supply, such as PPFD. V_X is the space (volume) occupied by B_X.

For example, in the case of *Abies alba*, a raised V_{cmax} (area based maximum rate of carboxylation reaction of rubisco) as a response to increased light availability is fully explained by morphological changes (increased thickness of palisade parenchyma; Grassi and Bagnaresi 2001). However, the competitive advantage of its competitor *Picea abies* in exploiting high irradiance in the case of canopy gaps was related to both morphological and physiological plasticity (higher partitioning of leaf nitrogen to rubisco and bioenergetics; Grassi and Bagnaresi 2001), exemplifying that both aspects have to be studied to understand the competitive relationship between the two species.

Following the initial idea of Küppers (1989) and similar to the above factors of Schwinning (1996), Grams et al. (2002) defined functional efficiencies that relate resource investments to resource gains to quantify woody plant competitiveness. Thus, competitiveness was analysed through several cost/benefit-ratios expressing efficiency in space occupation, i.e. resource investment per unit of occupied space (crown or soil volume) or efficiency in space exploitation, i.e. resource gains per occupied space or biomass investment. Using this concept to analyse an intense above-ground competitive situation between juvenile beech and spruce trees, Kozovits et al. (2005b) identified space occupation (not resource gains) to be the crucial factor for the competitive success of a juvenile tree. For both juvenile and mature trees those efficiency ratios were found to quantitatively relate to the competitive success of trees (Grams et al. 2002; Kozovits 2005a,b; Reiter et al. 2005).

Cost/benefit ratios in resource uptake might vary for physiological and morphological responses in resource uptake. It is generally assumed that physiological plasticity is energetically less 'expensive' than morphological plasticity; however this may not always be the case (Aerts 1999; Hodge 2006). Several factors influence the energetic costs of physiological plasticity, including resource availability in space and time, and the degree to which uptake is enhanced through physiological plasticity. In addition, many studies have been conducted in pots or mesocosms where resources are evenly available or continuously supplied, artificially influencing resource availability (R) in equations above and, thus, limiting the ecological significance of such studies. Morphological plasticity may be more energetically efficient than physiological plasticity if tissues are long-lived and exploit a sustained source of resource. Physiological plasticity may be more efficient when resource availability is transient and requires rapid accommodation to utilize the available resource. For example, nitrate reductase is an inducible enzyme that allows plants to chemically reduce NO_2 in leaves in response to NOx deposition (Schmidt et al. 1991). Relative to constitutive enzymes, substrate inducible enzymes are only synthesized when the substrate is available, conserving plant energy (Tingey and Andersen 1991).

4.2 The whole-tree perspective: above-ground/below-ground interactions in resource competition

The vast majority of competition studies have examined herbaceous or agronomic species and not trees. In contrast to herbaceous plants, woody plants contain large amounts of structural biomass, which on one hand are essential for the spatial arrangement of biomass involved in resource capture (see section 3.2) but on the other hand require additional resources for support. Therefore, in addition to the two cost/benefit-parameters for quantifying plant competitiveness mentioned above, a third parameter, the efficiency of "running costs" for support of structural biomass, was introduced for studies on woody plant competition (Grams et al. 2002; Kozovits et al. 2005b). In many cases resources needed for supporting the structure involved are reflecting the whole-tree perspective. Transpiratory costs of an occupied crown volume or respiratory costs of below-ground structural biomass exemplify this above-/below-ground link in tree competitiveness (cf. Kozovits et al. 2005b; Reiter et al. 2005). This is one reason why analysis of interplant competition should take both above- and below-ground aspects into account.

It is difficult to make generalities about the relative importance of above- versus below-ground competition due to the complex nature of multiple stress interactions. For example, Coomes and Grubb (2000) evaluated a series of trenching experiments and found that light limitation alone limits seedling growth in forests on moist, nutrient rich soils, but competition for below-ground resources becomes important for seedling growth on infertile soils and drier regions.

Some plastic responses above ground can exacerbate below-ground resource limitation. For example, some species have difficulty tolerating both shade and drought stress, since shade adaptation leading to increased leaf area (with higher SLA) may lead to increased water loss and increased desiccation (Coomes and Grubb 2000). Cahill (1999) also found interaction between above- and below-ground competition, and the degree of interaction varied with productivity level such that no interaction occurred in the unfertilized blocks while a positive interaction occurred in the fertilized block. With fertilization, root competition decreased a plants ability to compete in size-asymmetric competition for light.

As noted previously, above-ground resource limitation also has been found to influence below-ground morphological responses to competition. Beech seedlings growing under *Pinus sylvestris* canopies with varying irradiance levels were studied to examine morphological shifts in both leaves and roots in response to light levels and interspecific competition (Curt et al. 2005). Seedlings grown under shade showed lower specific root lengths (SRL) and higher SLA. Bilbrough and Caldwell (1995) found that *Agropyron desertorum* roots increased SRL in N-enriched patches, and the magnitude of the response was decreased by shading. They also found an interaction such that relative growth rates of roots (RGR) in patches were reduced more than 50% by shading, while RGR in unenriched patches was unaffected by shading.

As might be expected, several controlled studies have shown that below-ground plasticity in response to increased nutrient limitation decreases as light levels increase (e.g. Machado et al. 2003). Because roots and symbionts require carbohydrates for growth and metabolism, it is important to consider resource limitations above ground when evaluating below-ground competition. For example, even though some hardwood species respond to soil resource availability in deeply shaded understories (Machado et al. 2003), many plant species will respond little to nutrient supply when severely shaded (Coomes and Grubb 1998). Hence, low resource availability above ground may limit a plants capacity to respond to resource limitation below ground. However, trees, which can have significant stored reserves, may exhibit greater plasticity than non-woody species in response to soil resource limitation even when above ground competition is high.

Taking the above arguments together, analysis of plant competition for resources should consider both above- and below-ground aspects of plant competitiveness.

5 Conclusions

Morphological and physiological plasticity above- and below ground in response to resource limitation is well documented. Our first hypothesis that when competition is size-asymmetrical, then allocation changes leading to morphological shifts are more important than physiological adjustments in obtaining limited resources is generally supported for above-ground competition. Although they may be crucial for tree survival in the shade, physiological adjustments appear to be less effective in competition for light than the ability to occupy new space and shading competitors at the same time. However, while competition for light is mechanistically size-asymmetric, situations exist where light capture can be size-symmetric as morphological aspects of the competing trees can influence size-symmetry.

Despite the relative importance of morphological shifts in response to above-ground asymmetric competition, physiological plasticity also is widespread and important. Whether or not hypothesis 2 is supported by the literature, i.e. physiological plasticity is more important when competition for resources is size-symmetric, is difficult to determine above ground for the following reasons. First, in most cases size-symmetrical competition above ground has not been documented or well studied since light is the most often limiting resource (e.g. size-asymmetrical). In addition, physiological changes rarely occur in the absence of morphological shifts, making the relative contribution of each difficult to quantify. Within canopies of individual trees it is not uncommon to find spatial heterogeneity in plasticity, depending on position and micro-environment.

Collectively, there is evidence that resource competition below ground is both size-symmetric and size-asymmetric and that plants exhibit morphological and physiological plasticity to increase resource capture. Although there is some evidence in support of hypotheses 1 and 2, at this time there is insufficient evidence to confirm or refute our hypotheses regarding the relative importance of morphological and physiological plasticity below ground (Fig. 1), particularly in trees. In addition, we can not confirm the long stated assumption that below-ground competition is size-symmetric. In particular, for less mobile ions and heterogeneous soils, below-ground competition might be size-asymmetric.

Evaluating the relative importance of morphological and physiological shifts in response to below-ground competition is very difficult due to the spatial and temporal complexity of resource distribution below ground. The vast majority of studies have manipulated individual resources, when in reality we know that multiple soil resources may be limiting simultaneously. In addition, resource availability can vary on temporal and spatial scales that are difficult to measure. Finally, similar to shoot responses, individual roots of root systems can respond to micro-environments that differ in resource availability, making whole-root and whole-plant generalizations difficult, especially given the hidden, inaccessible nature of roots. Although additional studies will shed light on the relative importance of morphological and physiological plasticity below ground, spatial and temporal complexity of soil resources suggest that both morphological and physiological adjustments are necessary strategies for plant competitiveness in the field. Additional research is needed and should focus on competition under natural soil conditions with heterogeneous resource distribution and should incorporate the use of whole-tree process models to help examine the importance of morphological and physiological shifts in terms of costs and benefits with respect to resource capture.

Acknowledgements

The information in this document has been subjected to EPA peer and administrative review, and it has been approved for publication as an EPA document. Mention of trade names or commercial products does not constitute endorsement or recommendation for use. T.E.E.G. was supported by SFB 607 *"Growth and Parasite Defense – Competition for Resources in Economic Plants from Agronomy and Forestry, Project B5"* by the "Deutsche Forschungsgemeinschaft" (DFG).

References

Aber JD, Melillo JM, Nadelhoffer KJ, McClaugherty CA, Pastor J (1985) Fine root turnover in forest ecosystems in relation to quantity and form of nitrogen availability—a comparison of two methods. Oecologia 66:317–321

Aerts R (1999) Interspecific competition in natural plant communities: mechanisms, trade-offs and plant-soil feedbacks. J Exp Bot 50:29–37

Aerts R, Boot RGA, Vanderaart PJM (1991) The relation between aboveground and below-ground biomass allocation patterns and competitive ability. Oecologia 87:551–559

Andersen CP (2003) Source-sink balance and carbon allocation below ground in plants exposed to ozone. New Phytol 157:213–228

Anten NPR, Hirose T (1998) Biomass allocation and light partitioning among dominant and subordinate individuals in *Xanthium canadense* stands. Ann Bot 82:665–673

Anten NPR, Hirose T (1999) Interspecific differences in above-ground growth patterns result in spatial and temporal partitioning of light among species in a tall-grass meadow. J Ecol 87:583–597

Balandier P, Collet C, Miller JH, Reynolds PE, Zedaker SM (2006) Designing forest vegetation management strategies based on the mechanisms and dynamics of crop tree competition by neighbouring vegetation. Forestry 79:3–27

Ballaré CL, Sanchez RA, Scopel AL, Casal JJ, Ghersa CM (1987) Early detection of neighbor plants by phytochrome perception of spectral changes in reflected sunlight. Plant Cell Environ 10:551–557

Ballaré CL, Sanchez RA, Scopel AL, Ghersa CM (1988) Morphological responses of *Datura ferox* L. seedlings to the presence of neighbors—their relationships with canopy microclimate. Oecologia 76:288–293

Ballaré CL, Scopel AL, Sanchez R (1990) Far-red radiation reflected from adjacent leaves: an early signal of competition in plant canopies. Science 247:329–332

Bartelink HH (1997) Allometric relationships for biomass and leaf area of beech (*Fagus sylvatica* L.). Ann Sci For 54:39–50

Bazzaz FA, Harper JL (1976) Relationship between plant weight and numbers in mixed populations of *Sinapsis alba* (L.) Rabenh and *Lepidium sativum* L. J Appl Ecol 13:211–216

Bazzaz FA, Miao SL (1993) Successional status, seed size, and responses of tree seedlings to CO_2, light, and nutrients. Ecology 74:104–112

Berntson GM, Wayne PM (2000) Characterizing the size dependence of resource acquisition within crowded plant populations. Ecology 81:1072–1085

Bilbrough CJ, Caldwell MM (1995) The effects of shading and N-status on root proliferation in nutrient patches by the perennial grass *Agropyron desertorum* in the field. Oecologia 103:10–16

Blair B (2001) Effect of soil nutrient heterogeneity on the symmetry of belowground competition. Plant Ecol 156:199–203

Broadmeadow MSJ, Griffiths H, Maxwell C, Borland AM (1992) The carbon isotope ratio of plant organic material reflects temporal and spatial variations in CO_2 within tropical forest formations in Trinidad. Oecologia 89:435–441

Buchmann N, Kao WY, Ehleringer JR (1996) Carbon dioxide concentrations within forest canopies—variation with time, stand structure, and vegetation type. Global Change Biol 2:421–432

Burton AJ, Pregitzer KS, Hendrick RL (2000) Relationships between fine root dynamics and nitrogen availability in Michigan northern hardwood forests. Oecologia 125:389–399

Cahill JF (1999) Fertilization effects on interactions between above- and belowground competition in an old field. Ecology 80:466–480

Cahill JF, Casper BB (2000) Investigating the relationship between neighbor root biomass and belowground competition: field evidence for symmetric competition belowground. Oikos 90:311–320

Caldwell MM, Manwaring JH, Jackson RB (1991) Exploitation of phosphate from fertile soil microsites by three Great Basin perennials when in competition. Funct Ecol 5:757–764

Caldwell MM, Manwaring JH, Durham SL (1996) Species interactions at the level of fine roots in the field: Influence of soil nutrient heterogeneity and plant size. Oecologia 106:440–447

Callaway RM, Aschehoug ET (2000) Invasive plants versus their new and old neighbors: a mechanism for exotic invasion. Science 290:521–523

Canham CD (1988) Growth and canopy architecture of shade-tolerant trees—response to canopy gaps. Ecology 69:786–795

Canham CD, LePage PT, Coates KD (2004) A neighborhood analysis of canopy tree competition: effects of shading versus crowding. Can J For Res 34:778–787

Cannell MGR, Morgan J, Murray MB (1988) Diameters and dry weights of tree shoots: effects of Young's modulus, taper, deflection and angle. Tree Physiol 4:219–231

Carpenter SB, Smith ND (1981) A somparative study of leaf thickness among southern Appalachian hardwoods. Can J Bot 59:1393–1396

Casper BB, Jackson RB (1997) Plant competition underground. Annu Rev Ecol Syst 28:545–570

Casper BB, Cahill JF (1996) Limited effects of soil nutrient heterogeneity on populations of Abutilon theophrasti (Malvaceae). Am J Bot 83:333–341

Chapin FS, Bloom AJ, Field CB, Waring RH (1987) Plant-responses to multiple environmental-factors. Bioscience 37:49–57

Chapman SK, Langley JA, Hart SC, Koch GW (2006) Plants actively control nitrogen cycling: uncorking the microbial bottleneck. New Phytol 169:27–34

Choudhury NK, Behera RK (2001) Photoinhibition of photosynthesis: role of carotenoids in photoprotection of chloroplast constituents. Photosynthetica 39:481–488

Clarkson DT (1985) Factors affecting mineral nutrient acquisition by plants. Annu Rev Plant Physiol Plant Mol Biol 36:77–115
Coomes DA, Grubb PJ (1998) Responses of juvenile trees to above- and belowground competition in nutrient-starved Amazonian rain forest. Ecology 79:768–782
Coomes DA, Grubb PJ (2000) Impacts of root competition in forests and woodlands: a theoretical framework and review of experiments. Ecol Monographs 70:171–207
Craine JM (2005) Reconciling plant strategy theories of Grime and Tilman. J Ecol 93:1041–1052
Curt T, Coll L, Prevosto B, Balandier P, Kunstler G (2005) Plasticity in growth, biomass allocation and root morphology in beech seedlings as induced by irradiance and herbaceous competition. Ann For Sci 62:51–60
Demmig-Adams B, Adams WW (1992) Photoprotection and other responses of plants to high light stress. Annu Rev Plant Physiol Plant Mol Biol 43:599–626
Eissenstat DM, Wells CE, Yanai RD, Whitbeck JL (2000) Building roots in a changing environment: implications for root longevity. New Phytol 147:33–42
Ellenberg H (1996) Vegetation Mitteleuropas mit den Alpen, 5th edn. Ulmer-Verlag, Stuttgart
Ericsson T (1995) Growth and shoot—root ratio of seedlings in relation to nutrient availability. Plant Soil 169:205–214
Farrar JF, Jones DL (2000) The control of carbon acquisition by roots. New Phytol 147:43–53
Fitter A, Williamson L, Linkohr B, Leyser O (2002) Root system architecture determines fitness in an Arabidopsis mutant in competition for immobile phosphate ions but not for nitrate ions. Proc R Soc Lond Ser B-Biol Sci 269:2017–2022
Fitter AH, Heinemeyer A, Staddon PL (2000) The impact of elevated CO_2 and global climate change on arbuscular mycorrhizas: a mycocentric approach. New Phytol 147:179–187
Fransen B, de Kroon H, Berendse F (2001) Soil nutrient heterogeneity alters competition between two perennial grass species. Ecology 82:2534–2546
Friend AL, Eide MR, Hinckley TM (1990) Nitrogen stress alters root proliferation in Douglas fir seedlings. Can J For Res 20:1524–1529
Gilbert IR, Jarvis PG, Smith H (2001) Proximity signal and shade avoidance differences between early and late successional trees. Nature 411:792–795
Givnish TJ (1988) Adaptation to sun and shade—a whole-plant perspective. Aust J Plant Physiol 15:63–92
Gleeson SK, Tilman D (1992) Plant allocation and the multiple limitation hypothesis. Am Nat 139:1322–1343
Graham JH, Leonard RT, Menge JA (1981) Membrane-mediated decrease in root exudation responsible for phosphorus inhibition of vesicular-arbuscular mycorrhiza formation. Plant Physiol 68:548–552
Grams TEE, Kozovits AR, Reiter IM, Winkler JB, Sommerkorn M, Blaschke H, Häberle K-H, Matyssek R (2002) Quantifying competitiveness in woody plants. Plant Biol 4:153–158
Granados J, Körner C (2002) In deep shade, elevated CO_2 increases the vigor of tropical climbing plants. Global Change Biol 8:1109–1117
Grassi G, Bagnaresi U (2001) Foliar morphological and physiological plasticity in *Picea abies* and Abies alba saplings along a natural light gradient. Tree Physiol 21:959–967
Grier CC, Vogt KA, Lee KM, Teskey RO (1985) Factors affecting root production in subalpine forests of the Northwestern United States. In: Turner H, Tranquillini W (eds) Establishment and tending of subalpine Forest: Research and Management. Proceedings of the 3rd IUFRO Workshop, 1984, Eidgenössische Anstalt für Forstliches Versuchswessen, Ber 270, pp 143–149
Grime JP (1977) Evidence for the existence of three primary stategies in plants and its relevance to ecological and evolutionary theory. Am Nat 111:1169–1194
Grime JP (1979) Plant strategies and vegetation processes. Wiley, Chichester
Grote R, Reiter IM (2004) Competition-dependent modelling of foliage biomass in forest stands. Trees-Struct Funct 18:596–607
Hättenschwiler S, Körner C (2000) Tree seedling responses to in situ CO_2-enrichment differ among species and depend on understorey light availability. Global Change Biol 6:213–226

Hirose T, Werger MJA (1995) Canopy structure and photon flux partitioning among species in a herbaceous plant community. Ecology 76:466–474
Hodge A (2006) Plastic plants and patchy soils. J Exp Bot 57:401–411
Hoflacher H, Bauer H (1982) Light acclimation in leaves of the juvenile and adult life phases of ivy (*Hedera helix*). Physiol Plant 56:177–182
Ingestad T, Agren GI (1991) The influence of plant nutrition on biomass allocation. Ecol Appl 1:168–174
Jackson RB, Manwaring JH, Caldwell MM (1990) Rapid physiological adjustment of roots to localized soil enrichment. Nature 344:58–60
Johnson MG, Phillips DL, Tingey DT, Storm MJ (2000) Effects of elevated CO2, N-fertilization, and season on survival of ponderosa pine fine roots. Can J For Res 30: 220–228
Johnson MG, Tingey DT, Phillips DL, Storm MJ (2001) Advancing fine root research with minirhizotrons. Environ Exp Bot 45:263–289
Jones DL, Hodge A, Kuzyakov Y (2004) Plant and mycorrhizal regulation of rhizodeposition. New Phytol 163:459–480
King JS, Pregitzer KS, Zak DR (1999) Clonal variation in above- and below-ground growth responses of *Populus tremuloides* Michaux: influence of soil warming and nutrient availability. Plant Soil 217:119–130
King JS, Albaugh TJ, Allen HL, Buford M, Strain BR, Dougherty P (2002) Below-ground carbon in put to soil is control led by nutrient availability and fine root dynamics in loblolly pine. New Phytol 154:389–398
Koch KE (1996) Carbohydrate-modulated gene expression in plants. Annu Rev Plant Physiol Plant Mol Biol 47:509–540
Kozovits AR, Matyssek R, Blaschke H, Göttlein A, Grams TEE (2005a) Competition increasingly dominates the responsiveness of juvenile beech and spruce to elevated CO_2 and/or O_3 concentration throughout two subsequent growing seasons. Global Change Biol 11:1387–1401
Kozovits AR, Matyssek R, Winkler JB, Göttlein A, Blaschke H, Grams TEE (2005b) Aboveground space sequestration determines competitive success in juvenile beech and spruce trees. New Phytol 167:181–196
Küppers M (1985) Carbon relations and competition between woody species in central European hedgerow. IV. Growth form and partitioning. Oecologia 66:343–352
Küppers M (1989) Ecological significance of above-ground architectural patterns in woody plants—a question of cost-benefit relationships. Trends Ecol Evol 4:375–379
Latowski D, Grzyb J, Strzalka K (2004) The xanthophyll cycle–molecular mechanism and physiological significance. Acta Physiol Plant 26:197–212
Leuschner C, Hertel D, Coners H, Büttner V (2001) Root competition between beech and oak: a hypothesis. Oecologia 126:276–284
Lichtenthaler HK, Burkard G, Kuhn G, Prenzel U (1981) Light-induced accumulation and stability of chlorophylls and chlorophyll-proteins during chloroplast development in radish seedlings. Z Naturforsch 36:421–430
Lilleskov EA, Fahey TJ, Horton TR, Lovett GM (2002) Belowground ectomycorrhizal fungal community change over a nitrogen deposition gradient in Alaska. Ecology 83: 104–115
Machado JL, Walters MB, Reich PB (2003) Below-ground resources limit seedling growth in forest understories but do not alter biomass distribution. Ann For Sci 60:319–330
Majdi H, Damm E, Nylund JE (2001) Longevity of mycorrhizal roots depends on branching order and nutrient availability. New Phytol 150:195–202
Matyssek R, Agerer R, Ernst D, Munch JC, Osswald W, Pretzsch H, Priesack E, Schnyder H, Treutter D (2005) The plant's capacity an regulating resource demand. Plant Biol 7:560–580

Messier C, Doucet R, Ruel JC, Claveau Y, Kelly C, Lechowicz MJ (1999) Functional ecology of advance regeneration in relation to light in boreal forests. Can J For Res 29:812–823

Muth CC, Bazzaz FA (2002) Tree canopy displacement at forest gap edges. Can J For Res 32:247–254

Muth CC, Bazzaz FA (2003) Tree canopy displacement and neighborhood interactions. Canadian J For Res 33:1323–1330

Niinemets U, Kull O (1995) Effects of light availability and tree size on the architecture of assimilative surface in the canopy of *Picea abies*—variation in needle morphology. Tree Physiol 15:307–315

Peters NK, Frost JW, Long SR (1986) A plant flavone, luteolin, induces expression of rhizobium-meliloti nodulation genes. Science 233:977–980

Philipson JJ, Coutts MP (1977) Influence of mineral-nutrition on root development of trees. 2. Effect of specific nutrient elements on growth of individual roots of sitka spruce. J Exp Bot 28:864–871

Poorter H, Navas ML (2003) Plant growth and competition at elevated CO_2: on winners, losers and functional groups. New Phytol 157:175–198

Pregitzer KS, Hendrick RL, Fogel R (1993) The demography of fine roots in response to patches of water and nitrogen. New Phytol 125:575–580

Pregitzer KS, Zak DR, Curtis PS, Kubiske ME, Teeri JA, Vogel CS (1995) Atmospheric CO_2, soil-nitrogen and turnover of fine roots. New Phytol 129:579–585

Rajaniemi TK (2003) Evidence for size asymmetry of belowground competition. Basic Appl Ecol 4:239–247

Read DJ, Perez-Moreno J (2003) Mycorrhizas and nutrient cycling in ecosystems—a journey towards relevance? New Phytol 157:475–492

Reiter IM, Häberle K-H, Nunn AJ, Heerdt C, Reitmayer H, Grote R, Matyssek R (2005) Competitive strategies in adult beech and spruce: space-related foliar carbon investment versus carbon gain. Oecologia 146:337–349

Robinson D (1994) The responses of plants to nonuniform supplies of nutrients. New Phytol 127:635–674

Robinson D, van Vuuren MMI (1998) Responses of wild plants to nutrient patches in relation to growth rate and life form. In: Lambers H, Poorter H, van Vuuren MMI (eds) Inherent variation in plant growth. Physiological mechanisms and ecological consequences. Backhuys Publishers, Leiden, Netherlands, pp 237–257

Schmidt B, Strack D, Weidner M (1991) Nitrate reductase in needles, roots and trunk wood of spruce trees [Picea-Abies (L) Karst]. Trees-Struct Funct 5:215–226

Schulze E-D, Küppers M, Matyssek R (1986) The roles of carbon balance and branching pattern in the growth of woody species. In: Givnish TJ (ed) On the economy of plant form and function. Cambridge University Press, Cambridge, pp 585–602

Schwinning S (1996) Decomposition analysis of competitive symmetry and size structure dynamics. Ann Bot 77:47–57

Schwinning S, Weiner J (1998) Mechanisms determining the degree of size asymmetry in competition among plants. Oecologia 113:447–455

Sims DA, Pearcy RW (1994) Scaling sun and shade photosynthetic acclimation of *Alocasia macrorrhiza* to whole plant performance. 1. Carbon balance and allocation at different daily photon flux densities. Plant Cell and Environ 17:881–887

Smith SE, Read DJ (1997) Mycorrhizal symbiosis, 2nd edition. Academic Press, London

Stancioiu PT, O'Hara KL (2006) Morphological plasticity of regeneration subject to different levels of canopy cover in mixed-species, multiaged forests of the Romanian Carpathians. Trees-Struct Funct 20:196–209

Suzuki A (2002) Influence of shoot architectural position on shoot growth and branching patterns in *Cleyera japonica*. Tree Physiol 22:885–890

Tilman D (1985) The resource-ration hypothesis of plant succession. Am Nat 125:827–852

Tilman D (1988) Plant strategies and the dynamics and function of plant communities. Princeton University Press, Princeton

Tinker PB, Nye PH (2000) Solute movements in the rhizosphere. Blackwell Scientific, Oxford

Tingey DT, Andersen CP (1991). The physiological basis of differential plant sensitivity to changes in atmospheric quality. In: Taylor GE, Clegg, MT, Pitelka, LF (eds) Ecological genetics, terrestrial vegetation and anthropogenic changes in the atmosphere. Springer-Verlag, New York, pp 209–234

Tremmel DC, Bazzaz FA (1995) Plant architecture and allocation in different neighbourhoods: implications for competitive success. Ecology 76:262–271

Umeki K (1997) Effect of crown asymmetry on size-structure dynamics of plant populations. Ann Bot 79:631–641

Vanninen P, Mäkelä A (2005) Carbon budget for Scots pine trees: effects of size, competition and site fertility on growth allocation and production. Tree Physiol 25:17–30

Wallenda T et al. (2000) Nitrogen uptake processes in roots and mycorrhizas. In: Schulze ED (ed) Ecological Studies 142: Carbon and nitrogen cycling in European forest ecosystems. Springer-Verlag, Berlin, Heidelberg, New York, pp 122–143

Walker TS, Bais HP, Grotewold E, Vivanco JM (2003). Root exudation and rhizosphere biology. Plant Physiol 132:44–51

Wayne PM, Bazzaz FA (1997) Light acquisition and growth by competing individuals in CO_2-enriched atmospheres: consequences for size structure in regenerating birch stands. J Ecol 85:29–42

Weiner J (1985) Size hierarchies in experimental populations of annual plants. Ecology 66:743–752

Weiner J (1990) Asymmetric competition in plant-populations. Trends Ecol Evol 5:360–364

Weiner J (2004) Allocation, plasticity and allometry in plants. Persp Plant Ecol Evol Syst 6:207–215

Weiner J, Fishman L (1994) Competition and allometry in *Kochia scoparia*. Ann Bot 73:263–271

Weiner J, Thomas SC (1986) Size variability and competition in plant monocultures. Oikos 47:211–222

Weiner J, Thomas SC (1992) Competition and allometry in three species of annual plants. Ecology 72:648–656

Weiner J, Wright DB, Castro S (1997) Symmetry of below-ground competition between *Kochia scoparia* individuals. Oikos 79:85–91

Wilson SD, Tilman D (1991) Components of plant competition along an experimental gradient of nitrogen availability. Ecology 72:1050–1065

Wilson SD, Tilman D (1993) Plant competition and resource availability in response to disturbance and fertilization. Ecology 74:599–611

Witkowski ETF, Lamont BB (1991) Leaf specific mass confounds leaf density and thickness. Oecologia 88:486–493

Woolfolk WTM, Friend AL (2003) Growth response of cottonwood roots to varied NH_4: NO_3 ratios in enriched patches. Tree Physiol 23:427–432

Würth MKR, Winter K, Körner C (1998) *In situ* responses to elevated CO_2 in tropical forest understorey plants. Funct Ecol 12:886–895

Young TP, Hubbell SP (1991) Crown asymmetry, treefalls, and repeat disturbance of broad-leaved forest gaps. Ecology 72:1464–1471

Zak DR, Pregitzer KS, Curtis PS, Vogel CS, Holmes WE, Lussenhop J (2000) Atmospheric CO2, soil-N availability, and allocation of biomass and nitrogen by Populus tremuloides. Ecol Appl 10:34–46

Zak DR, Tilman D, Parmenter RR, Rice CW, Fisher FM, Vose J, Milchunas D, Martin CW (1994) Plant production and soil microorganisms in late-successional ecosystems: a continental-scale study. Ecology 75:2333–2347

Thorsten E.E. Grams
Ecophysiology of Plants
Department of Ecology
Technische Universität München
Am Hochanger 13
85354 Freising-Weihenstephan
Germany
e-mail: grams@wzw.tum.de

Christian P. Andersen
US Environmental Protection Agency
Western Ecology Division
National Health and Environmental
Effects Research Laboratory
200 SW 35th St.
Corvallis Oregon 97333 USA
e-mail: andersen.christian@epamail.epa.gov

Explaining Variation in Fine Root Life Span

Michael S. Peek

1 Introduction

Fine roots represent an integral component to overall plant performance. As the belowground analogue to leaves (Eissenstat et al. 2000), fine roots are responsible for the uptake of water and nearly all essential soil nutrients, thereby generating competitive situations responsible for decreasing overall plant performance (Casper and Jackson 1997). Therefore, plants have evolved mechanisms to ensure adequate resource capture in competitive situations. Phenological mechanisms may modify root birth and death rates (Steingrobe et al. 2001), physiological mechanisms include increased uptake rates (Jackson et al. 1990; Caldwell et al. 1992) while morphological changes may increase absorptive area (Jackson and Caldwell 1989; Caldwell et al. 1991; Larigauderie and Richards 1994).

The strong competition belowground between fine roots can come at a considerable carbon expense for plants. Average estimates across many biomes suggest that net primary production (NPP) for fine roots approaches 33% of total global annual NPP (Gill and Jackson 2000). The largest variability in belowground carbon costs exists in temperate grasslands ranging between 24% and 87% of total NPP (Sims and Singh 1978). Some of this variability may be due to soil nutrient availability because in the nutrient poorest systems, below-ground net primary production (BNPP) often exceeds aboveground production (Caldwell and Camp 1974; Caldwell et al. 1977; Martinez et al. 1998), while in nutrient rich environments, root turnover can be much less (Gill and Jackson 2000). Despite the importance of fine roots, we are only beginning to elucidate some of the factors regulating fine root production and turnover (Eissenstat et al., 2000; Anderson et al., 2003).

While fine root life span is defined differently than production and turnover, it is directly related to both of these commonly used measurements. Therefore, life span is an integral component of ecosystem level processes such as carbon and nutrient cycling. Despite recent reviews on fine root production, turnover and life span (Eissenstat and Yanai 1997, 2002; Eissenstat et al. 2000; Laurenroth and Gill 2003), many questions still remain

Progress in Botany, Vol. 68

regarding the variation in fine root life span both within and across species. For example, what are the relative influences of life form (e.g. deciduous trees versus evergreen trees) and environment on root longevity? Certainly the former has implications for community composition and ecosystem function, while the latter is directly relevant to global change issues. The purpose of this review is to explore some generalizations regarding life form and environment on the life span of fine roots ($\leq$1 mm in diameter). While doing so I will explore the degree to which variation in root lifespan can be attributed to species and environmental conditions (Eissenstat and Yanai 2002). Additionally, methodological differences will be discussed.

2 Fine root estimation methodology

2.1 Indirect estimation

Estimating fine root longevity can be difficult due to obvious observational limitations. Therefore, indirect methods to obtain estimates of root life span are often employed. Estimating root turnover (yr^{-1}) is the common metric of these methods, which can then be inverted to give a life span estimate. The most common method for obtaining turnover is root coring, either sequentially or using ingrowth cores (Eissenstat and Yanai 2002). For this method, turnover is calculated by dividing production ($kg\ ha^{-1}\ yr^{-1}$) by the average, minimum or maximum standing crop ($kg\ ha^{-1}$). Less common methods include nitrogen (N) or carbon (C) budgeting. The former yields estimates of root turnover by the difference between annual net mineralization rate and N uptake (Nadelhoffer et al. 1985) while C budgeting calculates root turnover by the difference between soil respiration and root respiration plus above ground litterfall (Raich and Nadelhoffer 1989). Finally, isotopic tracer techniques estimate turnover by measuring the dilution of the element at intervals after label introduction (Caldwell and Camp 1974; Gaudinski et al. 2001). While all have their respective advantages and disadvantages (Eissenstat and Yanai 2002), all estimate longevity without directly tracking root births and deaths.

2.2 Direct estimation

Direct observation of roots through rhizotrons, or root observation windows, offers the advantage of following individual roots through time. Initially large root boxes (on the order of m^2) were used to track roots growing on transparent windows (Fernandez and Caldwell 1975). As technology progressed, these windows were greatly reduced in size to the current several mm^2, which are more commonly referred to as minirhizotrons (Johnson et al. 2001). Additionally, computer and software technology has advanced to ease collection and processing of vast amounts of root images. The major advantage of this technique is that direct observation allows the tracking of individual roots or root segments from inception to disappearance. Other advantages include: sampling at depths of greater than 1 m (Peek et al. 2006), while coring rarely exceeds the top 0.2 m of soil (Eissenstat and Yanai 2002); no damage to intact roots and the integrity of the soil environment is maintained; and other useful measures can be obtained such as root architecture (Milchunas et al. 2005), individual root growth and expansion (Johnson et al. 2001), and neighbour analysis (Wells and Eissenstat 2001).

While the appeal of tracking individual roots through time is attractive, monitoring roots in this way does have numerous disadvantages. First, functionality of roots often cannot be gleaned. However, it is certainly assumed that thinner diameter fine roots primarily serve in nutrient and water uptake while thicker roots are primarily for anchorage and transport (Eissenstat et al. 2000). Root death sometimes can be related to visible colour changes in roots (Comas et al. 2000). In their study, white roots were clearly alive and black roots were demonstrated to be dead while roots that had disappeared indicated complete decay. However, roots exhibiting various shades of brown appeared to be in a continuum between reduced metabolic activity and death when tested with vital stains and respiration measurements (Wells and Eissenstat 2001; Comas et al. 2000). Furthermore, in arid systems or other systems where decomposition is slow, senescing or dead roots can persist for long periods of time (Moretto and Distel 2003; Peek et al. 2005). Additionally, standing crop and productivity estimates are difficult to obtain with two dimensional images and coring must be done in conjunction with digital imaging (Johnson et al. 2001). Although the invasiveness of tube installation is minimal and short lived, significant amounts of time may be required for the roots to return to steady state conditions (Peek et al. 2006), and the presence of the tube may create an artificial environment that results in preferential flow of nutrients and/or water and even the material of the tube itself may alter root functioning (Withington et al. 2003).

2.3 Direct and indirect comparison

Indirect estimates may overestimate fine root longevity (Luo 2003; Trumbore and Gaudinski 2003). This can be attributed to differences in the methods used to infer longevity. Direct methods often estimate median life span, recognizing that root life spans are highly skewed right, whereas indirect methods calculate means (or mean residence times) using the methods described above (Trumbore and Gaudinski 2003). Furthermore, methodological differences create situations where fine roots may have been born and died during the sampling interval, therefore overestimating life span. Certainly for isotopic labelling techniques this is also true, as the only roots that are sampled are the longest lived.

In an attempt to quantify whether indirect measures overestimate life span compared to direct estimates, I examined direct techniques of estimating fine root life span using the rhizotron technique with the indirect method of soil coring, either sequential cores or ingrowth (Table 1; Fig. 1). Literature values were obtained for species of various life forms, where each life form was represented by each method with the exception of Forbs (Table 1). Indirect measurements, revealed significantly longer life spans (419±48 days), than the direct observations (234±38; $F_{1,80}=8.2$, $P=0.005$). This result is robust since the variability across species and life form was less than the comparison between methods, resulting in the significantly different means. Furthermore, five species existed, *Fagus sylvatica, Picea abies, Pinus resinosa, Pseudotsuga menziesii, Schixachyrium scoparium,* where estimates were obtained by both direct and indirect methods. Four of the five estimates were longer for the indirect techniques compared to the direct technique (Fig. 2). Further support is provided in a recent study by Hendricks et al. (2006), where they compared all indirect techniques (ingrowth and sequential coring and elemental budgeting) with the direct minirhizotron technique and concluded that ingrowth coring and sequential coring yielded lower production estimates which would correspond to longer lived roots, assuming a constant standing crop. Reconciling the overestimation of indirect methods and determining if minirhizotrons provide true estimates of longevity will be challenging and may require more comparative studies like Hendricks et al. (2006) or incorporate sophisticated modelling techniques (Eissenstat and Yanai 2002; Luo 2003; Majdi and Andersson 2005).

2.4 Predicting fine root life span

Predicting fine root longevity from secondary correlates may prove to be equally difficult. Fine roots are often considered to be the belowground analogue of leaves and many parallels have been made regarding the structural, physiological and phenological patterns between roots and leaves (Eissenstat et al. 2000; Craine and Lee 2003; West et al. 2003). For example, high rates of photosynthesis relate to high leaf N concentrations, high

Table 1. Fine root life spans (*MLS*, median or mean) obtained from literature values either from direct observation using the rhizotron technique or indirectly using ingrowth cores. Each species is assigned to one of six growth form categories, C_3 grass, C_4 grass, deciduous trees, evergreen trees, forbs or legumes

Species	Method	MLS	Reference
C_3 grasses			
Festuca rubra	Direct	49	Pärtel and Wilson 2001
Lolium perenne	Direct	44	Watson et al. 2000
Lolium perenne	Direct	13	"
Bromus tectorum	Direct	51	Peek et al. 2005
Agropyron desertorum	Direct	45	"
Lolium perenne	Direct	98	Van der Krift and Berendse 2002
Arrhenatherum elatius	Direct	280	"
Molinia cerulea	Direct	371	"
Nardus striata	Direct	406	"
Agropyron repens	Indirect	761	Tjoelker et al. 2005
Agrostis scabra	Indirect	105	"
Koeleria cristata	Indirect	484	"
Poa pratensis	Indirect	709	"
Stipa spartea	Indirect	1173	"
Elymus pycnanthus	Indirect	102	Bouma et al. 2002
C_4 grasses			
Bouteloua gracilis	Direct	320	Gill et al. 2002
Bouteloua gracilis	Direct	180	"
Aristida stricta	Direct	777	West et al. 2003
Schizachyrium scoparium	Direct	374	"
Andropogon gerardii	Indirect	710	Tjoelker et al. 2005
Bouteloua curtipendula	Indirect	494	"
Calamovilfa longifolia	Indirect	740	"
Panicum virgatum	Indirect	600	"
Schizachyrium scoparium	Indirect	1118	"
Sorghastrum nutans	Indirect	1409	"
Spartina anglica	Indirect	365	Bouma et al. 2002
Deciduous trees			
Malus domestica	Direct	30	Eissenstat et al. 2000
Citrus spp.	Direct	300	"
Prunus persica	Direct	110	Wells et al. 2002b
Prunus persica	Direct	140	"
Prunus avium	Direct	18	Black et al. 1998
Acer pseudoplatanus	Direct	31	"
Populus canadensis	Direct	38	"
Populus tristis	Direct	149	Coleman et al. 2000
Acer saccharum	Direct	346	Hendrick and Pregitzer 1992
Populus deltiodes	Direct	438	Kern et al. 2004
Populus deltiodes	Direct	511	"

Table 1. *Continued*

Species	Method	MLS	Reference
Quercus robur	Direct	77	Ponti et al. 2004
Fraxinus oxyphylla	Direct	75	"
Various hardwoods	Direct	42	Pregitzer et al. 1993
Various hardwoods	Direct	314	Tierney and Fahey 2001
Malus domestica	Direct	14	Head 1966
Populus spp.	Direct	21	"
Acer platanoides	Direct	591	Withington et al. 2006
Acer pseudoplatanus	Direct	902	"
Fagus sylvatica	Direct	208	"
Quercus robur	Direct	358	"
Tilia cordata	Direct	234	"
Larix deciduas	Direct	402	"
Corylys Americana	Indirect	430	Tjoelker et al. 2005
Quercus macrocarpa	Indirect	360	"
Various hardwoods	Indirect	369	Fahey and Hughes 1994
Fagus sylvatica	Indirect	720	Van Praag et al. 1988
Evergreen trees			
Picea sitchensis	Direct	63	Black et al. 1998
Pinus resinosa	Direct	291	Coleman et al. 2000
Pinus taeda	Direct	181	King et al. 2002
Quercus ilex	Direct	67	López et al. 2001
Picea abies	Direct	450	Majdi 2001
Picea abies	Direct	730	Majdi and Andersson 2005
Picea abies	Direct	365	"
Picea abies	Indirect	730	"
Picea abies	Indirect	406	"
Picea abies	Indirect	261	Ostonen et al. 2005
Pinus ponderosa	Direct	74	Johnson et al. 2000
Pinus resinosa	Direct	169	Zeleznik and Dickmann 2004
Juniperus osteosperma	Direct	368	Peek et al. 2005
Abies alba	Direct	412	Withington et al. 2006
Picea abies	Direct	256	"
Pinus nigra	Direct	281	"
Pinus sylvestris	Direct	245	"
Pseudotsuga menziesii	Direct	591	"
Pseudotsuga menziesii	Indirect	190	Santantionio and Grace 1987
Pinus resinosa	Indirect	692	Haynes and Gower 1995
Picea abies	Indirect	720	Van Praag et al. 1988
Forbs			
Achillea millefolium	Indirect	241	Tjoelker et al. 2005
Anemone cylindrica	Indirect	135	"
Asclepias syriaca	Indirect	768	"
Asclepias tuberose	Indirect	135	"

Table 1. *Continued*

Species	Method	MLS	Reference
Asclepias verticillata	Indirect	201	"
Aster ericoides	Indirect	203	"
Aster nova-angliae	Indirect	73	"
Liastris aspera	Indirect	113	"
Penstemon grandiflorus	Indirect	265	"
Pontentilla arguta	Indirect	153	"
Rudbekia serotina	Indirect	191	"
Solidago nemoralis	Indirect	42	"
Solidago rigida	Indirect	267	"
Solidago speciosa	Indirect	433	"
Legumes			
Trifolium repens	Direct	42	Watson et al. 2000
Trifolium repens	Direct	12	"
Medicago sativa	Direct	131	Goins and Russelle 1996
Desmodium canadense	Indirect	150	Tjoelker et al. 2005
Lespediza capitata	Indirect	302	"
Lupinus perennis	Indirect	32	"
Petalostemon purpureum	Indirect	562	"
Petalostemon villosum	Indirect	56	"

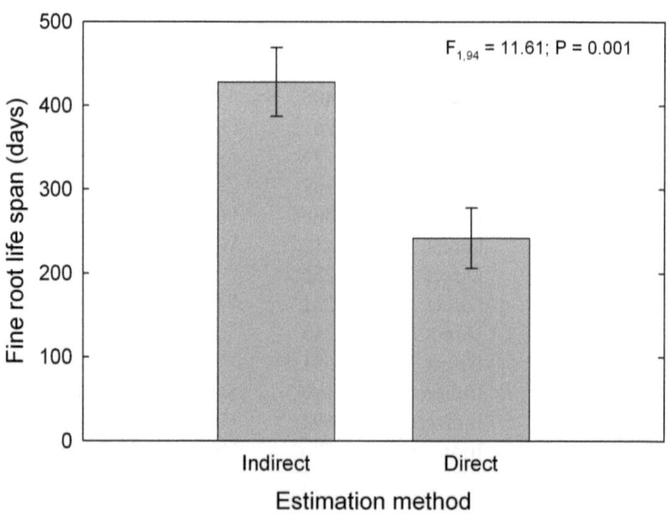

Fig. 1. Comparison of estimation method for determining fine root life span. Estimations were obtained from data in Table 1 and were broken into two categories: Indirect where life span was calculated from turnover rates and; Direct where estimates were obtained from direct root measurement from rhizotron techniques. Least squares means ± 1 SE are presented

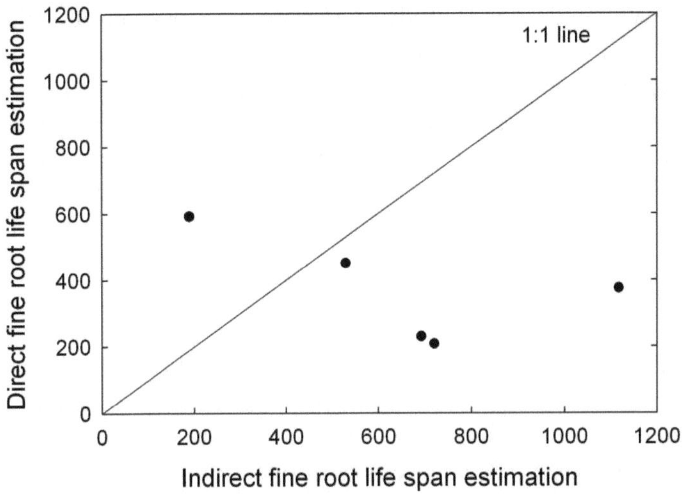

Fig. 2. Comparison of 5 species with estimates of direct and indirect fine root life spans. The five species are *Fagus sylvatica, Picea abies, Pinus resinosa, Pseudotsuga menziesii, Schixachyrium scoparium. Picea abies* had four direct and indirect estimates, therefore averaged into a single value for each method

specific leaf area and short leaf longevity. Likewise, high tissue density, lower maintenance respiration, and larger diameter have been associated with longer-lived roots (Pregitzer et al. 1997, 1998; Eissenstat et al. 2000). Additionally, faster growing species tend to invest less in structural tissues and have shorter leaf and root lifespans (Ryser, 1996). Consequently, fine root lifespan may be directly related to tissue construction costs and the return on the carbon investment (Yanai et al. 1995) and even be related to leaf life span directly given the similarities in suites of traits for many plant species (Chapin et al. 1993).

While the general consensus on leaf traits correlating with leaf life span (Reich et al. 1999) is well established, the evidence for this in fine roots still needs empirical support. There are only a handful of studies that have examined such traits in roots to find correlates for fine root longevity. Craine and Lee (2003) and Tjoelker et al. (2005) examined root and leaf traits of 24 and 39 grassland and savannah species, respectively and found low CO_2 exchange, low specific leaf and root lengths and low C:N ratios associated with long lived roots and leaves. Both of these studies examined species in natural environments across a range of soil moisture and nutrient availabilities. Two studies have examined leaf and root traits in common garden experiments (Ryser 1996; Withington et al. 2006). Ryser (1996) showed that in five different grass species, growth rate and tissue longevity were inversely correlated. While in a 4-year study of 11 temperate tree species, Withington et al. (2006) found similar results of low C:N root ratios with longer lived roots, however, there was no direct relationship between leaf longevity and root longevity, suggesting that leaf traits and/or growth rate may not be good predictors of root longevity. Furthermore, better predictors were found to correlate with root life span, such as root exodermis thickening.

3 Environment

Some of the variability in root life spans in the soil can also be associated with environmental factors. There are broad-scale patterns, such as the association between longer-lived roots and resource-poor environments, with the notion that longer life is necessary in order to insure that construction costs are met by the return of water and nutrients for the plant (Eissenstat and Yanai 1997). Local conditions such as temperature, moisture and nutrients are linked with root length growth, root mortality (Hendrick and Pregitzer 1993) and decomposition (Silver and Miya 2001). Experimental manipulations that increased water and nitrogen indicated stimulation of fine root production (Pregitzer et al. 1993), but the effect on lifespan was quite variable (see Pärtel and Wilson 2001). However, the question still remains, under changing conditions, what are the likely factors that will influence root mortality, and ultimately lifespan?

A promising way to determine whether there are environmental correlates with root death is to conduct a survival analysis using a Cox proportional hazard regression model (Cox 1972). This model allows continuous and discrete variables to be entered into the model in order to determine the risk of mortality of a root at any given point in time. This model is also attractive due to the nature of the data collected. In most studies, roots are present at the start of experiment, at the end, or both, so that the actual life spans cannot be determined. These data are considered to be censored (Allison 1995; Black et al. 1998). Current maximum likelihood estimation methods are robust for this treatment (Lee 2004). The major advantage is that any number of environmental variables (e.g. soil moisture; soil nutrient availability; temperature) can be measured and correlated with root mortality. The hazard risk in the Cox proportional hazard regression refers to the instantaneous risk of root disappearance, or mortality, at time *t*. Negative parameter estimates for a given covariate indicate that increasing values of the covariate are associated with a decreasing risk of mortality. Alternatively, a positive parameter estimate indicates that increasing values of the covariate correspond to an increasing risk of mortality. Another useful statistic obtained in this analysis is the risk or hazard ratio, which can quantify the proportional risk as a function of a one-unit increase in the covariate. For quantitative covariates, the risk ratio can be subtracted from 1 and multiplied by 100 to get the percent change in the hazard for each one-unit increase or decrease in the continuous variable (Allison 1995).

Only a handful of studies have utilized the Cox proportional hazard model to identify factors that may influence root death, and studies with strict environmental covariates are even fewer. The first to model this on roots was Wells and Eissenstat (2001) where they documented a significant decrease in the risk of root mortality with increasing root diameter. Also, the presence of neighbouring roots significantly increased the risk of roots dying. Further studies in savannah bunchgrasses demonstrated that root cohort production also influences root mortality such that shorter lived roots were produced under seasonal patterns of lower water availability (West et al. 2003).

Two studies have examined the effects of soil moisture on risk of mortality. Anderson et al. (2003) provided supplemental water to grape vines and found mixed effects probably due to other unmeasured environmental factors relating to seasonality, while Peek et al. (2006) examined fine root dynamics of three cold desert species of different life form, a perennial shrub, a perennial grass, and an annual grass. In all but one case in 1 year, decreases in soil moisture were associated with an increase in root disappearance risk. These authors spoke not of mortality, but of root disappearance due to the limitation of many minirhizotron studies where functionality cannot be gleaned from roots. Nevertheless, modelling environmental variables in this way will identify important variables that might shorten or lengthen root life span. It is likely, however, that a combination of several factors influences fine root life span (Anderson et al. 2003). And these factors may interact with life form and other local edaphic factors to exhibit different life spans across species and across the same species growing in different environments. For example, in deserts, water is most limiting, its presence or absence may have the largest influence on fine root longevity, while in more mesic systems, nutrients may play a greater role (Nadelhoffer et al. 1985). The environment component controlling fine root lifespan is exhibited in a study where identical varieties of *Lolium perenne* and *Trifolium repens* were grown in two very different climates, the United Kingdom and in Italy. Both species exhibited significantly longer lived roots in the UK environment, with a lower mean temperatures and higher soil nutrients; though the specific environmental mechanism for the observed differences could not be determined (Watson et al. 2000).

4 Life history/growth form

Faster growing species tend to invest less in structural tissues and have shorter tissue life spans (Ryser 1996). Therefore, predictions based on growth form, or tissue structure may be useful in predicting fine root longevity. However, from published studies, fine root lifespan is highly variable across species, ranging from weeks (Black et al. 1998) to a few years (Eissenstat and Yanai 1997) in similar life forms (e.g. trees). Even within a species, fine root lifespan can range from a few days to many months depending on the timing of cohort production and other characteristics (Anderson et al. 2003). Yet there is only one comparison of fine root longevity across different life forms (Tjoelker et al. 2005). In that study, 39 grassland and savannah species were examined and negative correlations were found when correlating the log of fine root life span and both mass-based fine root respiration and N concentration. Furthermore, they directly compared leaf longevity and root longevity and found no significant correlation, similar to that of Withington et al. (2006).

In order to examine the life form question using similar designations as Tjoelker et al. (2005), I found 96 estimates of fine root life span for 75 different species from published estimates (Table 1, Fig. 3). I looked at average or median fine root longevity using two methods, direct estimation using the rhizotron technique and indirectly using either sequential coring or ingrowth cores as described above. Growth forms were divided into six categories, C_3 grasses, C_4 grasses, deciduous trees, evergreen trees, forbs and legumes. Both methods were represented in all categories with the exception of the forbs, where only indirect estimates were found. The estimates of life span were then analysed using a one way ANOVA to test for growth form differences in life span. The data were log transformed to satisfy homogeneity of variances, but least squared means are presented from the original data. I found that C_4 grasses and evergreen trees do not have significantly different fine root life spans, but C_4 grasses have greater fine root longevity than all other life forms ($F_{5,90}=4.2$, $P=0.002$; Fig. 3). While one may predict evergreen trees to have the longest lived roots based on their growth form and leaf longevity, it was surprising to find C_4 grasses with the largest overall mean of 644 days. Furthermore, deciduous trees were not significantly different from forbs, C_3 grasses or legumes. This signifies the large amount of variation across species that exists when estimating fine root life span.

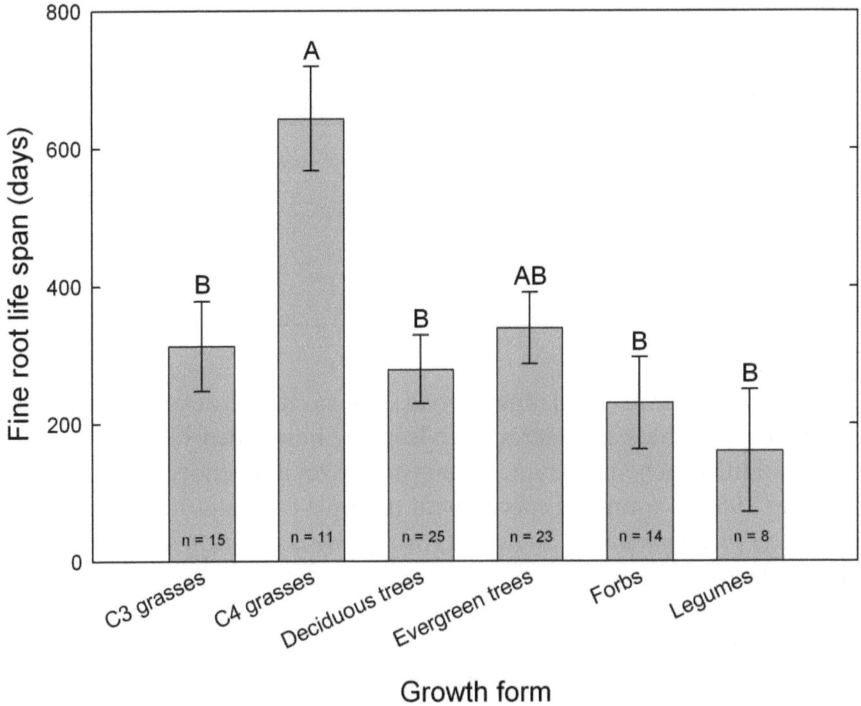

Fig. 3. Fine root life span estimates for six growth forms estimated from literature values in Table 1. Either median or mean values were given for fine roots (≤1 mm diameter). Least squares means (±1 SE) are reported with different letters indicating significantly different means using a Tukey adjustment on log-transformed values

5 Interactions between life form and environment

Certainly the large degree of variation in fine root life span is a result of the large number of variables affecting it. For example, studies and reviews have shown root age (Wells et al. 2002a), fine root order (Wells and Eissenstat 2003), diameter (Wells and Eissenstat 2001), temperature (Pregitzer et al. 2000), nutrients (Nadelhoffer 2000), CO_2 (Tingey et al. 2000), phenology/cohort (West et al. 2003) and growth from (Tjoelker et al. 2005) all influence fine root longevity. A key question remains: what is the relative proportion of variation attributable to each? In an attempt to answer this question, I examined different life span estimates for the same species regardless of study conditions and conducted an ANOVA for the fixed effect of species. By partitioning the variability between species, error and total, I was able to determine how much of the variation is due to species effects, and how much can be attributable to other factors (e.g. age, diameter, environment, measurement technique). This

Table 2. Analysis of variance table to examine within species variability for different estimates of fine root life span on the same species

Source of variation	*df*	SS	MS	*F*	*P* (*F*)
Species	12	1,404,703	117,058	1.79	0.1173
Error	21	1,373,861	65,422		
Total	33	2,778,564			
r^2=0.51	CV=75				

technique is similar to calculating a broad sense heritability estimate in genetics, where researchers are interested in learning how much variability is due to the additive genetic variance against the total phenotypic variability (Falconer 1981). I found 13 species with multiple estimates for fine root life span, (*Acer pseudoplatanus, Bouteloua gracilis, Fagus sylvatica, Lolium perenne, Malus domestica, Picea abies, Pinus resinosa, Populus deltiodes, Prunus persica, Pseudotsuga menziesii, Quercus robur, Schizachyrium scoparium, Trifolium repens*). Roughly half (r^2= 51) of the variability in the estimate for life span for these 13 species was due to species effects, while the remainder can be attributed to other external factors (Table 2). While a limited number of species are used here due to the lack of literature values, this still emphasizes the importance both of species and or life form controls versus external factors. While this still doesn't address the discrepancies between life forms, it suggests that a significant degree of variation in life form is due to genetic factors and advances in functional genomics may yield insights into how life span is controlled and selected upon (Majdi et al. 2005).

6 Conclusions

Fine root life span remains quite variable both within and across species. Methodological difficulties in estimating life span have yet to be resolved. Direct measures seem to be the most promising avenue, but studies are few due to the expense of minirhizotrons and most are truncated, leaving a large number of roots still alive at the end of the experiment. However, appropriate statistical modelling techniques should be used to handle these data. A more troubling concern is the discrepancy between direct and indirect techniques for life span estimation. Solutions may be found in complex modelling procedures. Tissue properties (e.g. respiration rates, C:N ratios) seem to be the best overall predictor of fine root life span, but the studies that have addressed this are purely correlational. In order for this to be a robust predictor, studies must move from general patterns to manipulation of tissue density and its affect on life span to ensure a functional prediction. The environment and species effects on fine root longevity explain roughly equal amounts of variation. Such a large component of the variability being environmental highlights many difficulties when trying to assess longevity. A powerful analysis tool, the Cox proportional hazard regression, while widely used in other disciplines, is underused in root research. The power to identify key environmental variables that influence root mortality should prove useful in explaining variability across many different systems. Certainly the greatest contribution to understand roots is the generation of significantly more estimates of life span across many different life forms and for the same species in different environments. Only then can a comprehensive assessment of the roles of life form, measurement techniques and environment be assessed.

Acknowledgements

Initial ideas for this work began while a post-doctoral fellow at the Ecology Center at Utah State University under funding from the National Science Foundation (DEB-9807097) to MM Caldwell. Funding was also provided by the Center for Research in the College of Science and Health at William Paterson University of New Jersey. I thank the editors and I. Forseth, J. Leffler and M. Robson for comments on earlier versions of this manuscript.

References

Allison PD (1995) Survival analysis using the SAS system: a practical guide. BBU Press, SAS Institute, Cary, N.C., USA

Anderson LJ, Comas LH, Lasko AN, Eissenstat DM (2003) Multiple risk factors in root survivorship: a 4-year study in Concord grape. New Phytol 158:489–501

Black KE, Harbron CG, Franklin M, Atkinson D, Hooker JE (1998) Differences in root longevity of some tree species. Tree Physiol 18:259–264

Bouma TJ, Hengst K, Koustaal BP, van Soelen J (2002) Estimating root lifespan of two grasses at contrasting elevation in a salt marsh by applying vitality staining on roots from ingrowth cores. Plant Ecol 165:235–245

Caldwell MM, Camp LB (1974). Belowground productivity of two cool desert communities. Oecologia 17:123–130

Caldwell MM, White RS, Moore RT, Camp LB (1977) Carbon balance, productivity, and water use of cold-winter desert shrub communities dominated by C_3 and C_4 species. Oecologia 29:275–300

Caldwell MM, Manwaring JH, Jackson RB (1991) Exploitation of phosphate from fertile soil microsites by three Great Basin perennials when in competition. Funct Ecol 5:757–764

Caldwell MM, Dudley LM, Lilieholm B (1992) Soil solution phosphate, root uptake kinetics and nutrient acquisition: implications for a patchy soil environment. Oecologia 89:305–309

Casper BB, Jackson RB (1997) Plant competition underground. Annu Rev Ecol Syst 28: 545–570
Chapin FS III, Autumn K, Pugnaire F (1993) Evolution of suites of traits in response to environmental stress. Am Nat 142:S78–S92
Coleman MD, Dickson RE, Isebrands JG (2000) Contrasting fine-root production, survival and soil CO_2 efflux in pine and poplar plantations. Plant Soil 225:129–139
Comas LH, Eissenstat DM, Lakso AN (2000) Assessing root death and root system dynamics in a study of grape canopy pruning. New Phytol 147:171–178
Cox D (1972) Regression models and life tables. J R Stat Soc 34:187–220
Craine JM, Lee WG (2003) Covariation in leaf and root traits for native and non-native grasses along an altitudinal gradient in New Zealand. Oecologia 134:471–478
Eissenstat DM, Yanai R (1997) The ecology of root lifespan. Adv Ecol Res 27:1–60
Eissenstat DM, Yanai RD (2002) Root Lifespan, efficiency, and turnover. In: Waisel Y, Eshel A, Kafkafi U (eds) Plant roots: the hidden half, 3rd edn. Marcel Dekker, New York, pp 221–238
Eissenstat DM, Wells CE, Yanai RD, Whitbeck JL (2000) Building fine roots in a changing environment: implications for root longevity. New Phytol 147:33–42
Fahey TJ, Hughes JW (1994) Fine root dynamics in a northern hardwood forest ecosystem, Hubbard Brook Experimental Forest, NH. J Ecol 82:533–548
Falconer DS (1981) Introduction to quantitative genetics. 2nd edn. Longman, N.Y.
Fernandez OA, Caldwell MM (1975) Phenology and dynamics of root growth of three cool semi-desert shrubs under field conditions. J Ecol 63:703–714
Gaudinski JB, Trumbore SE, Davidson EA, Cook AC, Markewitz D, Richter DD (2001) The age of fine-root carbon in three forests of the eastern United States measured by radiocarbon. Oecologia 129:420–429
Gill RA, Jackson RB (2000) Global patterns of root turnover for terrestrial ecosystems. New Phytol 147:13–31
Gill RA, Burke IC, Lauenroth WK, Milchunas DG (2002) Longevity and turnover of roots in the shortgrass steppe: influence of diameter and depth. Plant Ecol 159:241–251
Goins GD, Russelle MP (1996) Root demography in alfalfa (*Medicago stativa* L.). Plant Soil 185:281–291
Haynes BE, Gower ST (1995) Belowground carbon allocation in unfertilized and fertilized red pine plantations in northern Wisconsin. Tree Physiol 15:317–325
Head GC (1966) Estimating seasonal changes in the quantity of white unsuberized root on fruit trees. J Hort Sci 41:197–206
Hendrick RL, Pregitzer KS (1992) The demography of fine roots in a northern hardwood forest. Ecology 73:1094–1104
Hendrick RL, Pregitzer KS (1993) The dynamics of fine root length, biomass, and nitrogen content in two northern hardwood ecosystems. Can J For Res 23:2507–2520
Hendricks JJ, Hendrick RL, Wilson CA, Mitchell RJ, Pecot SD, Guo D (2006) Assessing the patterns and controls of fine root dynamics: an empirical test and methodological review. J Ecol 94:40–57
Jackson RB, Caldwell MM (1989) The timing and degree of root proliferation in fertile-soil microsites for three cold-desert perennials. Oecologia 81:149–153
Jackson RB, Manwaring JH, Caldwell MM (1990) Rapid physiological adjustment of roots to localized soil enrichment. Nature 344:58–60
Johnson MG, Phillips DL, Tingey DT, Storm MJ (2000) Effects of elevated CO_2, N-fertilization, and season on survival of ponderosa pine fine roots. Can J For Res 30:220–228
Johnson MG, Tingey DT, Phillips DL, Storm MJ (2001) Advancing fine root research with minirhizotrons. Environ Exp Bot 45:263–289
Kern CC, Friend AL, Johnson JMF, Coleman MD (2004) Fine root dynamics in a developing Populus deltoids plantation. Tree Physiol 24:651–660
King JS, Albaugh TJ, Allen HL, Buford M, Strain BR, Dougherty P (2002) Below-ground carbon input to soil is controlled by nutrient availability and fine root dynamics in loblolly pine. New Phytol 154:389–398

Larigauderie A, Richards JH (1994) Root proliferation characteristics of seven perennial arid-land grasses in nutrient-enriched microsites. Oecologia 99:102–111

Lauenroth WK, Gill RA (2003) Turnover of root systems. In: de Kroon H,d Visser EJW (eds) Root ecology. Ecological studies 168, Springer, Berlin, pp 61–90

Lee PM (2004) Bayesian statistics: an introduction. Halstead Press, N.Y., 360 pp.

López B, Sabaté S, Gracia CA (2001) Fine-root longevity of *Quercus ilex*. New Phytol 151:437–441

Luo Y (2003) Uncertainties in interpretation of isotope signals for estimation of fine root longevity: theoretical considerations. Global Change Biol 9:1118–1129

Majdi H (2001) Changes in fine root production and longevity in relation to water and nutrient availability in a Norway spruce stand in northern Sweden. Tree Physiol 21:1057–1061

Majdi H, Andersson P (2005) Fine root production and turnover in a Norway spruce stand in Northern Sweden: effects of nitrogen and water manipulation. Ecosystems 8:191–199

Majdi H, Pregitzer K, Morén AS, Nylund JE, Ågren GI (2005) Measuring fine root turnover in forest ecosystems. Plant Soil 276:1–8

Martinez F, Merino O, Martin A, Martin DG, Merino J (1998) Belowground structure and production in a Mediterranean sand dune shrub community. Plant Soil 201:209–216

Milchunas DG, Morgan JA, Mosier AR, LeCain DR (2005) Root dynamics and demography in shortgrass steppe under elevated CO_2, and comments on minirhizotron methodology. Global Change Biol 11:1837–1855

Moretto AS, Distel RA (2003) Decomposition of and nutrient dynamics in leaf litter and roots of *Poa ligularis* and *Stipa gyneriodes*. J Arid Environ 55:503–514

Nadelhoffer KJ (2000) The potential effects of nitrogen deposition on fine-root production in forest ecosystems. New Phytol 147:131–139

Nadelhoffer KJ, Aber JD, Melillo JM (1985) Fine roots, net primary production and nutrient availability: a new hypothesis. Ecology 66:1377–1390

Ostonen I, Lõhmus K, Pajuste K (2005) Fine root biomass, production and its proportion of NPP in a fertile middle-aged Norway spruce forest: comparison of soil core and ingrowth core methods. For Ecol Manag 212:264–277

Pärtel M, Wilson SD (2001) Root and leaf production, mortality and longevity in response to soil heterogeneity. Funct Ecol 15:748–753

Peek MS, Leffler AJ, Ivans CY, Ryel RJ, Caldwell MM (2005) Fine root distribution and persistence under field conditions of three co-occurring Great Basin species of different life form. New Phytol 165:171–180

Peek MS, Leffler AJ, Hipps L, Ivans S, Ryel RJ, Caldwell MM (2006) Significant root turnover and relocation in the soil profile in response to seasonal soil moisture variation in a natural stand of Utah Juniper (*Juniperus osteosperma*). Tree Physiol (in press)

Ponti F, Minotta G, Cantoni L, Bagnaresi U (2004) Fine root dynamics of pedunculate oak and narrow-leaved ash in a mixed-hardwood plantation in clay soils. Plant Soil 259:39–49

Pregitzer KS, Hendrick RL, Fogel R (1993) The demography of fine roots in response to patches of water and nitrogen. New Phytol 125:575–580

Pregitzer KS, Kubiske ME, Yu CK, Hendrick RL (1997) Relationships among root branch order, carbon, and nitrogen in four temperate species. Oecologia 111:302–308

Pregitzer KS, Laskowski MJ, Burton AJ, Lessard VC, Zak DR (1998) Variation in sugar maple root respiration with root diameter and soil depth. Tree Physiol 18:665–670

Pregitzer KS, King JS, Burton AJ, Brown SE (2000) Responses of tree fine roots to temperature. New Phytol 147:105–115

Raich JW, Nadelhoffer KJ (1989) Belowground carbon allocation in forest ecosystems: global trends. Ecology 70:1346–1354

Reich PB, Walters MB, Ellsworth (1999) Leaf life-span in relation to leaf, plant, and stand characteristics among diverse ecosystems. Ecol Monogr 62:365–392

Ryser P (1996) The importance of tissue density for growth and life span of leaves and roots: a comparison of five ecologically contrasting grasses. Funct Ecol 10:717–723

Silver WL, Miya RK (2001) Global patterns in root decomposition: comparisons of climate and litter quality effects. Oecologia 129:407–419

Sims PL, Singh JS (1978) The structure and function of ten western North American grasslands. III. Net primary production, turnover and efficiencies of energy capture and water use. J Ecol 66:573–597

Steingrobe B, Schmid H, Claassen N (2001) Root production and root mortality of winter barley and its implication with regard to phosphate acquisition. Plant Soil 237:239–248

Tierney GL, Fahey TJ (2001) Evaluating minirhizotron estimates of fine root longevity and production in the forest floor of a temperate broadleaf forest. Plant Soil 229:167–176

Tingey DT, Phillips DL, Johnson MG (2000) Elevated CO_2 and conifer roots: effects on growth, life span and turnover. New Phytol 147:87–103

Tjoelker MG, Craine JM, Wedin D, Reich PB, Tilman D (2005) Linking leaf and root trait syndromes among 39 grassland and savannah species. New Phytol 167:493–508

Trumbore SE, Gaudinski JB. 2003. The secret lives of roots. Science 302:1344–1345

Van der Krift TAJ, Berendse F (2002) Root life spans of four grass species from habitats differing in nutrient availability. Funct Ecol 16:198–203

Van Praag HJ, Sougnex-Remy S, Weissen F, Carletti G (1988) Root turnover in a beech and spruce stand of the Belgian Ardennes. Plant Soil 105:87–103

Watson CA, Ross, JM, Bagnaresi U, Minotta GF, Roffi F, Atkinson D, Black KE, Hooker JE (2000) Environment-induced modifications to root longevity in *Lolium perenne* and *Trifolium repens*. Ann Bot 85:397–401

Wells CE, Eissenstat DM (2001) Marked differences in survivorship among apple roots of different diameters. Ecology 82:882–892

Wells CE, Eissenstat DM (2003) Beyond the roots of young seedlings: the influence of age and order on fine root physiology. J Plant Growth Reg 21:324–334

Wells CE, Glenn DM, Eissenstat DM (2002a) Changes in the risk of fine-root mortality with age: a case study in peach, *Prunus persica* (Rosaceae). Am J Bot 89:79–87

Wells CE, Glenn DM, Eissenstat DM (2002b) Soil insects alter fine root demography in peach (*Prunus persica*). Plant Cell Environ 25:431–439

West JB, Espeleta JF, Donovan LA (2003) Root longevity and phenology differences between two co-occurring savanna bunchgrasses with different leaf habits. Funct Ecol 17:20–28

Withington JM, Elkin AD, Bulaj B, Olesinski J, Tracy KN, Bouma TJ, Oleksyn J, Anderson LJ, Modrzynski J, Reich PB, Eissenstat DM (2003) The impact of material used for minirhizotron tubes for root research. New Phytol 160:533–544

Withington JM, Reich PB, Oleksyn J, Eissenstat DM (2006) Root structure and lifespan are largely independent of leaf structure and lifespan in a common garden comparison of eleven tree species. Ecol Monogr (in press)

Yanai RD, Fahey TJ, Miller SL (1995) Efficiency of nutrient acquisition by fine roots and mycorrhizae. In: Smith WK, Hinckley TM (eds) Resource physiology of conifers. Academic Press, San Diego, pp 75–103

Zeleznik JD, Dickmann DI 2004 Effects of high temperatures on fine roots of mature red pine (*Pinus resinosa*) trees. For Ecol Manag 199:395–409

Michael S. Peek
Department of Biology
William Paterson University
300 Pompton Road
Wayne, NJ 07470 USA
peekm@wpunj.edu
Tel.: +1-973-7202247
Fax: +1-973-7202338

Subject Index

MIX
Papier aus verantwortungsvollen Quellen
Paper from responsible sources
FSC® C105338

If you have any concerns about our products,
you can contact us on
ProductSafety@springernature.com

In case Publisher is established outside the EU,
the EU authorized representative is:
Springer Nature Customer Service Center GmbH
Europaplatz 3, 69115 Heidelberg, Germany

Printed by Libri Plureos GmbH
in Hamburg, Germany